I0065976

Medicinal and Natural Product Chemistry

Medicinal and Natural Product Chemistry

Edited by **Allegra Smith**

NY RESEARCH
P R E S S

New York

Published by NY Research Press,
23 West, 55th Street, Suite 816,
New York, NY 10019, USA
www.nyresearchpress.com

Medicinal and Natural Product Chemistry
Edited by Allegra Smith

© 2016 NY Research Press

International Standard Book Number: 978-1-63238-482-9 (Hardback)

This book contains information obtained from authentic and highly regarded sources. Copyright for all individual chapters remain with the respective authors as indicated. All chapters are published with permission under the Creative Commons Attribution License or equivalent. A wide variety of references are listed. Permission and sources are indicated; for detailed attributions, please refer to the permissions page and list of contributors. Reasonable efforts have been made to publish reliable data and information, but the authors, editors and publisher cannot assume any responsibility for the validity of all materials or the consequences of their use.

The publisher's policy is to use permanent paper from mills that operate a sustainable forestry policy. Furthermore, the publisher ensures that the text paper and cover boards used have met acceptable environmental accreditation standards.

Trademark Notice: Registered trademark of products or corporate names are used only for explanation and identification without intent to infringe.

Printed in the United States of America.

Contents

Preface

The world is advancing at a fast pace like never before. Therefore, the need is to keep up with the latest developments. This book was an idea that came to fruition when the specialists in the area realized the need to coordinate together and document essential themes in the subject. That's when I was requested to be the editor. Editing this book has been an honour as it brings together diverse authors researching on different streams of the field. The book collates essential materials contributed by veterans in the area which can be utilized by students and researchers alike.

Medicinal and natural product chemistry is generally considered as a discipline concerned with biological molecules but it is essentially the chemistry of carbon compounds. A large number of organic compounds are either derived or completely synthesized in laboratories. This book discusses many such advanced compounds, their manufacturing processes, structures and properties. Some of the topics included are drug design to combinatorial synthesis and parallel synthesis, applications of compound collections, compound preparation, molecular diversity assessments, etc. The ever growing need for advanced technology is the reason that has fueled the research in this field in recent times. Such selected concepts that redefine medicinal and natural product chemistry have been presented in this book. The extensive content presented in the book provides the readers with a thorough understanding of this subject. It will serve as a resource guide for professionals, researchers and students engaged in this field.

Each chapter is a sole-standing publication that reflects each author's interpretation. Thus, the book displays a multi-facetted picture of our current understanding of applications and diverse aspects of the field. I would like to thank the contributors of this book and my family for their endless support.

Editor

1

Polyphenolic Composition and Evaluation of Antioxidant Activity, Osmotic Fragility and Cytotoxic Effects of *Raphiodon echinus* (Nees & Mart.) Schauer

Antonia Eliene Duarte [1,2], Emily Pansera Waczuk [2], Katiane Roversi [3],
Maria Arlene Pessoa da Silva [4], Luiz Marivando Barros [1,2], Francisco Assis Bezerra da Cunha [1,2],
Irwin Rose Alencar de Menezes [5], José Galberto Martins da Costa [6], Aline Augusti Boligon [7],
Adedayo Oluwaseun Ademiluyi [2,8], Jean Paul Kamdem [2,9], João Batista Teixeira Rocha [2,*] and
Marilise Escobar Burger [3,*]

Academic Editor: Derek J. McPhee

[1] Centro de Ciências Biológicas e da Saúde-CCBS, Departamento de Ciências Biológicas,
    Universidade Regional do Cariri (URCA), Pimenta, Crato CEP 63.100-000, CE, Brazil;
    duarte105@yahoo.com.br (A.E.D.); lmarivando@hotmail.com (L.M.B.); cunha.urca@gmail.com (F.A.B.C.)
[2] Programa de Pós-Graduação em Bioquímica Toxicológica, Departamento de Bioquímica e Biologia Molecular,
    Universidade Federal de Santa Maria, Santa Maria 97105-900, RS, Brazil; memypw@yahoo.com.br (E.P.W.);
    aoademiluyi@futa.edu.ng (A.O.A.); kamdemjeanpaul2005@yahoo.fr (J.P.K.)
[3] Departamento de Fisiologia e Farmacologia, Universidade Federal de Santa Maria, Santa Maria 97105-900,
    RS, Brazil; katianeroversi@gmail.com
[4] Laboratório de Botânica Aplicada, Departamento de Ciências Biológicas, Universidade Regional do
    Cariri (URCA), Pimenta, Crato CEP 63.100-000, CE, Brazil; arlene.pessoa@urca.br
[5] Laboratório de Farmacologia e Química Molecular, Departamento de Química Biológica,
    Universidade Regional do Cariri, Pimenta, Crato CEP 63.100-000, CE, Brazil; irwin.alencar@urca.br
[6] Laboratório de Pesquisas de Produtos Naturais, Departamento de Química Biológica,
    Universidade Regional do Cariri, Crato CEP 63.105.000, CE, Brazil; galberto.martins@urca.br
[7] Laboratório de Fitoquímica, Departamento de Farmácia Industrial, Universidade Federal de Santa Maria,
    Santa Maria 97105-900, RS, Brazil; alineboligon@yahoo.com.br
[8] Functional Foods and Nutraceutical Unit, Department of Biochemistry, Federal University of Technology,
    P.M.B. 704, Akure 340001, Nigeria
[9] Departamento de Bioquímica, Instituto de Ciências Básica da Saúde, Universidade Federal do Rio Grande
    do Sul, Porto Alegre CEP 90035-003, RS, Brazil; jpkamdem@gmail.com
*   Correspondence: jbtrocha@yahoo.com.br (J.B.T.R.); mariliseeb@yahoo.com.br (M.E.B.)

Abstract: *Raphiodon echinus* (*R. echinus*) is used in Brazilian folk medicine for the treatment of inflammation, coughs, and infectious diseases. However, no information is available on the potential antioxidant, cytotoxicity and genotoxicity of this plant. In this study, the polyphenolic constituents, antioxidant capacity and potential toxic effects of aqueous and ethanolic extracts of *R. echinus* on human erythrocytes and leukocytes were investigated for the first time. *R. echinus* extracts showed the presence of Gallic, chlorogenic, caffeic and ellagic acids, rutin, quercitrin and quercetin. Aqueous and ethanolic extracts of *R. echinus* exhibited antioxidant activity in DPPH radical scavenging with $IC_{50} = 111.9$ µg/mL (EtOH extract) and $IC_{50} = 227.9$ µg/mL (aqueous extract). The extracts inhibited Fe^{2+} (10 µM) induced thiobarbituric acid reactive substances (TBARS) formation in rat brain and liver homogenates. The extracts (30–480 µg/mL) did not induce genotoxicity, cytotoxicity or osmotic fragility in human blood cells. The findings of this present study therefore suggest that the therapeutic effect of *R. echinus* may be, in part, related to its antioxidant potential. Nevertheless, further *in vitro* and *in vivo* studies are required to ascertain the safety margin of its use in folk medicine.

Keywords: *Raphiodon echinus*; antioxidant activity; phenolic acids; HPLC-DAD

1. Introduction

Some medicinal plants used in folk medicine can cause toxicity to humans and also exhibit carcinogenicity and genotoxicity [1,2]. Therefore, toxicological studies of plant extracts used in traditional medicine are highly recommended, as it is part of the procedures that contribute to the standardization of phytopharmaceuticals [3,4].

The genus *Raphiodon* (*Lamiaceae*) is represented by only one species, *Raphiodon echinus* (*R. echinus*), which is common to Eastern Brazil and typical of the "caatinga" (semi-arid vegetation) [5]. It is a prostrate herb with aromatic leaves and long pedunculate spherical heads with bright purple flowers, found in the states of Bahia, Pernambuco, Paraíba, Ceará and Minas Gerais. The infusion of the leaves of *R. echinus* is used in Brazilian folk medicine for the treatment of inflammation, coughs and infectious diseases. Studies have shown that *R. echinus* exhibits antimicrobial [6], anti-inflammatory and analgesic activities [7]. These biological properties are generally attributed (at least in part) to the antioxidant activity. However, to the best of our knowledge, there are no reports on the antioxidant activity of this plant extract.

Free radicals are thought to be important mediators of tissues injury under pathological conditions [8]. Byproducts of lipid peroxidation (LPO) have been shown to decrease cell membrane fluidity, inactivation of membrane-bound enzymes and loss of essential fatty acids [9], resulting in increased osmotic fragility of the cell [10] Consequently, the use of plant extracts or compounds that can act as "physical barriers" to prevent free radicals generation from important sites (e.g., cell membranes), or able to inhibit the propagation of LPO are of utmost importance.

Given that there is limited literature information regarding the biological activities of *R. echinus*, especially its antioxidant activity, and no information on its potential toxic effects to human blood cells, the present study, therefore, aimed at investigating for the first time the antioxidant capacity, iron chelating activity, cytotoxicity, and genotoxicity of *R. echinus* leaf extracts (aqueous and ethanolic) in human leukocytes, as well as its effect on osmotic fragility in human erythrocytes. Furthermore, polyphenolic constituents that may be at least, in part, responsible for the beneficial effects of *R. echinus* extracts were characterized using high performance liquid chromatography coupled to diode array detector (HPLC-DAD). This study is particularly important in view of the fact that it provides supportive information on the use of this plant in traditional medicine.

2. Results and Discussion

The lack of scientific evidence for the biological activities and safety profile of plant extracts used in traditional medicine have generated considerable concern in the scientific community. In fact, it is imperative to isolate those plants that can represent serious public health problem. In the present study, we investigated for the first time the potential antioxidant activity of *R. echinus* leaves extracts as well as its potential cytotoxic and genotoxic effects in human leukocytes. In addition, the influence of *R. echinus* on human erythrocytes and the polyphenolic constituents of the leaves extracts were characterized and reported for the first time.

2.1. HPLC Characterization of the Polyphenolic Constituents of Aqueous and Ethanolic Extracts of R. echinus Leaves

The HPLC profile of the aqueous and ethanolic extracts of the leaves of *R. echinus* revealed the presence of polyphenolic constituents, which appeared with the following elution profile/order: Gallic acid (Rt = 7.12 min, peak 1), chlorogenic acid (Rt = 19.34 min, peak 2), caffeic acid (Rt = 22.61 min, peak 3), ellagic acid (Rt = 29.73 min, peak 4), rutin (Rt = 36.12 min, peak 5), quercitrin (Rt = 43.95 min, peak 6) and quercetin (Rt = 50.03 min, peak 7) (Figure 1A,B). However, quercitrin appeared to be absent in the aqueous extract (Figure 1A, Table 1) and present in the ethanolic extract of *R. echinus* (Figure 1B, Table 1). The identification of these constituents was made by comparing their retention time and UV spectra of the peaks in the samples with those of authentic reference samples or isolated compounds. Based on our results, ellagic acid appeared to be the major component of both extracts with 79.13 and 63.18 mg/g in ethanolic and aqueous extract, respectively. In contrast, quercetin (9.87 mg/g; representing about 0.98% of the aqueous extract) and quercitrin (7.12 mg/g; representing 0.81% of the ethanolic extract) were the two components detected in the smallest quantities in *R. echinus* extracts (Table 2).

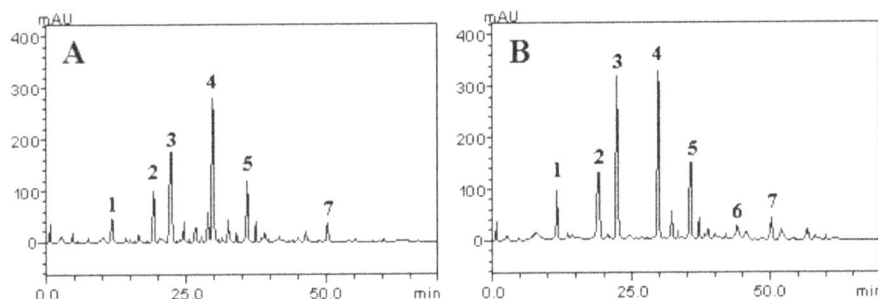

Figure 1. HPLC-DAD chromatograms of aqueous (**A**) and ethanolic (**B**) extracts of the leaves of *Raphiodon echinus* (*R. echinus*): Gallic acid (peak 1), chlorogenic acid (peak 2), caffeic acid (peak 3), ellagic acid (peak 4), rutin (peak 5), quercitrin (peak 6) and quercetin (peak 7). Calibration curve for Gallic acid: $y = 12574x + 1307.8$ ($r = 0.9999$); chlorogenic acid: $y = 11953x + 1278.2$ ($r = 0.9995$); caffeic acid: $y = 11976x + 1187.0$ ($r = 0.9996$); ellagic acid: $y = 13169x + 1346.8$ ($r = 0.9999$); quercitrin: $y = 12473x + 1187.5$ ($r = 0.9991$); rutin: $y = 12814x + 1189.3$ ($r = 0.9999$) and quercetin: $y = 12537x + 1375.6$ ($r = 0.9994$). All chromatography operations were carried out at ambient temperature and in triplicate.

Table 1. Schedule of evaluation of oxidation or chelation of Fe^{2+}/Fe^{3+} by plant extracts.

Time	Sequence of Addition	Reading at 510 nm
0 min	Extract (30–120 µg/mL) FeSO$_4$ (110 µM)	-
10 min	*Ortho*-phenanthroline (0.25%)	-
10 min	Immediately after mixing with *Ortho*-phenanthroline	First reading (0 min)
20 min	-	Second reading (10 min)
30 min	-	Third reading (20 min)
30 min	Ascorbic acid (AA, final concentration, 5 mM)	-
35 min	-	First reading after AA (25 min)
40 min	-	Second reading after AA (30 min)
50 min	-	Third reading after AA (40 min)

Table 2. Phenolics and flavonoids composition of *R. echinus* extracts.

Compounds	Aqueous Extract/mg·g^{-1} (%)	Ethanolic Extract/mg·g^{-1} (%)	LOD/µg·mL^{-1}	LOQ/µg·mL^{-1}
Gallic acid	11.59 ± 0.01 [a] (1.15)	25.03 ± 0.01 [a] (2.50)	0.011	0.037
Chlorogenic acid	25.07 ± 0.02 [b] (2.50)	31.94 ± 0.02 [b] (3.19)	0.009	0.035
Caffeic acid	40.19 ± 0.02 [c] (4.01)	76.45 ± 0.02 [c] (7.64)	0.026	0.090
Ellagic acid	63.18 ± 0.01 [d] (6.31)	79.13 ± 0.03 [c] (7.91)	0.017	0.056
Rutin	26.50 ± 0.03 [b] (2.65)	35.84 ± 0.02 [b] (3.58)	0.024	0.080
Quercitrin	-	7.12 ± 0.01 [d] (0.81)	0.035	0.118
Quercetin	9.87 ± 0.03 [a] (0.98)	12.37 ± 0.01 [e] (1.23)	0.019	0.063

Results are expressed as mean ± standard deviations (SD) of three determinations. Different letters in the same column indicate significant difference by Tukey test at $p < 0.01$. LOD: limit of detection, LOQ: limit of quantification.

2.2. Antioxidant Activity

2.2.1. Scavenging Effect of *R. echinus* Extracts on DPPH Radical

Aqueous and ethanolic (EtOH) extracts of *R. echinus* exhibited antioxidant activity against DPPH radical in a concentration-dependent manner (Figure 2). According to the calculated IC_{50} values, the ethanolic extract exhibited stronger DPPH radical scavenging activity than aqueous extract, which was about two times higher than that of aqueous extract (Figure 2).

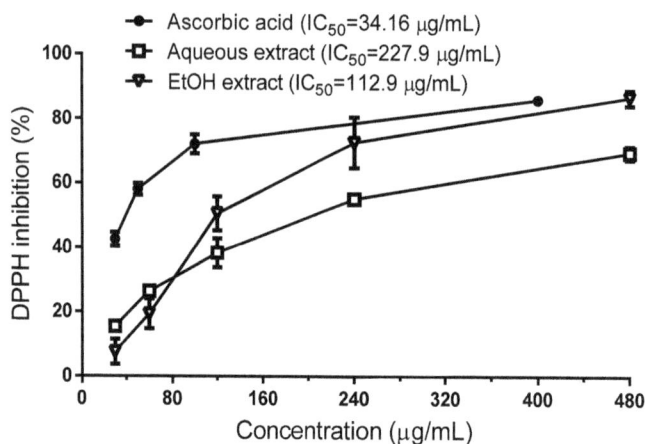

Figure 2. Quenching of DPPH radicals by aqueous and ethanolic extracts from the leaves of *R. echinus*. Data are expressed as mean $\pm$ SEM of $n = 4$ independent experiments.

In this assay, the bleaching of the DPPH coloration is an indication of the free radical scavenging capacity of the samples. Our results revealed that the ethanolic extract showed greater antioxidant activity than aqueous extract (IC_{50} = 112.9 μg/mL *vs.* 227.9 μg/mL). However, this activity was three times lower than that of ascorbic acid (IC_{50} = 34.16 μg/mL), used as a standard. The higher antioxidant capacity of the ethanolic extract compared to the aqueous extract can be explained by its higher polyphenolic contents. In line with this, the total phenolic content of both extracts assayed by the Folin–Ciocalteu method revealed that ethanolic extract exhibit higher total phenolic content (TPC) than aqueous extract (Table 3). Numerous studies have described a positive correlation between the antioxidant activity and phenolic content [11,12]. Although we did not perform such correlation calculations, the results obtained here clearly demonstrate that ethanolic extract, which exhibited high TPC, has a stronger antioxidant activity by DPPH radical scavenging.

Table 3. Total phenolic content of *R. echinus* extracts.

	R. echinus	
	Aqueous Extract	**Ethanolic Extract**
Total phenolics (mg GAE/g dry extract)	173.0 $\pm$ 0.07	389.1 $\pm$ 0.04

Results are expressed in milligram Gallic acid equivalent (GAE) per gram of dry extract; $n = 3$.

2.2.2. Effect of *R. echinus* Extracts on Fe^{2+} Induced Lipid Peroxidation (LPO) in the Rat Brain and Liver Homogenates

Fe^{2+} induced a significant increase in TBARS production in brain homogenates ($p < 0.05$; Figure 3). However, the aqueous (Figure 3A) and ethanolic (Figure 3B) extracts of *R. echinus* significantly reduced the LPO in concentration-dependent manner both under basal and iron-stimulated conditions (Figure 3A,B).

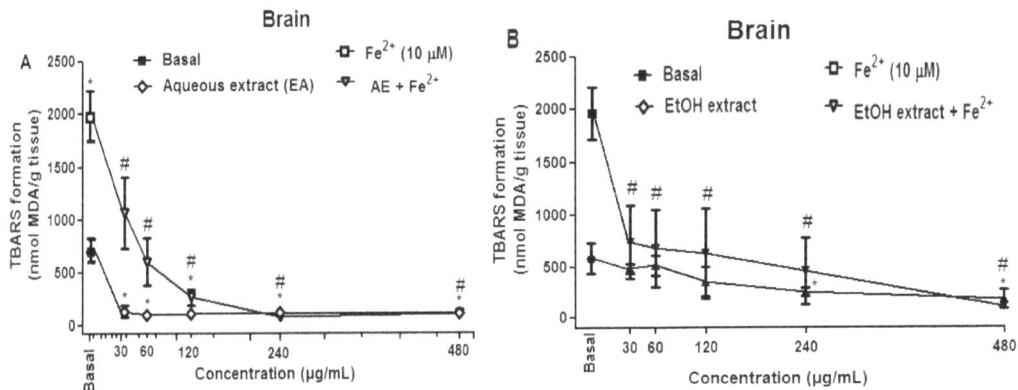

Figure 3. Effect of aqueous (**A**) and ethanolic (**B**) extracts of the leaves of *R. echinus* on lipid peroxidation induced by iron in rat brain homogenates. The homogenate was incubated for 1 h with Fe^{2+} (10 μM) in the presence or absence of different concentrations of the extracts. Values represent the mean of $n = 3$ independent experiments performed in duplicate $\pm$ SEM. *: $p < 0.05$ *vs.* basal and #: $p < 0.05$ *vs.* Fe^{2+}.

Similar to that observed with brain homogenates, crude extracts of *R. echinus* inhibited LPO induced by Fe^{2+} in rat liver homogenates (Figure 4). In contrast, ethanolic extract (EtOH extract) did not inhibited TBARS formation under basal condition (Figure 4B), while aqueous extract (AE) inhibited TBARS production under basal conditions (Figure 4A).

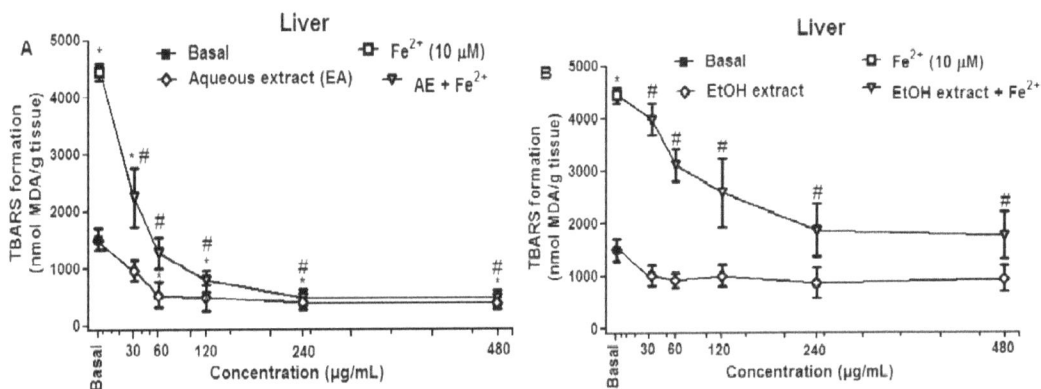

Figure 4. Effect of aqueous (**A**) and ethanolic (**B**) extracts of the leaves of *R. echinus* on lipid peroxidation induced by iron in rat liver homogenates. The homogenate was incubated for 1 h with Fe^{2+} (10 μM) in the presence or absence of different concentrations of the extracts. Values represent the mean of $n = 3$ independent experiments performed in duplicate $\pm$ SEM. *: $p < 0.05$ *vs.* basal and #: $p < 0.05$ *vs.* Fe^{2+}.

The results demonstrated that aqueous and ethanolic extracts from the leaves of *R. echinus* exhibited protective effects against Fe^{2+} induced lipid peroxidation (LPO) in rat brain homogenates. Although iron (II) is an essential element for life, free Fe^{2+} in biological systems, however, can be toxic [13] and its levels have been shown to be increased in many neurological disorders including Alzheimer's and Parkinson's diseases [14–16]. The results of the current study demonstrated that both extracts showed antioxidant activity against Fe^{2+} induced LPO in rat brain and liver homogenates at all the concentrations tested. Nevertheless, the ability of these extracts to inhibit LPO could be attributed (at least in part) to the capacity of their chemical constituents to chelate/inactivate Fe^{2+}, thereby, preventing or reducing reactive oxygen species generation. The HPLC fingerprint of these extracts revealed the presence of phenolic acids (Gallic, ellagic, chlorogenic and caffeic acids) and flavonoids (quercetin, quercitrin and rutin), compounds that are known scavengers and inhibitors of LPO [17,18]. Of particular importance, is the finding which revealed that chlorogenic and caffeic acids could inhibit free radicals formation through several mechanisms including the reaction complex formation with iron ions in the reactions of hydrogen peroxide with iron (II), hydrogen peroxide with ferric iron and 3-hydroxyanthranilic acid [19]. In both tissues, aqueous extract showed highest antioxidant activity against Fe^{2+} induced LPO by significantly reducing TBARS formation at much lower concentrations when compared to the ethanolic extract. At basal conditions, ethanolic extract did not have any effect on TBARS formation in the liver, while it did in the brain at the higher concentrations tested (240 and 480 µg/mL).

2.2.3. Iron Chelating Potential of *R. echinus* Extracts

The incubation of aqueous and ethanolic extracts with Fe^{2+} caused a decreased in the absorbance at 510 nm with the effect most apparent from 30 to 120 µg/mL (Figure 5A,B). The addition of ascorbic acid caused only a modest increase in the absorbance at 510 nm and this was a little more apparent for the aqueous extract (Figure 5A). The most plausible interpretation of these results is that the extracts can chelate Fe^{2+} and accelerate the oxidation of Fe^{2+} to Fe^{3+}. However, this Fe^{3+} was only partially or not released from the complex, as otherwise ascorbic acid should have reduced Fe^{3+} to Fe^{2+}, resulting in the formation of the complex between Fe^{2+} and *ortho*-phenanthroline. However, this occurred only to a limited extent.

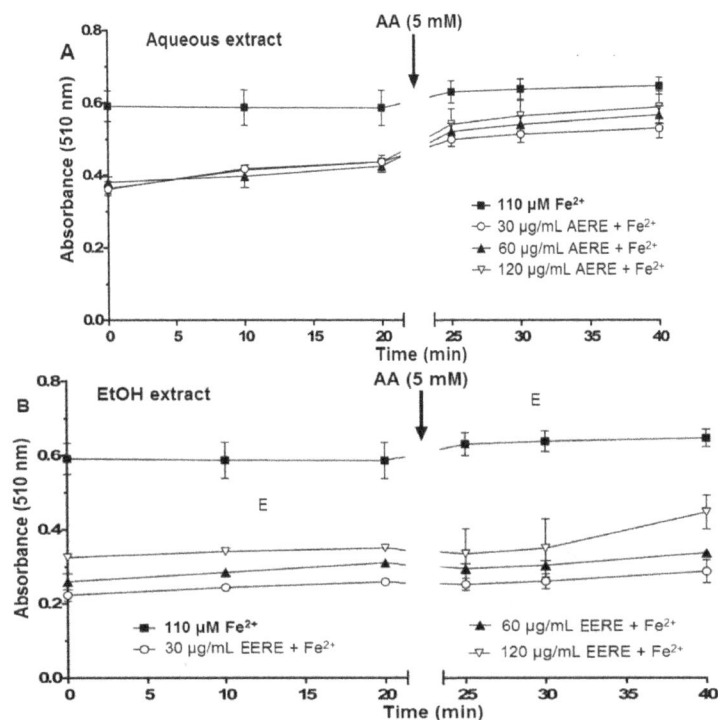

Figure 5. Oxidation of Fe^{2+} by aqueous (**A**) and ethanolic (**B**) extracts from the leaves of *R. echinus* (1–60 µg/mL). The extracts (30–120 µg/mL) were incubated with $FeSO_4$ (110 µM) for 10 min. Then, *ortho*-phenanthroline was added and the absorbance of the reaction mixture was measured at 0, 10 and 20 min following its addition. After the last reading (at 20 min), 5 mM ascorbic acid (AA) was added to the reaction mixture, and the absorbance was read again after 5 min (at 25 min), 10 min (at 30 min) and 20 min (at 40 min) (see Table 1 for details). Values represent the mean ± SEM of three independent experiments performed in duplicate. AERE, aqueous extract of *R. echinus*; EERE, ethanolic extract of *R. echinus*.

To verify whether the decrease in the absorbance in the presence of aqueous (Figure 5A) or ethanolic (Figure 5B) extracts was caused by the chelation or oxidation of Fe^{2+}, ascorbic acid (AA) was added to the reaction medium after 20 min of incubation of Fe^{2+} with the extracts and

$ortho$-phenanthroline. The objective was to reduce Fe^{3+} that could have being formed during the reaction back to Fe^{2+}. In the presence of aqueous extract (Figure 5A) or ethanolic extract (Figure 5B), the addition of AA to the reaction mixture caused a partial increase in the absorbance after 5, 10 and 20 min of incubation, which is an indication of stimulated oxidation of Fe^{2+} by the extracts.

Earlier reports have shown the ability of antioxidants to chelate or deactivate transition metals, thus preventing them from initiating lipid peroxidation and oxidative stress via metal-catalyzed reaction [20]. Chelation of transition metals such as iron can be viewed as a preventive antioxidant mechanism. In the present study, both extracts chelate Fe^{2+} partially and possibly converted it to Fe^{3+}. However, Fe^{3+} was only partially released from the plant extract-iron complex as evidenced by the small increase in the absorbance after addition of the reducing agent, ascorbic acid (AA) to the reaction medium. Furthermore, if there was an appreciable free Fe^{3+} pool, this should have been reduced back to Fe^{2+} in the presence of ascorbic acid. However, this was very modest. Consequently, it is possible to presume that the observed inhibition of Fe^{2+} induced LPO in the presence of the extracts was a result of both direct interaction with free radicals and via chelation of Fe^{2+}/Fe^{3+} species.

2.3. Cytotoxicity Effect of R. echinus Extracts on Human Leukocytes

The cytotoxicity of *R. echinus* extracts was investigated in human leukocytes using the Trypan blue exclusion method. In this method, dead cells allow passage of trypan blue into the cytoplasm due to loss of membrane selectivity [9,21]. Here, visual analysis of the cells using a Neubauer chamber allowed counting the number of dead (trypan blue positive) and living (trypan blue negative) cells. Exposure for 3 h of human leukocytes to aqueous and ethanolic extracts of *R. echinus* did not modify cell viability, when compared to the control (Figure 6A,B). Cell viability of treated leukocytes was generally greater than 90%, confirming the absence of cytotoxicity of the crude extracts of *R. echinus*. The potential protective effects of both extracts against H_2O_2 + azide induced cytotoxicity was also investigated. H_2O_2 (2 mM) + azide (1 mM) significantly decreased leukocytes viability when compared to control ($p < 0.05$; Figure 6C,D). However, aqueous (Figure 6C) and ethanolic (Figure 6D) extracts of *R. echinus* did not blunt the cytotoxicity caused by H_2O_2 + azide (Figure 6C,D).

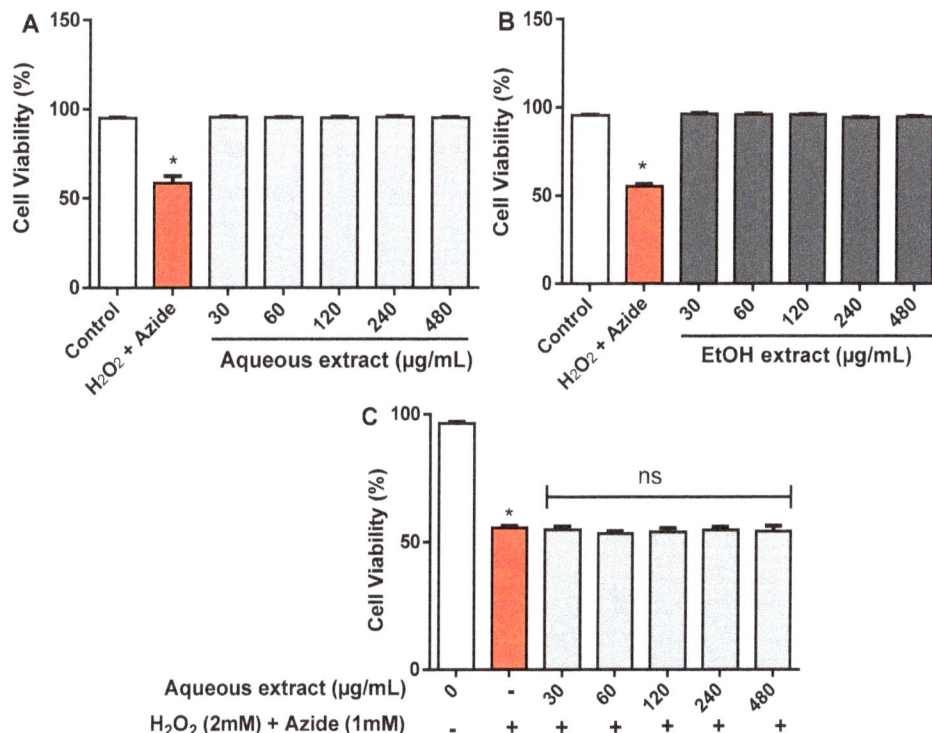

Figure 6. *Cont.*

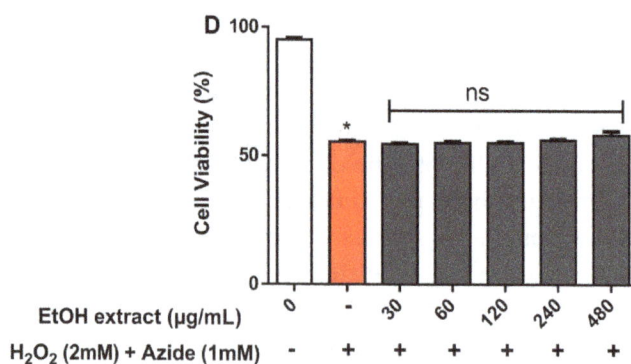

Figure 6. Effect of *R. echinus* leaves extracts on human leukocytes in the absence (**A,B**) and presence (**C,D**) of H_2O_2 (2 mM) + Azide (1 mM). The signs (−) and (+) indicate the absence and presence of the mixture H_2O_2 (2 mM) + Azide (1 mM), respectively. Results are expressed as percentage of control. Values are the means of n = 3 independent experiments performed in triplicate ± SEM. *: $p < 0.05$ *vs.* control.

2.4. Effect of R. echinus Extracts on Osmotic Fragility of Human Erythrocytes

The influence of aqueous and EtOH extracts from the leaves of *R. echinus* on human erythrocytes osmotic fragility at different salt concentrations (0%–0.9%) is depicted in Figure 7. The results demonstrated no significant difference between treated and untreated erythrocytes at different salt concentrations in comparison to their respective control ($p > 0.05$; Figure 7A,B).

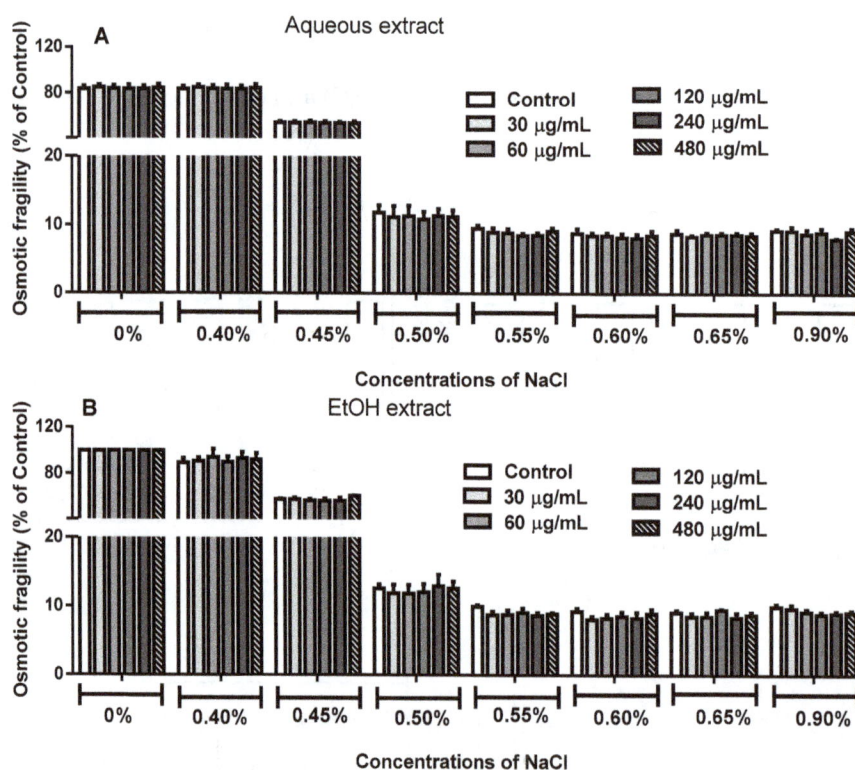

Figure 7. Osmotic fragility of human erythrocytes treated with aqueous (**A**) and ethanolic (**B**) extracts of the leaves of *R. echinus*. Untreated and treated erythrocytes were added to different concentrations of NaCl (0%–0.9%) and incubated for 20 min. The absorbance of the supernatant was measured and the hemolysis in each tube was expressed as percentage of control. Each bar represents the mean of n = 3 independent experiments performed in triplicate ± SEM.

Osmotic fragility assay is widely used to verify the toxicity of chemicals (and plant extracts) and environmental pollutants on membrane integrity of erythrocytes [9,22,23]. Despite the fact that erythrocytes are well equipped with several biological mechanisms to defend against free radical induced LPO, erythrocyte membrane, however, is prone to oxidative stress, particularly because of the high oxygen tension in the blood and high polyunsaturated fatty acid content [24,25]. Here, the cytotoxic effect of the leaves extracts of *R. echinus* was investigated on human erythrocytes, so as to provide primary information on the interaction between their phytoconstituents and erythrocytes membrane [26,27]. Our results demonstrated that 3 h treatment of human erythrocytes with *R. echinus* extracts did not affect erythrocytes membrane integrity when compared to the control. These results indicate that the chemical constituents of the extracts did not cause gross changes in membrane protein and lipid structure of the cells.

2.5. Effect of R. echinus Extracts on DNA Damage

As depicted in Figure 8, methyl methanesulfonate (MMS), used as positive control, caused a dramatic increase in DNA damage when compared with the control ($p < 0.001$). However, no significant difference was found in the DNA damage index (DI) between human leukocytes treated with aqueous (Figure 8A) and EtOH (Figure 8B) extracts, when compared to control ($p > 0.05$).

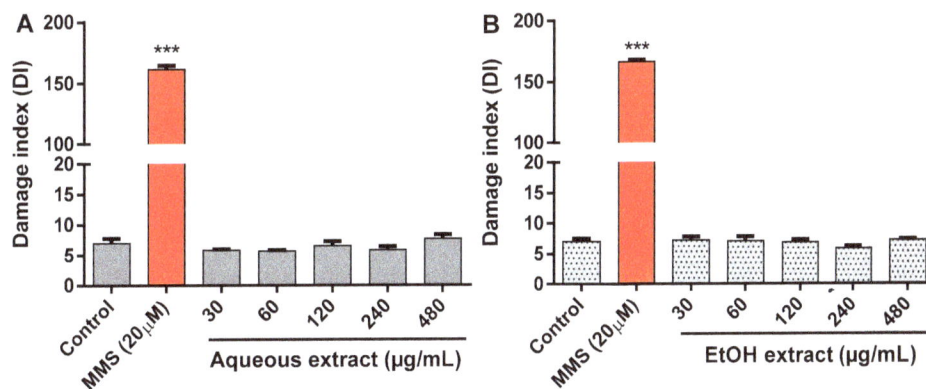

Figure 8. DNA damage of leukocytes treated with the aqueous (**A**) and ethanolic (**B**) leaves extracts of *R. echinus*. Methyl methanesulfonate (MMS) was used as positive control. The results are expressed as means $\pm$ SEM of $n = 3$ independent experiments. ***: $p < 0.001$ *vs.* control.

Comet assay (single-cell gel electrophoresis) is widely used to access DNA damage and repair in individual cells. Studies have indicated that comet assay is a useful tool for the detection of genotoxicity thus, making this assay an important biomarker in toxicological studies [27,28]. In the present study, the potential genotoxic effect of the extracts components at the level of DNA was evaluated in human leukocytes. Our data showed that the leaf extracts of *R. echinus* are not genotoxic. This suggests that the phytoconstituents of these extracts do not/or have not interacted with the DNA. According to Galloway *et al.* [29,30], a situation where DNA damage occurs without a concomitant cytotoxicity is of greater concern. Here, we found no DNA damage and no cytotoxicity. Although it is difficult to extrapolate *in vitro* findings to *in vivo* exposure; nevertheless, we can speculate that the popular use of both extracts may not cause overt toxicity.

3. Experimental Section

3.1. Chemicals

All chemicals used including solvents were of analytical grade. Methanol, acetic acid, Gallic acid, caffeic acid, ellagic acid and chlorogenic acid were purchased from Merck (Darmstadt, Germany). Gallic, chlorogenic, caffeic, and ellagic acids, quercetin, rutin, and quercitrin, were acquired from

Sigma Chemical Co. (St. Louis, MO, USA). 1,1-diphenyl-2-picrylhydrazyl (DPPH), ascorbic acid, malonaldehydebis-(dimethyl acetal) (MDA), thiobarbituric acid (TBA), sodium azide and hydrogen peroxide (H_2O_2) were purchased from Sigma Chemical Co.

3.2. Plant Material

The leaves *R. echinus* (Nees and Mart) Shauer were collected in Padre Cicero, Crato-Ceará (7°22′S; 39°28′W, 492 m above sea level), Brazil, in January 2014. The plant material was identified by Maria Arlene Pessoa da Silva, and deposited in the Herbarium Caririense Dárdano de Andrade—Lima, Universidade Regional do Cariri (URCA), with the number 7347. To obtain the aqueous extract of *R. echinus*, 300 g of crushed leaves was mixed with 2000 mL of hot water and the mixture was allowed to stand for three days. The mixture was filtered and the filtrate was lyophilized to obtain a dark green solid (3.3% yield). However, for the ethanolic extract, the fresh crushed leaves (300 g) were macerated with 1500 mL of 92% ethanol for three days. On the third day, the suspension was filtered and the filtrate containing the solvent was evaporated under reduced pressure and lyophilized to obtain 1.04% yield of the dry material. The prepared extracts (aqueous and ethanolic) were stored in the freezer until being tested. It should be stressed that the aqueous and ethanolic extracts of *R. echinus* were used in this study on the basis of its popular use in the form of infusion, and also because the roots of the plant are soaked in ethanol prior to use.

3.3. Quantification of Compounds by HPLC-DAD

Reverse phase chromatographic analyses were carried out under gradient conditions using C_{18} column (4.6 mm × 150 mm) packed with 5 μm diameter particles; the mobile phase was water containing 1% formic acid (A) and methanol (B), and the composition gradient was: 13% of B until 10 min and changed to obtain 20%, 30%, 50%, 60%, 70%, 20% and 10% B at 20, 30, 40, 50, 60, 70 and 80 min, respectively, following the method described by Pereira *et al.* [31], with slight modifications. Aqueous and ethanolic extracts of *R. echinus* as well as the mobile phase were filtered through 0.45 μm membrane filter (Millipore, Billerica, MA, USA) and then degassed by ultrasonic bath prior to use. The extracts were analyzed at concentration of 20 mg/mL for the presence or absence of Gallic, chlorogenic, caffeic and ellagic acids, and rutin, quercitrin and quercetin. These compounds were identified by comparing their retention time and UV absorption spectra with those of the commercial standards. The flow rate was 0.7 mL/min, injection volume 40 μL and the wavelength were 257 nm for Gallic acid, 327 nm for chlorogenic, caffeic and ellagic acids, and 365 nm for quercetin, quercitrin and rutin. Stock solutions of standards references were prepared in the HPLC mobile phase at a concentration range of 0.030–0.250 mg/mL for quercetin, quercitrin and rutin; and 0.035–0.300 mg/mL for ellagic, Gallic, caffeic and chlorogenic acids. Chromatography peaks were confirmed by comparing its retention time with those of reference standards and by DAD spectra (200 to 600 nm). All the chromatography operations were carried out at ambient temperature and in triplicates. The limit of detection (LOD) and limit of quantification (LOQ) were calculated based on the standard deviations of the responses and their slopes using three independent analytical curves. LOD and LOQ were calculated as 3.3 and 10 σ/S, respectively, where σ is the standard deviation of the response and S is the slope of the calibration curve.

3.4. Determination of Total Phenolics

The determination of total phenolic content was performed by the Folin–Ciocalteu method, as described by Kamdem *et al.* [32], with some modifications [33]. Briefly, 0.5 mL of 2 N Folin–Ciocalteu reagent was added to a 1 mL of ethanolic or aqueous extract of *R. echinus* (0.15 mg/mL) and the mixture was allowed to stand for 5 min before the addition of 2 mL of 20% Na_2CO_3. The tubes were then allowed to stand at room temperature for 10 min and the absorbance was measured against water at 730 nm using Shimadzu-UV-1201 (Shimadzu, Kyoto, Japan) spectrophotometer. The phenolic content was expressed in milligram of Gallic acid equivalent per gram of dry extract (mg GAE/g of dry extract). The standard curve obtained with Gallic acid was $Y = 29.315 \times X - 0.04$ ($r^2 = 0.9997$).

3.5. Antioxidant Activity

3.5.1. DPPH Radical Scavenging Activity

The radical scavenging ability of the aqueous and ethanolic extracts of R. echinus was performed using the stable free radical DPPH (1,1-diphenyl-2-picrylhydrazyl) as described by Kamdem et al. [32], with some modifications. Briefly, 50 µL of aqueous and ethanolic extracts at different concentrations (30–480 µg/mL) were mixed with 100 µL of freshly prepared DPPH solution (0.3 mM in ethanol). Then, the plate was kept in the dark at room temperature for 30 min. The reduction in the DPPH radical was measured by monitoring the decrease of absorption at 517 nm using a microplate reader (SpectraMax, Sunnyvale, CA, USA). Ascorbic acid was used as standard compound (i.e., positive control). The DPPH radical scavenging capacity was measured using the following equation:

$$\% \text{ inhibition } = 100 - (A_{sample} - A_{blank})/A_{control} \times 100 \qquad (1)$$

where A_{sample} is the absorbance of the tested sample with DPPH; A_{blank}, the absorbance of the test tube without adding the DPPH and $A_{control}$, is the absorbance of the DPPH solution.

3.5.2. Production of TBARS from Animal Tissues

The production of thiobarbituric acid reactive substances (TBARS) was measured as described by Ohkawa et al. [34], and modified by Barbosa-Filho et al. [35]. The rats were killed by decapitation. The whole brain and the liver were quickly removed, placed on ice and weighed. The tissues were immediately homogenized in cold 10 mM Tris-HCl, pH 7.4 (1:10, w/v for the liver and 1:5, w/v for brain) and centrifuged at 3600 rpm for 10 min. The pellet was discarded and the supernatant was used to perform the assay. Aliquots of brain or liver homogenates (20 µL) was incubated with 10 µM $FeSO_4$ in the presence or absence of extracts (30–480 µg/mL) at 37 °C for 1 h, to induce lipid peroxidation. Subsequently, 40 µL of sodium dodecyl sulphate (8.1%), 100 µL of acetic acid/HCl (pH 3.4) and 100 µL of 0.6% thiobarbituric acid (TBA) were added and incubated at 100 °C for 1 h. After cooling, the samples were centrifuged for 2 min at 6000 rpm and the absorbance of the supernatant was read at 532 nm using an ELISA plate reader (SpectraMax). This work was carried out in accordance with the Guidelines of the Ethical Committee of UFSM and approved by the institutional review board of UFSM (076.2012-2).

3.5.3. Iron Chelating Activity of R. echinus Extracts

The chelating capacity of aqueous and ethanolic extracts of the leaves of R. echinus was determined according to the modified method of Kamdem et al. [27]. The reaction mixture containing 58 µL of saline solution (0.9%, w/v), 45 µL Tris-HCl (0.1 M, pH, 7.5), 27 µL of extracts (30–120 µg/mL) and 36 µL of 110 µM FeSO4 was incubated for 10 min at 37 °C. Then, 34 µL of 1,10-phenanthroline (0.25%, w/v) was added and the absorbance of the orange coloured complex formed was measured at 0, 10 and 20 min at 510 nm (against blank solutions of the samples) using microplate reader (SpectraMax). The same procedure was performed for the control (i.e., Fe^{2+}), but without the extract. To ascertain the chelating potential of the extracts, we determined the potential reduction of any Fe^{3+} (that might be formed during the incubation periods) by adding the reducing agent, ascorbic acid (to give a final concentration of 5 mM) o the reaction mixture. The absorbance was then determined after 5, 10 and 20 min following ascorbic acid addition. This is because the extracts could be oxidizing Fe^{2+} to Fe^{3+}, leading to a decrease in absorbance that was not related to Fe^{2+} chelation. To summarize, the schedule for the evaluation of Fe^{2+} chelation or oxidation by the extracts is presented in Table 1.

3.6. Blood Collection and Preparation of Human Leukocytes and Erythrocytes

Heparinized venous blood was obtained from healthy volunteer donors at the hospital of the Federal University of Santa Maria (UFSM), Santa Maria-RS, Brazil (age 26 ± 9). The study was approved by the Ethical Committee of UFSM and registered under the protocol number 0089.0.243.000-07. Human leukocytes and erythrocytes were obtained as previously reported [27,35]. The leukocytes were separated by differential sedimentation rate using 5% dextran and subsequent adjustment of samples to 2×10^6 leukocytes/mL with Hank's buffered saline solution (HBSS)/heparin (5.4 mM KCl, 0.3 mM Na_2HPO_4, 0.4 mM KH_2PO_4, 4.2 mM $NaHCO_3$, 1.3 mM $CaCl_2$, 0.5 mM $MgCl_2$, 0.6 mM $MgSO_4$, 137 mM NaCl, 10 mM D-glucose, 10 mM Tris-HCl, and heparin 15 IU/mL, adjusted to pH 7.4). However, the erythrocytes were separated by centrifuging the blood sample at 2000 rpm for 5 min at room temperature. The plasma was aspirated and the cell pellet was washed three times with phosphate buffered saline (6.1 mM, pH 7.4, containing 150 mM NaCl).

3.7. Determination of Erythrocytes Osmotic Fragility

The influence of the extracts on osmotic fragility of erythrocytes was estimated by measuring their resistance to hemolysis in increasing concentrations of salt solutions as modified by Barbosa-Filho *et al.* [35]. Five hundred microliters of erythrocytes, 100 µL of various concentrations of aqueous and ethanolic extracts (30–480 µg/mL) and 900 µL of phosphate buffer saline (PBS) (6.1 mM, pH 7.4, containing 150 mM NaCl) were pre-incubated for 3 h at 37 °C. After incubation, samples were mixed and centrifuged at 2500 rpm for 10 min and the supernatant was discarded. The erythrocytes were washed twice with PBS, centrifuged at 2500 rpm for 2 min and the supernatant discarded. Treated and untreated erythrocytes (7.5 µL) were then incubated with 1.5 mL of varying concentrations (0%–0.9%) of NaCl, pH 7.4 for 20 min. The samples were homogenized and centrifuged at 2000 rpm for 5 min. The supernatant obtained from each Eppendorf was transferred to microplate and the lysis of erythrocytes was monitored by measuring the absorbance of hemoglobin content in the supernatants at 540 nm using microplate reader (SpectraMax). The results were expressed as percentage of the control.

3.8. Genotoxicity by the Comet Assay

The procedure described by Collins [36] was used with some modifications as described by Kamdem *et al.* [27]. Leukocytes were isolated as described earlier and treated with different concentrations of aqueous and ethanolic extracts of *R. echinus* (30–480 µg/mL) for 3 h. The comet assay was performed according to the following steps: (1) 15 µL of leukocyte suspension (2×10^6 leukocytes/mL) was mixed with low-melting agarose; (2) added to 90 mL of 0.75% LMP agarose (w/v), mixed, and placed on a microscope slide precoated with normal melting point agarose (1% w/v); (3) a coverslip was added and the samples allowed to solidify at 4 °C; (4) coverslips were removed and slides were placed in a lysis solution (2.5 M NaCl; 100 mM EDTA; 8 mM Tris-HCl; 1% Triton X-100, pH 10–10.5), where they remained for 24 h protected from light; (5) after lysis, the slides were placed in bucket containing neutralizing solution (400 mM Tris-HCl; pH 7.5) for 15 min; (6) then, they were incubated in electrophoretic solution (300 mM NaOH, 1 mM EDTA, pH 13.5) for 20 min at 4 °C under the condition of 25 V, 300 mA, 7W; (7) the slides were washed three times in distilled water and allowed to dry at room temperature; (8) the slides were rehydrated for 3 min in distilled water and fixed for 10 min in 15% trichloroacetic acid, 5% zinc sulfate, and 5% glycerol, then, washed three times in distilled water and allowed to dry at room temperature; (9) the slides were stained with 5% sodium carbonate, 0.1% ammonium nitrate, 0.1% silver nitrate, 0.25% tungstosilicico acid and 0.15% formaldehyde; (10) staining was stopped with 1% acetic acid and air dried sheets; and, finally, (11) slides were visualized and scored according to tail length into five class (from class 0: undamaged, without a tail; to class 4: maximum damage, comet with no heads) under blind conditions by at least two individuals.

DNA damage was presented as DNA damage index (DI), which is based on the length of migration. The DI was calculated from cells in different damage classes as follows:

$$DI = 1 \times n1 + 2 \times n2 + 3 \times n3 + 4 \times n4 \tag{2}$$

where, n1–n4 indicates the number of cells with level 1–4 of damage. Methyl methanesulfonate (MMS) (20 µM) was used as positive control and the negative control with distilled water.

3.9. Cytotoxicity by the Trypan Blue Exclusion Method (Cell Viability)

The cell viability was determined using the Trypan blue exclusion method, which assumes that nonviable cells are stained blue [37]. Briefly, 2.5 µL of the extracts (30–480 µg/mL) was added to 497.5 µL of leukocytes suspension and incubated in the presence or absence hydrogen peroxide (2mM) + azide (1 mM), at 37 °C for 3 h. The azide was used to inhibit the activity of catalase in the cell and consequently detect the toxicity induced by H_2O_2 used as positive control. Thereafter, 50 µL leukocytes was mixed with 50 µL of 0.4% Trypan blue and allowed to stand for 5 min at room temperature. From the mixture, an aliquot of 10 µL was checked microscopically for viability using a hemocytometer. The viability was as expressed as viable cells in percent of the total cells.

3.10. Statistical Analysis

Values were expressed as mean ± standard error of mean (S.E.M). All the analyses were performed using one-way analysis of variance (ANOVA), except for iron chelation, where data were treated with two-way ANOVA followed by Bonferroni post-test. The results were considered significantly different when $p < 0.05$.

4. Conclusions

In conclusion, the results presented in this study demonstrated for the first time the *in vitro* antioxidant activity of extracts from the leaves of *R. echinus*, which was evidenced by their potential to scavenge the DPPH radical and inhibit lipid peroxidation in rat brain and liver homogenates. *R. echinus* extracts was neither cytotoxic nor genotoxic to human leukocytes and erythrocytes, respectively, which give indirect support of the popular use of this plant in traditional Brazilian medicine. However, *in vivo* investigation to ascertain the safety of this plant and its extracts needs to be conducted.

Acknowledgments: Supported by CAPES, FAPERGS, CNPq, FINEP and Vitae Fundation. A.E.D. is particularly grateful to CAPES-DINTER for financial support. A.E.D. is a beneficiary of the CAPES Postgraduate (Doctoral) fellowship. J.P.K. acknowledges CAPES, TWAS and CNPq-TWAS for the financial support.

Author Contributions: M.E.B., J.B.T.R. and J.P.K. designed the research; J.G.M.C., I.R.A. and M.A.P.S. helped in the preparation, concentration and lyophilization of extracts; A.E.D., E.P.W., L.M.B. and K.R. performed the experiments and contributed to data collection; F.A.B.C. assisted in the statistical analysis; A.A.B. performed the HPLC analysis; A.O.A. determined the total phenolic content and edited the English; and A.E.D. and J.P.K. wrote the manuscript. J.B.T.R. and M.E.B. contributed to the revisions of the manuscript. All authors have read and approved the final version of the manuscript.

Conflicts of Interest: The authors declare no conflict of interest.

References

1. Schmeiser, H.H.; Stiborova, M.; Arlt, V.M. Chemical and molecular basis of the carcinogenicity of Aristolochia plants. *Curr. Opin. Drug Discov. Dev.* **2009**, *12*, 141–148.

2. Ekor, M. The growing use of herbal medicines: Issues relating to adverse reactions and challenges in monitoring safety. *Front. Pharmacol.* **2014**, *4*. [CrossRef] [PubMed]

3. Anantha, N.D. Approaches to pre-formulation R and D for phytopharmaceuticals emanating from herb based traditional Ayurvedic processes. *J. Ayurveda Integr. Med.* **2013**, *4*, 4–8. [CrossRef] [PubMed]

4. Waczuk, E.P.; Kamdem, J.P.; Abolaji, A.O.; Meinerz, D.F.; Caeran Bueno, D.; do Nascimento Gonzaga, T.K.S.; do Canto Dorow, T.S.; Boligon, A.A.; Athayde, M.L.; da Rocha, J.B.T.; *et al. Euphorbia tirucalli* aqueous extract induces cytotoxicity, genotoxicity and changes in antioxidant gene expression in human leukocytes. *Toxicol. Res.* **2015**, *4*, 739–748.

5. Vásquez, G.D.; Harley, R.M. Flora de Grão-Mogol, Minas Gerais: Labiatae. *Bol. Bot. Univ. São Paulo* **2004**, *22*, 193–204. [CrossRef]

6. Souza, A.A.; Rodrigues, S.A. Antimicrobial activity of essential oil of *Raphiodon echinus* (Nees. e Mart) Shauer. *Rev. Biol. Farm.* **2012**, *7*, 12–17.

7. Menezes, F.S.; Kaplan, M.A.K.; Cardoso, G.L.C.; Pereira, N.A. Phytochemical and pharmacological studies on *Rhaphiodon echinus*. *Fitoterapia* **1998**, *69*, 459–460.

8. Kehrer, J.P.; Klotz, L. Free radicals and related reactive species as mediators of tissue injury and disease: Implications for health. *Crit. Rev. Toxicol.* **2015**, *45*, 765–798. [CrossRef] [PubMed]

9. Ambali, S.F.; Abubakar, M.; Shittu, M.; Yaqub, L.S.; Anafi, S.B.; Abdullahi, A. Chlorpyrifos induced alteration of haematological parameters in Wistar rats: Ameriolative effect of zinc. *Res. J. Environ. Toxicol.* **2010**, *4*, 55–66.

10. Fetoui, H.; Gdoura, R. Synthetic pyrethroid increases lipid and protein oxidation and induces glutathione depletion in the cerebellum of adult rats: Ameliorative effect of vitamin C. Hum. *Exp. Toxicol.* **2012**, *31*, 1151–1160. [CrossRef] [PubMed]

11. Cheung, L.M.; Cheung, P.C.K.; Ooi, V.E.C. Antioxidant activity and total phenolics of edible mushroom extracts. *Food Chem.* **2003**, *81*, 249–255. [CrossRef]

12. Kotássková, E.; Sumczynski, D.; Mlcek, J.; Valásek, P. Determination of free and bound phenolics using HPLC-DAD antioxidante activity and *in vitro* digestibility of *Eragrostis tef. J. Food Compos. Anal.* **2016**, *46*, 15–21. [CrossRef]

13. Eugene, D.W. The hazards of iron loading. *Metallomics* **2010**, *2*, 732–740.

14. Zecca, L.; Casella, L.; Albertini, A.; Bellei, C.; Zucca, F.A.; Engelen, M.; Zadlo, A.; Szewczyk, G.; Zareba, M.; Sarna, T. Neuromelanin can protect against iron-mediated oxidative damage in system modeling iron overload of brain aging and Parkinson's disease. *J. Neurochem.* **2008**, *106*, 1866–1875. [PubMed]

15. Becerril-Ortega, J.; Bordji, K.; Fréret, T.; Rush, T.; Buisson, A. Iron overload accelerates neuronal amyloid-β production and cognitive impairment in transgenic mice model of Alzheimer's disease. *Neurobiol. Aging* **2014**, *35*, 2288–2301. [CrossRef] [PubMed]

16. Hassan, H.; Oliveira, C.S.; Noreen, H.; Kamdem, J.P.; Nogueira, C.W.; Rocha, J.B.T. Organoselenium compounds as potential neuroprotective therapeutic agents. *Curr. Org. Chem.* **2016**, *20*, 218–231. [CrossRef]

17. Chikezie, A.R.; Akuwudike, C.M. Membrane stability and methaemoglobin content of human erythrocytes incubated in aqueous leaf extract of *Nicotiana tabacum*. *Free Radic. Antioxid.* **2012**, *2*, 56–61. [CrossRef]

18. Pistón, M.; Machado, I.; Branco, C.S.; Cesio, V.; Heinzen, H.; Ribeiro, D.; Fernandes, E.; Chisté, R.C.; Freita, M. Infusion, decoction and of leaves from artichoke are effective scavengers of physiologically relevant ROS and RNS. *Food Res. Int.* **2014**, *64*, 150–156. [CrossRef]

19. Iwahashi, H. Inhibitory effects of caffeic acid on free-radical formation. In *Coffee in Health and Disease prevention*; Preedy, V.R., Ed.; Elsevier Inc.: Atlanta, GA, USA, 2014; Volume 88, pp. 803–812.

20. Oboh, G.; Puntel, R.L.; Rocha, J.B.T. Hot pepper (*Capsicum annuum*, tepin & *Capsicum Chinese*, habanero) prevents Fe^{2+}-induced lipid peroxidation in brain—*In vitro*. *Food Chem.* **2007**, *102*, 178–185.

21. Stoddart, M.J. Cell viability assays: Introduction. *Methods Mol. Biol.* **2011**, *740*, 1–6. [PubMed]

22. He, J.; Lin, J.; Li, J.; Zhang, J.H.; Sun, X.M.; Zeng, C.M. Dual effects of *Ginkgo biloba* leaf extract on human red blood cells. *Basic Clin. Pharmacol. Toxicol.* **2009**, *104*, 138–144. [CrossRef] [PubMed]

23. Ghorbel, L.; Maktouf, S.; Kallel, C.; Ellouze Chaabouni, S.; Boudawara, T.; Zeghal, N. Disruption of erythrocyte antioxidant defense system, hematological parameters, induction of pro-inflammatory cytokines and DNA damage in liver of co-exposed rats to aluminiun and acrylamide. *Chem. Biol. Interact.* **2015**, *236*, 31–40. [CrossRef] [PubMed]

24. Perrone, S.; Tataranno, M.L.; Stazzoni, G.; del Vecchio, A.; Buonocore, G. Oxidative injury in neonatal erythrocytes. *J. Matern. Fetal. Neonatal. Med.* **2012**, *25*, 104–108. [CrossRef] [PubMed]

25. Lang, E.; Lang, F. Mechanisms and pathophysiological significance of eryptosis, the suicidal erythrocyte death. Semin. *Cell. Dev. Biol.* **2015**, *39*, 35–42. [CrossRef] [PubMed]

26. Sharma, P.; Sharma, J.D. *In vitro* hemolysis of human erythrocytes by plant extracts with antiplasmodial activity. *J. Ethnopharmacol.* **2001**, *74*, 239–243. [CrossRef]

27. Kamdem, J.P.; Adeniran, A.; Boligon, A.A.; Klimaczewski, C.V.; Elekofehinti, O.O.; Hassan, W.; Ibrahim, M.; Waczuk, E.P.; Meinerz, D.F.; Athayde, M.L. Antioxidant activity, genotoxicity and cytotoxicity evaluation of lemon balm (*Melissa officinalis* L.) ethanolic extract: Its potential role in neuroprotection. *Ind. Crop. Prod.* **2013**, *51*, 26–34. [CrossRef]

28. Roy, P.; Mukherjee, A.; Giri, S. Evaluation of genetic damage in tobacco and arsenic exposed population of Southern Assam, India using buccal cytome and comet assay. *Ecotoxicol. Environ. Saf.* **2015**, *124*, 169–176. [CrossRef] [PubMed]

29. Galloway, S.M.; Miller, J.E.; Armstrong, M.J.; Bean, C.L.; Skopek, T.R.; Nichols, W.W. DNA synthesis inhibition as an indirect mechanism of chromosome aberrations: Comparison of DNA-reactive and non-DNA-reactive clastogens. *Mutat. Res.* **1998**, *400*, 169–186. [CrossRef]

30. Galloway, S.M. Cytotoxicity and chromosome aberrations *in vitro*: Experience in industry and the case for an upper limit on toxicity in the aberration assay. *Environ. Mol. Mutagen.* **2000**, *35*, 191–201. [CrossRef]

31. Pereira, R.P.; Boligon, A.A.; Appel, A.S.; Fachinetto, R.; Ceron, C.S.; Tanus-Santos, J.E.; Athayde, M.L.; Rocha, J.B.T. Chemical composition, antioxidant and anticholinesterase activity of *Melissa officinalis*. *Ind. Crops Prod.* **2014**, *53*, 34–45. [CrossRef]

32. Kamdem, J.P.; Stefanello, S.T.; Boligon, A.A.; Wagner, C.; Kade, I.J.; Pereira, R.P.; Preste, A.S.; Roos, D.H.; Waczuk, E.P.; Appel, A.S.; *et al. In vitro* antioxidant activity of stem bark of *Trichilia catigua* Adr. Juss. *Acta Pharm.* **2012**, *62*, 371–382. [CrossRef] [PubMed]

33. Boligon, A.A.; Pereira, R.P.; Feltrin, A.C.; Machado, M.M.; Janovik, V.; Rocha, J.B.T.; Athayde, M.L. Antioxidant activities of flavonol derivatives from the leaves and stem bark of Scutia buxifolia Reiss. *Bioresour. Technol.* **2009**, *100*, 6592–6598. [CrossRef] [PubMed]

34. Ohkawa, H.; Ohishi, N.; Yagi, K. Assay for lipid peroxides in animal tissues by thiobarbituric acid reaction. *Anal. Biochem.* **1979**, *95*, 351–358. [CrossRef]

35. Barbosa-Filho, V.M.; Waczuk, E.P.; Kamdem, J.P.; Lacerda, S.R.; Martin da Costa, J.C.; Alencar de Menesez, I.R.; Abolaji, A.O.; Boligon, A.A.; Athayde, M.L.; Rocha, J.B.T.; *et al.* Phytochemical constituents, antioxidant activity, cytotoxicity and osmotic fragility effects of caju (*Anacardium. microcarpum*). *Ind. Crop. Prod.* **2014**, *55*, 280–288. [CrossRef]

36. Collins, A.R. The comet assay for DNA damage and repair. Principles, application, and limitations. *Mol. Biotechnol.* **2004**, *26*, 249–261. [CrossRef]

37. Mischell, B.B.; Shiingi, S.M. *Selected Methods in Cellular Immunology*; WH Freeman Company: New York, NY, USA, 1980; pp. 1–469.

Benz[c,d]indolium-containing Monomethine Cyanine Dyes: Synthesis and Photophysical Properties

Eduardo Soriano [1,†], Cory Holder [1,†], Andrew Levitz [1] and Maged Henary [1,2,*]

Academic Editor: Pani Koutentis

1   Department of Chemistry, Georgia State University, 50 Decatur St., Atlanta, GA 30303, USA;
    esoriano1@gsu.edu (E.S.); cholder1@gsu.edu (C.H.); alevitz1@gsu.edu (A.L.)
2   Center for Diagnostics and Therapeutics, Georgia State University, Petit Science Center,
    100 Piedmont Ave SE, Atlanta, GA 30303, USA
*   Correspondence: mhenary1@gsu.edu
†   These authors contributed equally to this work.

Abstract: Asymmetric monomethine cyanines have been extensively used as probes for nucleic acids among other biological systems. Herein we report the synthesis of seven monomethine cyanine dyes that have been successfully prepared with various heterocyclic moieties such as quinoline, benzoxazole, benzothiazole, dimethyl indole, and benz[e]indole adjoining benz[c,d]indol-1-ium, which was found to directly influence their optical and energy profiles. In this study the optical properties *vs.* structural changes were investigated using nuclear magnetic resonance and computational approaches. The twisted conformation unique to monomethine cyanines was exploited in DNA binding studies where the newly designed sensor displayed an increase in fluorescence when bound in the DNA grooves compared to the unbound form.

Keywords: cyanine dye; unsymmetrical; synthesis; optical properties; DFT calculations; DNA grooves

---

1. Introduction

Polymethine dyes represent a class of organic molecules with absorption bands that cover a broad spectral range (430–1100 nm), larger than any other class of dye system [1]. Cyanine dyes consist of two terminal aza-heterocycles connected via an electron deficient polymethine bridge that allows for a push/pull system between the two heterocycles. The delocalization of electrons across this bridge causes them to exhibit long wavelength absorptions. In addition to the variable length of the conjugated system between the heterocycles, the heterocycles themselves can be altered which allows chemists to create dyes that possess ideal photophysical properties, such as high molar extinction coefficients ($>10^5$ M$^{-1}\cdot$cm^{-1}), tunable fluorescence intensities, and narrow absorption bands. Due to the diversity in function associated with this class of chromophore, an extensive number of cyanine dyes have been synthesized and developed for numerous applications in photographic processes and more recently as fluorescent probes for bio-molecular labeling and imaging [1–9].

As cyanine dyes have been shown to be highly modifiable for desirable properties such as solubility, permeability, and binding, these modifications can also cause changes in the dye's photophysical properties. Recently, the interpretation of the fluorogenic behavior of the monomethine cyanine dyes from *in silico* studies has been successfully used to design new fluorescent molecular rotors as viscosity sensors [10]. Two asymmetric dyes shown in Figure 1, thiazole orange (TO) and oxazole yellow (YO), are well known imaging probes in the biological sciences due to their enhanced photophysical properties which have been attributed to restricted torsional motion of the dye in the excited state upon binding to target a macromolecule (*i.e.*, nucleic acid structure, protein) [11–14]. TO absorbs and fluoresces at 501 nm and 525 nm, respectively, while YO absorbs and fluoresces at

491 nm and 509 nm, respectively [15]. The dimers of these compounds are also known imaging probes and shown in Figure 1. YOYO absorbs and fluoresces at 450 nm [16] and 510 nm, respectively, while TOTO absorbs and fluoresces at 513 and 530 nm, respectively [17,18]. Nonetheless, there is a lack of understanding of how the structure interplays with the optical performance (*i.e.*, extinction coefficient and fluorescence)—especially for those monomethine cyanines with red-shifted wavelengths [9,19–21]. Thus, it is important to understand how varying substituents and heterocycles would affect the optical properties of each dye.

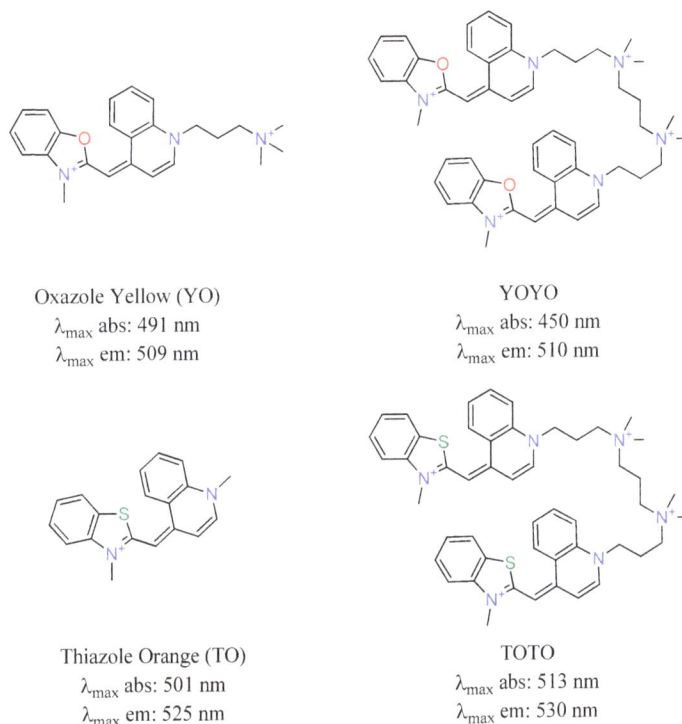

Oxazole Yellow (YO)
λ_{max} abs: 491 nm
λ_{max} em: 509 nm

YOYO
λ_{max} abs: 450 nm
λ_{max} em: 510 nm

Thiazole Orange (TO)
λ_{max} abs: 501 nm
λ_{max} em: 525 nm

TOTO
λ_{max} abs: 513 nm
λ_{max} em: 530 nm

Figure 1. Commercially available asymmetric monomethine cyanine dyes.

Imaging of macromolecules such as DNA by staining with fluorescent compounds is of great interest, therefore, expanding the options of available probes is vital to several areas of research spanning from medical diagnostics to genomics [22–38]. The synthesis of low cost, easy to manipulate systems for fast analysis is required [8]. Fluorescent detection has rapidly become one of the most widely used techniques due to its sensitivity and noninvasiveness [39]. Ethidium bromide has commonly been used for the detection of DNA, however it has mutagenic effects and poses other environmental concerns [40–42]. On the other hand, cyanine dyes are sensitive, safe and highly modifiable.

Recently, our group has synthesized a series of benz[*c,d*]indol-1-ium-containing monomethine cyanines with separate adjoining heterocyclic moieties which were found to directly influence the optical properties of the dye system [20]. In this report seven additional red-shifted monomethine cyanine dyes were synthesized and the structural influence on their fluorogenic properties was investigated by comparing the optical characteristics, examining the change in chemical shifts of methine proton and carbon NMR spectra, determining the energy profile through *in silico* approaches, as well as demonstrating that the dyes can be employed as DNA binding agents. The ability to use the theoretical calculations of optical properties for fluorophores, such as monomethine dyes could be useful for the development of the viscosity detection methods or bioimaging agents with desirable optical profiles.

2. Results and Discussion

2.1. Synthesis

Toward gaining better understanding of the relationship between various heterocyclic substitutions and changes in optical properties we began to rationally design and investigate the effect of altering the heterocyclic substitution on the photophysical characteristics of the dye systems. Two sets of monomethine cyanines were explored without altering the benz[c,d]indole heterocycle half of the dye. The first set possessing different heterocycles including 2-methylbenzothiazole, 2-methylbenzoxazole, 3,3-dimethylbenz[e]indole or 2-methylquinoline, respectively, and the second set containing the same 3,3-dimethylindole heterocycle, but with different substituents, one electron donating and one electron withdrawing, on the 5 position of the heterocyclic ring system.

The asymmetric red-shifted monomethine cyanine dyes were synthesized as shown in Scheme 1. The synthesis began with the alkylation of benz[c,d]indol-2(1H)-one (1) by reflux with iodobutane in acetonitrile. The alkylated amide 2 was then converted to the thioketone 3 under reflux with phosphorous pentasulfide in pyridine. The thioketone 3 was methylated to a thioether with iodomethane creating the key precursor, quaternary ammonium salt 4, which was used as one heterocycle. The second heterocycle was synthesized beginning with a Fischer indole synthesis by refluxing 4-substituted phenylhydrazine hydrochlorides 7 and 3-methyl-2-butanone in glacial acetic acid. The synthesized heterocyclic derivatives 8, 2-methylbenzothiazole, 2-methylbenzoxazole, 2,3,3-trimethylbenz[e]indole, and 2-methylquinoline were alkylated, respectively, with various alkyl halides in acetonitrile to form quaternary ammonium salts 5a–d and 9a–c, which acted as the second heterocycle for the final dyes. The two heterocycles were then connected by a condensation reaction in acetonitrile with a catalytic amount of triethylamine to afford final dyes 6a–d and 10a–c.

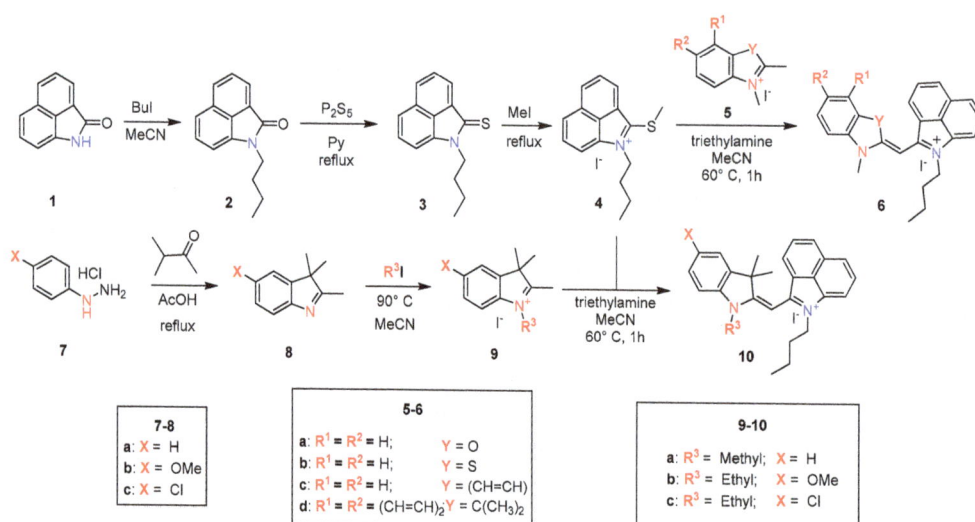

Scheme 1. Synthesis of Monomethine Dyes.

The reaction begins with the deprotonation of the methyl group at the 2 position of the heterocycle. This activated methylene group of the various heterocyclic salts 5a–d and 9a–c displaces the methyl sulfide moiety of 4 and results in the formation of the asymmetrical monomethine dyes 6a–d and 10a–c. After isolation, the dyes were characterized by HRMS, ^{1}H- and ^{13}C-NMR and their photophysical properties were investigated.

2.2. Optical Properties

Optical properties are shown in Table 1. Absorption for each dye was recorded in methanol and 9/1 glycerol/methanol solution. Many monomethine cyanines display multiple bands which are attributed to different vibronic bands of the same electronic transition [16]. Because the compounds did not fluoresce in methanol due the ability to freely rotate around the methine bridge in free flowing solvent, emission was recorded in a more viscous solvent, 9/1 glycerol/methanol solution. Representative UV-Vis spectra are shown in Figure 2. A symmetrical monomethine dye containing two benzothiazole heterocycles has a λ_{max} of 430 nm in ethanol [43]. It has been shown by Brooker *et al.*, that if the nitrogen containing heterocycles are not identical, or if the relative stabilities of the two resonance forms are different, the absorption would not be at the midpoint [44]. The substitution of one of these heterocycles with benz[c,d]indole shifts the λ_{max} over 100 nm to 555 nm as seen in **6b**. This was accounted for by the further conjugated electron deficient system in the benz[c,d]indole heterocycle [1,20,45,46]. The conjugated system has more electronegativity due to the oxygen atom in **6a** causing a blue shift of the λ_{max} to 498 nm [11,47]. While the compounds containing 3,3-dimethylindole have similar absorption maxima to the benzothiazole compounds, the addition of an extra benzene ring as seen in **6d** red shifts the λ_{max} to 585 nm due to the increased conjugation through the heterocycle. All of the dyes displayed molar extinction coefficients in the range of 24,000–38,000 $M^{-1}\cdot cm^{-1}$. The dye with a methoxy substituted indole heterocycle **10b** showed the lowest molar absorptivity at 24,800 $M^{-1}\cdot cm^{-1}$ due to the electron donating nature of the methoxy group introducing electron density back into the system [47,48]. Aggregation was ruled out by measuring absorption of **6b** as a representative compound at various concentrations (5–25 µM) and the results were presented in the Supplementary Materials (Figure S2G). Solvatochromic studies were performed on dye **6b** in five different solvents (ethanol, dimethyl formamide, dichloromethane, acetonitrile, and aqueous tris buffer) (Figure S2H). Less than 5 nm change in λ_{max} was observed. Such a small shift suggests that the electronic distribution of the ground state dye is virtually unaffected by the solvent polarity [47].

Table 1. Spectral Characteristics of Dyes **6a–d** and **10ac**.

Dye	λ_{abs} (nm) [a]	λ_{abs} (nm) [b]	$\lambda_{emission}$ (nm) [b]	Stokes Shift (nm) [b]	ε ($M^{-1}\cdot cm^{-1}$) [a]
6a	498	505	570	65	37600
6b	555	563	609	46	32300
6c	585	587	609	22	36500
6d	553	557	625	68	25300
10a	537	552	657	105	33300
10b	563	569	606	37	24800
10c	552	571	662	91	30100

[a] methanol [b] methanol/glycerol 9/1 (*v/v*).

It has been reported that the fluorescence of these compounds cannot be observed in methanol alone because of a high nonradiative rate of return of the excited molecule as previously reported with many monomethine cyanines [20,49–51]. However, when a viscous solution is used, the free rotation around the methine bridge is restricted and a fluorescence signal is observed as shown in Figure 2. Methanol (10%) was used in order to solubilize the compounds in the highly viscous glycerol. Fluorescence maxima ranged from 570 nm to 662 nm, almost reaching the near-infrared region. The benzoxazole containing dye **6a** had the highest fluorescence intensity followed by benzothiazole containing dye **6b**. The quinoline containing dye **6c** had the least fluorescence intensity due to alternative relaxation pathways [52]. The largest Stokes shift, greater than 100 nm, was observed for the dye with an indole based heterocycle, **10a**. Since the emission intensity was so low the Stokes shift reported could be slightly skewed due to low signal to noise. However, this finding is in agreement with red-shifted compounds previously synthesized by our group [20]. Large Stokes shifts are ideal for imaging applications as the excitation light is farther from the fluorescence signal of the compound [39,53].

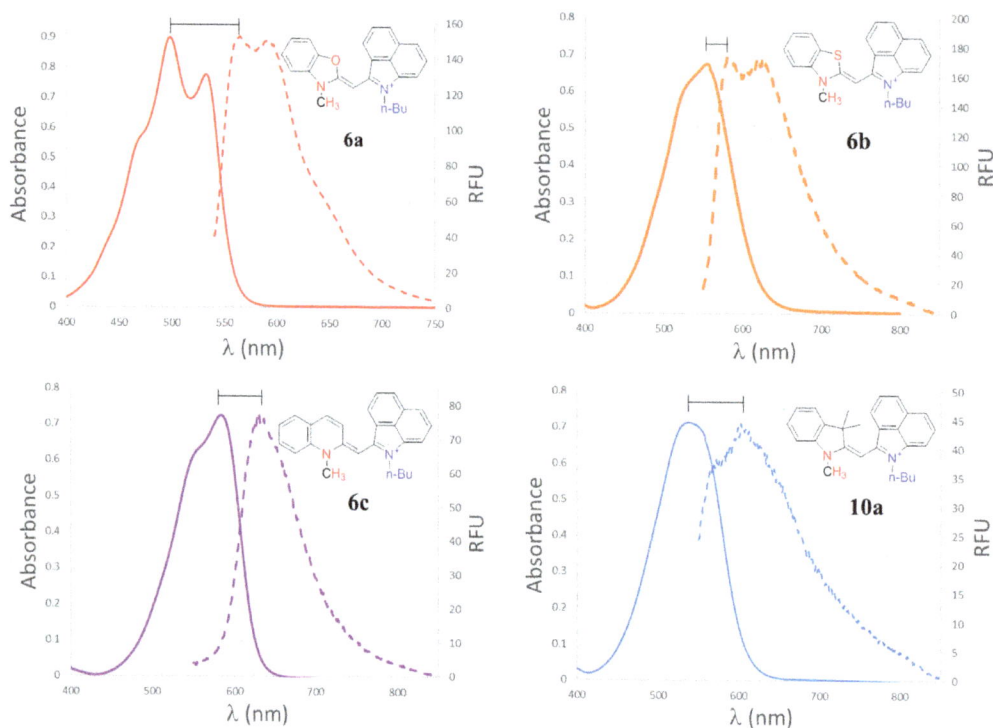

Figure 2. Absorbance (solid lines) and emission (dashed lines) in methanol/glycerol 9/1 spectra at 20 µM.

2.3. Computational Evaluations

The electronic spectra of the monomethine dyes were investigated to help elucidate the trends described above in the optical properties. As shown by the calculations in Figure 3, over the series of dyes when the geometry is planar both the HOMO and LUMO orbitals are spread evenly throughout the dye. When the dyes are twisted out of plane the HOMO orbitals are localized around the more conjugated system benz[c,d]indole heterocycle. The energy transitions in cyanine dyes have been shown to be a dominant π–π* transition [11,21], but if the dye assumes a twisted geometry the orbitals are not delocalized throughout the dye, as shown in Figure 3, and the system is not conjugated or planar [54].

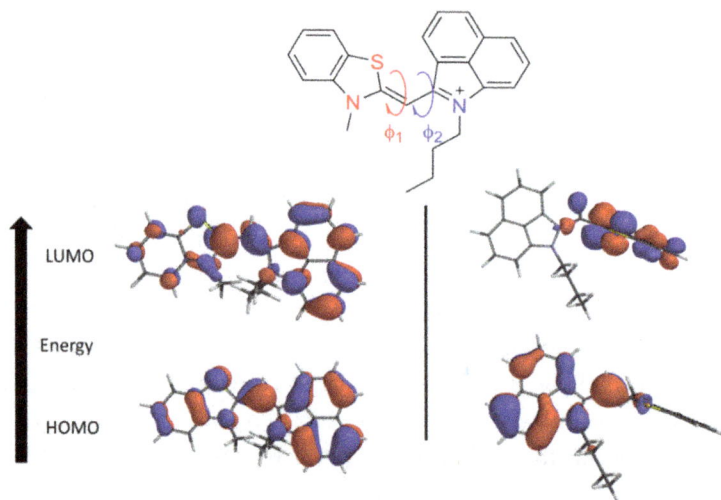

Figure 3. Frontier molecular orbitals of **6a** constrained in planar (**left**) and twisted (**right**) configurations.

The geometry was constrained to keep the molecule planar to observe trends in the HOMO–LUMO gaps for comparing with excitation energies. As shown in Figure 4, the energy gap between HOMO and LUMO of compound **6d** containing a benz[*e*]indole heterocycle is the lowest among the series of dyes at 2.06 eV. This finding is corroborated by the bathochromic absorbance maximum of the benz[*e*]indole compared to the Fischer indole, benzothiazole, and benzoxazole heterocycles which led to further delocalizing of the electrons and therefore stabilizing the orbitals.

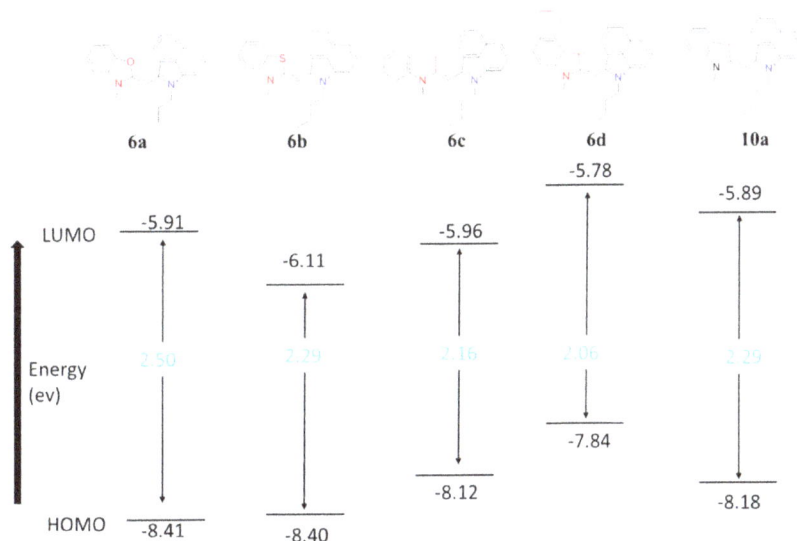

Figure 4. HOMO and LUMO orbital analysis of differing heterocycles in the monomethine system; energies (black), HOMO-LUMO gaps (blue).

The benzoxazole heterocycle in dye **6a** influenced the conjugated system shown by shifting the absorbance maximum to the blue. This dye **6a** shows the highest energy gap likely due to both the lone pair of electrons and electronegativity of the oxygen atom similar to dye **6b** with a sulfur containing benzothiazole heterocycle that had the second highest energy gap. Dye **10a** containing a 3,3-dimethylindole heterocycle had the same energy gap as **6b** with the benzothiazole heterocycle, but had higher energy.

The theoretical absorption λ_{max} values are plotted along with the experimental data as shown in Figure 5. Time-dependent density functional theory (TD-DFT) has been shown to work well for large conjugated molecules because the orbitals are obtained by solving the Kohn-Sham equation involving exchange and correlation (XC) terms [55]. Although a discrepancy gap is observed between the theoretical and experimental results, the observed trends in absorbance wavelength are almost the same with the calculated absorbance wavelength giving slightly blue-shifted values [47].

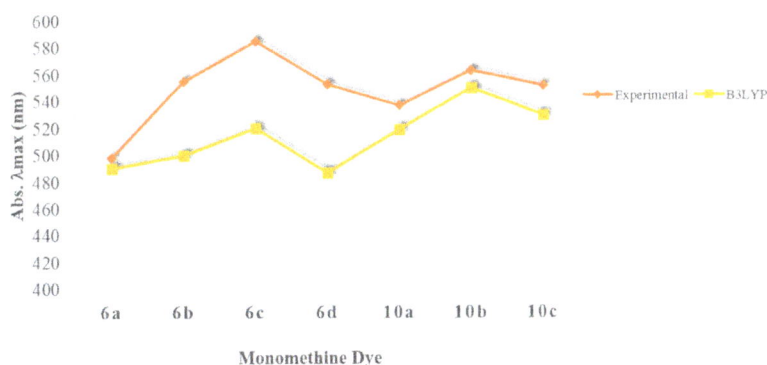

Figure 5. Experimental and Calculated λ_{max} values.

As shown in Figure 6 and Table 2, the observed change of the chemical shift of the methine-proton is most likely due to altering the electron density from the surrounding atoms.

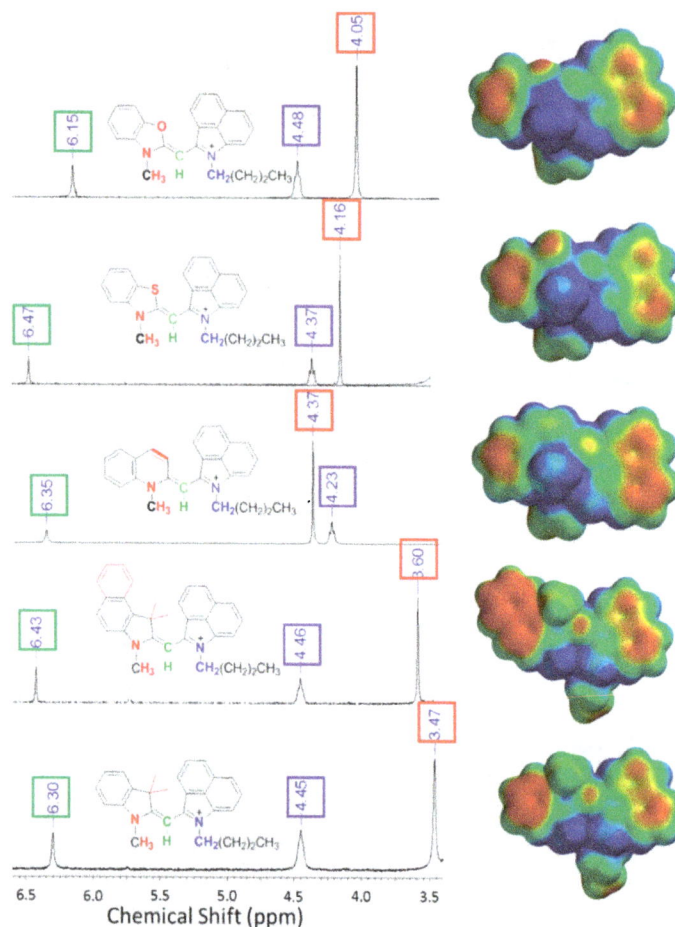

Figure 6. ^{1}H-NMR shift of *meso*-proton in DMSO-d_6 at 25 °C, Calculated EMP on the right.

Table 2. λ_{max}, NMR shifts, and computational charges of monomethine cyanine dyes.

	Heterocycle Included in Monomethine Dye	λ_{abs} (nm) exp.	λ_{abs} (nm) calc.	Charge of Methine Carbon	Methine Carbon Shift (ppm)	Methine Proton Shift (ppm)	N-CH₃ [1]H Shift (ppm)
6a	benzoxazole	498	490	−0.535	75.51	6.15	4.05
6b	benzothiazole	555	500	−0.421	87.40	6.47	4.16
6c	quinoline	585	520	−0.526	93.65	6.35	4.37
6d	benz(e)indole	553	487	−0.284	94.10	6.43	3.60
10a	3,3-dimethylindole	542	519	−0.328	82.78	6.30	3.47
	Substitution at the 5-position of heterocycle 10a						
10a	H	542	519	−0.328	82.78	6.31	-
10b	OMe	563	550	−0.316	83.44	6.23	-
10c	Cl	552	530	−0.344	83.81	6.29	-

Calculated values obtained via TD-DFT in vacuum, NMR run in DMSO-d_6 at 25 °C.

2.4. DNA Binding

It has been reported that a combination of a crescent shape complements the helical DNA minor groove, hydrogen bond donors and acceptors on the side of the molecule facing the DNA, a cationic center to enhance electrostatic interactions with negatively charged phosphate groups, and hydrophobic character from an extended fused heterocyclic structure allows for optimization of the compound for DNA minor groove interactions [56–59]. Dye **6b**, which is crescent shaped and has an overall hydrophobic structure, includes a sulfur on the side suggested by computational data to be facing the DNA (Figure 7) and contains delocalized positive charge throughout the polymethine chain; therefore, it was selected for DNA binding as a representative example of the series.

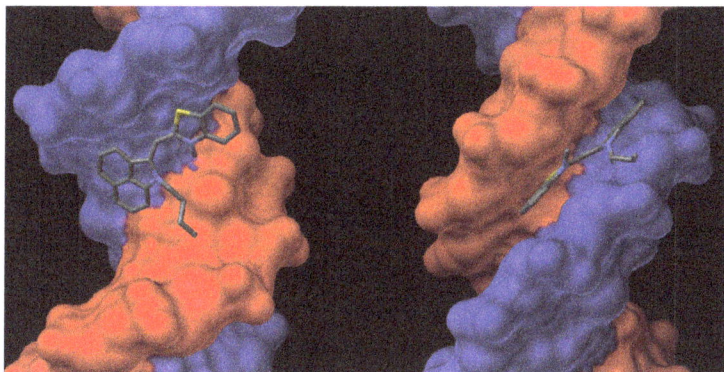

Figure 7. Dye **6b** with fixed torsion angles and planar geometry suggested to bind to the major (**left**) and minor (**right**) grooves of dsDNA by computational studies.

As presented in Figure 8, the fluorescence spectrum of **6b** in Tris-HCl buffer exhibits a particularly weak fluorescence spectrum with 2 local maxima at 565 nm and 630 nm. The 565 nm band is red shifted to 582 when ct-DNA is added and an increase in fluorescence is observed. Similar to the previously described enhancement in glycerol, a viscous solvent, this enhancement is also attributable to the fact that on excitation the inability to freely rotate around the methine bond due to binding does not allow for nonradiative deactivation of the ground state causing the dye to fluoresce. Using a double reciprocal plot, the binding constant, K_b, of **6b** was determined to be $1.0 \times 10^4 \ M^{-1}$ which is on par with similar monomethine cyanine dyes [8].

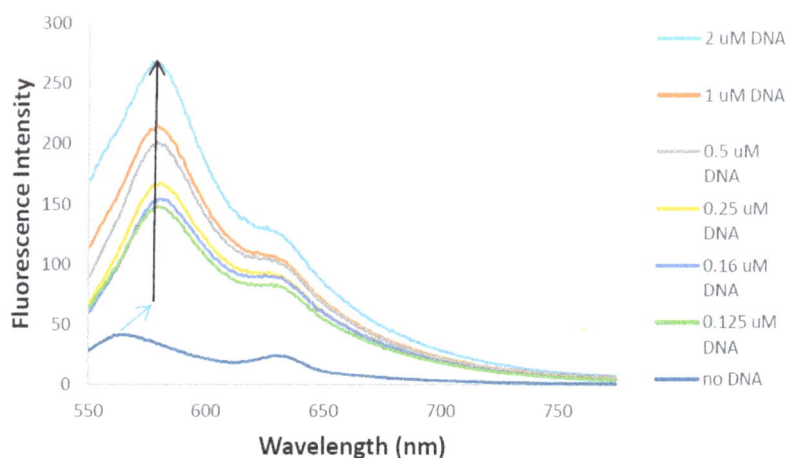

Figure 8. Emission spectra of dye **6b** (10 μM) in Tris-HCl buffer with and without ct-DNA (excitation wavelength 520 nm).

Although dye **6b** is structurally similar to TO (Figure 1), a known intercalating agent, it is intriguing to investigate interactions at the molecular level. Therefore, computational studies were conducted to get better insight on the mode of binding for these red shifted monomethines. The 264D (a dodecamer d(CGCAAATTTGCG)$_2$) was chosen from the Protein Data Bank as a representative model for dsDNA binding. Molecular docking was then performed on **6b** using Autodock (Figure 7). As it turns out, docking was achieved in both the minor and major grooves. Our computational data indicates higher propensity to bind in the minor groove based on relative scoring. Surprisingly,

6b did not display intercalation based on these computational studies. This could be due to the bulkiness of the benz[c,d]indole heterocycle. Further studies such as electrophoresis unwinding assays or crystallography can be conducted in the future to more accurately define the binding modes of these compounds.

3. Experimental

3.1. General Information

All chemicals and solvents were of American Chemical Society grade or HPLC purity and were used as received. HPLC grade methanol and glycerol were purchased from Sigma-Aldrich (St. Louis, MO, USA). All other chemicals were purchased from Fisher Scientific (Pittsburgh, PA, USA) or Acros Organics (Pittsburgh, PA, USA). The reactions were followed using silica gel 60 F_{254} thin layer chromatography plates (Merck EMD Millipore, Darmstadt, Germany). The ^{1}H-NMR and ^{13}C-NMR spectra were obtained using high quality Kontes NMR tubes (Kimble Chase, Vineland, NJ, USA) rated to 500 MHz and were recorded on an Avance spectrometer (Bruker, Billerica, MA; 400 MHz for ^{1}H and 100 MHz for ^{13}C) in DMSO-d_6 or CD$_3$Cl-d_3. High-resolution accurate mass spectra (HRMS) were obtained at the Georgia State University Mass Spectrometry Facility using a Q-TOF micro (ESI-Q-TOF) mass spectrometer (Waters, Milford, MA, USA). HPLC data was obtained using a Waters 2487 dual detector wavelength absorption detector with wavelengths set at 260 and 600 nm. The column used in LC was a Waters Delta-Pak 5 μM 100 Å 3.9 × 150 mm reversed phase C18 column, with a flow rate of 1mL/min employing a 5%–100% acetonitrile/water/0.1% formic acid gradient. All compounds tested were >95% pure. Infrared spectra (FT-IR) were obtained using a Spectrum 100 spectrometer (PerkinElmer, Duluth, GA, USA) (see Supplementary Materials). UV-Vis/NIR absorption spectra were recorded on a Cary 50 spectrophotometer (Varian, Palo Alto, CA, USA) interfaced with Cary WinUV Scan Application v3.00 using VWR disposable polystyrene cuvettes with a 1 cm pathlength. Laser Induced Fluorescence (LIF) emission spectra were acquired using Shimadzu RF-5301 Spectroflurophotometer (Shimadzu Corporation Analytical Instruments Division, Duisburg, Germany) interfaced to a PC with RF-5301PC software using Sigma-Aldrich disposable polystyrene fluorimeter cuvettes with a 1 cm pathlength. All spectral measurements were recorded at room temperature. The data analysis and calculations were carried out using Microsoft Excel (Microsoft Corporation, Redmond, WA, USA).

3.2. Synthesis

3.2.1. General Synthetic Procedure for the Indolium Salts 4 and 9a–c

Thioether **4** was previously synthesized by our group and others [20,60]. The substituted indoles **8** were synthesized as previously reported by our group and others [20,61]. Each individual compound **8** was dissolved in acetonitrile and alkyl halide was added. The reaction mixture was then refluxed for 12 h. Thin layer chromatography (TLC) was used to monitor the reaction progress using a mixture of 4:1 dichloromethane-hexanes. Upon cooling to room temperature, the quaternary ammonium salts **9a–c** were precipitated in diethyl ether and collected by vacuum filtration [36,62].

3.2.2. General Synthesis of the Monomethine Dyes

Thioether **4** and each quaternary ammonium salt **5a–d** and **9a–c**, respectively, were dissolved in acetonitrile and a catalytic amount of triethlyamine was added to the solution. The reaction mixture was refluxed at 60 °C for 1 h and monitored by UV-Vis. Upon cooling to room temperature, the corresponding dyes **6a–d** and **10a–c** were precipitated by adding diethyl ether. The solid was collected by vacuum filtration and triethylammonium salts were removed by washing with deionized water. The final dyes were purified via precipitation from methanol with diethyl ether.

1-Butyl-2-[(3-methyl-1,3-benzoxazol-2(3H)-ylidene)methyl]benzo[c,d]indolium iodide (**6a**); Yield 0.43 g, 69%; mp > 260 °C; ^{1}H-NMR (DMSO-d_6): δ ppm 0.95 (t, J = 7.1 Hz, 3H), 1.44–1.49 (m, 2H), 1.82–1.85 (m, 2H), 4.04 (s, 3H), 4.48 (t, J = 7.3 Hz, 2H), 6.14 (s, 1H), 7.55–7.67 (m, 3H), 7.73 (t, J = 8.6 Hz, 1H), 7.82–7.89 (m, 2H), 8.04 (t, J = 7.3 Hz, 1H), 8.15 (d, J = 7.1 Hz, 1H), 8.39 (d, J = 7.6 Hz, 1H), 9.17 (d, J = 7.6 Hz, 1H); ^{13}C-NMR (DMSO-d_6): δ ppm 14.3, 20.1, 30.2, 32.0, 75.5, 110.1, 112.3, 112.6, 126.6, 127.2, 129.7, 129.7, 130.3, 130.4, 131.8, 132.9, 141.1, 146.8, 155.6, 162.0; HRMS (ESI): Calcd for $C_{24}H_{23}N_2O^+$ m/z 355.1805, obsd m/z 355.1791.

1-Butyl-2-[(3-methyl-1,3-benzothiazol-2(3H)-ylidene)methyl]benzo[c,d]indolium iodide (**6b**); Yield 0.37 g, 57%; mp 249–251 °C; ^{1}H-NMR (DMSO-d_6): δ ppm 0.96 (t, J = 7.3 Hz, 3H), 1.43–1.49 (m, 2H), 1.75–1.92 (m, 2H), 4.16 (s, 3H), 4.37 (t, J = 7.2 Hz, 2H), 6.47 (s, 1H), 7.55 (d, J = 7.3Hz, 1H), 7.59–7.72 (m, 2H), 7.74–7.81 (m, 2H), 7.89 (t, J = 7.8 Hz, 1H), 8.04 (d, J = 8.3 Hz, 1H), 8.20 (d, J = 7.8 Hz, 1H), 8.32 (d, J = 8.1 Hz, 1H), 9.25 (d, J = 7.6 Hz, 1H); ^{13}C-NMR (DMSO-d_6): δ ppm 13.8, 19.7, 29.7, 35.4, 43.4, 87.0, 109.0, 115.0, 122.0, 123.6, 124.7, 126.8, 128.8, 129.2, 129.6, 129.7, 132.3, 141.0, 141.2, 154.0, 165.9; HRMS (ESI): Calcd for $C_{24}H_{23}N_2S^+$ m/z 371.1576, obsd m/z 371.1566.

1-Butyl-2-[(1-methylquinolin-2(1H)-ylidene)methyl]benzo[c,d]indolium iodide (**6c**); Yield 0.44 g, 69%; mp 225–227 °C; ^{1}H-NMR (DMSO-d_6): δ ppm 0.95 (t, J = 7.2 Hz, 3H), 1.40–1.54 (m, 2H), 1.79–1.85 (m, 2H), 4.25 (t, J = 7.3 Hz, 2H), 4.37 (s, 3H), 6.35 (s, 1H), 7.31 (d, J = 7.3 Hz, 1H), 7.55–7.62 (m, 2H), 7.65 (t, J = 7.7 Hz, 1H), 7.82 (t, J = 7.4 Hz, 1H), 8.07 (t, J = 7.7 Hz, 1H), 8.12 (d, J = 8.1 Hz, 1H), 8.21 (d, J = 7.8 Hz, 1H), 8.35 (d, J = 8.1 Hz, 2H), 8.58–8.71 (m, 2H); ^{13}C-NMR (DMSO-d_6): δ ppm 13.9, 19.7, 29.7, 42.9, 93.6, 106.4, 118.4, 120.0, 123.7, 127.5, 128.9, 129.5, 129.7, 130.4, 133.9, 141.2, 152.1, 157.0; HRMS (ESI): Calcd for $C_{26}H_{25}N_2^+$ m/z 365.2012, obsd m/z 365.1999.

1-Butyl-2-[(1,1,3-trimethyl-1,3-dihydro-2H-benzo[e]indol-2-ylidene)methyl]benzo[c,d]indolium iodide (**6d**); Yield 0.52 g, 72%; mp 190–192 °C; ^{1}H-NMR (DMSO-d_6): δ ppm 0.95 (t, J = 7.3 Hz, 3H), 1.45 (q, J = 7.3 Hz, 2H), 1.80–1.97 (m, 8H), 3.60 (s, 3H) 4.46 (t, J = 7.3 Hz, 2H), 6.43 (s, 1H), 7.60 (t, J = 7.5 Hz, 1H), 7.67–7.78 (m, 3H), 7.81 (d, J = 7.3 Hz, 1H), 7.84–7.93 (m, 3H), 8.14 (d, J = 8.0 Hz, 1H), 8.21 (d, J = 8.7 Hz, 1H), 8.35 (d, J = 8.2 Hz, 2H); ^{13}C-NMR (DMSO-d_6): δ ppm 13.7, 19.7, 25.2, 29.8, 43.8, 53.2, 54.9, 82.9, 110.4, 113.1, 122.9, 123.0, 124.1, 125.8, 127.7, 128.0, 128.6, 129.3, 129.6, 129.8, 130.0, 130.2, 130.3, 132.1, 132.3, 133.6, 140.8, 141.3, 156.5, 181.1; HRMS (ESI): Calcd for $C_{31}H_{31}N_2^+$ m/z 431.2482, obsd m/z 431.2469.

1-Butyl-2-[(1,1,3-trimethyl-1,3-dihydro-1H-indol-2-ylidene)methyl]benzo[c,d]indolium iodide (**10a**); Yield 0.42 g, 63%; mp 238–240 °C; ^{1}H-NMR (DMSO-d_6): δ ppm 0.95 (t, J = 7.08 Hz, 3H), 1.42–1.47 (m, 2H), 1.65 (s, 6H), 1.85–1.88 (m, 2H), 3.47 (s, 3H), 4.46 (t, J = 7.0 Hz, 2H), 6.31 (s, 1H), 7.44 (t, J = 7.3 Hz, 1H), 7.51–7.63 (m, 2H), 7.69–7.85 (m, 4H), 7.88–7.96 (m, 2H), 8.38 (d, J = 8.0 Hz, 1H); ^{13}C-NMR (DMSO-d_6): δ ppm 13.6, 19.6, 25.6, 29.7, 43.8, 45.7, 51.4, 82.9, 110.9, 113.3, 122.7, 123.2, 123.9, 126.4, 128.6, 129.1, 129.1, 129.5, 130.0, 130.2, 132.5, 140.1, 140.6, 143.9, 156.9, 179.5; HRMS (ESI): Calcd for $C_{27}H_{29}N_2^+$ m/z 381.2325, obsd m/z 381.2313.

1-Butyl-2-[(3-ethyl-5-methoxy-1,1-dimethyl-1,3-dihydro-1H-indol-2-ylidene)methyl]benzo[c,d] indolium iodide (**10b**); Yield 0.65 g, 90%; mp 187–189 °C; ^{1}H-NMR (DMSO-d_6): δ ppm 0.92 (t, J = 7.2 Hz, 3H), 1.14 (t, J = 6.7 Hz, 3H), 1.37–1.43 (m, 2H), 1.61 (s, 6H), 1.79–1.83 (m, 2H), 3.85 (s, 3H), 4.20 (q, J = 6.3 Hz, 2H), 4.39 (t, J = 6.1 Hz, 2H), 6.23 (s, 1H), 7.10 (d, J = 9.9 Hz, 1H), 7.40 (s, 1H), 7.58–7.67 (m, 2H), 7.70 (t, J = 7.9 Hz, 1H), 7.77–7.87 (m, 2H), 7.87–7.94 (m, 1H), 7.91 (d, J = 7.5 Hz, 1H), 8.30 (d, J = 8.0 Hz, 2H); ^{13}C-NMR (DMSO-d_6): δ ppm 13.3, 13.7, 19.5, 25.1, 29.7, 43.4, 45.6, 51.9, 56.0, 83.4, 109.5, 109.9, 113.8, 115.3, 122.4, 124.1, 127.3, 129.4, 129.5, 129.9, 130.1, 132.1, 134.8, 140.9, 142.8, 154.9, 158.9, 179.4; HRMS (ESI): Calcd for $C_{29}H_{33}N_2O^+$ m/z 425.2587, obsd m/z 425.2576.

1-Butyl-2-[(5-chloro-3-ethyl-1,1-dimethyl-1,3-dihydro-1H-indol-2-ylidene)methyl]benzo[c,d] indolium iodide (**10c**); Yield 0.25 g, 34%; mp 152–154 °C; ^{1}H-NMR (DMSO-d_6): δ ppm 0.91 (t, J = 7.2 Hz, 3H), 1.08 (t, J = 6.7 Hz, 3H), 1.37–1.43 (m, 2H), 1.63 (s, 6H), 1.81–1.84 (m, 2H), 4.16 (q, J = 6.7 Hz, 2H), 4.47 (t, J = 6.3 Hz, 2 H), 6.29 (s, 1H), 7.59 (d, J = 8.7 Hz, 1H), 7.67 (d, J = 8.7 Hz, 1H), 7.71–7.81 (m, 2H), 7.85–7.95 (m, 3H), 8.10 (d, J = 7.3 Hz, 1H), 8.39 (d, J = 8.2 Hz, 1H); ^{13}C-NMR (DMSO-d_6): δ ppm 13.0, 13.7,

19.4, 25.4, 29.9, 43.8, 45.7, 51.5, 83.8, 111.8, 115.5, 123.5, 123.7, 123.9, 128.4, 128.9, 129.2, 129.6, 130.1, 130.6, 133.2, 140.5, 140.8, 142.7, 156.8, 179.1; HRMS (ESI): Calcd for $C_{28}H_{30}N_2Cl^+$ m/z 429.2092, obsd m/z 429.2083.

3.3. Stock Solutions for Optical Measurements

Stock solutions were prepared by weighing the solid of each individual compound on a 5-digit analytical balance and adding solvent via class A volumetric pipette to make a 1.0 mM solution. The vials were vortexed for 20 s and then sonicated for 5 min to ensure complete dissolution. When not in use, the stock solutions were stored in a dark at 4 °C. For emission spectra in methanol/glycerol solutions the concentrations were prepared via the dilution of the stock solution in methanol followed by the addition of the appropriate volume of glycerol to achieve the desired concentrations.

3.4. Method of Determining Absorbance and Fluorescence

Stock solutions were used to prepare five dilutions of dyes with concentrations ranging from 5 to 25 μM using a class A volumetric pipette in order to maintain absorption between 0.1 and 1.0. The dye solutions were diluted ten-fold for fluorescence in order to minimize inner filter effect. The absorption spectra of each sample were measured in duplicate from 400 to 750 nm. Aggregation of **6b** was ruled out by measuring absorption at different concentrations (Figure S2G). Dye **6b** was tested for solvatochromic changes in absorption by dissolving the dye in five different solvents (ethanol, dimethyl formamide, dichloromethane, acetonitrile, and aqueous tris buffer) to observe any change in λ_{max} (Supplementary Materials Figure S2H). The emission spectra of each sample were measured in duplicate with a 530 nm excitation wavelength and slit widths of 5 nm for both excitation and emission. Emission spectra were corrected automatically by our developed method file used for reading the spectrofluorometer.

3.5. Computational Methods

The structure of each compound was first optimized using the TD-DFT method with the hybrid exchange-correlation functional, B3LYP/6-31G* basis set using *SPARTAN '14* (Wavefunction, Inc., Irvine, CA, USA) [63]. The torsional angles from the quaternary nitrogen to the α-carbon on the alternate heterocycle were restricted to 0° to get the calculated absorbance values, LUMO and HOMO orbitals, and electrostatic potential maps. The calculated LUMO and HOMO orbitals were obtained using a restricted hybrid HF-DFT SCF calculation performed with B3LYP/6-31G* basis set. The electrostatic potential maps were investigated for the optimized structures at HF/6-31G*. DNA docking studies were achieved using AutoDockTools 1.5.6 (Scripps Research Institute, La Jolla, CA, USA). Results of DNA docking study with dye **6b** under constraints were obtained by making all bonds within the dye to be non-rotatable and planar [64,65]. Polar and aromatic hydrogens were added to the DNA using GROMACS package [66] using GROMOS 53A6 force field [67] and Gasteiger Marsili charges [68]. A 78 × 70 × 64 grid box with a resolution of 0.375 Å was created encompassing the entire DNA using module AutoGrid 4.0. Dye **6b** was then added and simulations were preformed using Genetic Algorithm (GA).

3.6. DNA Binding Studies

A stock solution of **6b** (1×10^{-4} M) and ct-DNA type 1 (7.5×10^{-3} M) were prepared in ethanol and Tris-HCl buffer solution, respectively. Fluorescence titration with ct-DNA concentrations (0–200 mM) were made by mixing 35 μL **6b** solution with Tris-HCl buffer solution with and without ct-DNA to a total volume of 3500 μL in a fluorescence cuvette to make working solutions of 10 μM **6b**. Fluorescence spectra were measured in duplicate with excitation at 520 nm and slit widths of 10 nm for both excitation and emission.

4. Conclusions

A series of seven monomethine cyanines were synthesized in good yield with red-shifted absorbance properties in comparison to previously synthesized monomethine cyanine dyes. Although the benz[c,d]indolium containing monomethine cyanine dyes in this report are non-fluorescent in free flowing solvent, when the dyes are in a viscous environment their fluorescence becomes observable due to the restricted ability to rotate around the methine bridge. Computational methods outlined above were shown to be useful as a predictive tool for determining their optical properties. Dye **6b** was chosen as a representative example for DNA binding studies and was shown to bind DNA with an observable increase in fluorescence. Computational studies suggest it is binding the minor groove. Utilizing the described techniques these dyes could be developed as potential biological probes. Future studies will investigate how the different heterocycles and substituents affect binding to biological targets.

Acknowledgments: M.H. appreciates the Georgia State University Chemistry Department and Center for Diagnostics and Therapeutics for their Support. This study was supported by the Georgia State University Neuroscience Institute Brains and Behavior Seed Grant and Health Innovation Program Seed Grant. The authors thank Eric Owens and Vincent Martinez for editing.

Author Contributions: M.H. designed the research and all authors wrote the paper. E.S., C.H., and A.L. performed experiments. All authors discussed the results and commented on the manuscript.

Conflicts of Interest: The authors declare no conflict of interest.

References

1. Mishra, A.; Behera, R.K.; Behera, P.K.; Mishra, B.K.; Behera, G.B. Cyanines during the 1990s: A Review. *Chem. Rev.* **2000**, *100*, 1973–2012. [CrossRef] [PubMed]

2. Pisoni, D.S.; Todeschini, L.; Borges, A.C.A.; Petzhold, C.L.; Rodembusch, F.S.; Campo, L.F. Symmetrical and Asymmetrical Cyanine Dyes. Synthesis, Spectral Properties, and BSA Association Study. *J. Org. Chem.* **2014**, *79*, 5511–5520. [CrossRef] [PubMed]

3. Wada, H.; Hyun, H.; Vargas, C.; Gravier, J.; Park, G.; Gioux, S.; Frangioni, J.V.; Henary, M.; Choi, H.S. Pancreas-Targeted NIR Fluorophores for Dual-Channel Image-Guided Abdominal Surgery. *Theranostics* **2015**, *5*, 1–11. [CrossRef] [PubMed]

4. Hyun, H.; Wada, H.; Bao, K.; Gravier, J.; Yadav, Y.; Laramie, M.; Henary, M.; Frangioni, J.V.; Choi, H.S. Phosphonated Near-Infrared Fluorophores for Biomedical Imaging of Bone. *Angew. Chem. Int. Ed. Engl.* **2014**, *53*, 10668–10672. [CrossRef] [PubMed]

5. Hyun, H.; Owens, E.A.; Wada, H.; Levitz, A.; Park, G.; Park, M.H.; Frangioni, J.V.; Henary, M.; Choi, H.S. Cartilage-Specific Near-Infrared Fluorophores for Biomedical Imaging. *Angew. Chem. Int. Ed. Engl.* **2015**, *54*, 8648–8652. [CrossRef] [PubMed]

6. Njiojob, C.N.; Owens, E.A.; Narayana, L.; Hyun, H.; Choi, H.S.; Henary, M. Tailored Near-Infrared Contrast Agents for Image Guided Surgery. *J. Med. Chem.* **2015**, *58*, 2845–2854. [CrossRef] [PubMed]

7. Hyun, H.; Park, M.H.; Owens, E.A.; Wada, H.; Henary, M.; Handgraaf, H.J.M.; Vahrmeijer, A.L.; Frangioni, J.V.; Choi, H.S. Structure-inherent targeting of near-infrared fluorophores for parathyroid and thyroid gland imaging. *Nat. Med.* **2015**, *21*, 192–197. [CrossRef] [PubMed]

8. El-Shishtawy, R.M.; Asiri, A.M.; Basaif, S.A.; Rashad Sobahi, T. Synthesis of a new beta-naphthothiazole monomethine cyanine dye for the detection of DNA in aqueous solution. *Spectrochim. Acta A* **2010**, *75*, 1605–1609. [CrossRef] [PubMed]

9. Henary, M.; Levitz, A. Synthesis and applications of unsymmetrical carbocyanine dyes. *Dyes Pigment.* **2013**, *99*, 1107–1116. [CrossRef]

10. Silva, G.L.; Ediz, V.; Yaron, D.; Armitage, B.A. Experimental and Computational Investigation of Unsymmetrical Cyanine Dyes: Understanding Torsionally Responsive Fluorogenic Dyes. *J. Am. Chem. Soc.* **2007**, *129*, 5710–5718. [CrossRef] [PubMed]

11. Cao, J.; Wu, T.; Hu, C.; Liu, T.; Sun, W.; Fan, J.; Peng, X. The nature of the different environmental sensitivity of symmetrical and unsymmetrical cyanine dyes: An experimental and theoretical study. *Phys. Chem. Chem. Phys.* **2012**, *14*, 13702–13708. [CrossRef] [PubMed]

12. Deligeorgiev, T.G.; Gadjev, N.I.; Drexhage, K.-H.; Sabnis, R.W. Preparation of intercalating dye thiazole orange and derivatives. *Dyes Pigments* **1995**, *29*, 315–322. [CrossRef]

13. Thompson, M. Synthesis, Photophysical Effects, and DNA Targeting Properties of Oxazole Yellow-Peptide Bioconjugates. *Bioconjugate Chem.* **2006**, *17*, 507–513. [CrossRef] [PubMed]

14. Dähne, S.; Resch-Genger, U.; Wolfbeis, O.S. *Near-Infrared Dyes for High Technology Applications*; Daehne, S., Resch-Genger, U., Wolfbeis, O.S., Eds.; Kluwer: Dordrecht, The Netherland; Boston, MA, USA, 1998.

15. Nygren, J.; Svanvik, N.; Kubista, M. The interactions between the fluorescent dye thiazole orange and DNA. *Biopolymers* **1998**, *46*, 39–51. [CrossRef]

16. Joseph, M.J.; Taylor, J.C.; McGown, L.B.; Pitner, B.; Linn, C.P. Spectroscopic studies of YO and YOYO fluorescent dyes in a thrombin-binding DNA ligand. *Biospectroscopy* **1996**, *2*, 173–183. [CrossRef]

17. Kabatc, J. Multicationic monomethine dyes as sensitizers in two- and three-component photoinitiating systems for multiacrylate monomers. *J. Photochem. Photobiol. A* **2010**, *214*, 74–85. [CrossRef]

18. Hirons, G.T.; Fawcett, J.J.; Crissman, H.A. Toto and Yoyo—New Very Bright Fluorochromes for DNA Content Analyses by Flow-Cytometry. *Cytometry* **1994**, *15*, 129–140. [CrossRef] [PubMed]

19. Dietzek, B.; Brüggemann, B.; Persson, P.; Yartsev, A. On the excited-state multi-dimensionality in cyanines. *Chem. Phys. Lett.* **2008**, *455*, 13–19. [CrossRef]

20. Soriano, E.; Outler, L.; Owens, E.A.; Henary, M. Synthesis of Asymmetric Monomethine Cyanine Dyes with Red-Shifted Optical Properties. *J. Heterocycl. Chem.* **2015**, *52*, 180–184. [CrossRef]

21. Upadhyayula, S.; Nunez, V.; Espinoza, E.M.; Larsen, J.M.; Bao, D.D.; Shi, D.W.; Mac, J.T.; Anvari, B.; Vullev, V.I. Photoinduced dynamics of a cyanine dye: Parallel pathways of non-radiative deactivation involving multiple excited-state twisted transients. *Chem. Sci.* **2015**, *6*, 2237–2251. [CrossRef]

22. Sameiro, M.; Goncalves, T. Fluorescent Labeling of Biomolecules with Organic Probes. *Chem. Rev.* **2009**, *109*, 190–212.

23. Lavis, L.D.; Raines, R.T. Bright ideas for chemical biology. *ACS Chem. Biol.* **2008**, *3*, 142–155. [CrossRef] [PubMed]

24. Mann, S. Life as a nanoscale phenomenon. *Angew. Chem. Int. Ed. Engl.* **2008**, *47*, 5306–5320. [CrossRef] [PubMed]

25. Resch-Genger, U.; Grabolle, M.; Cavaliere-Jaricot, S.; Nitschke, R.; Nann, T. Quantum dots *versus* organic dyes as fluorescent labels. *Nat. Methods* **2008**, *5*, 763–775. [CrossRef] [PubMed]

26. Chen, A.K.; Cheng, Z.; Behlke, M.A.; Tsourkas, A. Assessing the sensitivity of commercially available fluorophores to the intracellular environment. *Anal. Chem.* **2008**, *80*, 7437–7444. [CrossRef] [PubMed]

27. Simeonov, A.; Jadhav, A.; Thomas, C.J.; Wang, Y.; Huang, R.; Southall, N.T.; Shinn, P.; Smith, J.; Austin, C.P.; Auld, D.S.; Inglese, J. Fluorescence spectroscopic profiling of compound libraries. *J. Med. Chem.* **2008**, *51*, 2363–2371. [CrossRef] [PubMed]

28. Longmire, M.R.; Ogawa, M.; Hama, Y.; Kosaka, N.; Regino, C.A.; Choyke, P.L.; Kobayashi, H. Determination of optimal rhodamine fluorophore for *in vivo* optical imaging. *Bioconjugate Chem.* **2008**, *19*, 1735–1742. [CrossRef] [PubMed]

29. Johnsson, N.; Johnsson, K. Chemical tools for biomolecular imaging. *ACS Chem. Biol.* **2007**, *2*, 31–38. [CrossRef] [PubMed]

30. Marti, A.A.; Jockusch, S.; Stevens, N.; Ju, J.; Turro, N.J. Fluorescent hybridization probes for sensitive and selective DNA and RNA detection. *Acc. Chem. Res.* **2007**, *40*, 402–409. [CrossRef] [PubMed]

31. Willis, R.C. Portraits of life, one molecule at a time. *Anal. Chem.* **2007**, *79*, 1785–1788. [CrossRef] [PubMed]

32. Hama, Y.; Urano, Y.; Koyama, Y.; Bernardo, M.; Choyke, P.L.; Kobayashi, H. A comparison of the emission efficiency of four common green fluorescence dyes after internalization into cancer cells. *Bioconjugate Chem.* **2006**, *17*, 1426–1431. [CrossRef] [PubMed]

33. Yuste, R. Fluorescence microscopy today. *Nat. Methods* **2005**, *2*, 902–904. [CrossRef] [PubMed]

34. Lichtman, J.W.; Fraser, S.E. The neuronal naturalist: Watching neurons in their native habitat. *Nat. Neurosci.* **2001**, *4*, 1215–1220. [CrossRef] [PubMed]

35. Selvin, P.R. The renaissance of fluorescence resonance energy transfer. *Nat. Struct. Biol.* **2000**, *7*, 730–734. [CrossRef] [PubMed]

36. Hu, H.; Owens, E.A.; Su, H.; Yan, L.; Levitz, A.; Zhao, X.; Henary, M.; Zheng, Y.G. Exploration of Cyanine Compounds as Selective Inhibitors of Protein Arginine Methyltransferases: Synthesis and Biological Evaluation. *J. Med. Chem.* **2015**, *58*, 1228–1243. [CrossRef] [PubMed]

37. Nanjunda, R.; Owens, E.; Mickelson, L.; Dost, T.; Stroeva, E.; Huynh, H.; Germann, M.; Henary, M.; Wilson, W. Selective G-Quadruplex DNA Recognition by a New Class of Designed Cyanines. *Molecules* **2013**, *18*, 13588–13607. [CrossRef] [PubMed]

38. Mapp, C.T.; Owens, E.A.; Henary, M.; Grant, K.B. Oxidative cleavage of DNA by pentamethine carbocyanine dyes irradiated with long-wavelength visible light. *Bioorganic Med. Chem. Lett.* **2014**, *24*, 214–219. [CrossRef] [PubMed]

39. Lakowicz, J.R. *Principles of Fluorescence Spectroscopy*, 3rd ed.; Springer: New York, NY, USA, 2006.

40. Kantor, G.J.; Hull, D.R. An effect of ultraviolet light on RNA and protein synthesis in nondividing human diploid fibroblasts. *Biophys. J.* **1979**, *27*, 359–370. [CrossRef]

41. Cerutti, P.A. Prooxidant states and tumor promotion. *Science* **1985**, *227*, 375–381. [CrossRef] [PubMed]

42. Marks, R. An overview of skin cancers. Incidence and causation. *Cancer* **1995**, *75*, 607–612. [CrossRef]

43. Krasnaya, Z.A.; Tret'yakova, E.O.; Kachala, V.V.; Zlotin, S.G. Synthesis of conjugated polynitriles by the reactions of β-dimethylaminoacrolein aminal and 1-dimethylamino-1,3,3-trimethoxypropane with 2-dicyanomethylene-4,5,5-trimethyl-3-cyano-2,5-dihydrofuran. *Mendeleev Commun.* **2007**, *17*, 349–351. [CrossRef]

44. Brooker, L.G.S.; Keyes, G.H.; Williams, W.W. Color and Constitution. V.1 The Absorption of Unsymmetrical Cyanines. Resonance as a Basis for a Classification of Dyes. *J. Am. Chem. Soc.* **1942**, *64*, 199–210. [CrossRef]

45. Williams, C.G. XXVI.—Researches on Chinoline and its Homologues. *Earth Environ. Sci. Trans. R. Soc. Edinb.* **1857**, *21*, 377–401. [CrossRef]

46. Yarmoluk, S.M.; Kovalska, V.B.; Losytskyy, M.Y. Symmetric cyanine dyes for detecting nucleic acids. *Biotech. Histochem.* **2008**, *83*, 131–145. [CrossRef] [PubMed]

47. Levitz, A.; Ladani, S.T.; Hamelberg, D.; Henary, M. Synthesis and effect of heterocycle modification on the spectroscopic properties of a series of unsymmetrical trimethine cyanine dyes. *Dyes Pigments* **2014**, *105*, 238–249. [CrossRef]

48. Murphy, S.; Yang, X.Q.; Schuster, G.B. Cyanine Borate Salts That Form Penetrated Ion-Pairs in Benzene Solution—Synthesis, Properties, and Structure. *J. Org. Chem.* **1995**, *60*, 2411–2422. [CrossRef]

49. Deligeorgiev, T.G.; Zaneva, D.A.; Kim, S.H.; Sabnis, R.W. Preparation of monomethine cyanine dyes for nucleic acid detection. *Dyes Pigments* **1998**, *37*, 205–211. [CrossRef]

50. Timcheva, I.I.; Maximova, V.A.; Deligeorgiev, T.G.; Gadjev, N.I.; Sabnis, R.W.; Ivanov, I.G. Fluorescence spectral characteristics of novel asymmetric monomethine cyanine dyes in nucleic acid solutions. *FEBS Lett.* **1997**, *405*, 141–144. [CrossRef]

51. Oster, G.; Nishijima, Y. Fluorescence and Internal Rotation: Their Dependence on Viscosity of the Medium1. *J. Am. Chem. Soc.* **1956**, *78*, 1581–1584. [CrossRef]

52. Potts, K.T. *Heteropentalenes, in Chemistry of Heterocyclic Compounds: Special Topics in Heterocyclic Chemistry*; Weissberger, A., Taylor, E.C., Eds.; John Wiley & Sons, Inc.: Hoboken, NJ, USA, 1977; Volume 30.

53. Escobedo, J.O.; Rusin, O.; Lim, S.; Strongin, R.M. NIR dyes for bioimaging applications. *Curr. Opin. Chem. Biol.* **2010**, *14*, 64–70. [CrossRef] [PubMed]

54. Kuhn, H. A Quantum-Mechanical Theory of Light Absorption of Organic Dyes and Similar Compounds. *J. Chem. Phys.* **1949**, *17*, 1198–1212. [CrossRef]

55. Bickelhaupt, F.M.; Baerends, E.J. Kohn-Sham Density Functional Theory: Predicting and Understanding Chemistry. *Rev. Comput. Chem.* **2000**, *15*, 1–86.

56. Cantor, C.R.; Schimmel, P.R. *Biophysical Chemistry*; W.H. Freeman: San Francisco, CA, USA, 1980.

57. Shaikh, S.A.; Ahmed, S.R.; Jayaram, B. A molecular thermodynamic view of DNA-drug interactions: A case study of 25 minor-groove binders. *Arch. Biochem. Biophys.* **2004**, *429*, 81–99. [CrossRef] [PubMed]

58. Chaires, J.B.; Ren, J.; Hamelberg, D.; Kumar, A.; Pandya, V.; Boykin, D.W.; Wilson, W.D. Structural selectivity of aromatic diamidines. *J. Med. Chem.* **2004**, *47*, 5729–5742. [CrossRef] [PubMed]

59. Fairley, T.A.; Tidwell, R.R.; Donkor, I.; Naiman, N.A.; Ohemeng, K.A.; Lombardy, R.J.; Bentley, J.A.; Cory, M. Structure, DNA minor groove binding, and base pair specificity of alkyl- and aryl-linked bis(amidinobenzimidazoles) and bis(amidinoindoles). *J. Med. Chem.* **1993**, *36*, 1746–1753. [CrossRef] [PubMed]

60. Sinha, S.H.; Owens, E.A.; Feng, Y.; Yang, Y.; Xie, Y.; Tu, Y.; Henary, M.; Zheng, Y.G. Synthesis and evaluation of carbocyanine dyes as PRMT inhibitors and imaging agents. *Eur. J. Med. Chem.* **2012**, *54*, 647–659. [CrossRef] [PubMed]

61. Nanjunda, R.; Owens, E.A.; Mickelson, L.; Alyabyev, S.; Kilpatrick, N.; Wang, S.; Henary, M.; Wilson, W.D. Halogenated pentamethine cyanine dyes exhibiting high fidelity for G-quadruplex DNA. *Bioorg. Med. Chem.* **2012**, *20*, 7002–7011. [CrossRef] [PubMed]

62. Narayanan, N.; Patonay, G. A New Method for the Synthesis of Heptamethine Cyanine Dyes: Synthesis of New Near-Infrared Fluorescent Labels. *J. Org. Chem.* **1995**, *60*, 2391–2395. [CrossRef]

63. Shao, Y.; Molnar, L.F.; Jung, Y.; Kussmann, J.; Ochsenfeld, C.; Brown, S.T.; Gilbert, A.T.B.; Slipchenko, L.V.; Levchenko, S.V.; O'Neill, D.P.; *et al.* Advances in methods and algorithms in a modern quantum chemistry program package. *Phys. Chem. Chem. Phys.* **2006**, *8*, 3172–3191. [CrossRef] [PubMed]

64. Morris, G.M.; Huey, R.; Lindstrom, W.; Sanner, M.F.; Belew, R.K.; Goodsell, D.S.; Olson, A.J. AutoDock4 and AutoDockTools4: Automated docking with selective receptor flexibility. *J. Comput. Chem.* **2009**, *30*, 2785–2791. [CrossRef] [PubMed]

65. Ricci, C.G.; Netz, P.A. Docking Studies on DNA-Ligand Interactions: Building and Application of a Protocol To Identify the Binding Mode. *J. Chem. Inf. Model.* **2009**, *49*, 1925–1935. [CrossRef] [PubMed]

66. Kulchin, Y.N.; Vitrik, O.B.; Kamenev, O.T.; Romashko, R.V. *Vector Fields Reconstruction by Fiber Optic Measuring Network*; SPIE: Bellingham, WA, USA; 2001; pp. 100–108.

67. Oostenbrink, C.; Soares, T.; van der Vegt, N.A.; van Gunsteren, W. Validation of the 53A6 GROMOS force field. *Eur. Biophys. J.* **2005**, *34*, 273–284. [CrossRef] [PubMed]

68. Gasteiger, J.; Marsili, M. Iterative partial equalization of orbital electronegativity—A rapid access to atomic charges. *Tetrahedron* **1980**, *36*, 3219–3228. [CrossRef]

3

Isolation of Terpenoids from the Stem of *Ficus aurantiaca* Griff and their Effects on Reactive Oxygen Species Production and Chemotactic Activity of Neutrophils

Shukranul Mawa, Ibrahim Jantan and Khairana Husain *

Academic Editor: Derek J. McPhee

Drug and Herbal Research Centre, Faculty of Pharmacy, Universiti Kebangsaan Malaysia (UKM), Jalan Raja Muda Abdul Aziz, Kuala Lumpur 50300, Malaysia; shuku_76@yahoo.com (S.M.); profibj@gmail.com (I.J.)
* Correspondence: khairana@ukm.edu.my

Abstract: Three new triterpenoids; namely 28,28,30-trihydroxylupeol (**1**); 3,21,21,26-tetrahydroxy-lanostanoic acid (**2**) and dehydroxybetulinic acid (**3**) and seven known compounds; *i.e.*, taraxerone (**4**); taraxerol (**5**); ethyl palmitate (**6**); herniarin (**7**); stigmasterol (**8**); ursolic acid (**9**) and acetyl ursolic acid (**10**) were isolated from the stem of *Ficus aurantiaca* Griff. The structures of the compounds were established by spectroscopic techniques. The compounds were evaluated for their inhibitory effects on polymorphonuclear leukocyte (PMN) chemotaxis by using the Boyden chamber technique and on human whole blood and neutrophil reactive oxygen species (ROS) production by using a luminol-based chemiluminescence assay. Among the compounds tested, compounds **1–4**, **6** and **9** exhibited strong inhibition of PMN migration towards the chemoattractant *N*-formyl-methionyl-leucyl-phenylalanine (fMLP) with IC_{50} values of 6.8; 2.8; 2.5; 4.1; 3.7 and 3.6 μM, respectively, comparable to that of the positive control ibuprofen (6.7 μM). Compounds **2–4**, **6**, **7** and **9** exhibited strong inhibition of ROS production of PMNs with IC_{50} values of 0.9; 0.9; 1.3; 1.1; 0.5 and 0.8 μM, respectively, which were lower than that of aspirin (9.4 μM). The bioactive compounds might be potential lead molecules for the development of new immunomodulatory agents to modulate the innate immune response of phagocytes.

Keywords: *Ficus aurantiaca* Griff; terpenoids; immunomodulatory; neutrophils; chemotaxis; reactive oxygen species

1. Introduction

Medicinal plant products have long been used in traditional medicines for the treatment of many immunological disorders. Their therapeutic effects may be due to their effects on the immune system [1]. Many herbs such as *Andrographis paniculata*, *Allium sativum*, *Trigonella foenum graecum*, *Pouteria cambodiana*, *Centella asiatica*, *Asparagus racemosus*, *Baliospermum montanum*, *Curcuma longa*, *Panax ginseng*, *Phyllanthus debilis*, *Tinospora cordifolia*, and *Picrorhiza scrophulariiflora* have been reported to be able to modulate the immune system, including both adaptive and innate arms of the immune responses, exhibiting either immunostimulant or immunosuppressive effects [2–4]. Immunomodulation by these plant extracts and their active components provide new sources of lead molecules for development of natural immunomodulators for a variety of immunologic diseases. Assessment of the immunological effects of compounds is based on their selective activities on the different components of the immune system. In recent years, there was an increased interest in the search for natural immunomodulators from medicinal plants to substitute conventional therapy as they are considered to possess fewer side effects [5].

Ficus (family: Moraceae) is one of the largest genera of angiosperms, with more than 800 species of trees, shrubs, hemi-epiphytes and climbers in the tropics and sub-tropics worldwide [6]. Due to its high economic and nutritional values the genus is a vital hereditary source and also an important component of biodiversity of the rainforest ecosystem. It is also a good source of food for fruit eating animals in tropical area [7]. Chemical investigations of various *Ficus* species have shown the presence of flavonoids, coumarins, alkaloids, steroids, triterpenoids, simple phenols and salicylic acids from *F. benghalensis, F. carica, F. hirta, F. hispida, F. microcarpa, F. nymphaeifolia, F. ruficaulis* and *F. septica* [8,9]. Some of these plants exhibited a wide range of biological activities such as anti-inflammatory [10–12], antioxidant, hypolipidemic and hypoglycemic activities [13]. Some *Ficus* species are well known in Asia as medicinal plants and are widely used in folk medicines for the treatment of flu, malaria, tonsillitis, bronchitis and rheumatism [14]. One of the most biologically active species of *Ficus* is *F. carica* which has been reported to have some 21 traditional and current uses in various ethnopharmacological practices [15].

Ficus aurantiaca Griff is an evergreen tree that can grow up to 9 m. It is widespread in the lowland forests at Kelantan, Terengganu, Perak, Pahang, Selangor, Melaka and Johor in Malaysia, where it is locally known as "tengkuk biawak" or "akar tengkuk biawak hitam". Traditionally, the plant parts have been used for the treatment of various ailments including headache, wound and toothache [16]. There is little investigation on the chemical composition and biological properties of *F. aurantiaca*. Our preliminary study on the stem extract of this plant had demonstrated that it inhibited *in vitro* chemotactic migration as well as suppressed the release of reactive oxygen species by granulocytes (unpublished data). The aims of the present study were to isolate and identify the constituents of the stem of *F. aurantiaca* and evaluate their effects on reactive oxygen species (ROS) production and chemotactic activity of neutrophils.

2. Results and Discussion

2.1. Characterization of the Isolated Compounds

Successive separations of *n*-hexane, ethyl acetate and methanol extracts using silica gel chromatography afforded three new triterpenoids, 28,28,30-trihydroxylupeol (**1**), 3,21,21,26-tetrahydroxylanostanoic acid (**2**) and dehydroxybetulinic acid (**3**) along with five known triterpenoids, taraxerone (**4**), taraxerol (**5**), stigmasterol (**8**), ursolic acid (**9**), acetyl ursolic acid (**10**), one sesquiterpenoid, herniarin (**7**) and one diterpenoid, ethyl palmitate (**6**). The structures of the new compounds are shown in Figure 1. The structures of the known compounds were elucidated by the combination of ESIMS, ^{1}H- and ^{13}C-NMR spectral data and comparison of their spectral data with literature values [17–23].

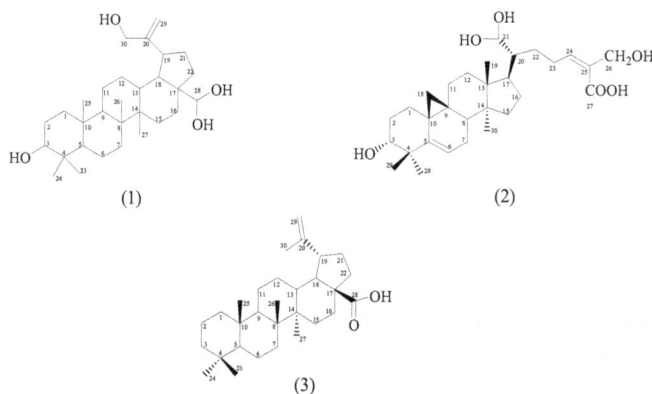

Figure 1. New terpenoids from *F. aurantica*.

Compound **1** (20 mg) was isolated as a white powder with a melting point range of 222–224 °C. The IR spectrum showed a band at 3311 cm^{-1} due to the presence of hydroxyl groups (-OH). C-H stretching vibrations were observed at 2920 cm^{-1} while the methyl bending vibration was at 1411 cm^{-1}. The HRESIMS displayed a molecular ion at m/z 474.1465 [M]$^{+}$, corresponding to a molecular formula of $C_{30}H_{50}O_4$ (M = 474.1465). The ^{1}H-NMR spectrum exhibited singlets indicating the presence of five methyl groups, deshielded methine protons at δ_H 2.04 (H-18) and 2.38 (H-19) and two olefinic protons at δ_H 4.69 and 4.57. The corresponding ^{13}C-NMR spectrum displayed 30 carbon signals, including five methyls, twelve methylenes, seven methines and six quaternary carbons. Two methine carbons at δ_C 80.1 (C-3) and 95.1 (C-28) were typical for C-OH and two sp^2 carbons at δ_C 150.9 (C-20) and 109.3 (C-29), indicated the presence of one double bond. Another methylene carbon at δ_C 69.2 (C-30) was characteristic for hydroxyl (C-OH). The ^{1}H- and ^{13}C-NMR data of **1** were very similar to those reported for lupeol [24], the main difference being the presence of four hydroxyl groups instead of only one. Two methyl group protons at C-28, were substituted by two hydroxyl groups, and in the C-30 methyl group, a proton was also ubstituted by a hydroxyl group. The positions of the hydroxyl groups were established by the HMQC correlation of the hydroxyl (-OH) protons, δ_H 2.37, 2.04, 2.29 and 2.39 with the δ_C 80.1 (C-3) 69.2 (C-30) and 95.1 (C-28) carbon signals, respectively.

The connectivity between protons and carbons established by the HMQC and HMBC spectra indicated that the two olefinic protons were attached to carbon at δ_C 109.3 (C-29), the two adjacent methyl protons resonating as a singlet at δ_H 0.79 (s, 3H, H-23) and 0.83 (s, 3H, H-24) attached to a quaternary carbon at δ_C 39.6 (C-4) were attributed to methine carbon δ_C 55.6 (C-5) and hydroxyl methine carbon δ_C 80.1 (C-3) by ^3J correlation, respectively. Meanwhile another methyl proton at δ_H 0.94 (s, 3H, H-25) displayed ^3J correlations with methylene carbon at δ_C 38.1 (C-1) and methine carbon at δ_C 50.3 (C-9). Another two groups of methyl protons were observed at δ_H 1.01 (s, 3H, H-26) and 1.07 (s, 3H, H-27), showed ^3J correlation with methylene carbon at δ_C 34.1(C-7) and δ_C 27.4 (C-15), respectively. One hydroxyl methine proton at δ_H 3.39 (dd, 1H, J = 10.8, 4.8 Hz, H-3) showed a ^2J correlation with methylene carbon δ_C (C-2). Another three methine protons were visible at δ_H 1.34 (m, 1H, H-13), 2.04 (m, 1H, H-18) and 2.38 (d, 1H, J = 7.2 Hz, H-19) whereas at δ_H 1.34 (m, 1H, H-13) showed ^2J correlation with methylene carbon at δ_C 25.1 (C-12) and methine carbon at δ_C 47.9 (C-18) and proton at δ_H 2.04 (m, 1H, H-18) and 2.38 (d, 1H, J = 7.2 Hz, H-19) showed ^2J correlation with methine carbon at δ_C 48.2 (C-19) and quaternary carbon at δ_C 150.9 (C-20), respectively. Based on these ^{1}H and ^{13}C-NMR spectral data of **1**, summarized in Table 1, and the comparison with the literature values of lupeol [24], **1** was identified as 28,28,30-trihydroxylupeol. Selected C-H correlations in **1** established by HMBC are shown in Figure 2. In the genus *Ficus*, lupeol was previously isolated from *F. carica* [25], *F. micricarpa* [26] and *F. benjamina* [9], while lupeol acetate was isolated from *F. microcarpa* [26]. This is the first report of a hydroxy derivative of lupeol in *Ficus* species.

Table 1. ^{1}H- and ^{13}C-NMR spectroscopic data (600 MHz, CDCl$_3$) for triterpenoids **1–3**.

Positions	Compound 1		Compound 2		Compound 3	
	δ_C	δ_H (J in Hz)	δ_C	δ_H (J in Hz)	δ_C	δ_H (J in Hz)
1	38.1	1.28 (d, 4H, J = 6.6)	29.8	0.94, (m, 2H)	36.1	1.41 (m, 1H) 1.31 (m,1H)
2	27.3	1.68 (d, 4H, J = 3.0)	25.7	1.01 (m, 2H)	28.2	1.42 (m, 1H) 1.32 (m, 1H)
3	80.1	3.39 (dd, 1H, J = 10.8, 4.8)	80.1	3.44 (1H, t, J = 18.0), 4.68 (OH)	36.7	1.42 (m, 1H) 1.33 (m, 1H)
4	39.6	-	43.0	-	30.8	-
5	55.6	1.32 (d, 1H, J = 7.2)	151.0	-	59.6	1.34 (m, 1H)
6	17.7	1.28 (m, 4H, J = 6.6)	109.3	4.79 (t, 1H, J = 6.0)	18.6	1.65 (m, 2H)
7	34.1	1.28 (m, 4H)	31.9	1.77 (m, 2H)	37.4	1.46 (m, 1H)
8	40.9	-	50.3	1.48 (m, 1H)	40.1	-

Table 1. *Cont.*

Positions	Compound 1		Compound 2		Compound 3	
	δ_C	δ_H (*J* in Hz)	δ_C	δ_H (*J* in Hz)	δ_C	δ_H (*J* in Hz)
9	50.3	0.88 (m, 10H)	48.2	-	51.9	1.56 (m, 1H)
10	37.3	-	39.9	-	36.7	-
11	21.0	1.61 (m, 4H)	23.7	1.34 (m, 2H)	20.6	1.54 (m, 1H) 1.29 (m, 1H)
12	25.1	1.68 (m, 4H)	22.7	1.31 (m, 2H)	25.2	1.55 (m, 1H) 1.29 (m, 1H)
13	37.9	1.34 (m, 1H)	42.8	-	37.7	1.24 (m, 1H)
14	42.8	-	40.9	-	42.9	-
15	27.4	1.26 (m, 4H)	34.5	1.57 (m, 2H)	39.7	1.49 (m, 1H) 1.00 (m, 1H)
16	35.5	1.26 (m, 4H)	29.4	1.63 (m, 2H)	40.4	1.86 (m, 1H) 1.25 (m,1H)
17	43.0	-	55.6	1.48 (m,1H)	59.6	-
18	47.9	2.04 (m, 1H)	27.4	1.03 (s, 2H)	46.5	1.68 (m, 1H)
19	48.2	2.38 (d, 1H, *J* = 7.2)	17.7	1.04 s, 1H	37.2	2.09 (m, 1H)
20	150.9	-	35.3	1.68 (m, 1H)	150.9	-
21	29.7	1.88–1.94 (m, 2H)	95.1	4.80 (m, 1H)	29.2	1.60 (m, 6H)
22	39.9	1.34 (m, 4H)	35.5	1.18 (m, 2H)	23.0	1.75 (t, 2H, *J* = 5.4)
23	14.5	0.79 (s, 3H)	25.0	1.35 (m, 2H)	22.1	0.83 (s, 6H)
24	18.0	0.83 (s, 3H)	150.9	4.57 (m, 1H)	16.6	0.90 (s, 3H)
25	15.9	0.94 (s, 3H)	109.3	-	15.4	0.76 (s, 3H)
26	16.1	1.01(s, 3H)	69.2	4.03 (d, 1H, *J* = 12)	17.6	0.89 (s, 3H)
27	14.1	1.07 (s, 3H)	173.8	-	13.9	0.73 (s, 3H)
28	95.1	4.79 (t, 1H, *J* = 6 Hz)	14.1	0.80 (s, 3H)	183.5	-
29	109.3	4.69 (br s, 1H) 4.57 (br s, 1H)	15.9	0.85 (s, 1H)	116.3	5.32 (brs, 1H)
30	69.2	5.14 (m, 1H, H-30) 5.19 (m, 1H, H-30)	16.5	0.92 (q, 3H)	20.1	1.10 (s, 3H)

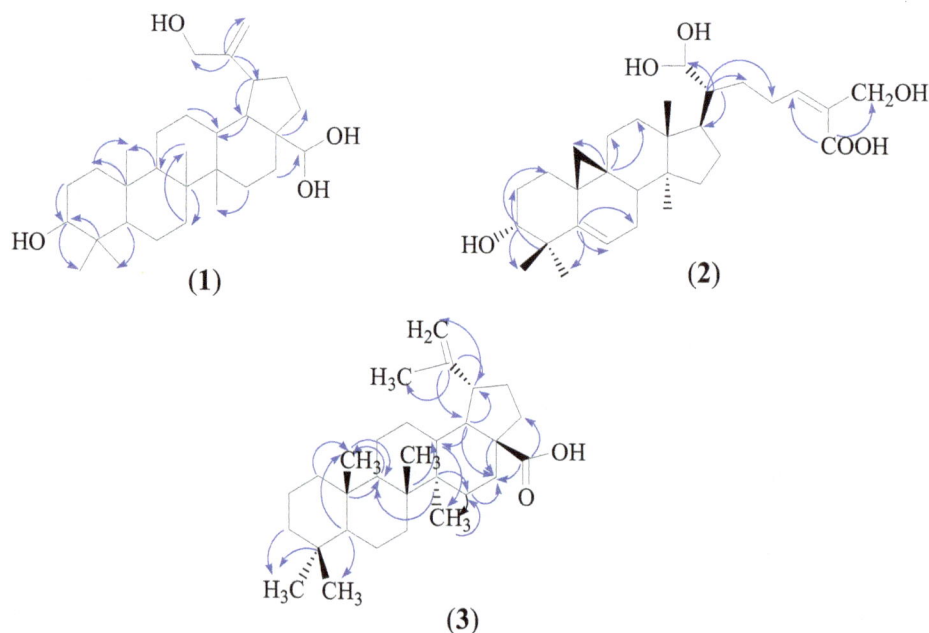

Figure 2. Selected HMBC correlations (C→H) of compounds **1–3**.

Compound **2** was isolated as a white powder with a melting point range of 150–155 °C. The IR spectrum displayed absorption bands for a carboxylic acid group (3215 cm^{-1}, 1678 cm^{-1}) and a keto function at 1735 cm^{-1} and unsaturation at 1643 cm^{-1}. The HRESIMS showed a molecular ion at *m/z* 502.3290 [M]$^+$, 479.1023 [M − 23]$^+$, 453.0885 [M − CH$_3$OH − 17]$^+$ and 311.1563 [M − side chain

($C_8H_{14}O_5$) − 1]$^+$. Based on HRESIMS data its molecular formula was determined to be $C_{30}H_{46}O_6$ and the molecular weight was established as 502.3290 [M]$^+$. The ^{1}H-NMR spectrum displayed characteristic signals for methyl protons at δ_H 0.80 (s, 3H, H-28), 0.85 (s, 1H, H-29), 0.92 (q, 3H, H-30) and 1.04 (s, 1H, H-19). One proton downfield signal appeared at δ_H 4.79 (t, 1H, J = 6.0 Hz, H-6) and was attributed to the vinylic H-6 proton. Another proton resonating as a triplet at δ_H 3.44 (1H, t, J = 18.0 Hz) was ascribed to a β-oriented H-3 carbinol proton. The ^{13}C-NMR spectrum showed thirty carbon signals, including four methyls, ten methylenes, six methines and eight quaternary carbon signals. One methylene carbon at δ_C 69.2 and two methine carbons at δ_C 80.1 and δ_C 95.1 were typical for C-OH moieties. Important signals were observed for a carboxylic acid carbon at δ_C 173.8 (C-27), vinylic carbons at δ_C 151.0 (C-5) and δ_C 109.3 (C-6) carbinol carbons at δ_C 80.1 (C-3) and methyl carbons between δ_C 17.7–14.1. The ^{1}H- and ^{13}C-NMR data of **2** were compared to the reported data of other lanostanoic acids [22,27].

Proton and carbon connectivity established by HMQC and HMBC spectral analysis indicated that four methyl protons at δ_H 0.80, 0.80, 0.92 and 1.04 were connected to the C-28, C-29, C-30 and C-19 tertiary methyl protons, respectively, all attached to the saturated carbons. Two methyl groups resonating as a singlet were observed at δ_H 0.80 (s, 3H, H-28), 0.85 (s, 1H, H-29) and were attached to quaternary carbon at δ_C 43.0 (C-4). Methyl protons at δ_H 0.80 (s, 3H, H-28) and 0.85 (s, 1H, H-29) displayed ^3J correlations with quaternary carbon at δ_C 151.0 (C-5) and the carbinol carbon at δ_C 80.1 (C-3). Meanwhile two methylene protons observed at δ_H 0.94, (m, 2H, H-1) and 1.01 (m, 2H, H-2) showed ^3J and ^2J correlations, respectively, with the same hydroxyl carbinol carbon at δ_C 80.1 (C-3). Another two methyl groups positioned at δ_H 1.04 (s, 1H, H-19) and 0.92 (q, 3H, H-30) were attached to quaternary carbons at δ_C 42.8 (C-13) and δ_C 40.9 (C-14), respectively. Three methylene proton signals at δ_H 1.34 (m, 2H, H-11), 1.31 (m, 2H, H-12) and 1.03 (s, 2H, H-18) were observed to show ^2J, ^3J and ^2J correlations, respectively, with the quaternary carbon at δ_C 48.2 (C-9). One carbinol proton appearing as multiplet at δ_H 4.80 was attributed to C-21 and showed a ^2J correlation with the methine carbon at δ_C 35.3 (C-20). The COSY spectrum revealed the presence of two sets of methyl groups at δ_H 0.80 (s, 3H, H-28) and δ_H 0.85 (s, 1H, H-29) coupled with each other and attached to a quaternary carbon at C-4 (δ_C 43.0) and another set of methyl protons at δ_H 0.92 (q, 3H, H-30) coupled with a methylene proton at δ_H 1.57 (m, 2H, H-15). The ^{1}H- and ^{13}C-NMR spectral data are summarized in Table 1. The C-H correlations established by HMBC are shown in Figure 2. Based on these evidences and comparisons with the spectral data of lanostene type triterpenoic acids [22,27], the structure of **2** was identified as 3,21,21, 26-tetrahydroxylanostanoic acid.

Compound **3** was isolated as a white powder with a melting point range of 285–290 °C. The UV spectrum showed an absorption at v_{max} 254 nm. The IR spectrum displayed absorption bands for carboxylic acid hydroxyl groups at v_{max} 2924 cm^{-1} (C-OH, acid), carboxylic acid carbonyl carbon at 1732 cm^{-1} (C=O, acid) and olefins at 1648 (C=C). There is no absorption band for an alcoholic hydroxyl (-OH, alcohol). The ESIMS (positive mode) showed a molecular ion at m/z 441.3663 [M + H]$^+$ and 463.3474 [M + Na]$^+$ while ESIMS (negative mode) showed a molecular ion at m/z 439.3759 [M − H]$^-$ and 879.7791 [2M − H]$^-$. HRESIMS (positive mode) displayed a molecular ion at m/z 457.2734 [M + OH]$^+$. Its molecular formula was thus deduced to be $C_{30}H_{48}O_2$ (calculated 440.3650 [M$^+$] for $C_{30}H_{48}O_2$). The ^{1}H-NMR spectrum of **3** revealed signals for five tertiary methyls at δ_H 0.83 (s, 6H, H-23), 0.90 (s, 3H, H-24), 0.76 (s, 3H, H-25), 0.89 (s, 3H, H-26), 0.73 (s, 6H, H-27), a vinyl methyl at δ_H 1.10 ppm (s, 3H, H-30) and an *exo*-methylene proton at δ_H 5.32 ppm (brs, 1H, H-29).

The ^{13}C-NMR spectrum showed six methyl groups at δ_C 22.1 (C-23), 16.6 (C-24), 15.4 (C-25), 17.6 (C-26), 13.9 (C-27), and 20.1 ppm (C-30), an *exo*-methylene group at δ_C 150.9 (C-20) and 116.3 (C-29) and a carboxylic acid group at δ_C 183.5 (C-28). In addition the ^{13}C-NMR spectrum showed twelve methylene carbons, five methine carbons and six quaternary carbon signals. The ^{1}H- and ^{13}C-NMR data of **3** were very similar to those reported for betulinic acid [23]. The difference lies only the presence of a hydrogen instead a hydroxyl at C-3. These NMR spectral data indicated that compound **3** is a pentacyclic lupeol-type triterpenoic acid without a secondary hydroxyl-bearing carbinol carbon at the C-3 position. Proton and carbon connectivity was determined by HMQC and HMBC spectral analyses. Two methyl protons resonating as a singlet observed at δ_H 0.90 (H-24) and 0.83 (H-23) were attached to C-4 (δ_C 30.8) and correlated with carbons at δ_C 36.7 (C-3) and 59.6 (C-5), respectively, by a ^3J correlation. One of these methyl protons also showed a ^2J correlation with the quaternary carbon at C-4 (δ_C 30.8). Another methyl group at δ_H 0.76 (H-25) displayed ^3J correlations to a methylene carbon at δ_C 36.1 (C-1) and two methyne carbons at δ_C 59.6 (C-5) and 51.9 (C-9). An *exo*-methylene proton at δ_H 5.32 (H-29) resonating as a broad singlet attached to a quaternary carbon at δ_C 150.9 (C-20) was correlated with a methine carbon at δ_C 37.2 (C-19). Two methyl groups at δ_H 0.73 (s, 3H, H-27) and 1.10 (s, 3H, H-30) resonating as a singlet were attached to a quaternary carbon at δ_C 42.9 (C-14) and 150.9 (C-20), respectively, whereas a methyl proton at δ_H 0.73 (s, 3H, H-27) showed ^3J correlations with the methylene carbon at δ_C 39.7 (C-15), methine carbon δ_C 37.7 (C-13) and the proton at δ_H 1.10 (s, 3H, H-30) showed a ^2J correlation with quaternary carbon at δ_C 150.9 (C-20). Two methylene protons at δ_H 1.86 (m, 1H, 16a) and 1.25 (m, 1H, 16b) displayed ^2J correlations with δ_C 39.7 (C-15) and δ_C 59.6 (C-17). These methylene protons also showed ^3J correlations with the methine carbon at δ_C 46.5 (C-18) and the quaternary carboxylic carbon at δ_C 183.5 (C-28). Similarly the carboxylic acid carbon at δ_C 183.5 (C-28) was correlated with methylene proton at δ_H 1.75 (t, 2H, J = 5.4 Hz, H-22) by ^3J correlation. Two methine protons at δ_H 1.68 (m, 1H, H-18) and 2.09 (m, 1H, H-19) were observed to show ^3J and ^2J correlations with the quaternary carbon at δ_C 150.9 (C-20) and the proton at δ_H 2.09 (m, 1H, H-19) also showed a ^2J correlation with δ_C 46.5 (C-18). In the same way the methine proton at δ_H 1.24 (m, 1H, H-13) was correlated with the methine carbon at δ_C 46.5 (C-18) and the quaternary carbon at δ_C 40.1 (C-8) by ^2J and ^3J correlations, respectively. Another methine proton at δ_H 1.56 (m, 1H, H-9) showed ^2J and ^3J correlations with the quaternary carbon at δ_C 36.7 (C-10) and 42.9 (C-14), respectively.

The COSY spectrum revealed the presence of one set of methylene protons at δ_H 5.32 (brs, 1H, H-29) attached to C-29 (δ_C 116.3) were coupled to each other and with one set of methyl proton δ_H 1.10 (s, 3H, H-30). Based on the above ^{1}H-NMR, ^{13}C-NMR information as well as the HMBC, HSQC and COSY data and comparison with the literature values of betulinic acid [23], the structure of **3** was identified as dehydroxybetulinic acid. The ^{1}H- and ^{13}C-NMR spectral data of **3** are summarized in Table 1. The positions of carbons and protons were established by 2D-NMR data (HMBC, HSQC and COSY) and the C-H correlations established by HMBC are shown in Figure 2. In the genus Ficus, betulinic acid and acetylbetulinic acid and betulonic acid were previously isolated from *F. microcarpa* [28]. This is the first report of dehydroxybetulinic acid from a *Ficus* species.

2.2. Chemotactic Activity

The cell viability test was performed using trypan blue to determine the toxicity of the compounds on immune cells at different concentrations. The elevated cell viability indicated that the compounds were nontoxic to immune cells and were able to modulate the cellular immune response of phagocytes. Cells were viable (>92%) at the concentrations of 6.25 and 100 µg/mL of the extracts after incubation for 2 h. All the compounds were tested for chemotaxis activity at five different concentrations (10, 5, 2.5, 1.25 and 0.625 µg/mL). Distance migrated by cell was specified in µm. The percentage of inhibition and IC_{50} values were calculated comparing the distance travelled by negative control and samples. The average distance travelled by the negative control (DMSO and HBSS, 1:1 ratio) was 15.8 µm while that for the positive control, ibuprofen, was 4.8 µm. Ibuprofen was used as a positive control as it was found to be the most effective drug in a study to determine the effect of selected nonsteroidal anti-inflammatory drug (NSAIDs) in blocking the migration of PMNs [29,30]. The results showed that DMSO did not affect the movement of cells. Seven compounds showed high percentage of inhibition (70% to 80%), significantly different to the control ($p < 0.05$). Among the ten compounds, seven showed strong inhibitory activities with dose-dependent effects on the migration of PMNs towards the chemo attractant (fMLP). Compounds **1–4**, **6**, **7** and **9** exhibited strong (more than 70%) inhibition and the IC_{50} values of these compounds were comparable to that of ibuprofen. The percentage of inhibitions of

PMNs chemotaxis for the active compounds is shown in Figure 3. The IC_{50} values of active compounds with positive control are shown in Table 2.

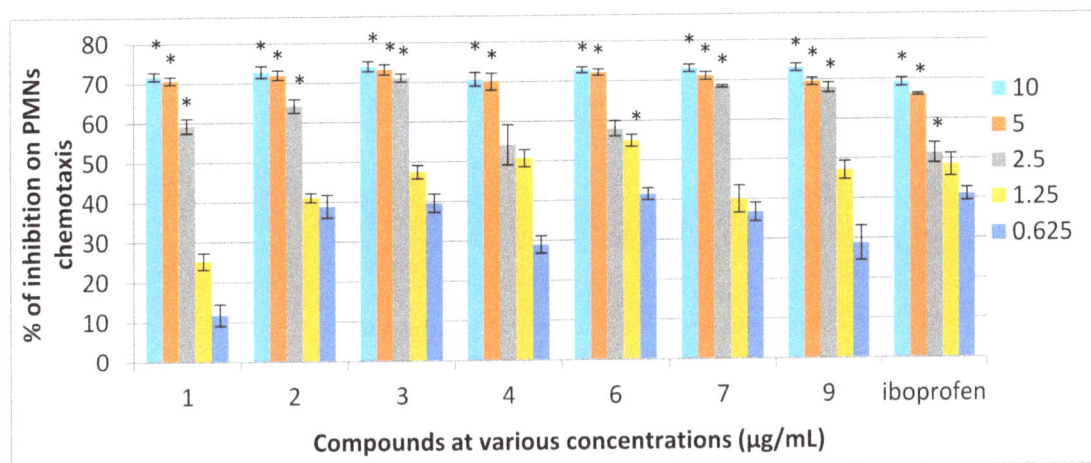

Figure 3. Dose dependent percentage of inhibition of compounds of *F. aurantiaca* on PMNs chemotaxis. Data are mean $\pm$ SEM ($n = 3$). * $p < 0.05$ is significant difference compared with the respective control determined by one-way ANOVA followed by Tukey's test.

Table 2. IC_{50} values of compounds isolated from *F. aurantiaca*.

Compounds	IC_{50} Value (μM)		
	ROS		Chemotaxis
	PMNs	WB	
28,28,30-Trihydroxylupeol (**1**)	9.2 ± 0.5	11.5 ± 0.05	6.8 ± 0.1
3,21,21,26-Tetrahydroxylanostanoic acid (**2**)	0.9 ± 0.03	0.1 ± 0.3	2.8 ± 0.1
Dehydroxybetulinic acid (**3**)	0.9 ± 0.1	1.7 ± 0.1	2.5 ± 0.1
Taraxerone (**4**)	1.3 ± 0.1	4.6 ± 0.08	4.1 ± 0.5
Taraxerol (**5**)	-	-	-
Ethyl palmitate (**6**)	1.1 ± 0.05	0.7 ± 0.05	3.7 ± 0.2
Herniarin (**7**)	0.5 ± 0.5	1.6 ± 0.1	8.2 ± 0.2
Stigmasterol (**8**)	-	-	-
Ursolic acid (**9**)	0.8 ± 0.5	1.3 ± 0.8	3.6 ± 0.2
Acetylursolic acid (**10**)	-	-	-
Aspirin (positive control)	9.4 ± 0.5	11.6 ± 0.3	-
Ibuprofen (positive control)	-	-	6.7 ± 0.5

2.3. Reactive Oxygen Species (ROS) Inhibitory Activity of Isolated Pure Compounds on Human Whole Blood (WB) and PMNs

Compounds **1–10** were evaluated for their effects on the oxidative burst of PMNs. Of the ten compounds, **1–4**, **6**, **7** and **9** were strongly active against PMNs. The IC_{50} values of these compounds were lower than that of aspirin which was used as a positive control. The result showed that the compounds inhibited ROS generation during the metabolic phase of phagocytises in a dose-dependent manner as the concentrations of samples increased the percentage of inhibition was also increased. Further evaluation of compounds **1–10** for their effects on oxidative burst on human whole blood showed that compounds **1–4**, **6**, **7** and **9** exhibited remarkable activity for luminol-enhanced chemiluminescence. In addition, compounds **1–3**, **6**, **7** and **9** showed strong inhibition on both PMNs and whole blood whereas inhibition of PMNs was higher than that of whole blood. The dose dependent ROS inhibitory effects on the oxidative burst of human whole blood (b) and PMNs (a) for active compounds are shown in Figure 4 and their IC_{50} values are shown in Table 2.

(a)

(b)

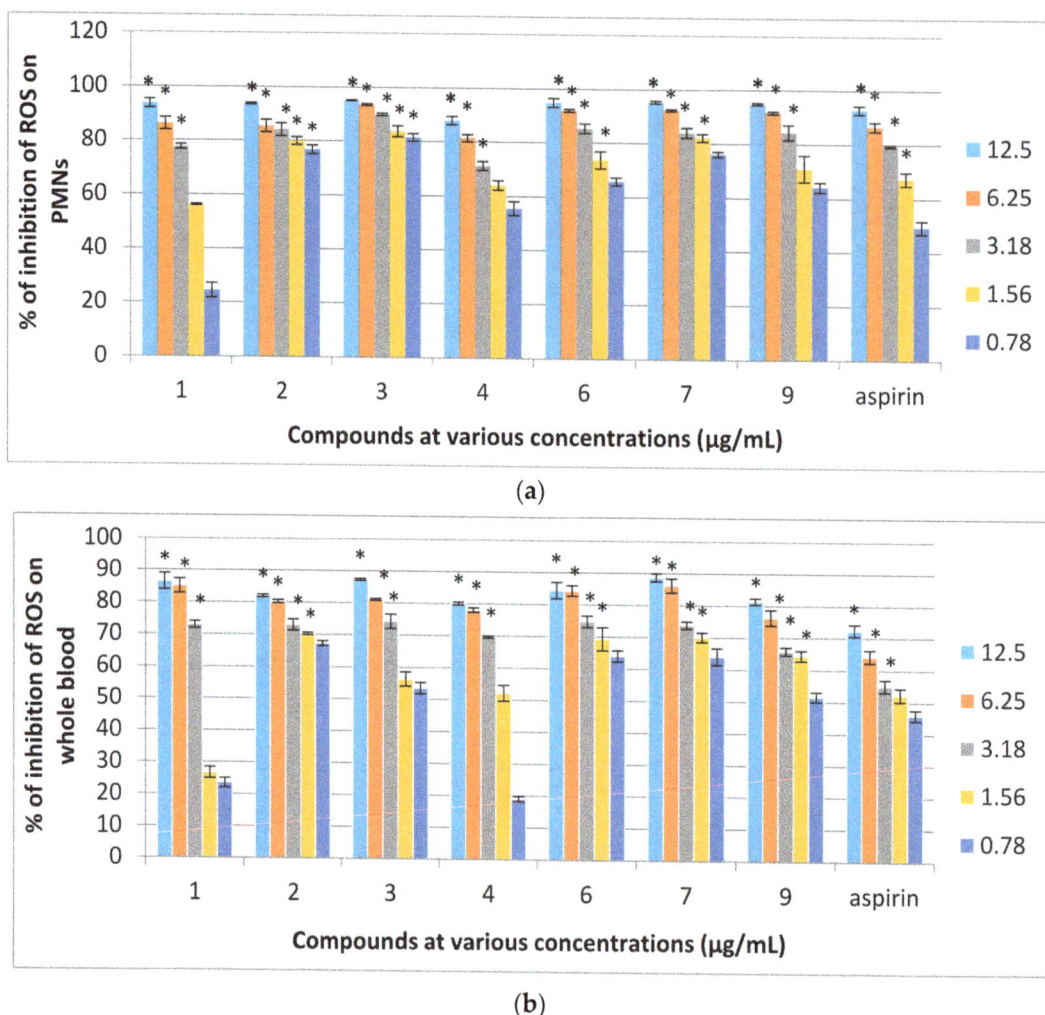

Figure 4. (a) Dose dependent percentage of inhibition of ROS inhibitory activity of compounds isolated from *F. aurantiaca* on PMNs and (b) whole blood assayed by luminol amplified chemiluminescence. Data are mean $\pm$ SEM ($n = 3$).* $p < 0.05$ is significant difference compared with the respective control determined by one-way ANOVA followed by Tukey's test.

The potency of triterpenoids from *F. aurantiaca* as inhibitors of chemotaxis and ROS production of neutrophils is in agreement with many previous studies. Triterpenoids, especially pentacyclic triterpenes like compounds **1** and **3** were implicated in studies on the mechanisms of action and pharmacological effects of many medicinal plants used in folk medicine against diseases related to the immune system such as anti-inflammatory, antiviral, antimicrobial, antitumoral agents, as well as immunomodulating effects. Several of them are implicated in the resolution of immune diseases, although their effects have not always been clearly correlated [31]. Oleanolic acid and its related compounds such as ursolic acid and betulinic acid are known to be anti-inflammatory and immuno-suppressive through their reduction of relevant cytokines such as interleukin-1 (IL-1) interleukin-6 (IL-6), and tumor necrosis factor-α (TNF-α), as well as their effect on the classic pathway of complement activation though the inhibition of C3 convertase. Triterpenoic acid such as oleanolic acid and 3-*epi*-katonic acid also inhibited adenosine deaminase, an enzyme which is found at increased levels in various immune diseases [31,32]. The immunosuppressive effects of *Tripterygium wilfordii* might also involve triterpenoids, as well as compounds such as oleanolic acid, 3-*epi*-katonic acid, terpenoic acid. Eight cycloartanes isolated from *Astragalus melanophrurius* (Fabaceae) were found to show interesting immunomodulatory activity in an isolated human lymphocyte stimulation test [32].

The effects of a mixture of triterpenes from *Quillaja saponaria* (Rosaceae) on the production of IL-1 and IL-6, as well as their role in the activation of antigen-presenting cells (APC), a prerequisite for the development of immune responses have been reported [33].

Triterpenes of diverse structural types are widely distributed in prokaryotes and eukaryotes. The physiological function of these compounds is generally supposed to be a chemical defense against pathogens. Triterpenes act against certain pathogens causing human and animal diseases, such as imflammation. This may be primarily due to the hydrophobic nature of most of the compounds [34]. In this study, among the eight triterpenes, there are two lupine-type skeletons (compounds **1** and **3**), two taraxerane-type (**4** and **5**), two ursane-type (**9** and **10**), one lanostane-type skeleton (**2**) and one stigmastane skeleton (**8**). A variety of terpenoids with oleanane-, ursane-, taraxerane-, lupine- and friedelane-type skeletons previously identified in many higher plants can act as precursors for many biomarkers found in biological screening. Oleanane, ursane and lupane triterpenes with carboxylic acid groups and alcoholic derivatives of oleanane, ursane and lupane are active against inflammation. However, it is difficult to identify exact molecular motifs, largely spread among these terpenes and implicated in their anti-inflammatory action. Some of the terpenoids act as plant hormones regulating different physiological roles, but some secondary metabolites is concerned in host defence and in the protection of the plant or animal from possible pathogens. Recent research in the field of the regulation of innate immunity from insects to mammals has established the existence of a previously unexpected protection in the pathways (receptors, kinases and effector molecules) that are involved in this process [35]. We suggest that the general analysis of the relationship between chemical structure and immunomodulatory activity of the isolated triterpenes may be related to the presence of an oxygenated group at C-3 and carboxyl group at C-28.

3. Experimental Section

3.1. General Information

Melting points were determined with an electrothermal apparatus (digital series) and were uncorrected. Ultraviolet (UV) spectra were recorded with a Shimadzu UV-1601 Spectrophotometer (Shimadzu Corp., Tokyo, Japan), for ethanol and methanol solutions in 1 cm Quartz cells. IR spectra were recorded on a GX FTIR instrument (Perkin-Elmer Corp., Waltham, MA, USA) using potassium bromide pellets and sodium chloride cells. NMR spectral analyses were carried out with a FT-NMR 600 MHz Cryoprobe spectrometer (Bruker, Basel, Switzerland). ^{1}H-^{1}H COSY, ^{1}H-^{13}C HSQC and ^{1}H-^{13}C HMBC were obtained with the usual pulse sequence and data processing was performed with the ACD lab software (Version 12, Advanced Chemistry Development, Toronto, ON, Canada). Mass spectra ESIMS and HRESIMS were measured on Micro TOF-Q mass spectrometer (Bruker, Basel, Switzerland). All experiments were carried out at Universiti Kebangsaan Malaysia (UKM). Analytical Thin Layer Chromatography (TLC) was carried out on pre-coated silica gel 60 GF 254 (20 × 20) TLC plate (Merck, art 5554, Darmstadt, Germany). The plates were visualized under ultraviolet light (λ_{254} nm, model UVGL-58) and by charring the compounds after spraying the plates with 10% sulphuric acid or Dragendorff reagent. Vacuum liquid chromatography was carried out over silica gel type H (10–0 µm, Sigma Chemical Co., St. Louis, MO, USA). The conventional column chromatography was done by using silica gel 60 (230–400 mesh ASTM, Merck, art 9385) or Sephadex LH20 (GE Healthcare Bio-Sciences AB, Uppsala, Sweden). Radial chromatography (centrifugal chromatography) was performed on a Chromatotron (Analtech Inc., Newark, Denmark) using silica gel 60 GF 254 containing gypsum (TLC grade) (Merck, art. 7749). Organic solvents such as hexane, ethyl acetate, chloroform, diethyl ether and methanol were purchased from Merck.

3.2. Plant Material

Ficus aurantiaca Griff was collected from Banting, Selangor, Malaysia and a voucher specimen (SM2109) was identified and deposited at the Herbarium of Universiti Kebangsaan Malaysia (UKM), Bangi, Malaysia.

3.3. Extraction and Isolation

Extraction and isolation of compounds from *F. aurantiaca* was carried out by modification of the method described by Rukachaisirikul [36]. The plant material was dried at room temperature and its air-dried stems were ground. The ground powder (196 g) was sequentially extracted thrice with *n*-hexane, ethyl acetate and methanol (500 mL each) by soaking at least for 48 h at room temperature for each time. After filtration each of the solvent extracts was evaporated to dryness under reduced pressure using a rotary evaporator to yield crude n-hexane (7.8 g), ethyl acetate (4.0 g) and methanol (10.0 g) extracts, respectively. The separation of hexane, ethyl acetate and methanol extracts was carried out using vacuum liquid chromatography (VLC), column chromatography and Chromatotron over silica gel, respectively.

The *n*-hexane crude extract (6.0 g) of *F. aurantiaca* was chromatographed using VLC on silica gel 60 GF$_{254}$ (TLC grade) and hexane–ethyl acetate (10:0 to 0:10, *v/v*) as solvent system to give sixteen fractions (F1–F16) that were analyzed using TLC and combined into five fractions (Fa1–Fa5) based on their TLC profiles. For further isolation and purification, each fraction was subjected to column chromatography (CC) using silica gel 60 (230–400 mesh ASTM). Fa1 (1.234 g) was subjected to CC on silica gel as stationary phase and hexane–ethyl acetate (10:0 to 0:10, *v/v*) of increasing polarity as mobile phase to obtain seven subfractions (Fa1A–Fa1G). Subfractions Fa1A and Fa1B were subjected to silica gel CC eluted with hexane–ethyl acetate (9:1 to 5:5 *v/v*) and further recrystallized from 100% *n*-hexane to obtain compound 6. Fraction Fa3 (2.135 g) was purified by silica gel CC eluted with hexane–ethyl acetate (9:1 to 0:10 *v/v*) to obtain compound 7.

The ethyl acetate crude extract (3.0 g) of *F. aurantiaca* was subjected to silica gel CC eluted with hexane–ethyl acetate (5:5 to 0:10, *v/v*) and ethyl acetate–methanol (10:0 to 6:4, *v/v*) to obtain eighteen fractions (F1–F18). Based on the TLC profiles, these eighteen fractions were combined into six fractions (Faa–Faf). Fractions Faa (322.1 mg) and Fab (173.2 mg) were separated and purified using repeated CC to obtain compounds 1 and 4, respectively. In the same way 5, 2 and 8 were obtained from fractions Fad (321.7 mg), Fae (523.2 mg) and Faf (457.2 mg), respectively. Similarly the methanol extract of *F. aurantiaca* (8.0 g) was fractionated using VLC on silica gel 60 GF$_{254}$ using ethyl acetate-methanol of increasing polarity as eluent and the fractions were purified by silica gel CC to obtain subfractions FF1–FF10. Subfractions FF2 (567.0 mg), FF5 (234.0 mg) and FF7 (638.0 mg) was run on a Chromatotron eluted with ethyl acetate–methanol (10:0 to 2:8, *v/v*) to yield compounds 9, 3 and 10, respectively. To summarize, compounds 6 and 7 were obtained from the *n*-hexane extract; compounds 1, 4, 5, 2 and 8 were isolated from ethyl acetate extract while compounds 9, 3 and 10 were isolated from the methanol extract. The structure elucidation of the isolated compounds was performed using UV, IR, ESIMS and NMR spectral data analyses and comparison with appropriate literature values.

28,28,30-Trihydroxylupeol (1): White powder (20.0 mg), melting point 222–224 °C, UV λ_{max} (C_2H_5OH) nm: 272, 210 (log ε 3.4, 2.2). IR λ_{max} ($CHCl_3$) cm^{-1}: 3310, 2920, 2158, 2042, 1976, 1611, 1411, 1021. ^{1}H-NMR ($CDCl_3$): δ_H 0.79 (s, 3H, H-23), 0.83 (s, 3H, H-24), 0.94 (s, 3H, H-25), 1.01 (s, 3H, H-26), 1.07 (s, 3H, H-27), 0.88 (m, 10H, H-9), 1.26 (m, 4H, H-15, H-16), 1.28 (m, 4H, J = 6.6 Hz, H-6, H-7), 1.32 (m, 1H, J = 7.2 Hz, H-5), 1.34 (m, 4H, H-22, H-13), 1.61 (m, 4H, H-11), 1.68 (m, 4H, J = 3.0 Hz, H-2, H-12), 1.88–1.94 (m, 2H, H-21), 2.04 (m, 1H, H-18), 2.38 (d, 1H, J = 7.2 Hz, H-19), 3.39 (dd, 1H, J = 10.8, 4.8 Hz, H-3), 4.57 (br s, 1H, H-29), 4.69 (br s, 1H, H-29), 4.79 (t, 1H, J = 6.0 Hz, H-28), 5.14 (m, 1H, H-30) 5.19 (m, 1H, H-30). ^{13}C-NMR ($CDCl_3$): δ_C 14.1(C-27), 14.5 (C-23), 15.9 (C-25), 16.1 (C-26), 17.7 (C-6), 18.0 (C-24), 21.0 (C-11), 25.1 (C-12), 27.3 (C-2), 27.4 (C-15), 29.7 (C-21), 34.1 (C-7), 35.5 (C-16), 37.3 (C-10), 37.9 (C-13), 38.1 (C-1), 39.6 (C-4), 39.9 (C-22), 40.9 (C-8), 42.8 (C-14), 43.0 (C-17), 47.9 (C-18), 48.2

(C-19), 50.3 (C-9), 55.6 (C-5), 69.2 (C-30), 80.1(C-3), 95.1 (C-28), 109.3 (C-29), 150.9 (C-20). ESIMS m/z: 463.3393 $[M + H - 12]^+$, 931.7436 $[2M + H - 18]^+$, HRESIMS m/z: 474.1465 $[M]^+$, calculated 474.1465 for molecular formula $C_{30}H_{50}O_4$.

3,21,21,26-Tetrahydroxylanostanoic acid (**2**): White powder (17.2 mg), melting point 150–155 °C, UV v_{max} (C_2H_5OH) nm: 272, 207 (log ε 2.3, 2.8). IR v_{max} ($CHCl_3$) cm^{-1}: 3215, 2915, 2850, 2134, 1985, 1735, 1678, 1642, 1472, 1120. ESIMS (negative mode) m/z: 502.3290 $[M]^+$, 479.1023 $[M - 23]^+$, 453.0885 $[M - CH_3OH - 17]^{+/}$ and 311.1563 $[M - side\ chain\ (C_8H_{14}O_5) - 1]^+$, HRESIMS (positive mode) m/z: 503.3570 $[M + H]^+$, calculated 502.3290 for molecular formula $C_{30}H_{50}O$. ^{1}H-NMR ($CDCl_3$): δ_H 0.80 (s, 3H, H-28), 0.85 (s, 1H, H-29), 0.92 (q, 3H, H-30), 0.94 (m, 2H), 0.99 (s, 2H, H-18), 1.01(m, 2H, H-2), 1.03 (s, 1H, H-19), 1.18 (m, 2H, H-22), 1.31 (m, 2H, H-12), 1.34 (m, 2H, H-11), 1.35 (m, 2H, H-23), 1.48 (m, 1H, H-8, H-17), 1.57 (m, 2H, H-15), 1.63 (m, 2H, H-16), 1.68 (m, 1H, H-20), 1.77 (m, 2H, H-7), 3.44 (t, 1H, H-3), 4.03 (d, 1H, $J = 12.0$ Hz, H-26), 4.57 (m, 1H, H-24), 4.79 (t, 1H, $J = 6.0$ Hz, H-6), 4.80 (m, 1H, H-21), 2.25 (C-21), 2.30 (C-21), 4.68 (C-3), 4.69 (C-26). ^{13}C-NMR ($CDCl_3$): δ_C 29.8 (C-1), 25.7 (C-2), 80.1 (C-3), 43.0 (C-4), 151.0 (C-5), 109.3 (C-6), 31.9 (C-7), 50.3 (C-8), 48.2 (C-9), 39.9 (C-10), 23.7 (C-11), 22.7 (C-12), 42.8 (C-13), 40.9 (C-14), 34.5 (C-15), 29.4 (C-16), 55.6 (C-17), 27.4 (C-18), 17.7 (C-19), 35.3 (C-20), 95.1 (C-21), 35.5 (C-22), 25.0 (C-23), 150.9 (C-24), 109.3 (C-25), 69.2 (C-26), 173.8 (C-27), 14.1 (C-28), 15.9 (C-29), 16.5 (C-30).

Dehydroxybetulinic acid (**3**): White powder (17.4 mg) with a melting point 285–290 °C. UV λ_{max} (C_2H_5OH) nm: 195. IR v_{max}($CHCl_3$) cm^{-1}: 2924, 2238, 2158, 1732 1648, 1463, 1248 and 971 cm^{-1}. ESIMS (positive mode) m/z: 441.3663 $[M + H]^+$ 463.3474 $[M + Na]^+$and ESIMS (negative mode) m/z: 439.3759 $[M - H]^-$, 879.7791 $[2M - H]$. HRESIMS (positive mode) m/z: 457.2734 $[M + OH]^+$, calculated 440.3787 for molecular formula $C_{30}H_{48}O_2$. ^{1}H-NMR ($CDCl_3$): δ_H 1.41 (m, 1H, H-1), 1.31 (m, 1H, H-1), 1.42 (m, 1H, H-2), 1.32 (m, 1H, H-2), 1.42 (m, 1H, H-3), 1.33 (m, 1H, H-3), 1.34 (m, 1H, H-5), 1.65 (m, 2H, H-6), 1.46 (m, 1H, H-7), 1.56 (m, 1H, H-9), 1.54 (m, 1H, H-11), 1.29 (m, 1H, H-11), 1.55 (m, 1H, H-12), 1.29 (m, 1H, H-12), 1.24 (m, 1H, H-13), 1.49 (m, 1H, H-15), 1.00 (m, 1H, H-15), 1.80 (m, 1H, H-16), 1.25 (m, 1H, H-16), 1.68 (m, 1H, H-18), 2.09 (m, 1H, H-19), 1.60 (m, 6H, H-21), 1.75 (t, 2H, $J = 5.4$ Hz, H-22), 0.83 (s, 6H, H-23), 0.90 (s, 3H, H-24), 0.76 (s, 3H, H-25), 0.89 (s, 3H, H-26), 0.73 (s, 6H, H-27), 5.32 (br s, 1H, H-29), 1.10 (s, 3H, H-30). ^{13}C-NMR ($CDCl_3$): δ_C 36.1 (C-1), 28.2 (C-2), 36.7 (C-3), 30.8 (C-4), 59.6 (C-5), 18.6 (C-6), 37.4 (C-7), 40.1 (C-8), 51.9 (C-9), 37.7 (C-10), 20.6 (C-11), 25.2 (C-12), 37.7 (C-13), 42.9 (C-14), 39.7 (C-15), 40.4 (C-16), 59.6 (C-17), 46.5 (C-18), 37.2 (C-19), 150.9 (C-20), 29.2 (C-21), 23.0 (C-22), 22.1 (C-23), 16.6 (C-24), 15.4 (C-25), 17.6 (C-26), 13.9 (C-27), 183.5 (C-28), 116.3 (C-29), 20.1 (C-30).

3.4. Chemicals, Reagents and Equipment

Serum opsonized zymosan A (*Saccharomyses cerevisiae* suspensions and serum), luminol (3-aminophthalhydrazide), phosphate buffer saline (PBS), Hanks Balance Salt Solution (HBSS^{++}), ficoll, N-formyl-methionylleucyl-phenilalanine (fMLP), tryphan blue, phorbol 12-myristate 13-acetate (PMA), dimethyl sulfoxide (DMSO), acetylsalicylic acid (purity 99%) and ibuprofen (purity 99%) were purchased from Sigma. Haematoxylin and xylene were obtained from BDH (London, UK). Chemiluminscence measurements were carried out on a Luminoscan Ascent luminometer (Thermo Scientific, Loughborough, UK). A Boyden chamber with a 3 and 5 μm polycarbonate membrane filter separating the upper and lower compartments was purchased from Neuro probe (Cabin John, Montgomery County, MD, USA).

3.5. Isolation of Polymorphonnuclear Leucocytes (PMNs)

Human blood was collected from a healthy volunteer fasted for at least 8 h. PMNs were isolated by Ficol-gradient separation following the published method [37]. The use of human blood in this study was approved by the Human Ethics Committee, Universiti Kebangsaan Malaysia Medical

Centre (HUKM), Cheras with permission (number FF-220-2008). Cell counts were performed using a haemocytometer.

3.6. Cell Viability

Cell viability tests were performed using trypan blue dye exclusion method with the modification of a published procedure [38]. The neutrophils (1×10^6 mL^{-1}) were incubated with 6.25 or 100 µg/mL of pure compounds in triplicate at 37 °C for 1 to 2 h. The cell death was indicated by the blue dye uptake. The percentage of cell viability was considered from the total cell counts.

3.7. Chemiluminescence Assay

Luminol enhanced chemiluminescence assay was carried out using the modification of the published method [39]. In brief, 25 µL of human whole blood or 25 µL of isolated PMN cells were suspended in $HBSS^{++}$ into each well of 96-well microplate. The plate was incubated with 25 µL of tested compounds and aspirin at five different concentrations (12.5, 6.25, 3.18, 1.56 and 0.78 µg/mL) of each sample for 50 min at 37 °C in a luminoscan while 25 µL of $HBSS^{++}$ was used as the negative control. The cells were induced with 25 µL of serum opsonized zymosan (SOZ) followed by 25 µL of luminol into each well. Then $HBSS^{++}$ solution was added into each well to make the final volume of 200 µL. Aspirin was used as a positive control while the negative control contained zymosan, luminol, DMSO (0.5%), $HBSS^{++}$ and cells. The percentage of inhibition was calculated by the measurement of RLU (reading luminometer unit) of peak and total integral values with repeated scans.

3.8. Chemotaxis Assay

The assay was carried out using the modified 48-well Boyden chamber method [40]. The Boyden chamber contains 48 wells with a diameter of 8 µm and is divided into two compartments by a filter separation. In brief, 25 µL of chemo attractant, fMLP (10^{-8} M, diluted with chemo attractant buffer solution) was added to the lower compartment of the Boyden chamber. PMN cell suspension (45 µL) with 5 µL of test compounds and ibuprofen at five different concentrations (6.25, 12.5, 25, 50, and 100 µg/mL) were added to the upper compartment of the chamber whereas the negative control contained 45 µL of PMN cell suspension and 5 µL of chemo attractant buffer. Ibuprofen was used as a positive control. The final concentrations of test compounds and ibuprofen in the wells were 0.625, 1.25, 2.5, 5 and 10 µg/mL. The chamber was incubated in 5% carbon dioxide incubator for 1 h at 37 °C. After incubation the polycarbonate membrane (where the migrated cells were remained) was stained with PBS, methanol (99.5%), haematoxylin, distilled water, ethanol (70% and 95.8%) and xylene respectively. The distance of cell's migration was measured using a low power microscope with magnifications 40×.

3.9. Statistical Analysis

Data were analyzed using One-way ANOVA, *post hoc* Tukey's test. $p < 0.05$ was considered to be statistically significant using the SPSS 17.0 (SPSS Inc., Chicago, IL, USA) statistical software. All values were characterized as mean ± SEM. Probit programme was used to determine the IC_{50} values for active compounds.

4. Conclusions

In conclusion, a phytochemical investigation of the stem of *F. aurantiaca* afforded three new triterpenoids—28,28,30-trihydroxylupeol (**1**), 3,21,21,26-tetrahydroxylanostanoic acid (**2**) and dehydroxybetulinic acid (**3**) together with five known triterpenoids: taraxerone (**4**), taraxerol (**5**), stigmasterol (**8**), ursolic acid (**9**), acetyl ursolic acid (**10**), one coumarin, herniarin (**7**) and one unsaturated hydrocarbon, ethyl palmitate (**6**). This is the first report on the phytochemical investigation and biological effects of compounds from *F. aurantiaca* on PMN chemotaxis and ROS inhibitory activity of human whole blood and PMNs. Compounds **1–3**, **6**, **7** and **9** showed the strongest inhibitory activities for both chemiluminescence and chemotaxis whereas the IC_{50} values were lower than those of the standard drugs used. This study provided evidence that *F. aurantiaca* is a source of new immunomodulatory compounds to modulate the innate immune response of phagocytes and further studies are needed to investigate the mechanisms of their immuno-modulatory responses.

Acknowledgments: This work was supported by the grant UKM-GGPM-TKP-058-2010, UKM-Pharmacy-03-FRGS 0029-2010 and UKM-DIPM-006-2011. Special thanks to the Ministry of Higher Education Malaysia, for CSFP.

Author Contributions: S.M., I.J., and K.H. designed research; S.M. and K.H. performed research and analyzed the data; S.M., I.J. and K.H. wrote the paper, S.M. and I.J. performed on reactive oxygen species production and chemotactic activity of neutrophils. All authors read and approved the final manuscript.

Conflicts of Interest: The authors declare that they have no conflict of interests.

References

1. Vander, N.J.M.; Klerx, J.P.; Vandijk, H.; de Silva, K.T.; Labadie, R.P. Immunomodulatory activity of an aqueous extract of *Azadirachta indica* stem bark. *J. Ethnopharmacol.* **1987**, *19*, 125–131.

2. Diwanay, S.; Chitre, D.; Patwardhan, B. Immunoprotection by botanical drugs in cancer chemotherapy. *J. Ethnopharmacol.* **2004**, *90*, 49–55.

3. Gautam, M.; Diwanay, S.; Gairola, S.; Shinde, Y.; Patki, P.; Patwardhan, B. Immuno adjuvantpotential of *Asparagus racemosus* aqueous extract in experimental system. *J. Ethnopharmacol.* **2004**, *91*, 251–255. [CrossRef] [PubMed]

4. Jayathirtha, M.G.; Mishra, S.H. Preliminary immunomodulatory activities of methanol extracts of *Eclipta alba* and *Centella asiatica*. *Phytomedicine* **2004**, *11*, 361–365. [CrossRef] [PubMed]

5. Diasio, R.B.; LoBuglio, A.F. Immunomodulators: Immunosuppressive agents and immunostimulants. In *Goodman and Gilman's The Pharmacological Basis of Therapeutics*, 9th ed.; Hardman, J.G., Limbird, L.E., Molinoff, P.B., Ruddon, R.W., Eds.; McGraw-Hill: New York, NY, USA, 1996; pp. 1291–1307.

6. Fordin, D.G. History and concepts of big plant genera. *Taxon* **2004**, *53*, 753–776. [CrossRef]

7. Ronsted, N. Phylogeny, biogeography and ecology of *Ficus* section *Malvanther* (Moraceae). *Mol. Phylogenet. Evol.* **2008**, *48*, 12–22. [CrossRef] [PubMed]

8. Lansky, E.P.; Paavilainen, H.M.; Pawlus, A.D.; Newman, R.A. *Ficus* spp. (fig): Ethnobotany and potential as anticancer and anti-inflammatory agents. *J. Ethnopharmacol.* **2008**, *119*, 195–213. [CrossRef] [PubMed]

9. Simo, C.C.F.; Simeon, F.K.; Poumale, H.M.P.; Simo, I.K.; Ngadjui, B.T.; Green, I.R.; Krohn, K. Benjaminamide: A new ceramide and other compounds from the twigs of *Ficus benjamin* (Moraceae). *Biochem. Syst. Ecol.* **2008**, *36*, 238–243. [CrossRef]

10. Li, R.W.; Leach, D.N.; Myers, S.P.; Lin, G.D.; Leach, G.J.; Waterman, P.G. A new anti-inflammatory glucoside from *Ficus racemosa* L. *Planta. Med.* **2004**, *70*, 421–426. [PubMed]

11. Mandal, S.C.; Maity, T.K.; Das, J.; Das, B.P.; Pal, M. Anti-inflammatory evaluation of *Ficus racemosa* Linn. Leaf extract. *J. Ethnopharmacol.* **2000**, *72*, 87–92. [CrossRef]

12. Sackyfio, A.C.; Lugeleka, O.M. The anti-inflammatory effect of a crude aqueous extract of the root bark of *Ficus elastica* in the rat. *Arch. Int. Pharmacodyn. Ther.* **1986**, *281*, 169–176.

13. Shukla, R.; Gupta, S.; Gambhir, J.K.; Prabhu, K.M.; Murthy, P.S. Antioxidant effect of aqueous extract of the bark of *Ficus benghalensis* in hyper cholesterolaemic rabbits. *J. Ethnopharmacol.* **2004**, *92*, 47–51. [CrossRef] [PubMed]

14. Burkill, I.H. *A Dictioanary of the Economic Products of the Malays Panisular*; Ministry of Agriculture and Co-Operatives: Kuala Lumpur, Malaysia, 1966.

15. Mawa, S.; Husain, K.; Jantan, I. *Ficus carica* L. (Moraceae): Phytochemistry, traditional uses and biological activities. *Evid. Based Complement. Altern. Med.* **2013**. [CrossRef] [PubMed]

16. Kamaruddin, M.S.; Latif, A. *Tumbuhan Ubatan Malaysia Selangor*; Pusat Pengurusan Penyelidikan: Bangi, Malaysia, 2002; Volume 1.

17. Himanshu, J.; Arun, B.J.; Hemlata, S.; Gururaja, M.P.; Prajwal, R.S.; Subrahmanyam, A.V.S.; Satyanarayana, D. Fatty acids from *Memecylonum bellatum* (Burm). *Asian J. Res. Chem.* **2009**, *2*, 178–180.

18. Ibrahim, A.; Najmuldeen, A.; Hamid, A.H.; Mohamad, K.; Khalijah, A.; Mehran, F.N.; Kamal, A.K.; Mat Ropi, M.; Hiroshi, M. Steroids from *Chisocheton tomentosus*. *Malays. J. Sci.* **2011**, *30*, 144–153.

19. Jamal, A.K.; Yacoob, W.A.; Din, L.B. Triterpenes from the root bark of *Phyllanthus columnaris*. *Aust. J. Basic Appl. Sci.* **2009**, *3*, 1428–1431.

20. Mahdi, A.; Amirhossein, S.; Mehrda, I. Synthesis and purification of 7-prenyloxycoumarins and herniarin as bioactive natural coumarins. *Iran. J. Basic Med. Sci.* **2009**, *1*, 63–69.

21. Mawa, S.; Ikram, M.S. Chemical constituents of *Garcinia prainiana*. *J. Sains Malays.* **2012**, *41*, 585–590.

22. Shahid, H.A.; Ali, M.; Kamran, J.N. New manglanostanoic acid from the stem bark of *Megifera indica* var. "Fazli". *J. Soudi Chem. Soc.* **2014**, *18*, 561–565.

23. Uddin, G.; Waliullah, M.; Siddiqui, B.S.; Alam, M.; Sadat, A.; Ahmad, A.; Uddin, A. Chemical constituents and phytotoxicity of solvent extracted fractions of stem bark of *Grew optiva* Drummond ex Burret. *Middle East J. Sci. Res.* **2011**, *8*, 85–91.

24. Supaluk, P.; Puttirat, S.; Rungrot, C.; Somsak, R.; Virapong, P. New bioactive triterpenoids and antimalarial activity of *Diospyros rubra* lec. *EXCLI J.* **2010**, *9*, 1–10.

25. Ahmed, W.; Khan, A.Q.; Malik, A. Two triterpenes from the leaves of *Ficus carica*. *Planta Med.* **1988**, *54*. [CrossRef] [PubMed]

26. Chiang, Y.M.; Kuo, Y.H. Novel triterpenoids from the arial parts of *Ficus microcarpa*. *J. Org. Chem.* **2002**, *67*, 7656–7661. [CrossRef] [PubMed]

27. Ali, M.; Shuaib, M.; Naqvi, K.J. New lanostane type triterpenes from the oleo-gum resin of *Commiphora myrrha* (NEES) ENGL. *Int. J. Pharm. Pharm. Sci.* **2014**, *6*, 372–375.

28. Chiang, Y.M.; Chang, J.Y.; Kuo, C.C.; Kuo, Y.H.; Chang, C.Y. Cytotoxic triterpenes from the arial roots of *Ficus microcarpa*. *Phytochemistry* **2005**, *66*, 495–501. [CrossRef] [PubMed]

29. Spisani, S.; Vanzini, G.; Traniello, S. Inhibition of human leucocytes locomotion by anti-inflammatory drugs. *Experientia* **1979**, *35*, 803–804. [CrossRef] [PubMed]

30. José-Luis, R. Effects of triterpenes on the immune system. *J. Ethnopharmacol.* **2010**, *128*, 1–14.

31. Brinker, A.M.; Ma, J.; Lipsky, P.E.; Raskin, I. Medicinal chemistry and pharmacology of genus Tripterygium (Celastraceae). *Phytochemistry* **2007**, *68*, 732–766. [CrossRef] [PubMed]

32. Calis, I.; Yürüker, A.; demir, D.; Wright, A.D.; Sticher, O.; Luo, Y.D.; Pezzuto, J.M. Cycloartane triterpene glycosides from the roots of *Astragalus melanophrurius*. *Planta Med.* **1997**, *63*, 183–186. [CrossRef] [PubMed]

33. Behboudi, S.; Morein, B.; Villacres-Eriksson, M. *In vitro* activation of antigen presentin cells (APC) by defined composition of *Quillaja saponaria* molina triterpenoids. *Clin. Exp. Immunol.* **1996**, *105*, 26–30. [CrossRef] [PubMed]

34. De las, H.B.; Rodríguez, B.; Boscá, L.; Villar, A.M. Terpenoids: Sources, structure elucidation and therapeutic potential in inflammation. *Curr. Top. Med. Chem.* **2003**, *3*, 53–67.

35. Shashi, B.M.; Sucharita, S. Advances in triterpenoid research, 1990–1994. *Phytochemistry* **1997**, *44*, 1185–1236.

36. Rukachaisirikul, V.; Ritthiwigrom, T.; Pinsa, A.; Sawangchote, P.; Taylor, W.C. Xanthones from the bark of *Garcinia nigrolineata*. *Phytochemistry* **2003**, *64*, 1149–1156. [CrossRef]

37. Demirkiran, O.; Mesaik, M.A.; Beynek, H.; Abbaskhan, A.; Iqbal, C.M. Cellular reactive oxygen species inhibitory constituents of *Hypericum thasium*. *Phytochemistry* **2009**, *70*, 244–249. [CrossRef] [PubMed]

38. Koko, W.S.; Mesaik, M.A.; Yousof, S.; Galal, M.; Choudary, M.I. *In vitro* immonumodulating properties of selected Sudanese medicinal plants. *J. Ethnopharmacol.* **2008**, *118*, 26–34. [CrossRef] [PubMed]

39. Haklar, G.; Ozveri, E.S.; Yuksel, M.; Aktan, A.; Yalcin, A.S. Different kinds of reactive oxygen and nitrogen species were detected in colon and breast tumors. *Cancer Lett.* **2001**, *165*, 219–224. [CrossRef]

40. Sacerdote, P.; Massi, P.; Panerai, A.E.; Parolaro, D. *In vivo* and *in vitro* treatment with the synthetic cannabinoid CP55,940 decreases the *in vitro* migration of macrophages in the rat: involvement of both CB1 and CB2 receptors. *J. Neuroimmunol.* **2000**, *109*, 155–163. [CrossRef]

4

Cytotoxic and Antifungal Constituents Isolated from the Metabolites of Endophytic Fungus DO14 from *Dendrobium officinale*

Ling-Shang Wu [1,†], Min Jia [2,†], Ling Chen [2], Bo Zhu [1], Hong-Xiu Dong [1], Jin-Ping Si [1], Wei Peng [3,*] and Ting Han [1,2,*]

Academic Editor: Derek J. McPhee

[1] Nurturing Station for the State Key Laboratory of Subtropical Silviculture, Zhejiang A & F University, Lin'an 311300, China; shang2002012@163.com (L.-S.W.); aurora0119@163.com (B.Z.); dhx0103@126.com (H.-X.D.); lssjp@163.com (J.-P.S.)
[2] Department of Pharmacognosy, School of Pharmacy, Second Military Medical University, Shanghai 200433, China; jm7.1@163.com (M.J.); m15201916813@163.com (L.C.)
[3] College of Pharmacy, Chengdu University of Traditional Chinese Medicine, Chengdu 610075, China
[*] Correspondence: pengwei002@126.com (W.P.); than927@163.com (T.H.)

[†] These authors contributed equally to this work.

Abstract: Two novel cytotoxic and antifungal constituents, (4S,6S)-6-[(1S,2R)-1, 2-dihydroxybutyl]-4-hydroxy-4-methoxytetrahydro-2H-pyran-2-one (**1**), (6S,2E)-6-hydroxy-3-methoxy-5-oxodec-2-enoic acid (**2**), together with three known compounds, LL-P880γ (**3**), LL-P880α (**4**), and Ergosta-5,7,22-trien-3b-ol (**5**) were isolated from the metabolites of endophytic fungi from *Dendrobium officinale*. The chemical structures were determined based on spectroscopic methods. All the isolated compounds **1–5** were evaluated by cytotoxicity and antifungal effects. Our present results indicated that compounds **1–4** showed notable anti-fungal activities (minimal inhibitory concentration (MIC) ≤ 50 μg/mL) for all the tested pathogens including *Candida albicans*, *Cryptococcus neoformans*, *Trichophyton rubrum*, *Aspergillus fumigatus*. In addition, compounds **1–4** possessed notable cytotoxcities against human cancer cell lines of HL-60 cells with the IC_{50} values of below 100 μM. Besides, compounds **1**, **2**, **4** and **5** showed strong cytotoxities on the LOVO cell line with the IC_{50} values were lower than 100 μM. In conclusion, our study suggested that endophytic fungi of *D. officinale* are great potential resources to discover novel agents for preventing or treating pathogens and tumors.

Keywords: *Dendrobium officinale*; endophytic fungi; cytotoxic activities; antifungal activities; metabolites

1. Introduction

Since the endophytic fungus which can produce taxol was reported by Stroble *et al.* [1], finding the suitable endophytic fungi to ferment and synthesize extensive active constituents has been considered as one of the effective way to resolve the resource shortage of some plant-derived compounds [2]. Endophytic fungi, microorganisms that reside in tissues of the host plant and can cause no apparent harm to the host plant during a certain phase in their life cycle [3], are known to produce some rare and novel natural agents with notable pharmacological activities including anti-tumor and anti-microbial, *etc.* [4]. However, only very few of them have been cultivated and screened for drugs.

Dendrobium officinale Kimura et Migo, is ranked "the first of the nine Chinese fairy herbs", which has been officially recorded in Chinese pharmacopoeia (Pharmacopoeia Committee of the

People's Republic of China 2010). *D. officinale* possesses great medicinal values for maintaining tonicity of stomach, promoting the body fluid production, reducing peripheral vascular obstruction, preventing the development of cataracts and enhancing the immune system, and has been commonly applied to anti-tumor, anti-aging, regulation of blood sugar, treatment of stomach disorders, *etc.* [5]. However, it has been listed as an endangered species and catalogued in the Chinese Plant Red Book since 1987 because limited natural resources and high demand threaten the survival of the species [6]. Therefore, protecting the wild resources of *D. officinale* becomes increasingly important for China. As a representative species of Orchidaceae, fungi are reported to play a critical role for seed germination and plants survival of *D. officinale* [7]. Therefore, exploiting the endophytic fungi in *D. officinale* is necessary, which can not only provide fungal resources for screening potential natural products but also lay a foundation of the further study on endophyte-host interaction. However, for endophytic fungi associated with medicinal orchids, especially in the *Dendrobium* genus, only a few have been explored.

In the course of our continuous search for plant–fungus associations and novel bioactive secondary metabolites [8] from endophyte cultures, we selected a fungus *Pestalotiopsis* sp. DO14 that can increase the content of main medicinal compounds (e.g., polysaccharides) of *D. officinale* from the shoot of *D. officinale* plants collected in Yandang Mountain, Zhejiang Province, People's Republic of China (unpublished). Through bioassay-oriented fractionation, two new monoterpenoids, (4S,6S)-6-[(1S,2R)-1,2-dihydroxypentyl]-4-hydroxy-4-methoxytetrahydro-2H-pyran-2-one (**1**) and (6S,2E)-6-hydroxy-3-methoxy-5-oxodec-2-enoic acid (**2**), together with three known compounds, LL-P880γ (**3**), LL-P880α (**4**) and Ergosta-5,7,22-trien-3β-ol (**5**), were isolated from the culture broth of *Pestalotiopsis* sp. DO14. We report herein the details of the isolation and identification of endophytes and compounds, and the evaluation for cytotoxic and antifungal activity of those isolated compounds.

2. Results and Discussion

2.1. Identification of the Endophytic Fungus

The phylogenetic tree (Figure 1) inferred from the ribosomal DNA ITS (Internal Transcribed Spacer) sequences indicated that the endophytic fungus DO14 was classified into the clade including *Pestalotiopsis clavispora* KJ677242, *P. mangiferae* KF155295, *P. microspora* KJ019328. Thus, the endophytic fungus DO14 was identified as a *Pestalotiopsis* sp. closely related to these three taxa with the ITS sequence similarity of 100.0%.

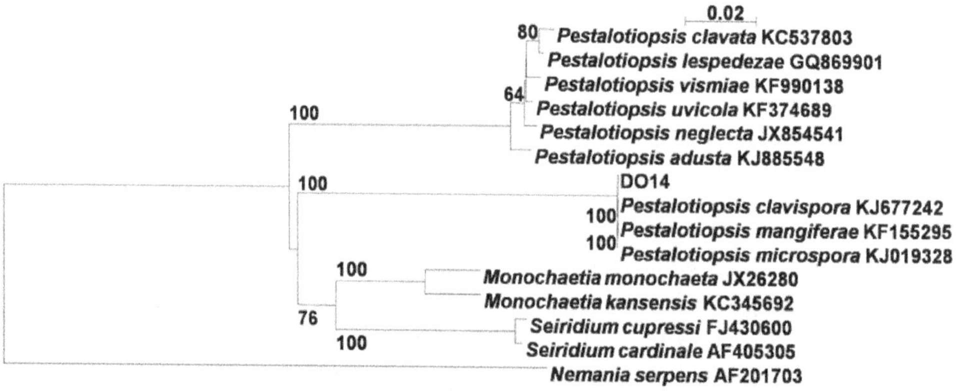

Figure 1. Phylogenetic tree of the endophytic fungus DO14 based on 5.8S and ITS regions sequences. Bootstrap values above 50% (1000 replicates) are shown at branches. *Nemania serpens* is used as an out-group.

2.2. Structural Determination of the Compounds

Compound **1** was obtained as yellow oil ($[\alpha]_D^{25}$ −14.3 (c 0.50, MeOH)) and analyzed for the molecular formula $C_{11}H_{20}O_6$ by HRESIMS $[M − H]^-$ at m/z 247.1186 (cald. 247.1182). The IR spectrum exhibited absorption bands for hydroxyl groups (3406 cm^{-1}) and carbonyl groups (1713 cm^{-1}). The ^{1}H-NMR spectra data exhibited signals for one-methyl groups (δ_H H 0.95, 3H, t, J = 7.3 Hz) and oxymethyl (δ_H H 3.36, 3H, s) (Table 1). The ^{13}C-NMR spectrum together with DEPT data resolved 11 carbon resonances attributable to one carbonyl group (δ_c C 171.9), two methyls, three sp^3 oxygenated methines, four sp^3 methylenes, and one sp^3 oxygenated quaternary carbons (Table 1). As one of the two degrees of unsaturation was consumed by one carbonyl group, the remaining degree of unsaturation required that compound **1** was monocyclic. The above-mentioned information was quite similar to that of co-isolated compound **3** reported from the same genus [9]. In comparison with compound **3**, the major differences of compound **1** were due to an additional oxygenated quaternary carbon (δ_c C 97.1) and one sp^3 methylene (δ_c C 41.6) instead of one tri-substituted double bonds (δ_c C 173.4 and 89.7), indicating that compound **1** was a derivative of compound **3**. HMBC correlations from H_3CO-4 (δ_H H 3.36) to C-4 (δ_c C 97.1) and from H_2-3 (δ_H H 2.85 and 2.82) to C-4, C-2 and C-5 assigned that the hydroxyl was connected at C-4 (Figure 2). The planar structure of compound **1** was further established by detailed interpretation of its 2D NMR data (Figure 3). The relative configuration of compound **1** was established by comparison of 1D NMR data with compound **3**. The absolute configuration of compound **3** was confirmed by exciton chirality method in previous work [10]. The stereochemistry of compound **3** was resolved by comparing optical rotation $[\alpha]_D^{25}$ −44.3 (c 1.30, MeOH) and the CD (Circular dichroism spectra) curve of compound **3** shows the cotton effects at 248 nm ($\Delta\varepsilon$ = −14.7), indicating that the configurations of C-6, C-1′, and C-2′ were S, S, and R. (Figure 4A). The absolute configuration of compound **1** was confirmed by the chemical correlation from compound **3** to compound **1** in methanol dropping with H_2O_2 and NaOH at room temperature. Therefore, the configurations of C-6, C-1′, and C-2′ in **1** were the same as **3**. As no convincing evidence was observed in the NOESY (nuclear Overhauser enhancement spectroscopy) spectrum to assign the configuration of 4-OH, the ^{1}H-NMR data of **1** was measured in $CDCl_3$ and C_5D_5N to obtain the pyridine-induced solvent shifts [11] (Figure 4B,C). The solvent shifts of H-6 ($\Delta\delta CDCl_3$ − C_5D_5N = 0.25) indicated that the 4-OH/H-6 was compound **1**, 3-diaxial-oriented. Thus, 4-OH was assigned in α-orientation and the configurations of C-4 was S. Therefore, compound **1** was determined to be (4S,6S)-6-[(1S,2R)-1,2-dihydroxypentyl]-4-hydroxy-4-methoxytetrahydro-2H-pyran-2-one.

Table 1. NMR data of compounds **1, 2**.

	1 [a]		2 [b]	
	δ_C, Type	δ_H Mult (J in Hz)	δ_C	δ_H Mult (J in Hz)
2	171.9, C		184.4, C	
3	41.6, CH_2	2.88, dd (18.5, 2.3) 2.85, d (18.5)	105.8, CH	5.63, s
4	97.1, C		167.1, C	
5	29.3, CH_2	2.44, dt (13.5, 2.3) 1.88, dd (13.5, 4.1)	36.5, CH_2	3.55, 2H, s
6	78.7, CH	4.74, m	204.6, C	
1′	66.8, CH	3.51, m	86.7, C	4.46, dd (7.8, 4.2)
2′	73.2, CH	3.61, m	30.7, CH_2	1.88, m 1.70, m
3′	33.7, CH_2	1.70, m 1.56, m	26.5, CH_2	1.40, 2H, m
4′	19.6, CH_2	1.43, 2H, m	22.3, CH	1.34, 2H, m
5′	14.3, CH_3	0.95, t (7.3)	13.8, CH_3	0.89, t (7.3)
CH_3O-4	49.2, CH_3	3.36, s	52.6, CH_3	3.76, s

[a] in CD_3OD. [b] in CDCl.

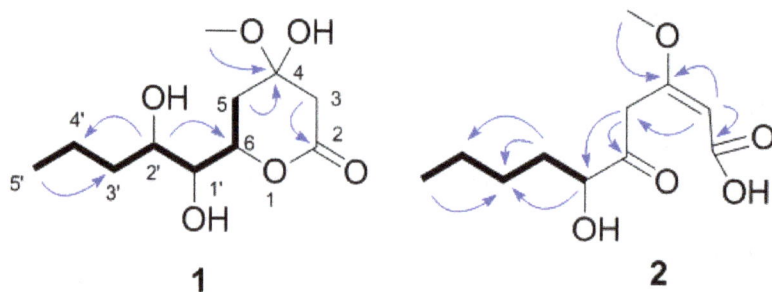

Figure 2. Selected ^{1}H–^{1}H COSY () and HMBC ($\rightarrow$) correlations of **1** and **2**.

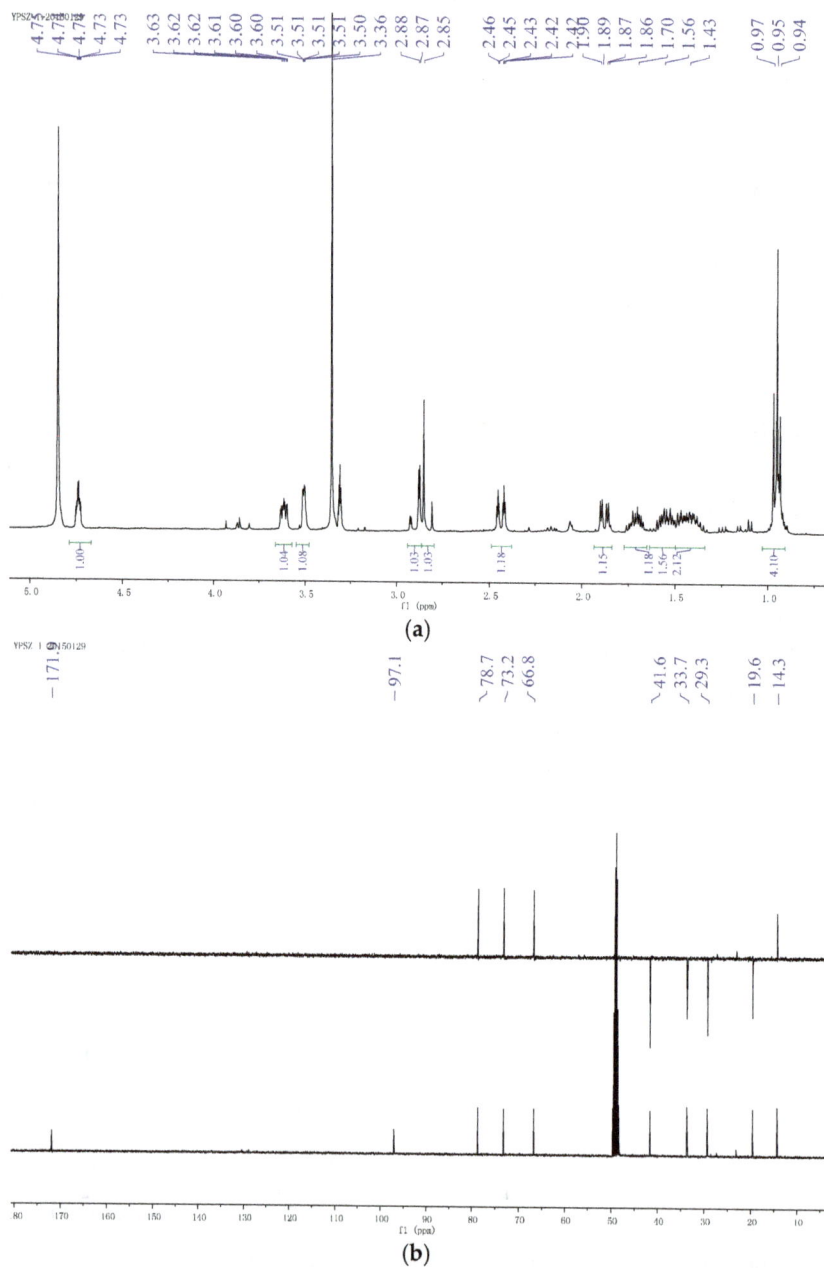

(a)

(b)

Figure 3. *Cont.*

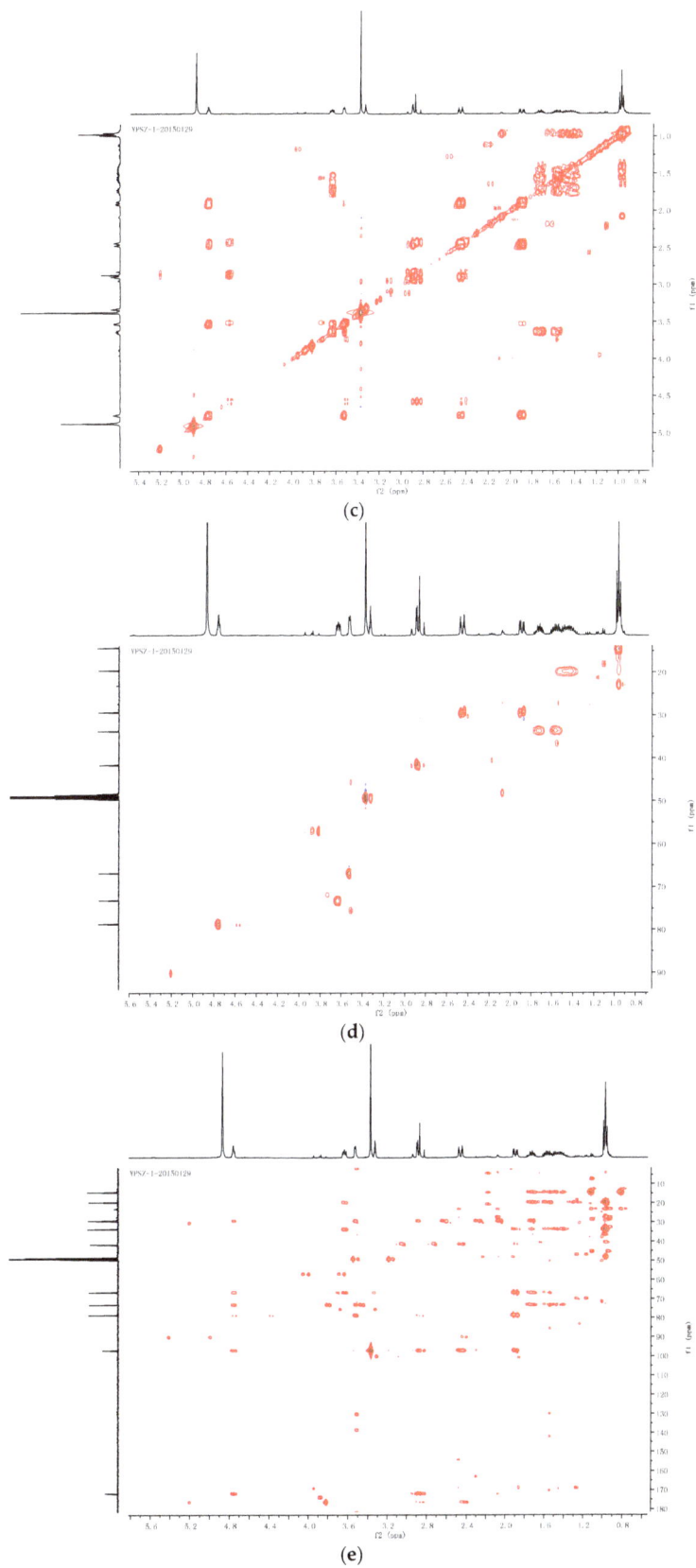

(c)

(d)

(e)

Figure 3. *Cont.*

(f)

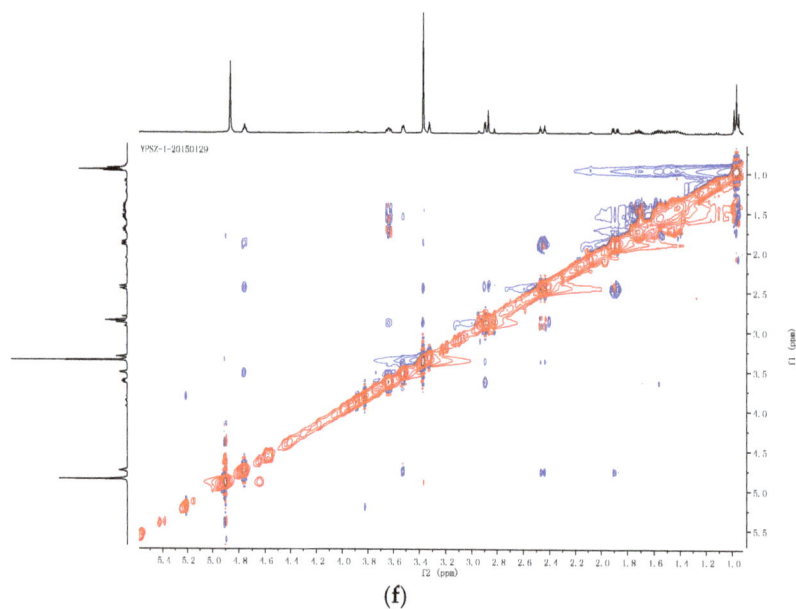

Figure 3. Supporting information of compound **1**. (**a**) ^{1}H-NMR of compound **1**; (**b**) ^{13}C-NMR and DEPT (Distortionless Enhancement by Polarization Transfer) of compound **1**; (**c**) HSQC (Heteronuclear Multiple-Quantum Correlation) of compound **1**; (**d**) ^{1}H-^{1}H COSY (correlated spectroscopy) of compound **1**; (**e**) HMBC (Heteronuclear Multiple Bond Correlation) of compound **1**; (**f**) NOESY of compound **1**.

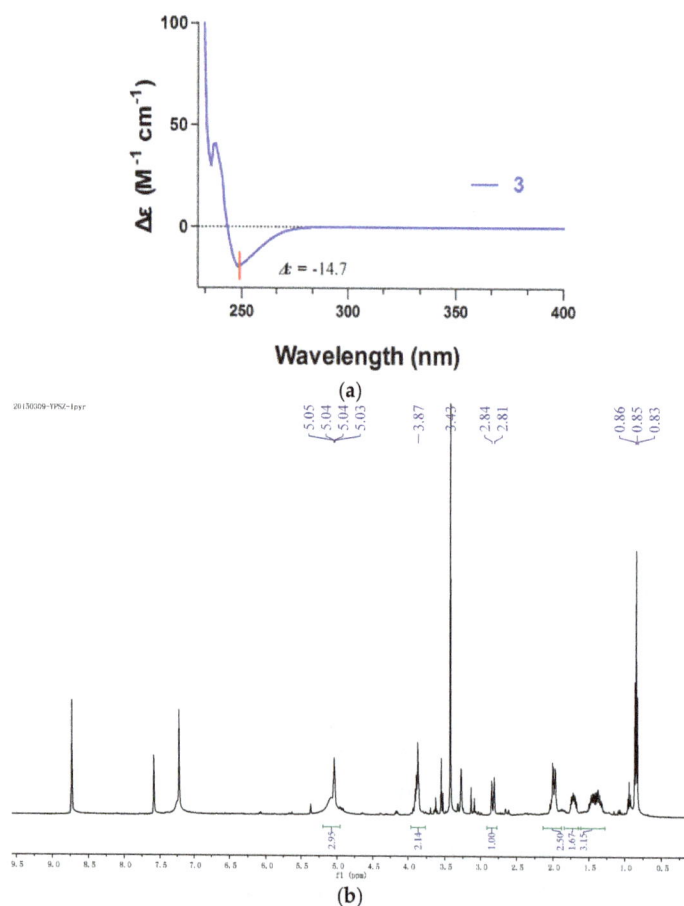

(a)

(b)

Figure 4. *Cont.*

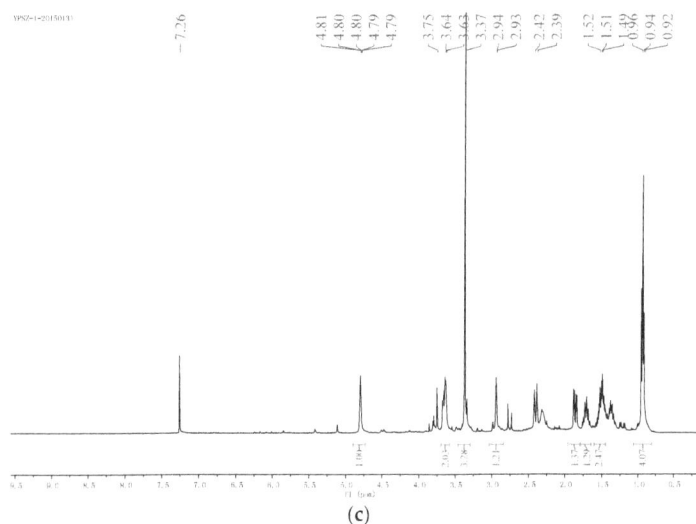

(c)

Figure 4. Absolute configuration of compound **1**. (**a**) CD spectrum of compound **3** in MeOH; (**b**) ^{1}H-NMR spectrum of compound **1** in C_5D_5N; (**c**) ^{1}H-NMR spectrum of compound **1** in $CDCl_3$.

Compound **2**, yellow oil ($[\alpha]_D^{25}$ +82.0 (*c* 1.30, MEOH)), had the molecular formula of $C_{11}H_{18}O_5$, as established by HRESIMS ion at m/z 253.1053 [M + Na]$^+$ (cald. 253.1052) and ^{13}C-NMR data. The IR spectrum exhibited absorption bands for hydroxyl groups (3416 cm^{-1}) and carbonyl groups (1746 and 1699 cm^{-1}). The ^{1}H-NMR spectra exhibited signals for one methyl groups (H 0.89, 3H, t, *J* = 7.2 Hz), one oxymethyl (H 3.76, 3H, s), and a series of aliphatic methylene or methine multiplets (Table 1). The ^{13}C-NMR spectrum, in combination with DEPT experiments, resolved 11 carbon resonances attributable to one ketone group (C 204.6), one carbonyl group (C 184.4), one tri-substituted double bond (C 167.1 and 105.8), two methyls, one sp^3 oxygenated methines, and four sp^3 methylenes (Table 1). As three degrees of unsaturation were consumed by one tri-substituted double bond, one kentone group, and one carbonyl group, then the structure of compound **2** should be a chain. The above-mentioned information was quite similar to that of co-isolated compound **4** reported from the same genus [12]. In comparison with compound **4**, the major differences of compound **2** were due to an additional ketone group and one carbonyl group indicating that compound **2** was a 2,6-hydrolysis derivative of compound **4**. HMBC correlations from H_3CO-4 (H 3.76) to C-4 (C 167.1) assigned that the methoxyl group was linked with C-4 (Figure 2). The planar structure of **1** was further established by detailed interpretation of its 2D NMR data (Figure 5). The relative configuration of compound **2** was established by comparison of 1DNMR data with compound **4**. The absolute configuration of compound **4** was confirmed by chemical synthesis in previous work [12]. The absolute configuration of compound **2** was confirmed by the chemical transformation from **4** to **2** in methanol added with 1 equiv NaOH at 60 °C. Therefore, compound **2** was determined as (6S,2E)-6-hydroxy-3-methoxy-5-oxodec-2-enoic acid.

Based on the NMR and MS data, compounds **3–5** were identified as LL-P880γ [10], LL-P880α [13], and Ergosta-5,7,22-trien-3β-ol [14], respectively (Figure 6).

2.3. Antifungal Activity

As can be seen form the Table 2, among the five compounds, compounds **1–4** all showed notable anti-fungal activities with the minimal inhibitory concentration (MIC) values no more than 50 μg/mL for all the tested fungi; interestingly, compounds **1** and **2** possess the strong activities with the MIC values no more than 25 μg/mL. In addition, our present result also demonstrated that compound **5** possessed moderate anti-fungal effects on the four tested fungi (MIC values were higher than 200 μg/mL).

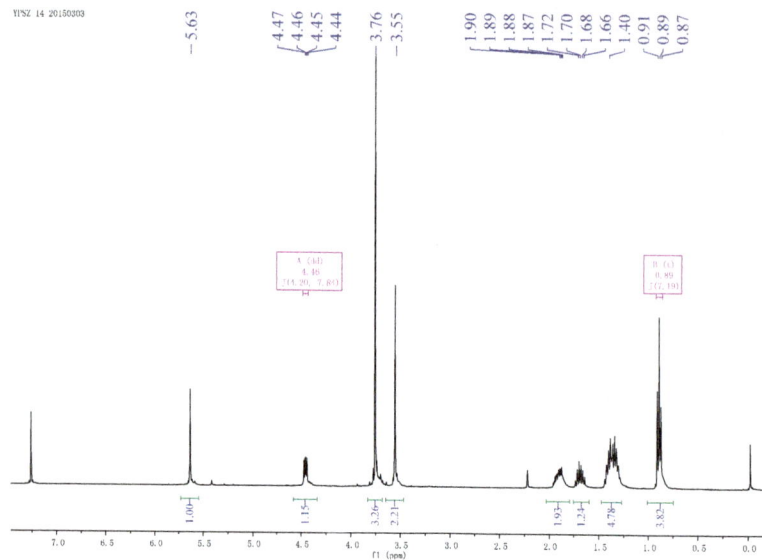

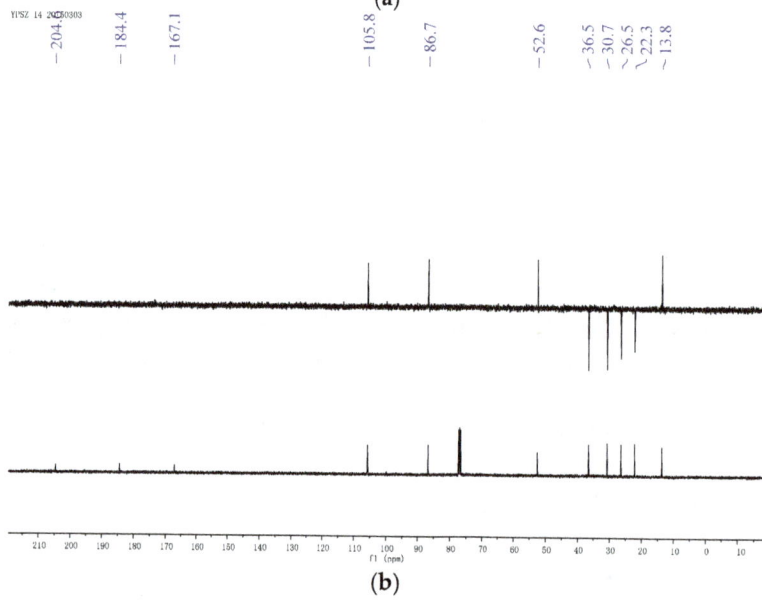

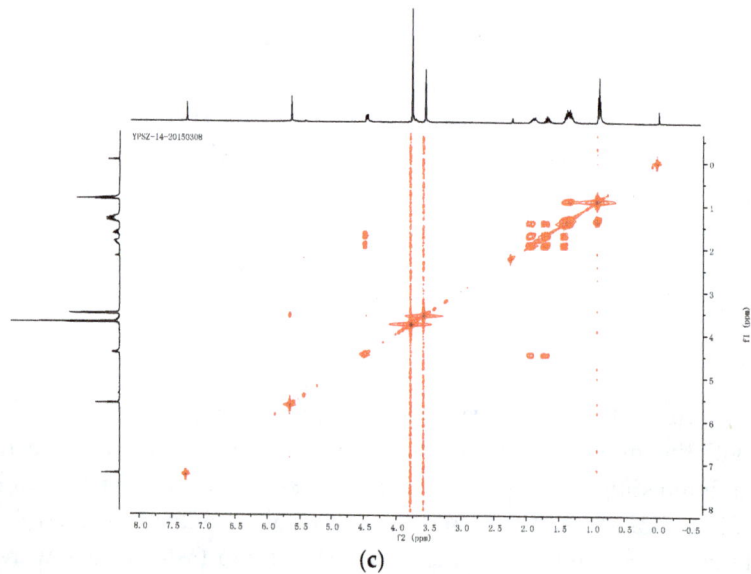

(a)

(b)

(c)

Figure 5. *Cont.*

(d)

(e)

(f)

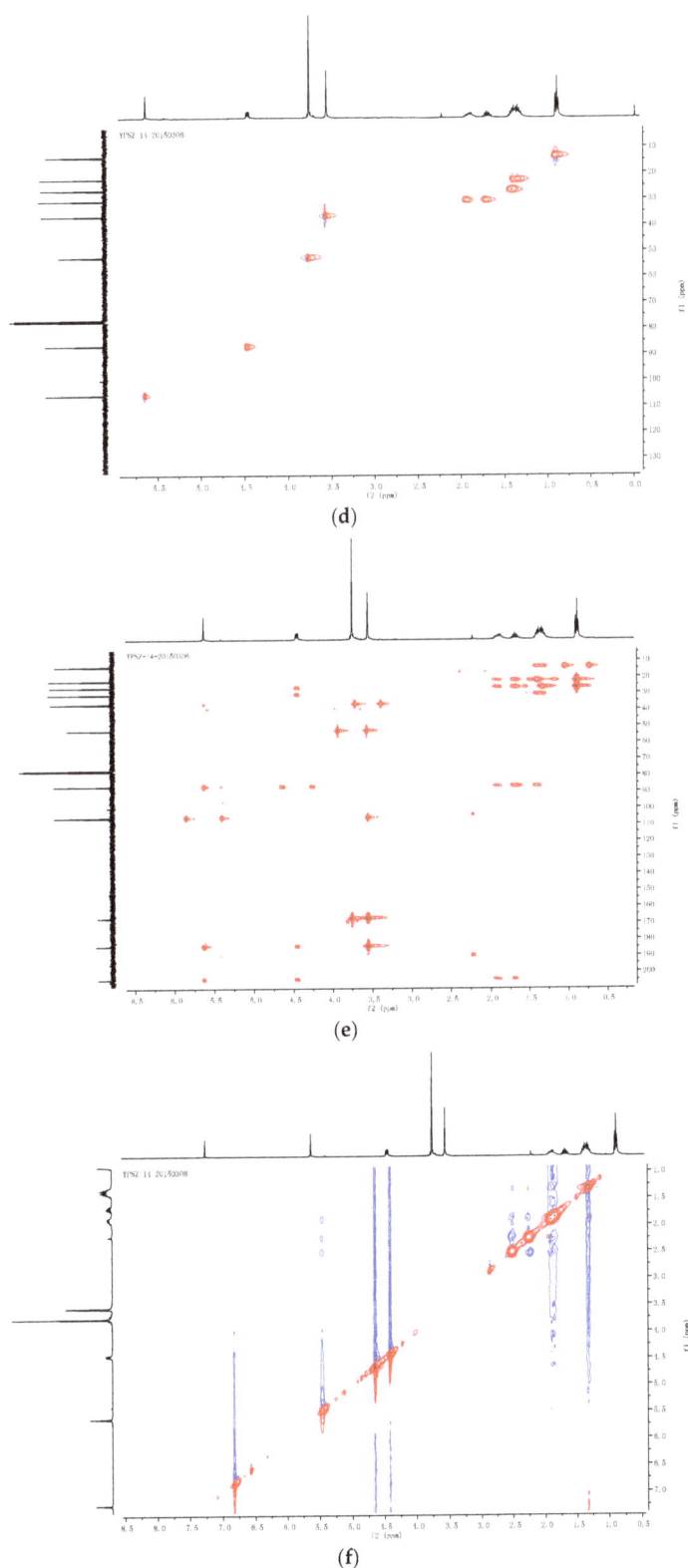

Figure 5. Supporting information of compound **2**. (**a**) ^{1}H-NMR of compound **2**; (**b**) ^{13}C-NMR and DEPT of compound **2**; (**c**) HSQC of compound **2**; (**d**) ^{1}H-^{1}H COSY of compound **2**; (**e**) HMBC of compound **2**; (**f**) NOESY of compound **2**.

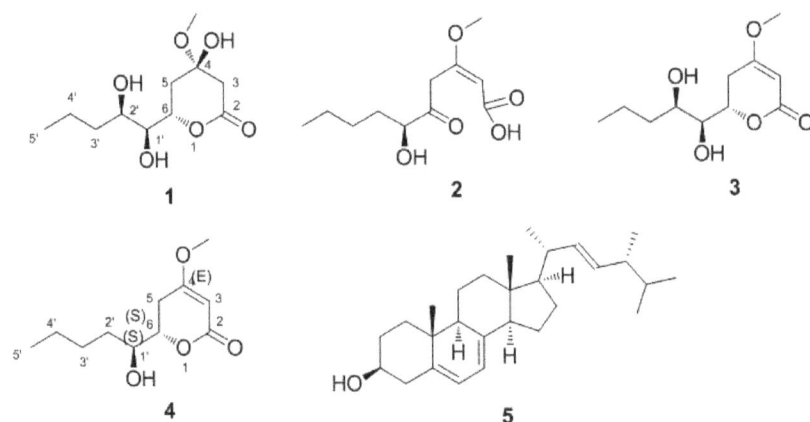

Figure 6. Structures of compounds 1–5.

Table 2. Anti-fungal effects of compounds 1–5.

	MIC (µg/mL)			
	C. albicans	*C. neoformans*	*T. rubrum*	*A. fumigatus*
1	6.25	3.13	25	25
2	12.5	12.5	6.25	3.13
3	12.5	50	50	50
4	6.25	3.13	50	25
5	>400	200	>400	>400
KTZ	0.0625	0.03125	0.5	1

Ketoconazole (KTZ) was used as positive control.

2.4. Cytotoxic Activity

From our present results showed in Table 3, compounds 1–4 showed significant cytotoxic activities on the MKN45, LOVO, A549 and HL-60 cancer cell lines with the IC_{50} values lower than 200 µM. In addition, the compounds 1–4 possessed notable cytotoxcities against human cancer cell lines of HL-60 cells with the IC_{50} values of below 100 µM. Besides, compounds 1, 2, 4 and 5 showed strong cytotoxies on the LOVO cell line with the IC_{50} values lower than 100 µM.

Table 3. Cytotoxic effects of compounds 1–5.

	IC_{50} (µM)				
	MKN45	LOVO	A549	HepG2	HL-60
1	104.76 ± 5.34	50.97 ± 1.87	157.02 ± 2.01	>200	15.24 ± 0.34
2	135.87 ± 6.15	41.91 ± 1.07	>200	>200	30.09 ± 0.98
3	125.87 ± 5.76	139.96 ± 5.76	182.92 ± 5.98	>200	64.87 ± 1.47
4	65.28 ± 1.98	68.88 ± 2.98	125.79 ± 4.07	191.68 ± 6.94	30.75 ± 1.65
5	>200	65.20 ± 1.37	>200	>200	171.54 ± 4.97
DOX	0.14 ± 0.002	0.06 ± 0.002	0.16 ± 0.004	0.18 ± 0.004	0.01 ± 0.001

Doxorubicin (DOX) was used as positive control.

2.5. Discussion

In terms of orchid–fungus relationships research, most works are concerning about the functional role of mycorrhizal fungi on seed germination and plants survival [15]; relatively little is known of endophytic nature of plant aboveground tissues and their roles in the establishment and growth of epiphytic orchids [16]. In recent years, non-mycorrhizal endophytes have been recorded and recognized [17]. Some researches indicated that non-mycorrhizal endophytes may function a previously underestimated way [18]. However, knowledge of orchids including *Dendrobium* associated non-mycorrhizal fungi is limited. We have previously investigated endophytic fungi isolated from leaves, stems and roots of *D. officinale* attached to nine tree species in Yandang Mountain of Zhejiang, China (unpublished). In order to select meaningful fungi that can promote the growth and contents of the host, 134 endophytic fungal taxa were isolated co-cultured with sterile plantlets of *D. officinale* one by one. As a result, *Pestalotiopsis* sp. DO14 could significantly improve the main medicinal compounds contents (e.g., polysaccharides) of *D. officinale*, which is therefore selected for the present study to understand chemical constituents and pharmacological activities of its metabolites.

Pestalotiopsis (Amphisphaeriaceae) species are distributed widely in tropical and temperate ecosystems as saprobes, pathogens, and endophytes of living plants [19]. Species of *Pestalotiopsis* have become a topic of research in many microbial-chemical and pharmacological laboratories because they contain structurally complex, biologically active metabolites. Xu *et al.* [20] reviewed 160 different compounds isolated from species of Pestalotiopsis containing alkaloids, steroids, sesquiterpenes, triterpenes, coumarins, chromones, simple phenols, phenolic acids, lactones, *etc.* Antitumor, antifungal, and antimicrobial activities were the most notable bioactivities of secondary metabolites isolated from this genus [21]. However, to our knowledge, there have been no monoterpenoids reported from this genus.

Previous studies indicated that anti-fungal and antitumor are the two major activities of compounds isolated form endophytic fungus *Pestalotiopsis* sp. [21]. In addition, Chen *et al.* reported that 4-(3′,3′-Dimethylallyloxy)-5-methyl-6-methoxyphthalide (DMMP) isolated from the endophytic fungus *Pestalotiopsis* sp. possessed significantly antitumor effect via mitochondrial extrinsic apoptotic pathway [22]. Furthermore, induction of apoptosis is one of the important mechanisms of taxol isolated from the *Pestalotiopsis* sp. [4]. Previous investigations reported that the destruction of the bacterial cell might be the possible antifungal activity of compounds isolated from the *Pestalotiopsis* sp. [23,24]. In this study, we found that the endophytic fungus *Pestalotiopsis* sp. DO14 produced different varieties of metabolite classes that were not yet reported from *Pestalotiopsis* species and that showed potent cytotoxic and anti-fungal activities. Compounds **1** and **2** were new members of the monoterpenoids metabolites with strong cytotoxic and antifungal activities, and they represent the first isolation of monoterpenoids derivative from the genus *Pestalotiopsis*. Compound **3** was first isolated from *Penicillium* citreo-viride as a pestalotin analogue, which was a gibberellin synergist. Compound **4** was first isolated from *Penicillium* sp. We found they both have strong cytotoxic and antifungal activities in this study. Compound **5** was ergosterol derivative. Ergosterol is the precursor of vitamin D2 which is very important for human health. Ergosterol and its derivatives comprise a big family in mushrooms [25]. Compound **5** was isolated from *Pestalotiopsis* for the first time. However, more works are needed to be devoted to systemically investigate the potential mechanisms of antitumor and antifungal activities of these compounds isolated in our present research.

3. Experimental Section

3.1. Isolation and Identification of the Endophytic Fungus

Healthy shoots of *D. officinale* plants were collected in Yandang Mountain, Zhejiang Province, PR China. Samples were immediately placed in plastic bags, labeled, and taken to the laboratory store at 4 °C for isolation of endophytic fungi within 48 h of collection. The samples were washed and then cut into 3 cm-long segments before surface-sterilization. Shoot segments were surface sterilized by using the method of Wu *et al.* [26]. Then, the shoots was cut into 1 cm-long segments, and placed on potato dextrose agar (PDA) media containing 50 mg/L penicillin and incubated at room temperature for 14 days. The hyphal tip was transferred to new PDA plates and incubated at 26 °C until the pure mycelium covered most of the plate.

The isolated endophytic fungus DO14 was identified according to its morphology and ITS sequences by using the universal primers ITS5 and ITS4 following the reported protocol [27]. Obtained ITS sequence was compared by Blast search with reference sequences at the GenBank and all sequences were aligned with CLUSTAL X software [28]. The phylogenetic tree was performed using the neighbor-joining method. Identification of sequences was according to Wu *et al.* [27]. The sequence was submitted to GenBank (accession No. KP050569). The fungal strain DO14 was deposited in the China Center for Type Culture Collection (CCTCC) as CCTCC M 2015180.

3.2. Fermentation and Compounds Isolation

The fermentation extracts of DO14 were prepared as reported [29]. DO14 was inoculated in Erlenmeyer flasks with potato dextrose broth (PDB) and incubated on a rotary shaker (180 rpm) for 7 days at 28 °C. Crude fermentation broths were filtered and blended for extraction with ethyl ether. After extracted by ethyl ether, 6.9 g crude extracts were obtained, and, subsequently, the extracts were subjected to silica gel column chromatography, eluting with gradient petroleum ether-acetone (30:1–1:1). Combination of similar fractions by using TLC (Thin Layer Chromatography) analysis, seven fractions (A–G) were afforded. Then, fraction C was purified by gel filtration on Sephadex LH-20 to afford compound **5** (36 mg). Fraction E was subject to column chromatography over silica gel, Sephadex LH-20, and preparative TLC to afford **1** (23 mg), **2** (19 mg), **3** (27 mg). Compound **4** (28 mg) were isolated from the fractions F in the same way (Figure 1).

3.3. Antifungal Assay

Antifungal activities of the compounds **1–5** were assayed on the four common pathogens (available in the Chinese Academy of Sciences) follows: *Candida albicans*, *Cryptococcus neoformans*, *Trichophyton rubrum*, *Aspergillus fumigatus*. The MIC was used to evaluate anti-fungal activities of the isolated compounds **1–5**, and the MIC assay was carried out according to the previous reported method [29]. Briefly, the sabouraud dextrose agar was used for fungal culture. Dilutions of the compounds **1–5** were prepared as follows: 400, 200, 100, 50, 25, 12.5, 6.25, 3.13 and 1.56 μg/mL. In addition, ketoconazole (KTZ) was used as positive control, and the concentrations were prepared as follows: 8, 4, 2, 1, 0.5, 0.25, 0.125, 0.0625 and 0.03125 μg/mL. In addition, Dimethyl sulfoxide (DMSO) at a concentration of 1% was used to enhance the compounds' solubility. Then, the pathogens $(1-5 \times 10^3$ CFU/mL) were seeded in the 96 well plates, and the total volume is 200 μL. MIC values were determined as the lowest samples' concentrations that prevent visible fungal growth at 35 °C after 24 h, 72 h and 168 h of incubation for *Monilia*, *Cryptococcus* and hyphomycete, respectively [20,30].

3.4. Cytotoxic Assay

Cytotoxic assay was determined by using the MTT [3-(4,5-dimethylthiazole-2-yl)-2,5-diphenylte trazoliumbromide] [31]. Briefly, five human cancer cell lines (available in the Chinese Academy of Sciences) were used: MKN45, LOVO, A549, HepG2, and HL-60. Briefly, cells with density of 1×10^5 cells/mL were cultured in 96-well plates for 24 h with 10% FBS DMEM medium. Subsequently, cells were treated with test samples at a series nine concentrations (0.5, 1, 5, 10, 20, 40, 60, 80, and 100 μg/mL) for 24 h. Then, 20 μL MTT (5 mg/mL) was added for 4 h, and after removing the medium, 150 μL DMSO was added into each well to dissolve blue formazan crystals. Finally, the optical density (OD) values were read at a wavelength of 570 nm on a micro-plate reader (Labsystems, WellscanMR-2). Cell proliferation inhibition (%) was determined according to the results of MTT assay, and IC_{50} values of the compounds on MKN45, LOVO, A549, HepG2, and HL-60 cell lines were calculated by LOGIT method. Each experiment was repeated three times.

4. Conclusions

Our results indicate that compounds isolated from the DO14 could be valuable candidates as potent tumor inhibitors and be beneficial in the therapy of cancer diseases. Our study also underscores that endophytic fungi of *D. officinale* are great potential resources to discover novel agents for preventing or treating pathogens and tumors. However, further investigations are still needed to study the extraction process and pharmacological action mechanisms of the active constituents in metabolites of endophytic fungi isolated from *D. officinale*.

Acknowledgments: This work was supported by the Zhejiang Open Foundation of the Most Important Forestry Level Subjects (No. KF201320) and Shanghai Municipal Committee of Science and Technology (Grant No. 14401902900).

Author Contributions: Conceived and designed the experiments: Ting Han, Wei Peng; Performed the experiments: Ling-Shang Wu, Min Jia, Bo Zhu, Ling Chen, Hong-Xiu Dong, Jin-Ping Si; Analyzed the data: Ting Han, Ling-Shang Wu; Contributed reagents/materials/analysis tools: Ling-Shang Wu, Min Jia, Jin-Ping Si; Wrote the paper: Wei Peng, Ting Han.

Conflicts of Interest: The authors declare no conflict of interest.

References

1. Stierle, A.; Strobel, G.; Stierle, D. Taxol and taxane production by *Taxomyces andreanae*, an endophytic fungus of Pacific yew. *Science* **1993**, *260*, 214–216. [CrossRef] [PubMed]

2. Dong, L.H.; Fan, S.W.; Ling, Q.Z.; Huang, B.B.; Wei, Z.J. Indentification of huperzine A-producing endophytic fungi isolated from *Huperzia serrata*. *World J. Microbiol. Biotechnol.* **2014**, *30*, 1011–1017. [CrossRef] [PubMed]

3. Yu, H.; Zhang, L.; Li, L.; Zheng, C.; Guo, L.; Li, W.; Sun, P.; Qin, L. Recent developments and future prospects of antimicrobial metabolites produced by endophytes. *Microb. Res.* **2010**, *165*, 437–449. [CrossRef] [PubMed]

4. Chen, L.; Zhang, Q.Y.; Jia, M.; Ming, Q.L.; Yue, W.; Rahman, K.; Qin, L.P.; Han, T. Endophytic fungi with antitumor activities: Their occurrence and anticancer compounds. *Crit. Rev. Microbiol.* **2014**, 1–20. [CrossRef] [PubMed]

5. Zhao, M.M.; Zhang, G.; Zhang, D.W.; Hsiao, Y.Y.; Guo, S.X. ESTs Analysis Reveals Putative Genes Involved in Symbiotic Seed Germination in *Dendrobium officinale*. *PLoS ONE* **2013**, *8*, e72705. [CrossRef] [PubMed]

6. Wang, H.; Fang, H.; Wang, Y.; Duan, L.; Guo, S. *In situ* seed baiting techniques in *Dendrobium officinale* Kimuraet Migo and *Dendrobium nobile* Lindl.: The endangered Chinese endemic *Dendrobium* (Orchidaceae). *World J. Microbiol. Biotechnol.* **2011**, *27*, 2051–2059. [CrossRef]

7. Tan, X.M.; Wang, C.L.; Chen, X.M.; Zhou, Y.Q.; Wang, Y.Q.; Luo, A.X.; Liu, Z.H.; Guo, S.X. *In vitro* seed germination and seedling growth of an endangered epiphytic orchid, *Dendrobium officinale*, endemic to China using mycorrhizal fungi (*Tulasnella* sp.). *Sci. Hortic.* **2014**, *165*, 62–68. [CrossRef]

8. Ming, Q.; Su, C.; Zheng, C.; Jia, M.; Zhang, Q.; Zhang, H.; Rahman, K.; Han, T.; Qin, L. Elicitors from the endophytic fungus Trichoderma atroviride promote Salvia miltiorrhiza hairy root growth and tanshinone biosynthesis. *J. Exp. Bot.* **2013**, *64*, 5687–5694. [CrossRef] [PubMed]

9. Kimura, Y.; Suzuki, A.; Tamura, S. ^{13}C-NMR spectra of pestalotin and its analogues. *Agric. Biol. Chem.* **1980**, *44*, 451–452. [CrossRef]

10. Mcgahren, W.J.; Ellestad, G.A.; Morton, G.O.; Kunstmann, M.P.; Mullen, P. A new fungal lactone, LL-P880 beta, and a new pyrone, LL-P880 gamma, from a *Penicillium* sp. *J. Org. Chem.* **1973**, *38*, 3542–3544. [CrossRef] [PubMed]

11. Dong, Z.; Gu, Q.; Cheng, B.; Cheng, Z.B.; Tang, G.H.; Sun, Z.H.; Zhang, J.S.; Bao, J.M.; Yin, S. Natural nitric oxide (NO) inhibitors from *Aristolochia mollissima*. *RSC Adv.* **2014**, *4*, 55036–55043. [CrossRef]

12. Masaki, Y.; Imaeda, T.; Kawai, M. Highly stereoselective synthesis and structural confirmation of a fungal metabolite, LL-P880 beta. *Chem. Pharm. Bull.* **1994**, *42*, 179–181. [CrossRef] [PubMed]

13. Ellestad, G.A.; Mcgahren, W.J.; Kunstmann, M.P. Structure of a new fungal lactone, LL-P880.alpha., from an unidentified *Penicillium species*. *J. Org. Chem.* **1972**, *37*, 2045–2047. [CrossRef] [PubMed]

14. Cai, H.; Liu, X.; Chen, Z.; Liao, S.; Zou, Y. Isolation, purification and identification of nine chemical compounds from *Flammulina velutipes* fruiting bodies. *Food Chem.* **2013**, *141*, 2873–2879. [CrossRef] [PubMed]

15. Oja, J.; Kohout, P.; Tedersoo, L.; Kull, T.; Kõljalg, U. Temporal patterns of orchid mycorrhizal fungi in meadows and forests as revealed by 454 pyrosequencing. *New Phytol.* **2015**, *205*, 1608–1618. [CrossRef] [PubMed]

16. Yuan, Z.L.; Chen, Y.C.; Yang, Y. Diverse non-mycorrhizal fungal endophytes inhabiting an epiphytic, medicinal orchid (*Dendrobium nobile*): Estimation and characterization. *World J. Microbiol. Biotechnol.* **2008**, *25*, 295–303. [CrossRef]

17. Bayman, P.; Otero, J.T. *Microbial Endophytes of Orchid Roots*; Springer: Berlin/Heidelberg, Germany, 2006; pp. 153–177.

18. Herre, E.A.; Mejía, L.C.; Kyllo, D.A.; Rojas, E.; Maynard, Z.; Butler, A.; van Bael, S.A. Ecological implications of anti-pathogen effects of tropical fungal endophytes and mycorrhizae. *Ecology* **2007**, *88*, 550–558. [CrossRef] [PubMed]

19. Maharachchikumbura, S.S.N.; Guo, L.D.; Chukeatirote, E.; Bahkali, A.H.; Hyde, K.D. Pestalotiopsis-morphology, phylogeny, biochemistry and diversity. *Fungal Divers.* **2011**, *50*, 167–187. [CrossRef]

20. Xu, L.L.; Han, T.; Wu, J.Z.; Zhang, Q.Y.; Zhang, H.; Huang, B.K.; Rahman, K.; Qin, L.P. Comparative research of chemical constituents, antifungal and antitumor properties of ether extracts of *Panax ginseng* and its endophytic fungus. *Phytomedicine* **2009**, *16*, 609–616. [CrossRef] [PubMed]

21. Xu, J.; Yang, X.; Lin, Q. Chemistry and biology of *Pestalotiopsis*-derived natural products. *Fungal Divers.* **2014**, *66*, 37–68. [CrossRef]

22. Chen, C.; Hu, S.Y.; Luo, D.Q.; Zhu, S.Y.; Zhou, C.Q. Potential antitumor agent from the endophytic fungus Pestalotiopsis photiniae induces apoptosis via the mitochondrial pathway in HeLa cells. *Oncol. Rep.* **2013**, *30*, 1773–1781. [PubMed]

23. Subban, K.; Subramani, R.; Johnpaul, M. A novel antibacterial and antifungal phenolic compound from the endophytic fungus Pestalotiopsis mangiferae. *Nat. Prod. Res.* **2013**, *27*, 1445–1449. [CrossRef] [PubMed]

24. Yang, X.L.; Zhang, S.; Hu, Q.B.; Luo, D.Q.; Zhang, Y. Phthalide derivatives with antifungal activities against the plant pathogens isolated from the liquid culture of Pestalotiopsis photiniae. *J. Antibiot.* **2011**, *64*, 723–727. [CrossRef] [PubMed]

25. Yaoita, Y.; Yoshihara, Y.R.; Machida, K.; Kikuchi, M. New Sterols from Two Edible Mushrooms, *Pleurotus eryngii* and *Panellus serotinus*. *Chem. Pharm. Bull.* **2002**, *50*, 551–553. [CrossRef] [PubMed]

26. Wu, L.; Han, T.; Li, W.; Jia, M.; Xue, L.; Rahman, K.; Qin, L. Geographic and tissue influences on endophytic fungal communities of *Taxus chinensis* var. *mairei* in China. *Curr. Microbiol.* **2013**, *66*, 40–48. [CrossRef] [PubMed]

27. Guo, L.D.; Huang, G.R.; Wang, Y.; He, W.H.; Zheng, W.H.; Hyde, K.D. Molecular identification of white morphotype strains of endophytic fungi from *Pinus tabulaeformis*. *Mycol. Res.* **2003**, *107*, 680–688. [CrossRef] [PubMed]

28. Thompson, J.D.; Gibson, T.J.; Plewniak, F.; Jeanmougin, F.; Higgins, D.G. The CLUSTAL_X Windows Interface: Flexible Strategies for Multiple Sequence Alignment Aided by Quality Analysis Tools. *Nucleic Acids Res.* **1997**, *25*, 4876–4882. [CrossRef] [PubMed]

29. You, F.; Han, T.; Wu, J.Z.; Huang, B.K.; Qin, L.P. Antifungal secondary metabolites from endophytic *Verticillium* sp. *Biochem. Syst. Ecol.* **2009**, *37*, 162–165. [CrossRef]

30. Peng, W.; Han, T.; Xin, W.B.; Zhang, X.G.; Zhang, Q.Y.; Jia, M.; Qin, L.P. Comparative research of chemical constituents and bioactivities between petroleum ether extracts of the aerial part and the rhizome of Atractylodes macrocephala. *Med. Chem. Res.* **2011**, *20*, 146–151. [CrossRef]

31. Wang, L.W.; Xu, B.G.; Wang, J.Y.; Su, Z.Z.; Lin, F.C.; Zhang, C.L.; Kubicek, C.P. Bioactive metabolites from *Phoma* species, an endophytic fungus from the Chinese medicinal plant *Arisaema erubescens*. *Appl. Microbiol. Biotechnol.* **2012**, *93*, 1231–1239. [CrossRef] [PubMed]

5

Synthesis and Properties of Bis-Porphyrin Molecular Tweezers: Effects of Spacer Flexibility on Binding and Supramolecular Chirogenesis

Magnus Blom [1], Sara Norrehed [1], Claes-Henrik Andersson [1], Hao Huang [1], Mark E. Light [2], Jonas Bergquist [1], Helena Grennberg [1] and Adolf Gogoll [1,*]

Academic Editor: M. Graça P. M. S. Neves

[1] Department of Chemistry-BMC, Uppsala University, Uppsala S-75123, Sweden;
magnus.blom@kemi.uu.se (M.B.); sara.norrehed@raa.se (S.N.); claes.henrik.andersson@gmail.com (C.-H.A.);
hao.huang@angstrom.uu.se (H.H.); jonas.bergquist@kemi.uu.se (J.B.); helena.grennberg@kemi.uu.se (H.G.)
[2] Department of Chemistry, University of Southampton, Highfield, Southampton SO17 1BJ, UK;
light@soton.ac.uk
* Correspondence: adolf.gogoll@kemi.uu.se

Abstract: Ditopic binding of various dinitrogen compounds to three bisporphyrin molecular tweezers with spacers of varying conformational rigidity, incorporating the planar enediyne (**1**), the helical stiff stilbene (**2**), or the semi-rigid glycoluril motif fused to the porphyrins (**3**), are compared. Binding constants $K_a = 10^4$–10^6 M^{-1} reveal subtle differences between these tweezers, that are discussed in terms of porphyrin dislocation modes. Exciton coupled circular dichroism (ECCD) of complexes with chiral dinitrogen guests provides experimental evidence for the conformational properties of the tweezers. The results are further supported and rationalized by conformational analysis.

Keywords: bisporphyrin tweezers; metalloporphyrins; porphyrinoids; host-guest chemistry; supramolecular chemistry; chirogenesis; chirality transfer; exciton coupled circular dichroism; conformational analysis

1. Introduction

Bisporphyrin molecular clips and tweezers are well studied systems for ditopic host–guest interactions [1,2]. In the majority of these compounds, two porphyrin chromophores are attached by a single bond to a usually conformationally flexible spacer. They have been used extensively to determine the absolute configuration of guests with a single stereogenic center, or to distinguish enantiomers [3–15]. We recently have shown that bisporphyrin tweezers also can be utilized for determination of the relative stereochemistry in molecules with several stereocenters, employing a semi-rigid bisporphyrin tweezer **3** incorporating a glycoluril spacer [16,17]. For further investigations, we required alternative tweezers with altered conformational flexibility, a key factor for guest affinity. Also, substitutes for **3** requiring a less demanding synthetic protocol are desirable. Therefore, we decided to replace the glycoluril spacer with enediyne and stiff stilbene spacers, respectively (Figure 1). These tweezers are expected to have restricted conformational flexibility, comparable to other tweezers with spacers composed of aromatic rings and ethyne segments [1]. Since several parameters are involved in host–guest binding, accurate predictions of ligand affinity are not always possible. However, bisporphyrins with flexible spacers are capable of strong binding to dinitrogen ligands. In typical examples, small aliphatic diamines were found to bind with $K_a = 10^3$–10^5 M^{-1} to a bisporphyrin with diphenylether spacer, with weaker binding for bulkier guests [5]. Much higher binding constants have been reported for the structurally more rigid DABCO (diazabicyclo[2.2.2]octane) with $K_a = 10^7–10^9$ M^{-1} [18,19]. The variation of binding constants with spacer length and flexibility has been explained by preorganization effects when binding rigid guests [20].

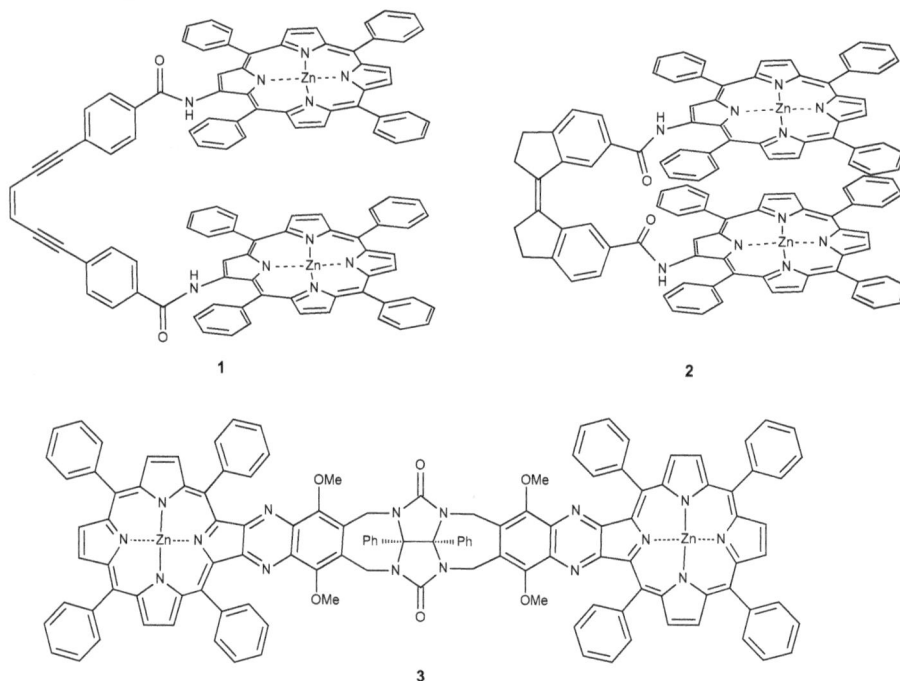

Figure 1. Bisporphyrin molecular tweezers with enediyne (**1**) and stiff stilbene (**2**) spacers, and the previously reported semi-rigid (**3**) with glycoluril spacer.

Bisporphyrin tweezers have been classified as belonging to three distinct categories regarding the conformational properties of their spacers: spacers with high conformational flexibility, spacers with conformational restrictions, and conformationally rigid spacers [1]. Rigid spacers may favor guest binding due to a preorganization effect, but also prevent binding of guests that cannot be accommodated by bitopic binding [21]. Flexible spacers allow more diverse binding via an induced fit [22–24]. The three bisporphyrins **1–3** discussed here present a more subtle conformational behavior.

In principle, four different types of dislocation of the two porphyrin units, some of which might be interdependent, may be distinguished (Figure 2).

For the previously studied glycoluril bisporphyrin tweezer **3**, we have observed conformational flexibility in terms of interporphyrin distance variation (Figure 2a), due to the conformational properties of the glycoluril spacer [16]. This manifested itself experimentally in variation of the hydrodynamic radius of the tweezer upon binding of various guests, as monitored by the diffusion coefficient of the host-guest complex. However, since the porphyrin units are attached to the spacer via two covalent bonds, lateral dislocation (Figure 2b) appears to be unlikely, and porphyrin rotation (Figure 2d) is impossible. There is, however, an option of porphyrin twisting (Figure 2c) via conformational changes of the seven-membered rings in the glycoluril backbone (vide infra). In contrast, both **1** and **2** might allow for lateral dislocation as well as twisting as the result of spacer bond distortion, in combination with porphyrin rotation around single bonds.

It occurred to us that detection of the induced circular dichroism (*i.e.*, exciton coupled circular dichroism, ECCD) [3–15] that results from binding a chiral diamine guest might provide a simple experimental verification of the distortion modes available to our tweezers (Table 1). Tweezer flexibility and guest geometry determine the sign and amplitude of the detected ECCD, and suitable mnemonics for reliable prediction of the effect are still under development [8,9]. The effect of rigidity variation on CD performance has been addressed for chiral bisporphyrin tweezers without bound guests [25], whereas we here investigate achiral bisporphyrin tweezers. Recently, Rath and co-workers have shown that chirogenesis in a bisporphyrin tweezers requires porphyrin twisting [26].

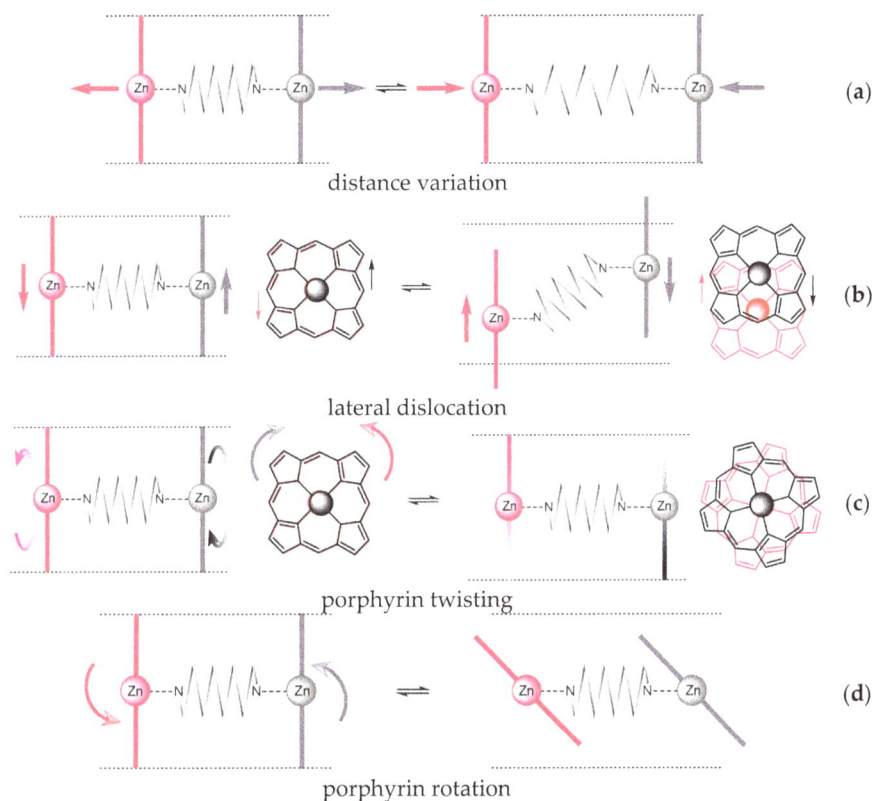

distance variation

lateral dislocation

porphyrin twisting

porphyrin rotation

Figure 2. Possible dislocations in bisporphyrin tweezers: (**a**) Variation of interporphyrin distance; (**b**) lateral displacement; (**c**) porphyrin twisting; (**d**) porphyrin rotation.

Table 1. Porphyrin dislocation modes (Figure 2) in bisporphyrin tweezers **1–3** [a].

	Dislocation	1	2	3
a	distance variation	+	(+)	+
b	lateral dislocation	+	inherent	−
c	porphyrin twisting	+	inherent	+
d	porphyrin rotation	+	+	−
	helical complex achievable?	+	inherent	−

[a] Dislocation mode capability is indicated as "+"(possible), "(+)" (requires high energy), "−" (impossible), "inherent" (present in the lowest energy conformation).

2. Results and Discussion

2.1. Synthesis

2.1.1. Synthesis of Porphyrin Units and Spacers

The synthesis of bisporphyrin **3** involved six steps from TPP (*meso*-tetraphenylporphyrin), and five steps for the glycoluril spacer, with a total yield of 4% from TPP [16]. For tweezers **1** and **2**, a synthesis scheme was devised involving eight steps (**1**) and six steps (**2**), respectively. Overall yields from TPP were 18% for *Z*-**1** and 19% for *Z*-**2**. Thus, the β-aminoporphyrin (**6**) [27] was prepared starting from the free-base *meso*-tetraphenylpophyrin (TPP) as shown in Scheme 1. Cu(II)TPP (**3**) was obtained in quantitative yield by metallation of TPP via refluxing with a solution of copper(II) acetate in dichloromethane-methanol. Mono-nitration with copper nitrate and acetic anhydride/acetic acid in chloroform [28–30] to afford CuTPPNO$_2$ (**4**, 80%) was followed by demetallation with sulphuric acid to give TPPNO$_2$ (**5**, 95%). Reduction of the nitro group with tin(II)chloride and hydrochloric acid yielded the target compound free-base aminoporphyrin **6**. The initial metallation with copper was performed because β-nitration of a copper metallated porphyrin results in vastly higher yields than nitration of free-base or zinc-metallated porphyrins [31]. Reduction of TPPNO$_2$ to TPPNH$_2$ with sodium borohydride in the presence of 10% palladium on activated carbon has been suggested as a faster alternative to tin(II)chloride [32], but in our hands it produced a complex mixture of products that was difficult to purify. The aminoporphyrin **6** is highly sensitive towards air and light, and therefore it was prepared immediately before use.

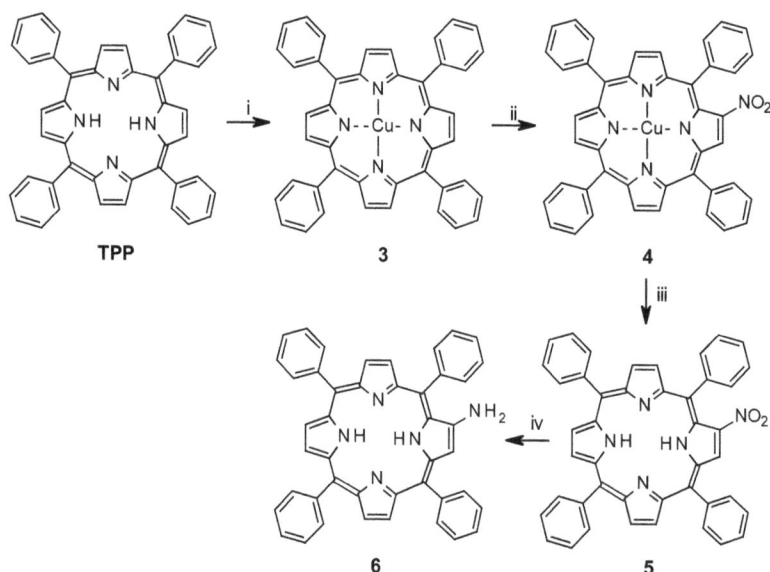

Scheme 1. Functionalization of tetraphenyl porphyrin. (i) Cu(OAc)$_2$·H$_2$O, CH$_2$Cl$_2$, MeOH, reflux, 2.5 h; (ii) Cu(NO$_3$)$_2$·3H$_2$O, acetic acid, acetic anhydride, CHCl$_3$, 35 °C, 5 h; (iii) H$_2$SO$_4$, CH$_2$Cl$_2$, 20 min r.t.; (iv) SnCl$_2$·2H$_2$O, HCl, CHCl$_3$ N$_2$-atm, dark, r.t., 3–4 days.

The synthetic route towards the enediyne spacer of **1** is described in Scheme 2. Microwave assisted esterification of 4-bromobenzoic acid (**7**) afforded 4-bromobenzoic acid methylester (**8**) in 72% yield. A microwave assisted Sonogashira coupling of **8** with trimethylsilyl acetylene yielded the 4-trimethyl silyl protected ethynylbenzoic acid methylester (**9**) almost quantitatively (98%) [33]. Deprotection of the TMS group was carried out with tetrabutylammonium fluoride, which after purification gave 4-ethynylbenzoic acid methylester in 90% yield (**10**) [34]. Methyl-4-[(Z)-6-(4-methoxycarbonylphenyl)hex-3-en-1,5-diynyl]benzoate (**11**) was obtained in 70% yield after a second Sonogashira coupling with Z-1,2-dichloroethylene [35]. Hydrolysis to dicarboxylic acid **12** (90% yield) was followed by quantitative conversion to its acid chloride **13** by oxalyl chloride, giving the enediyne spacer to be coupled to TPP-NH$_2$ **6**.

To obtain the stiff stilbene spacer of **2** (Scheme 3), 3-oxoindane-5-carboxylic acid (**14**) was converted to ethyl 3-oxoindane-5-carboxylate (**15**) via reflux in ethanol in the presence of hydrochloric acid (yield 95%), followed by a reductive McMurry coupling [36] to afford the usual mixture of the E and Z isomers of ethyl-3-(6-ethoxycarbonylindan-1-ylidene)indane-5-carboxylate **16** (E:Z = 3:1). Separation of E-**16** and enrichment of Z-**16** by recrystallization from ethanol, followed by chromatographic purification of Z-**16**, afforded pure isomers. Photoisomerization of E-**16** was used to produce more of Z-**16** [37,38]. Hydrolysis with sodium hydroxide in ethanol afforded dicarboxylic acid Z-**17** (94% yield). Z-**17** was quantitatively converted to the acid chloride (Z-**18**) with oxalyl chloride in dichloromethane. E- and Z-isomers can be distinguished via the $^3J_{HH}$ coupling between the olefinic protons, measured on their ^{13}C satellites (E-isomer: 11 Hz, Z-isomer: 5 Hz).

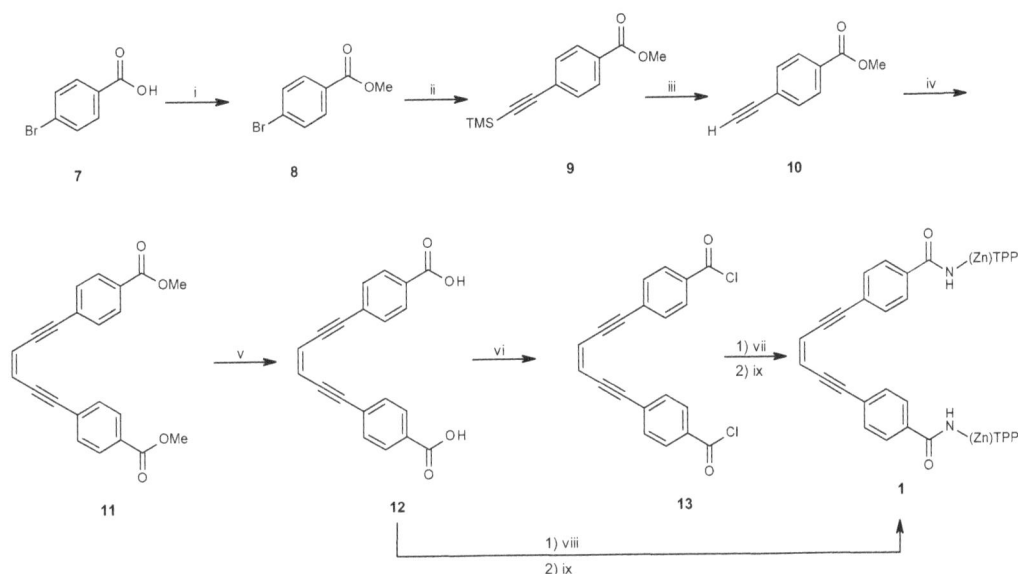

Scheme 2. Synthetic route towards enediyne spacer of **1**. (i) Trimethyl orthoacetate, Microwave 110 °C, 1 h; (ii) trimethyl silyl acetylene, Pd(PPh$_3$)$_2$Cl$_2$, CuI, Et$_2$NH, DMF, microwave 120 °C, 25 min; (iii) THF, TBAF, −20 °C, 3 h; (iv) 1,2-Z-dichloroethylene, Pd(PPh$_3$)$_2$Cl$_2$, CuI, Et$_2$NH, toluene, 0 °C, N$_2$-atm, 2 days; (v) NaOH, EtOH, reflux, 5 h; (vi) oxalyl chloride, CH$_2$Cl$_2$/THF, 0 °C, 1 h; (vii) TPPNH$_2$, CH$_2$Cl$_2$, r.t., 12 h; (viii) TPPNH$_2$, DCC, CH$_2$Cl$_2$, r.t., o.n.; (ix) Zn(OAc)$_2$·H$_2$O, CH$_2$Cl$_2$, MeOH, reflux, 30 min.

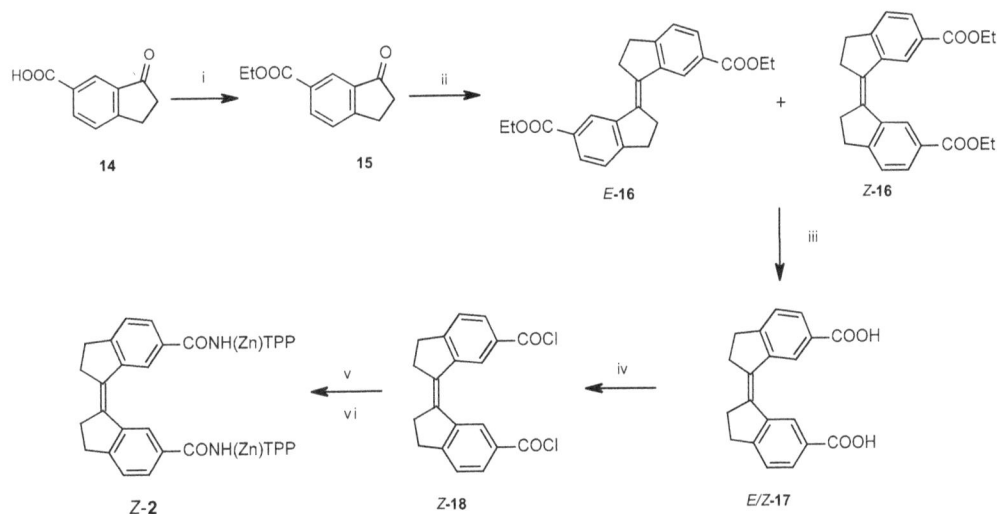

Scheme 3. Synthetic route towards stiff stilbene spacer of **2**. (i) EtOH, HCl, reflux, o.n.; (ii) TiCl$_4$, THF, Zn(s), reflux, 2 h, add **15**, reflux 12 h; (iii) NaOH, EtOH, reflux, 6 h; (iv) oxalyl chloride, CH$_2$Cl$_2$, r.t., 2 h; (v) TPPNH$_2$, CH$_2$Cl$_2$, r.t., o.n.; (vi) Zn(OAc)$_2$·H$_2$O, CH$_2$Cl$_2$, MeOH, reflux, 30 min.

Attempts to use 3-oxoindane-5-carboxylic acid **14** directly as substrate in the reductive coupling were unsuccessful. McMurry has previously discussed functional group compatibility in reductive carbonyl couplings using low-valent titanium reagents in a review [36]. Carboxylic acids were classified as "semi-compatible" for reductive coupling due to their propensity to slowly react with low valent titanium reagents (*i.e.*, TiCl$_3$/LiAlH$_4$). Consequently, their compatibility is largely limited to conditions where shorter reaction times are employed. Therefore, for the sterically crowded alkene **16**, where longer reaction time is required, a carboxylic acid substrate is less well suited.

2.1.2. Coupling of Spacers to β-Monoaminotetraphenylporphyrin

Coupling of the β-monoaminoporphyrin (**6**) to the enediyne spacer was attempted via the dicarboxylic acid (**12**) using DCC coupling reagent. Although successful, low yields were obtained (up to 14%). Other approaches with DCC/HOBT- and HATU-mediated coupling did not produce any isolable product. An alternative route via the acid chloride (**13**) afforded bisporphyrin tweezer **1** in 25% isolated yield after purification. This route was then also used for the coupling of **6** to acid chloride **18**, producing bisporphyrin tweezer **2** in 24% yield. Low yields in amide formation from aminoporphyrins have been reported previously [39] and are most likely due to the reduced nucleophilicity of the amino group, probably in combination with sterical effects. As metallation of the porphyrin moiety provides stabilization and simplified purification over silica, the crude coupling products were metallated prior to column chromatography. Still, cumbersome purification and solubility problems add to the low yields of these coupling reactions.

2.2. Conformational Analysis of Spacer and Tweezer Geometry

Indications for the envisioned different steric properties of the spacers in tweezers **1–3** were obtained from conformational analysis. Here, we focus on the spacer distortions required to generate the dislocation modes indicated in Figure 2. As expected, there are considerable differences between the three spacers. Regarding the lateral dislocation, *i.e.*, twisting of the two spacers attached to the double bond via hindered double bond rotation (Figure 3, right), tweezer **1** has a narrower profile than tweezer **2**, the latter also having a built-in twisting (*i.e.*, energy minima for a twisted alignment of the two fused ring units attached to the double bond with a 9° dihedral angle), separated by a small local energy maximum at 0° of 1.7 kJ/mol. For tweezer **3**, the energy profile for twisting was monitored by changing the dihedral angle between the two phenyl rings attached to the glycoluril unit (Figure 4). The resulting energy profile resembled closely that of Z-**1**.

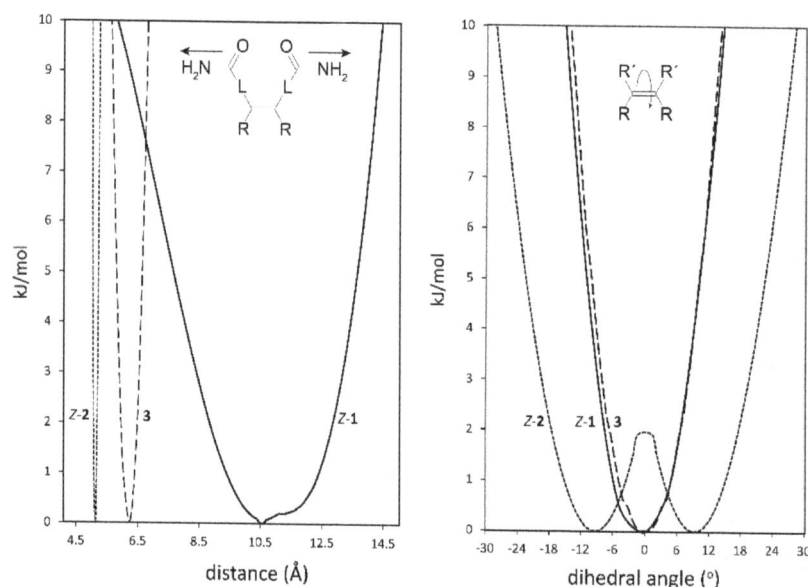

Figure 3. Energy profiles for spacer dihedral angle variation (**right**) and for spacer C=O–C=O distance variation (**left**) in bisporphyrin tweezers Z-**1** (-), Z-**2** (···) and **3** (—). L represents the spacer between the double bond and the amide group, e.g., –Ph–C≡C- in **1**. The two minima for tweezer **2** are at ±9°. For **3**, the Zn–Zn distance (**left**), and the dihedral angle between the two phenyl rings attached to the glycoluril unit (**right**, *cf.* Figure 4) in the complete tweezer are shown.

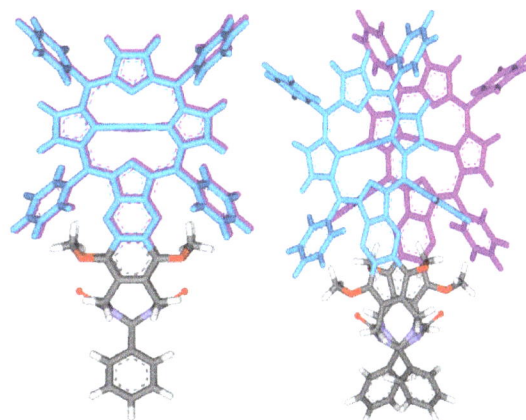

Figure 4. Effect of glycoluril spacer twisting in tweezer **3**, monitored via the Ph-C-C-Ph dihedral angle (viewed along the Ph-<u>C</u>-<u>C</u>-Ph bond, *cf.* Figure 3). **Left**: dihedral angle = 0°, **right**: dihedral angle = 30°.

Regarding the distance variation, relative conformational energies were calculated depending on the distance between the C=O carbons (Figure 3, left). Tweezer **1** has a very broad profile (*ca.* 10.2 Å wide, between 7.7 Å and 13.7 Å within a 5 kJ/mol span) and, thus, should be able to accommodate a variety of guests. In contrast, tweezer **2** shows a much narrower profile (*ca.* 0.2 Å wide, between 5.0 Å and 5.2 Å within a 5 kJ/mol span.

Attachment of the porphyrin units to the spacers results in a less clear cut picture, since interactions between the two porphyrins modulate the energy profiles. However, for tweezer **1** we again obtain a dihedral angle of 0° for the lowest energy conformation, with a Zn–Zn distance of 4.9 Å. For tweezer **2**, the lowest energy conformation has a dihedral angle of 6.5°, with a Zn–Zn distance of 6.2 Å. Tweezer **3** has previously been shown to have a comparatively shallow energy profile upon Zn–Zn distance variation, covering at least ≈10 Å, with a minimum for a Zn–Zn distance of 6.25 Å [16].

Regarding these energy profiles, it should be kept in mind that, according to Berova and co-workers [11], only host–guest conformers within an energy span of 10 kJ/mol above the minimum were considered as relevant contributors to the CD spectrum (*vide infra*).

We can summarize the tweezer conformational properties as follows. Tweezer **1** is capable of all four distortions indicated in Figure 2. This should enable binding of a large variety of diamine guests with similar binding constants. Porphyrin rotation is not necessary in order to accommodate sterically demanding guests. In tweezer **2**, distance variation is limited, and therefore larger guests can be accommodated only by increased spacer twisting and/or porphyrin rotation. Both of these introduce helicity. Tweezer **3** can apparently vary the interporphyrin distance and achieve porphyrin twisting via glycoluril conformational changes, which also might induce helicity.

2.3. *Conformational Analysis of Host-Guest Complex Geometry*

Conformational analysis results of tweezer complexes with dinitrogen guests (Scheme 4) are summarized in Scheme 5 and Table 2. They show a considerable variation of Zn–Zn distances in complexes involving tweezers **1** and **3**, and less in those with tweezer **3**. The latter accommodates larger but flexible guests by twisting (complexes with **21** and **22**), whereas tweezers **1** and **2** both show porphyrin rotation, or lateral dislocation (only **1**). Higher energy conformers did not exhibit any substantial deviations from these geometries. These structures should be interpreted with some caution, though. For example, the complex of 1,12-diaminododecane **22** with tweezer **1** gives the lowest energy for an in-out binding mode (Scheme 5), which is not in agreement with experimental evidence, such as ¹H-NMR data (*vide infra*).

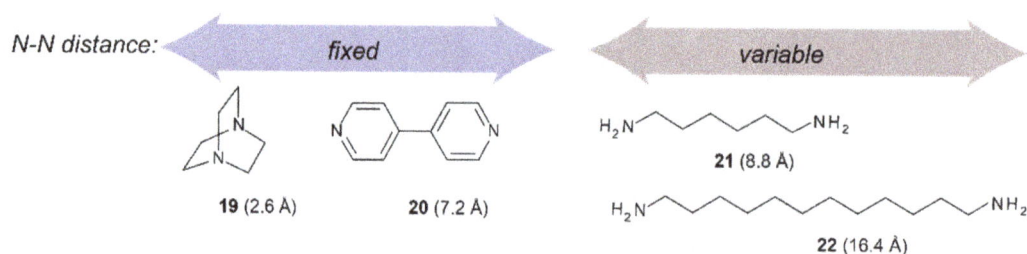

Scheme 4. Maximal distance between nitrogen atoms for the guest molecules used in this study.

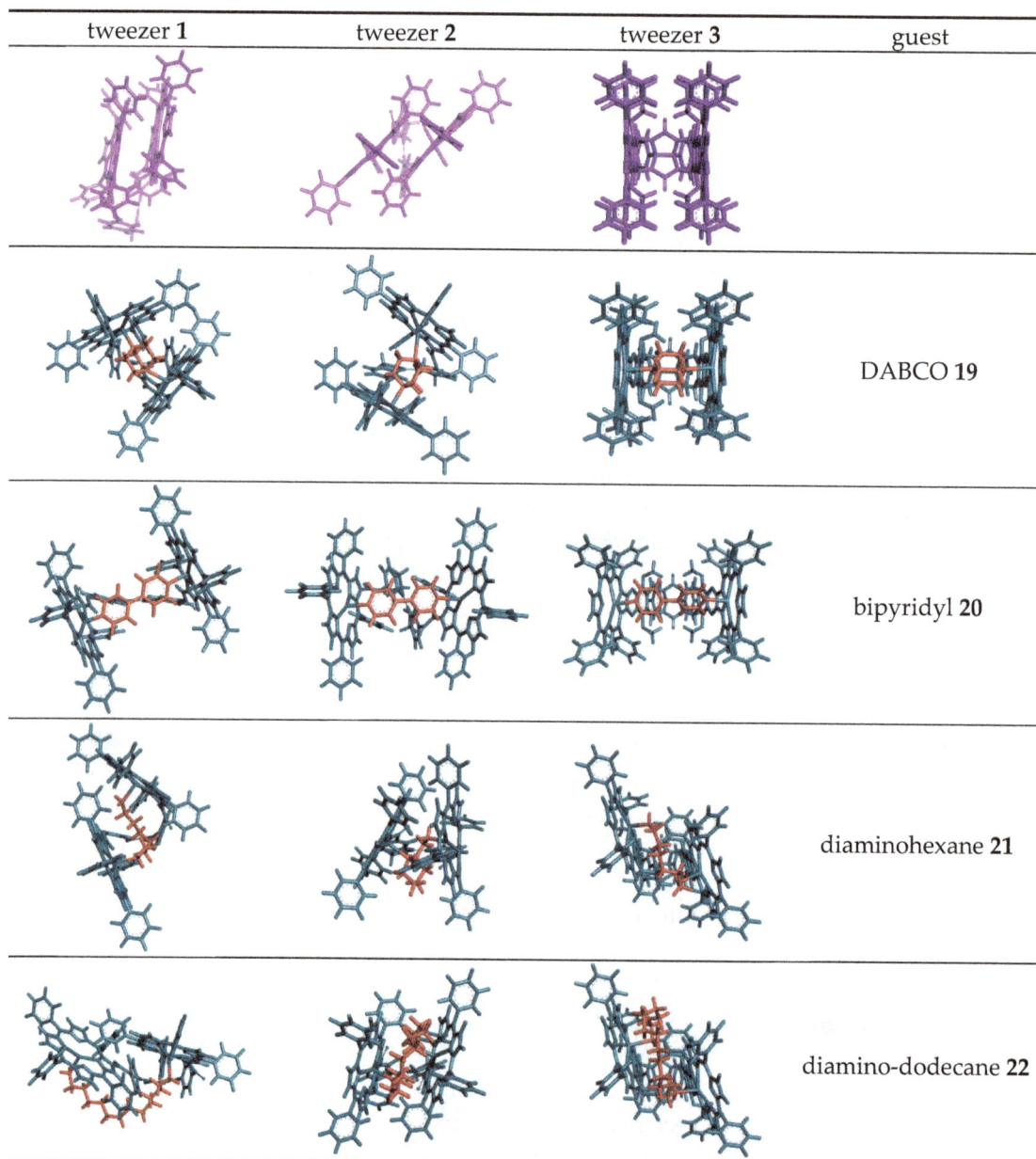

Scheme 5. Porphyrin dislocations in complexes of tweezers 1–3 (green color) with various dinitrogen guests (red color). Shown are the structures corresponding to the global minimum obtained in conformational search with the OPLS 2005 force field.

Table 2. Dihedral angles, spacer C=O–C=O distances, and Zn–Zn distances (Å) from conformational analysis of tweezers **1**–**3** with bound guests.

Guest	Host						
	Z-1			Z-2			3
	Dihedral Angle [a]	CO–CO Distance	Zn–Zn Distance	Dihedral Angle [a]	CO–CO Distance	Zn–Zn Distance	Zn–Zn Distance
(free tweezer)	0°	8.5 Å	4.9 Å	6.5°	4.9 Å	6.1 Å	6.3 Å [b]
DABCO **19**	0.2°	7.9 Å	7.0 Å	11.2°	5.4 Å	7.3 Å	7.3 Å
4,4′-bipyridyl **20**	0.1°	10.6 Å	11.3 Å	11.5°	5.9 Å	11.6 Å	11.5 Å
1,6-diaminohexane **21**	0.5°	7.1 Å	8.4 Å	10.4°	5.7 Å	5.8 Å	8.5 Å
1,12-diaminododecane **22**	c	c	c	10.5°	4.8 Å	6.1 Å	8.2 Å
Lys methylester **23**	0.3°	6.9 Å	9.9 Å	8.5°	5.3 Å	8.5 Å	11.3 Å

[a] Dihedral angle over spacer double bond; [b] Ref. [16]; [c] Minimizes to in-out complex (*cf.* Scheme 5).

2.4. X-ray Crystallography of the Stiff Stilbene Spacer

X-ray crystallographic analysis of diester **16**, a congener of the spacer of tweezer **2**, supports the results from conformational analysis, namely the effect of sterical interaction between the double bond substituents in the Z-isomer. Whereas the E-isomer is completely planar (dihedral angle between the aromatic carbons attached to the double bond = 180.0(1)°), the corresponding angle in the Z-isomer is 9.1(2)°. Transannular interaction between the aromatic rings results in an even larger angle when measured on the two carbonyl carbons, at 27.30(8)° (Figure 5).

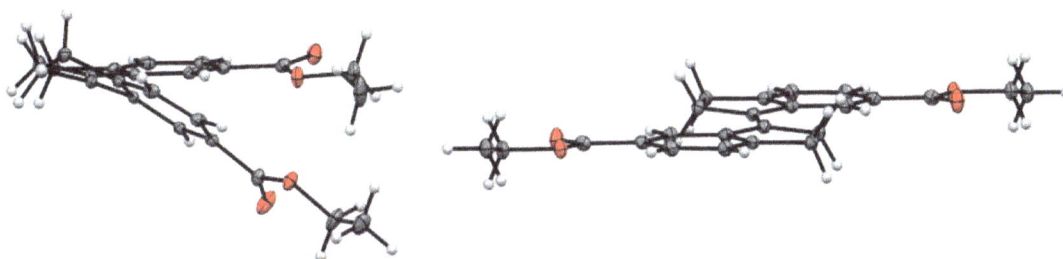

Figure 5. ORTEP view of Z-**16** (**left**) and E-**16** (**right**). Thermal ellipsoids are drawn at the 35% probability level.

This is in keeping with other 1,1′-biindanylidene derivatives. The geometry of the stiff stilbene unit appears to be predominantly dependent on steric interactions. All structures reported in the CCDC database are symmetrical with respect to the central double bond. In the E-isomers, the torsional angle between the two indane subunits (measured as the torsional angle involving the double bond and the two ortho positions of the fused phenyl rings) varies between 180.00(4)° for unsubstituted indanyl units [40] and 138.6(2)° for 2,2,2′,2′-tetramethyl-indanyl rings [41]. Also, ortho substituents on the phenyl rings result in deviations from the 180° angle, such as 154.39(7)° in (E)-7,7′-dimethyl-1,1′-biindanylidene [41]. For Z-isomers, compounds devoid of indane substituents with steric interactions show torsional angles in the range of approximately 24.90(9)° for (Z)-(1,1′)biindanylidene [42] and 18.7(1)° for (Z)-6,6′-dimethyl-1,1′-biindanylidene [43]. Substituents on the cyclopentyl ring, in particular dimethyl substitution such as in Z-2,2,2′,2′-tetramethyl-1,1′-biindanylidene (41.2(1)°) [43] or on the phenyl ring, such as ortho-methyl substituents in (Z)-4,4′,7,7′-tetra-methyl-1,1′-biindanylidene (41.4(1)°) [44], result in the largest deviations from this value. The diester **16** shows an undistorted geometry, with a torsional angle for E-**16** at 180.00(7)°, and for Z-**16** at 24.1(1)°, which is close to the values measured for unsubstituted stiff stilbenes (*i.e.*, approximately 20°) [37].

2.5. Binding of Dinitrogen Guests

2.5.1. UV-Vis Spectroscopy

Binding studies with tweezers **1–3** and a series of dinitrogen guests (Scheme 4), with both variable and fixed N−N distances, were performed to probe the impact of conformational flexibility on binding affinity. If we compare their maximal N−N distances with those obtained from the calculated Zn−Zn distances of their complexes with tweezer **3**, subtraction of twice the assumed Zn−N bond length (2.2 Å) [26] gives the N−N distance of bound guest, which is 2.62 Å (**19**), 6.76 Å (**20**), 7.84 Å (**21**) and 8.12 Å (**22**), respectively, indicating coiling of the flexible guests **21** and, in particular, **22**. As shown by pronounced isosbestic points (Figures 6 and 7), accompanied by red shifts of the Soret- and Q-bands, formation of a single, well defined complex is indicated (Table 3) [10,45,46] which, according to NMR data (*vide infra*), is a 1:1 complex (see Supplementary Materials).

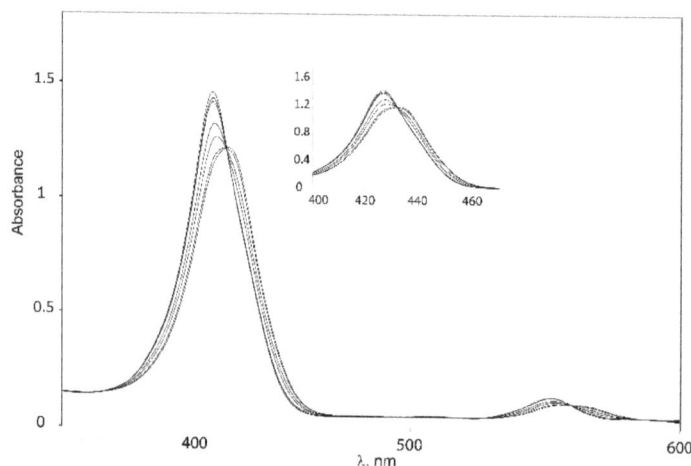

Figure 6. Isosbestic points for binding of 1,6-diaminohexane **21** to Z-**1**.

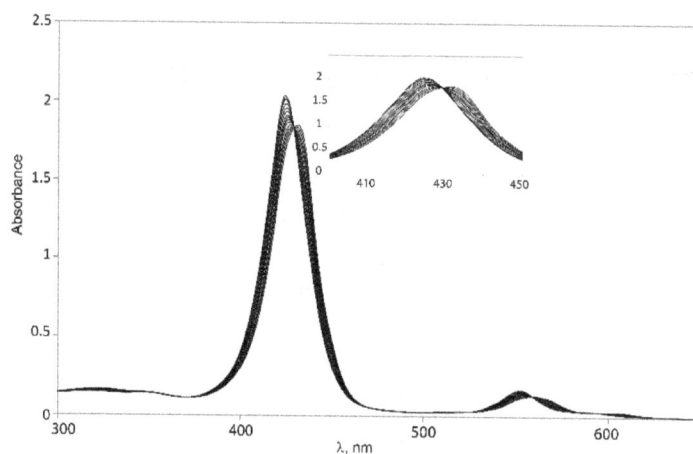

Figure 7. Isosbestic points for binding of 1,6-diaminohexane **21** to Z-**2**.

The consistent increase of the red shifts observed with larger N−N distances of the guests indicates the formation of 1:1 host–guest complexes [47]. Red shifts due to amine binding may be counteracted by the blue shift resulting from close proximity of two porphyrin rings [8,48]. Thus, the two guests with the shortest N−N distance (DABCO **19** and 4,4′-bipyridyl **20**) produce the smallest red shifts for their complexes. For **20**, the smaller red shift is also caused by the different electron density at its nitrogens, as indicated by the smaller red shift for the complex with Zn-TPP.

Complexes with glycoluril tweezer **3** show the smallest red shifts, which would indicate a closer proximity between the two porphyrin units than for the complexes with **1** and **2**. Another factor that could account for the smaller red shifts exhibited in the complexes with tweezer **3** is the degree of porphyrin rotation. Since tweezer **3** has no possibility of porphyrin rotation, the porphyrin rings should preserve their coplanarity with respect to the glycoluril backbone, and thus the blue shifts caused by π–π interactions between the porphyrin moieties should be larger than for **1** and **2** at the same Zn−Zn distances.

Binding constants of the investigated host–guest complexes (Table 4) show the expected effect of cooperativity when compared to the reference Zn-TPP, with K_a being up to three orders of magnitude larger for the tweezers. Amongst these, the largest binding constants are observed for **1**, followed by **3**, and with **2** showing the weakest binding. We rationalize this as follows. Tweezer **2** has the least flexible backbone, and therefore can accommodate guests only at the expense of distortion, primarily backbone twisting. The backbone of tweezer **1** is more flexible, and the cost of distortion upon guest accommodation is therefore low, resulting in larger binding constants. For the previously reported tweezer **3**, backbone distortion via conformational variation of the glycoluril rings also requires relatively low energies, giving high binding constants. Here, the possible distortions appear to be variation of interporphyrin distance and porphyrin twisting. Both **1** and **2** have an additional distortion available, *viz.* porphyrin rotation, which is not available to tweezer **3**. It should be emphasized that the larger flexibility of **1**, as compared to both **2** and **3**, which might result in a lesser degree of preorganization, is not manifested in the binding constants. Instead, it appears that tweezer **1** allows for an induced fit of guest molecules.

Table 3. UV-vis red shift of Soret bands upon binding ($\Delta\lambda = \lambda_{complex} - \lambda_{tweezer}$) for molecular tweezers Z-**1**, Z-**2** and **3** (CH$_2$Cl$_2$ solution, r.t.).

| Guest | ($\Delta\lambda = \lambda_{complex} - \lambda_{tweezer}$) Host | | | |
	Z-1	Z-2	3	Zn-TPP
DABCO **19** [a]	5	5	4 [b]	10
4,4'-bipyridyl **20**	7	7	5	8
1,6-diaminohexane **21**	9	9	10 [b]	10
1,12-diaminododecane **22**	10	10	10	10

[a] 1,4-Diazabicyclo[2.2.2]octane; [b] CHCl$_3$ solution.

Table 4. Binding constants (K_a, M^{-1}) (CH$_2$Cl$_2$ solution, r.t.) for molecular tweezers Z-**1**, Z-**2** and **3** with selected dinitrogen compounds, determined by UV-vis titration.

| Guest | Host K_a/M^{-1} | | | |
	Z-1	Z-2	3	Zn-TPP [a]
DABCO **19** [b]	2×10^4	4×10^5	3×10^{6} [c]	8×10^3
4,4'-bipyridyl **20**	4×10^6	2×10^5	2×10^6	-
1,6-diaminohexane **21**	6×10^6	5×10^5	2×10^{6} [c]	5×10^4
1,12-diaminododecane **22**	8×10^5	2×10^5	2×10^5	-

[a] Zn-meso-Tetraphenylporphyrin; [b] 1,4-Diazabicyclo[2.2.2]octane; [c] CHCl$_3$ solution.

It is interesting to note that both **1** and **3** show a 10 times lower binding constant for 1,12-diaminododecane **22** than for diaminohexane **21**, whereas **2** shows more similar binding constants for both guests. Obviously, the large guest **22** does not fit well into the binding cavity of any tweezer, despite coiling, and tweezer **2**, in particular, cannot vary its interporphyrin distance to improve binding, as is the case for **1** and **3**. For the smallest guest, DABCO **19**, tweezer **1** has the lowest binding constant, which is probably due to the difference between interporphyrin distance in the free tweezer and in the complex, resulting in only a small preorganization effect. In tweezer **1**, the Zn−Zn distance is 4.9 Å, as compared to 7.0 Å in the complex (Table 2). For tweezer **2**, the Zn−Zn distance increases only slightly from 6.2 Å (tweezer) to 7.3 Å (complex). However, these numbers have to be interpreted with some caution. Guest binding requires distortion of tweezer geometry (the source of the ECCD observed in CD spectroscopy, *vide infra*). As illustrated in Scheme 5 for binding of tweezers **1** and **2** to DABCO (**19**), for a rigid guest that itself cannot contribute to complex stability by conformational distortion and that has been used to test binding modes for bisporphyrins [49,50], geometries of free and complexed tweezers differ substantially.

2.5.2. NMR Spectroscopy

NMR titrations of tweezer **1** with 1,ω-diamino-n-alkanes were also consistent with ditopic binding, showing (n + 2)/2 signals at low chemical shifts (Figure 8). For 1:1 complexes, low temperature ^{1}H-NMR spectra showed an increase in signal number. While the broadening at 25 °C indicates an onset of dynamics, possibly guest dissociation, the observation of an increased signal number at lower temperatures is likely due to the presence of several conformers of the host–guest complex that are no longer exchanging. At reduced temperature, most CH_2 signals show a slightly decreased chemical shift (Figure 8), and TOCSY spectra show the presence of several isolated spin systems, each with the signal number expected for one ditopically bound diamine guest molecule (Figures 9 and 10).

A possible explanation for the larger number of guest signals for 1,6-diaminohexane **21** in the complex with stiff stilbene tweezer **Z-2** as compared to that with the enediyne tweezer **Z-1** might be the reduced inter-porphyrin distance flexibility imposed by the stiff stilbene spacer. In order for guest **21** to fit into the cavity, the tweezer sidewalls of **Z-2** have to twist and rotate in relation to each other, resulting in two non-equivalent porphyrin faces in contact with the guest. This type of signal splitting has been observed previously [49].

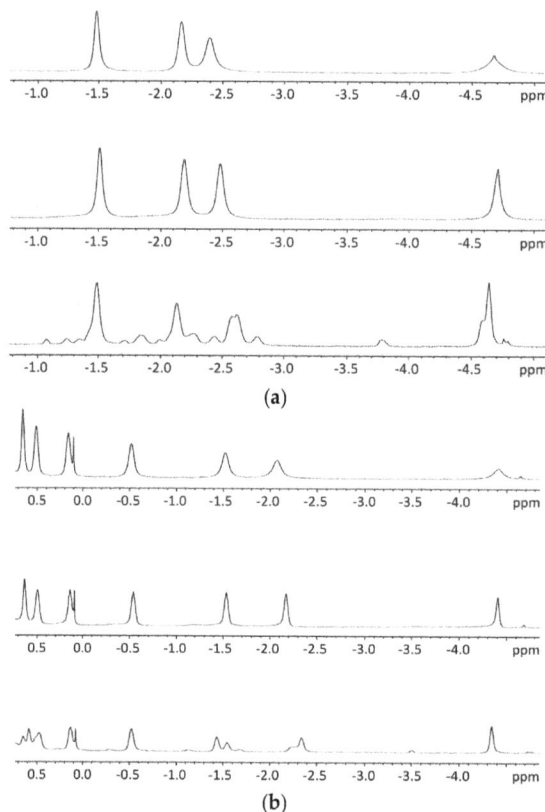

Figure 8. Variable temperature ^{1}H-NMR spectra (500 MHz), (**a**): 1:1 complex of enediyne tweezer **Z-1** and 1,6-diaminohexane **21** in CDCl$_3$. Top: 25 °C Middle: 0 °C Bottom: −55 °C; (**b**): 1:1 complex of enediyne tweezer **Z-1** and 1,12-diaminododecane **22** in CDCl$_3$. Top: 25 °C, Middle: 0 °C, Bottom: −55 °C.

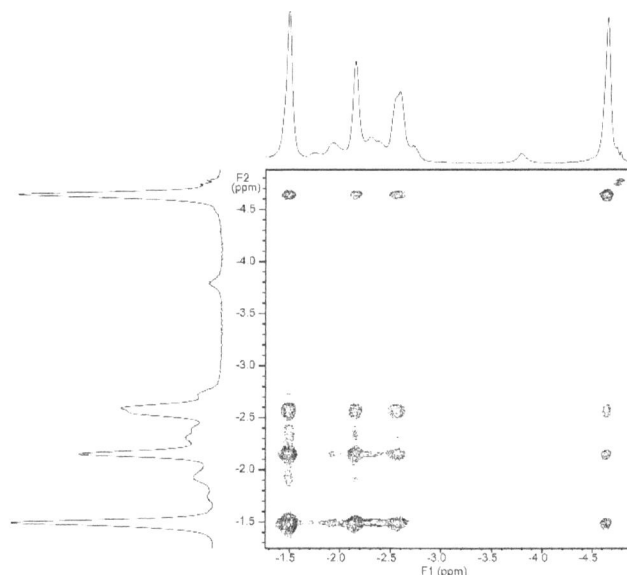

Figure 9. Expansion from TOCSY spectrum of a ≈1:1 complex of enediyne tweezer Z-**1** and 1,6-diaminohexane **21** (500 MHz, CDCl$_3$. solution, −40 °C).

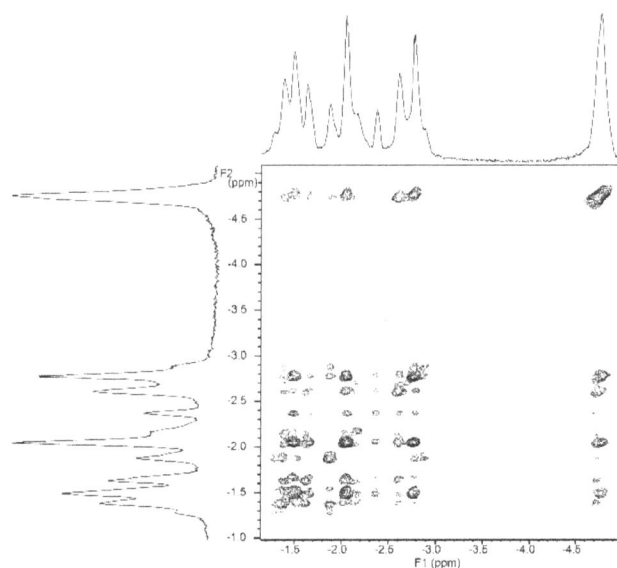

Figure 10. Expansion from TOCSY spectrum of a ≈1:1 complex of tweezer Z-**2** and 1,6-diaminohexane **21** (500 MHz, CDCl$_3$. solution, −40 °C).

2.5.3. Circular Dichroism

Further information on the binding properties of the three tweezers should be accessible by binding studies involving chiral diamino compounds, *i.e.*, by utilizing the chirogenesis that is the foundation of exciton coupled circular dichroism (ECCD). Parameters involved in this phenomenon have been extensively studied, and models to rationalize the effect have been proposed [3–15]. We found it tempting to test this effect by exposing the sterically flexible lysine methylester **23**, and the more rigid tryptophan methylester **24** (Figure 11) to tweezers **1–3**. Both esters have previously been shown to be capable of chirogenesis with bisporphyrin hosts [9,13,51]. While **23** constitutes a diamino ligand, **24** binds monotopically, but may give a strong CD response with suitable bisporphyrin tweezers.

As illustrated in Figure 12, there is indeed a relationship between the rigidity of the bisporphyrin tweezer and the observed CD signal. Tweezer **3** shows no chirality transfer from both **23** and **24**. In contrast, strong CD-signals for both **23** and **24** when binding to Z-**2** demonstrate a high degree of induced conformational helicity (Table 5). Tweezer Z-**1** on the other hand produces a slightly weaker ECCD than Z-**2** for the complex with the flexible chiral lysine methylester **23**, but none at all for tryptophan methylester **24**, despite succesful binding. The reason for this might be the already twisted spacer in Z-**2**, which favors unidirectional porphyrin twisting in the presence of a chiral guest, which is to some extent reminescent of bisporphyrins with Tröger's base as chiral spacer, which also have produced very strong CD signals [25]. The enediyne spacer, on the other hand, shows no twisting in the absence of a guest, and might bind a chiral guest without preference for porphyrin twisting in a particular direction, which would result in cancellation of the CD signal. Such a situation has been described by Rath and co-workers, based on X-ray crystallography [26]. However, there are also examples where tryptophan methylester **24** bound to bisporphyrin tweezers fails to give a CD signal at all [9]. Monotopic binding of **24** via its α-amino group is a likely reason for absence of chirality transfer to Z-**1**, which leaves the second (non-binding) porphyrin ring free to assume random orientations. This is supported by the absence of complexation between indole and Zn-TPP (see Supplementary Materials). Furthermore, similar observations for **24** bound to bisporphyrin tweezers with flexible pentanediol as opposed to tweezers with rigid melamine spacer have been reported [9].

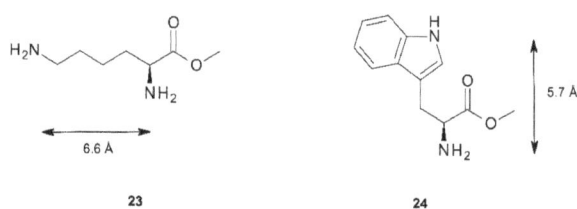

Figure 11. Chiral diamines **23** and **24** used in CD studies. The maximal N–N distances are indicated.

Table 5. CD signal amplitudes ($L \times mol^{-1} \times cm^{-1}$) for the complexes of tweezers Z-1 and Z-2 with chiral guests **23** and **24** at 20-fold excess. Measurements were performed at 25 °C for CH_2Cl_2 solutions at tweezer concentration of 33 μM.

Guest	Host					
	Z-1		Z-2		3	
	$\lambda/\Delta\varepsilon$	A_{CD}	$\lambda/\Delta\varepsilon$	A_{CD}	$\lambda/\Delta\varepsilon$	A_{CD}
23	460 nm (+50) 448 nm (−59)	+109	457 nm (+26) 448 nm (−99)	+125		-
24		-	451 nm (−391) 442 nm (+306)	−697		-

The absence of chirality transfer from both **23** and **24** to tweezer **3** is surprising, since twisting of the porphyrin units is possible (*cf.* Figure 4). However, if we inspect the calculated structures of the complexes between the three tweezers and these guests, there is a striking difference: If we compare the angle between lines bisecting the porphyrin units from the amide-attached pyrrol ring to the one on the opposite side (seen along the connecting line between their centers), this angle is close to 0° for the complexes of tweezer **3**. For complexes with tweezer **1**, the angle is 60°–70°, for those with tweezer **2** it is 35°–45° (Figure 13). This is reminiscent of rationalization of the ECCD produced by bisporphyrins by the effective transition moment in porphyrin derivatives [52].

Thus, for the purpose of chiral transfer, it seems to be beneficial with a tweezer exhibiting a more restricted inter-poprhyrin distance since this forces the porphyrins to translocate and rotate in order to accommodate the guest. It is interesting to note that tryptophan methylester with tweezer **2** produces a strong CD signal in dichloromethane solution, while previously typical bisporphyrins with flexible spacer used for this purpose have been reported to generate no CD signals in this solvent [9], requiring the use of less polar solvents such as methyl cyclohexane. Furthermore, tweezer **2** shows a CD signal already at low guest concentrations ($\geqslant$1:1 host:guest ratio), while typical systems require an appreciable excess of ligand ($\geqslant$6:1).

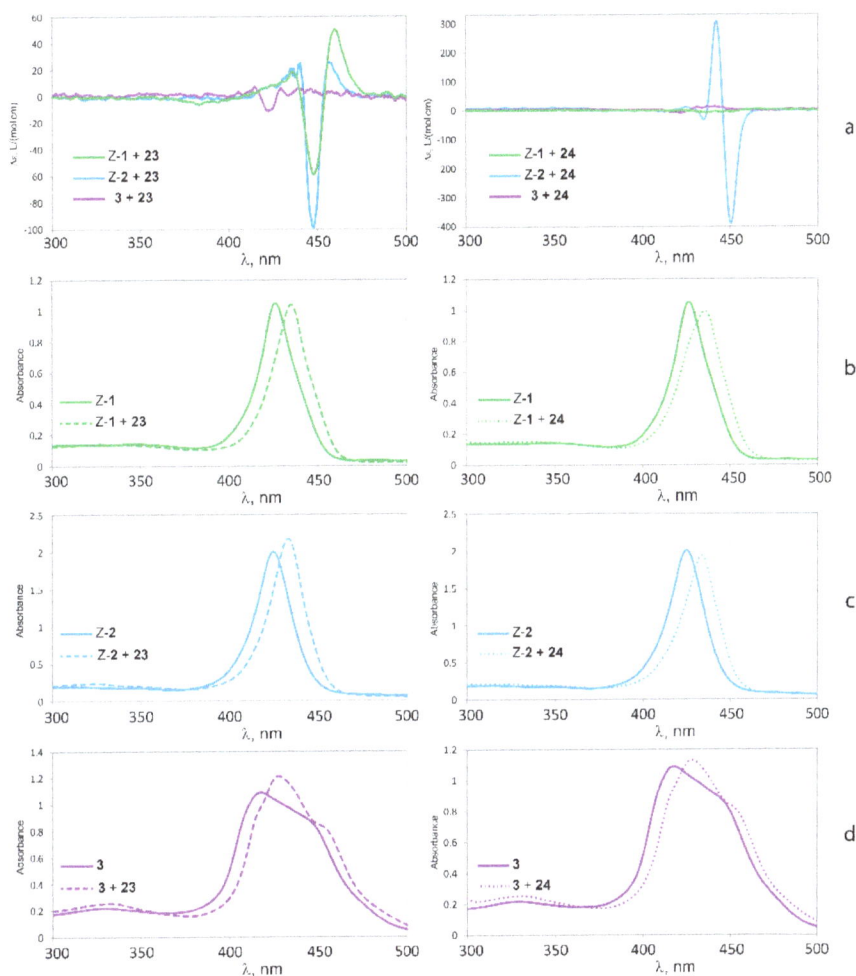

Figure 12. CD spectra (**a**) and UV-Vis spectra of L-lysine methylester **23** (**left**) and L-tryptophan methylester **24** (**right**) bound to tweezers **1** (**b**); **2** (**c**) and **3** (**d**).

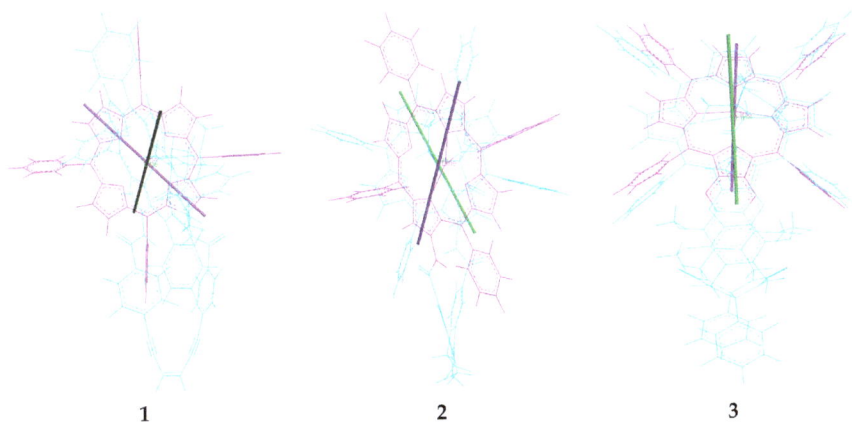

Figure 13. Alignment of porphyrin axes in the complexes with chiral guest lysine methylester **23**. Line positions are indicated for **23** with tweezers **1**, **2** and **3**, respectively.

3. Experimental Section

3.1. General

Commercially available compounds were used without purification. *meso*-Tetraphenylporphyrin was purchased from Porphyrin Systems GbR, Germany. Microwave heating was carried out in a Biotage Initiator microwave instrument using 10–20 mL Biotech microwave vials, applying microwave irradiation at 2.4 GHz, with a power setting up to 400 W, and an average pressure of 3–4 bar. Analytical TLC was performed using Merck precoated silica gel 60 F_{254} plates and for column chromatography Matrex silica gel (60 Å, 35–70 µm) was used. Melting points were determined using a Stuart Scientific melting point apparatus SMP10 and are uncorrected. Molecular structures for tweezers **1–3**, as well as their spacers, were calculated in MacroModel 9.9 with the OPLS-2005 force field [53] and a dielectric constant of 9.1. The coordinate scan option was used to obtain energy profiles for dihedral angle scan about the double bond of the alkene unit, or for alteration of distances between the two halves of the tweezer or spacer (Figure 3). Conformational analysis was performed with the lowest energy conformation from these scans as the starting structure. For bisporphyrins and host–guest complexes, Zn–N distances were initially constrained to 2.0 Å, resulting in final Zn–N distances of 2.1–2.3 Å [9,11]. In complexes of **24**, the ligand was constrained to ditopical binding (involving both nitrogens) to keep it within the bisporphyrin cleft.

^{1}H- and ^{13}C-NMR spectra were recorded on Varian Mercury Plus (^{1}H at 300.03 MHz, ^{13}C at 75.45 MHz), Varian Unity (^{1}H at 399.98 MHz, ^{13}C at 100.58 MHz), or Varian Unity Inova (^{1}H at 499.94 MHz, ^{13}C at 125.7 MHz) spectrometers at 25 °C unless noted otherwise. Chemical shifts are reported referenced to tetramethylsilane via the residual solvent signal (CDCl$_3$, ^{1}H at 7.26 and ^{13}C at 77 ppm; DMSO-d_6, ^{1}H at 2.50 and ^{13}C at 39.5 ppm). Signal assignments were derived from COSY [54,55], P.E.COSY [56], gHSQC [57], gHMBC [58], gNOESY [59], ROESY [60], and TOCSY [61] spectra.

Mass spectra were recorded on a GCQ/Polaris MS spectrometer using direct inlet interface (EI-MS), on an Advion Expression-L CMS with APCI interface, or on an Ultraflex II MALDI TOF/TOF (Bruker, Rheinstetten, Germany) mass spectrometer equipped with a gridless delayed extraction ion source, a 337-nm nitrogen laser, and a gridless ion reflector, using an α-cyano-4-hydroxycinnamic acid matrix (MALDI-MS). HR-MS were acquired using a Thermo Scientific LTQ Orbitrap Velos apparatus in direct nanospray infusion mode.

Circular dichroism spectra were recorded for solutions in CH$_2$Cl$_2$ on a JASCO J-810 spectropolarimeter from 300–500 nm using a 0.1 cm path length quartz cell. CD-spectra were measured in millidegrees and normalized into $\Delta\varepsilon_{max}$ (L·mol^{-1})/λ (nm) units. For comparison with literature data, molar ellipticities θ were converted to A$_{CD}$ values with A$_{CD}$ = θ/32.982.

UV-Vis spectra were recorded on a Varian Cary 3 Bio spectrophotometer using 5 mm or 10 mm quartz cuvettes. K$_a$ for diamines were determined by UV-Vis titration utilizing the iterative fitting program (in Matlab R2012b) recently published by P. Thordarson [62]. A 1:1 complexation model and global fitting to several datasets was applied. K$_a$ is calculated by (Equation (1)). The standard error (SE$_y$) is estimated by (Equation (2)). For NMR titrations, aliquots of freshly prepared guest solutions (CDCl$_3$, AlOx-filtered) were added to a solution of tweezer **1** or **2** in an NMR tube.

$$[HG] = \frac{1}{2}\left\{\left([G]_0 + [H]_0 + \frac{1}{K_a}\right) - \sqrt{\left([G]_0 + [H]_0 + \frac{1}{K_a}\right)^2 - 4[H]_0[G]_0}\right\} \tag{1}$$

$$SE_y = \sqrt{\frac{\sum(ydata - ycalc)^2}{N - k}} \tag{2}$$

CCDC 1437031 and 1437032 contain the supplementary crystallographic data for this paper. These data can be obtained free of charge via http://www.ccdc.cam.ac.uk/conts/retrieving.html

(or from the CCDC, 12 Union Road, Cambridge CB2 1EZ, UK; Fax: +44 1223 336033; E-Mail: deposit@ccdc.cam.ac.uk).

Synthesis of compounds **3**, **4**, **5**, **6**, **8**, **9**, **10**, **15**, **23** and **24**: See Supplementary Materials.

3.2. Syntheses

Methyl 4-[(Z)-6-(4-methoxycarbonylphenyl)hex-3-en-1,5-diynyl]benzoate (Z-**11**).   This protocol is a modification of one previously reported for Z-enediyne oligomers [35]. $Pd(PPh_3)_2Cl_2$ (51 mg, 0.064 mmol), *n*-butylamine (450 mg) and dry benzene (6 mL) were added to a 25 mL round bottomed flask.   The stirred mixture was cooled to 0 °C on an ice bath, before addition of methyl 4-(ethynyl)benzoate **10** (320 mg, 2.0 mmol) and (Z)-1,2-dichloroethylene (96 mg, 1.0 mmol). Subsequently, copper(I)iodide (34 mg, 0.18 mmol) was added and the mixture was allowed to stir at 0 °C for two hours and then at room temperature for two days. The dark reaction mixture was washed with 1 M HCl, (sat.) $NaHCO_3$ solution and brine before separating the phases. The aqueous phase was then extracted three times with diethyl ether, and the combined organic phases were dried over Na_2SO_4 before removal of solvent by reduced pressure. The resulting black solid was purified by column chromatography on silica using pentane as eluent giving the product Z-**11** as dark solid (240 mg, yield 70%). R_f = 0.42 (*n*-pentane/EtOAc = 5:1). ^{1}H-NMR (400 MHz, $CDCl_3$) δ = 8.02 (AA′BB′, 4H), 7.56 (AA′BB′, 4H), 6.16 (s, 2H, CH=CH), 3.93 (s, 6H, OCH_3). ^{13}C-NMR (100.6 MHz, DMSO-d_6) δ = 165.5, 131.7, 129.7, 129.6, 126.6, 120.7, 96.7, 90.1, 52.4. MS (EI): *m/z* = 344 ([M + H]$^+$), 313 ([M–CH_3O]$^+$). UV-vis: (CH_2Cl_2) $λ_{max}$: 345, 280 nm.

Methyl 4-[(E)-6-(4-methoxycarbonylphenyl)hex-3-en-1,5-diynyl]benzoate (E-**11**). ^{1}H-NMR (500 MHz, $CDCl_3$) δ = 8.01 (AA′BB′, 4H, Ar-H), 7.52 (AA′BB′, 4H, Ar-H), 6.33 (s, 2H, CH=CH), 3.92 (6H, OCH_3). ^{13}C-NMR (100.6 MHz, $CDCl_3$) δ = 166.4, 129.9, 129.6, 129.5, 127.3, 120.1, 94.5, 89.8, 52.3.

4-[(Z)-6-(4-carboxyphenyl)hex-3-en-1,5-diynyl]benzoic acid (Z-**12**). Dimethylester Z-**11** (0.5 g, 1.45 mmol) was dissolved in ethanol (100 mL) and NaOH (0.23 g, 5.8 mmol) was added. The mixture was refluxed for 1.5 h and followed by TLC. Water (150 mL) was added to the mixture, followed by a dropwise addition of HCl (12 M) upon which a fine-particle precipitate formed. The mixture was filtrated several times and the product Z-**12** was obtained as beige solid (0.409 g, 1.3 mmol, 90%). ^{1}H-NMR (400 MHz, d_6-DMSO) δ = 13.12 (br s, 2H, OH), 7.98 (AA′BB′, 4H, Ar-H) 7.63 (AA′BB′, 4H, Ar-H), 6.45 (s, 2H, CH=CH, $^3J_{HH}$ = 11 Hz, measured on the ^{13}C satellites). ^{13}C-NMR (100.6 MHz, DMSO-d_6) δ = 166.6, 131.6, 131.0, 129.7, 126.2, 120.6, 96.8, 89.8. IR (KBr): 3433 cm^{-1} (broad), 1689 cm^{-1}. HRMS, calcd. for $C_{20}H_{12}O_4$: *m/z* = 317.0808 [M + H]$^+$, found 317.0805.

4-[(Z)-6-(4-chlorocarbonylphenyl)hex-3-en-1,5-diynyl]benzoyl chloride (Z-**13**). Dicarboxylic acid Z-**12** (50 mg, 0.16 mmol) was dissolved in THF (10 mL) and 2 drops of DMF was added. The mixture was cooled to 0 °C and oxalyl chloride (0.1 mL, 1.1 mmol) added upon which gas formation was observed. The mixture was allowed to stir for 1 h after which solvents were evaporated to yield an orange solid (Z-**13**) that was used without further purification (57 mg, quant.) ^{1}H-NMR (400 MHz, $CDCl_3$) δ = 8.11 (AA′BB′, 4H), 7.60 (AA′BB′, 4H), 6.22 (s, 2H, CH=CH). MS (EI) *m/z*: 352 [M]$^+$, 317 [M − Cl]$^+$.

*Z-1,6-Bis-[4-(5,10,15,20-tetraphenylporphyrin-2-yl-ato)Zn(II)carbamoylphenyl]-hexa-1,5-diyn-3-ene (Z-**1**)*

Alternative Route 1

Dicarboxylic acid Z-**12** (25 mg, 0.08 mmol) was dissolved in dry and degassed THF (5 mL). DCC (40 mg, 0.19 mmol) was added and the mixture was allowed to stir for 5 min. The mixture was then added via syringe to a solution of TPPNH$_2$ (100 mg, 0.16 mmol) in dry and degassed THF (20 mL) under N$_2$-atm, dark, at 0 °C. The mixture was left to warm to r.t. and stir over night. Urea precipitate was filtered off and solvents evaporated to give a dark purple-greenish crude solid The crude solid was dissolved in CH_2Cl_2 (50 mL), MeOH (5 mL) and $Zn(OAc)_2 \cdot H_2O$ (0.5 g) was added. The mixture was refluxed for 20 min and allowed to cool to r.t. and solvents evaporated. The reaction residue was purified by column chromatography (eluent CH_2Cl_2) from which the pink fraction showed to contain product. Solvents were evaporated to give a purple solid of crude precursor to Z-**1** that was used directly for metallation (19 mg, 14%). MS (MALDI-TOF), calculated for $C_{108}H_{70}N_{10}O_2$, $[M + H]^+$, m/z = 1540. found: m/z = 1540.

Alternative Route 2

Dicarboxylic acid chloride Z-**13** (31 mg, 0.08 mmol) was dissolved in dry and degassed THF (3 mL), cooled to 0 °C and added via syringe to a solution of TPPNH$_2$ (141 mg, 0.22 mmol) in dry and degassed THF (20 mL) under N_2-atm in the dark. The mixture was left to warm to r.t. and stir over night. MeOH (10 mL) and Zn(OAc)$_2 \cdot$ H$_2$O (0.7 g) was added and left to stir at 50 °C for 3 h. Solvents were evaporated to give a crude dark purple solid that was purified by column chromatography (eluent CH_2Cl_2) to give a purple solid of Z-**1** (34 mg, 0.02 mmol, 25%). R_f = 0.1 (dichloromethane). ^{1}H-NMR (400 MHz, CDCl$_3$) δ = 9.76 (s, 2H, H-3'), 9.22 (br s, 2H, NH), 8.94 (d, J = 4.7 Hz, 2H, H-7'/8'), 8.91–8.89 (m, 6H, H-7'/8', H-12', H-13'), 8.85 (d, J = 4.7 Hz, 2H, H-17'), 8.47 (d, J = 4.7 Hz, 2H, H-18'), 8.33–8.29 (m, 3H, Ph-20), 8.24–8.15 (m, 12H, Ph), 7.87–7.84 (m, 5H, Ph-20) 7.78–7.68 (m, 20H, Ph), 7.64 (AA'BB', 4H, H-8), 7.45 (AA'BB', 4H, H-9), 6.27 (s, 2H, CH=CH). ^{13}C-NMR (100.6 MHz, CDCl$_3$) δ = 163.7 (CO), 152.0, 150.8, 150.6, 150.1, 150.0, 149.6, 149.1, 142.6, 141.7, 141.4, 140.5, 139.8, 134.2, 134.19, 134.17, 133.7, 132.8, 132.7, 132.3, 132.2, 131.8, 131.0, 129.0, 128.5, 127.6, 127.1, 126.7, 126.6, 126.2, 124.8, 122.0, 121.5, 121.0, 120.6 (C-3'), 120.2 (C-3), 117.2, 97.4 (C-1), 89.9 (C-2). UV-Vis (CH$_2$Cl$_2$) λ_{max}: 427 nm (ϵ = 6.4 × 10^5 M$^{-1} \cdot$ cm^{-1}), 552 nm (ϵ = 5.6 × 10^4 M$^{-1} \cdot$ cm^{-1}). MS (MALDI-TOF): m/z = 1667 $[M + H]^+$, 1689 $[M + Na]^+$. HRMS: m/z calculated for $C_{108}H_{70}N_{10}O_2$, $[M + 2H]^{2+}$: 834.2206, found 834.2032.

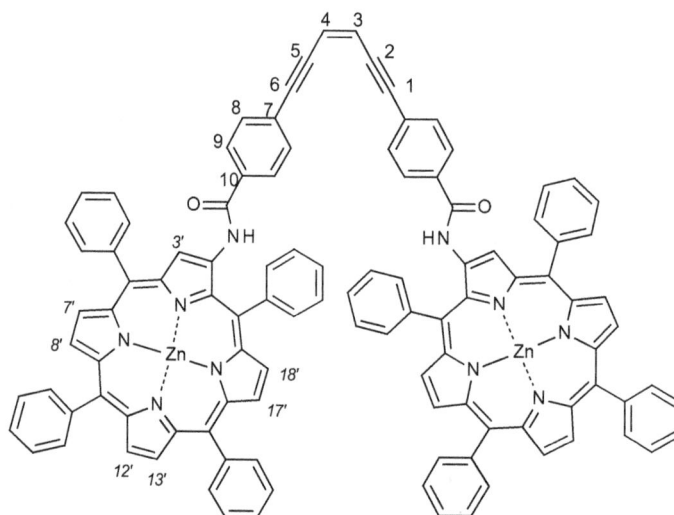

Ethyl 3-(6-ethoxycarbonylindan-1-ylidene)indane-5-carboxylate (**16**). To a suspension of TiCl$_4$ (1.394 g, 7.35 mmol) in dry THF (30 mL) solvent was slowly added Zn powder (0.96 g, 14.7 mmol) under an Argon atmosphere. The resultant deep green slurry was heated at reflux for 2 h. A THF solution (10 mL) of ethyl-3-oxoindane-5-carboxylate (**15**) (500 mg, 2.45 mmol) was added to the mixture in one portion and refluxed for 12 h. After all starting material was consumed (TLC), the mixture was quenched with saturated aqueous NH$_4$Cl and extracted with diethyl ether. The organic phase was filtered through MgSO$_4$ and solvents evaporated. The yellow crude product was purified by column chromatography (eluent CH$_2$Cl$_2$:pentane 2:3) from which E-**16** (pale yellow solid) and Z-**16** (yellow oil) were isolated (324 mg, 0.86 mmol, 70% (E:Z = 3:1).

Ethyl (3E)-3-(6-ethoxycarbonylindan-1-ylidene)indane-5-carboxylate (E-**16**). R_f = 0.73 (dichloromethane). ^{1}H-NMR (500 MHz, CDCl$_3$) δ = 8.26 (2H, d, J = 1.5 Hz, H-4), 7.91 (2H, dd, J = 1.5, 7.8 Hz, H-6), 7.37 (2H, dm, J = 7.8 Hz, H-7), 4.41 (4H, q, J = 7.1 Hz, OCH$_2$), 3.27 (4H, m, CH$_2$-2), 3.19 (4H, m, CH$_2$-1), 1.42 (6H, t, J = 7.1, OCH$_2$CH$_3$). E-configuration indicated by NOE between H-4 and H-2'. ^{13}C-NMR (100.6 MHz,

CDCl$_3$) δ = 167.0 (CO), 152.5 (C-7a), 143.2 (C-3a), 135.4 (C-3), 129.0 (C-5), 128.5 (C-6), 125.5 (C-4), 124.8 (C-7), 60.9 (OCH$_2$), 31.9 (CH$_2$-2), 31.2 (CH$_2$-1), 14.4 (CH$_2$CH$_3$). IR 1703 [ν(C=O)] cm^{-1}. MS m/z (EI): 376 [M + H]$^+$.

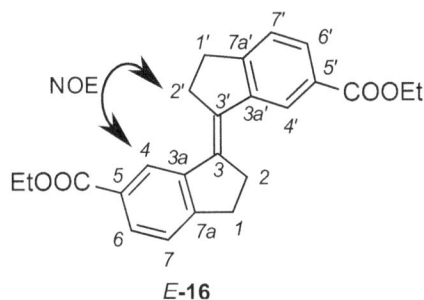

E-16

Ethyl (3Z)-3-(6-ethoxycarbonylindan-1-ylidene)indane-5-carboxylate (Z-**16**). R$_f$ = 0.43 (dichloromethane). ^{1}H-NMR (500 MHz, CDCl$_3$) δ = 8.76 (2H, d, J = 1.6 Hz, H-4), 7.89 (2H, dd, J = 1.6, 7.9 Hz, H-6), 7.35 (2H, dm, J = 7.9, H-7), 4.32 (4H, q, J = 7.1 Hz, OCH$_2$), 3.04 (4H, m, CH$_2$-1), 2.85 (4H, m, CH$_2$-2), 1.31 (6H, t, J = 7.1 Hz, OCH$_2$CH$_3$). ^{13}C-NMR (100.6 MHz, CDCl$_3$) δ = 166.6 (CO), 153.3 (C-7a), 140.4 (C-3a), 135.0 (C-3), 128.9 (C-6), 128.3 (C-5), 125.0 (C-7), 124.2 (C-4), 60.7 (COCH$_2$), 34.7 (CH$_2$-2), 30.7 (CH$_2$-1), 14.2 (CH$_2$CH$_3$); MS m/z (EI) 376 [M + H]$^+$.

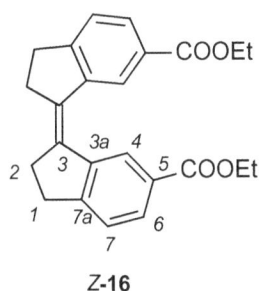

Z-16

Preparative isomerizations of *E*-**16** followed by chromatographic purification were used to get access to larger quantities of *Z*-**16**.

Photoisomerizations

Photoisomerizations were conducted for CDCl$_3$ solutions at 25 °C using an Oriel 1000 W Xe ARC light source equipped with a band pass filter 20BPF10-340 (for Z-**1** and Z-**11**) or 10BPF10-300 (for Z-**2** and Z-**16**) (Newport). Solutions were degassed by argon bubbling for 15 min prior to irradiation. As reaction vessels, 5 mm NMR-tubes, Type 5Hp, 178 mm were used. The course of the irradiation was followed by NMR-spectroscopy (Varian Unity Inova 500 MHz NMR spectrometer, ^{1}H at 499.9 MHz).

General procedure for hydrolysis of Z/E stiff stilbene diethyl ester 16 to 3-(6-carboxy-indan-1-ylidene)indane-5-carboxylic acid (**17**). Z- or E- stiff stilbene diethylester **16** (120 mg, 0.32 mmol), NaOH (150 mg, 3.75 mmol) and ethanol (20 mL) was added to a round-bottomed flask and the mixture was refluxed at 80 °C for 6 h. The solvent was removed in vacuo and the residue was dissolved in H$_2$O. The aqueous solution was acidified by addition of HCl (6 M) and the formed a yellow precipitate was collected by filtration (using a grade 3 filter paper). The solid product (**17**) was washed with H$_2$O and dried overnight.

*Z-Dicarboxylic acid Z-***17** (96 mg, 94%). ^{1}H-NMR (500 MHz, dmso-d_6, 25 °C) δ = 8.54 (2H, d, J = 1.5 Hz, H-4), 7.79 (2H, dd, J = 1.5, 7.8 Hz, H-6), 7.46 (2H, dm, J = 7.8 Hz), 3.02 (4H, m), 2.83 (4H, m). ^{13}C-NMR (100.6 MHz, dmso-d_6) δ = 167.6, 153.7, 140.33, 135.3, 129.2, 129.1, 125.9, 123.8, 34.7, 30.6.

E-Dicarboxylic acid *E*-**17** (83 mg, 82%). ^{1}H-NMR (500 MHz, dmso-d_6, 25 °C) δ = 8.17 (2H, d, *J* = 1.5 Hz, H-4), 7.84 (2H, dd, *J* = 1.5, 7.8 Hz, H-6), 7.48 (2H, d, *J* = 7.8 Hz, H-7), 3.18 (8H, m, CH$_2$CH$_2$). ^{13}C-NMR (75.5 MHz, dmso-d_6, 25 °C) δ = 167.9, 152.7, 143.1, 135.5, 129.7, 129.0, 125.6, 125.5, 31.8, 31.0. HRMS, calc. for C$_{20}$H$_{16}$O$_4$: *m/z* = 321.1121 [M + H]$^+$, found 321.1118.

(3Z)-3-(6-Chlorocarbonylindan-1-ylidene)indane-5-carbonyl chloride (**18**). Dicarboxylic acid *Z*-**17** (55 mg, 0.17 mmol) was added to a dried 25 mL round-bottomed flask together with dry CH$_2$Cl$_2$ (5 mL), oxalyl chloride (0.9 mL) and DMF (3 drops). The mixture was stirred at RT under nitrogen atmosphere for two hours. Volatile components were removed *in vacuo* and the residue was washed several times with dry Et$_2$O in which the product was soluble. The heterogeneous Et$_2$O solution was filtrated (using a grade 3 filter paper) to remove by-products and the filtrate was evaporated to leave the acid chloride *Z*-**18** as a pale yellow solid (yield: 45 mg, 74%). ^{1}H-NMR (500 MHz, CDCl$_3$, 25 °C) δ = 8.80 (2H, d, *J* = 1.8 Hz, H-4), 7.96 (2H, dd, *J* = 1.8, 8.0 Hz, H-6), 7.43 (2H, dm, *J* = 8.0 Hz, H-7), 3.10 (4H, m, CH$_2$), 2.90 (4H, m, CH$_2$).

(3Z)-N-(5,10,15,20-tetraphenylporphyrin-2-yl-ato)Zn(II)-3-[6-(5,10,15,20-tetraphenylporphyrin-2-yl-ato)Zn(II)-carbamoyl)indan-1-ylidene]indane-5-carboxamide (**2**). TPP-NH$_2$ **6** (210 mg, 0.33 mmol) was added to a dried round-bottomed flask and *Z*-stif-stilbene dicarboxylic acid chloride *Z*-**18** (42 mg, 0.13 mmol) dissolved in dry CH$_2$Cl$_2$ (20 mL) was added via syringe. Dry pyridine (0.2 mL) was added and the mixture was allowed to stir for three days under nitrogen atmosphere and with protection from light. The solvent was evaporated and the residue redissolved in CH$_2$Cl$_2$. The solution was washed with HCl (1 M, 50 mL) and brine (5%, 50 mL). The organic phase was then dried over Na$_2$SO$_4$ and the solvent evaporated. The crude compound was added to a flask together with Zn(OAc)$_2$·2H$_2$O (1 g), methanol (20 mL), CH$_2$Cl$_2$ (20 mL) and the mixture was stirred at 50 °C for one hour. The solvents were removed by evaporation and the crude metallated compound was purified by column chromatography (silica, CH$_2$Cl$_2$) followed by a subsequent silica column (CH$_2$Cl$_2$:methanol:acetic acid 96:4:1). The compound was further purified by washing several times with methanol to yield the target compound *Z*-**2** as a light/air sensitive purple solid (yield: 53 mg, 24%). R$_f$ = 0.13 (dichloromethane). ^{1}H-NMR (500 MHz, CDCl$_3$, 25 °C) δ = 9.25 (2H, s, H-3′), 8.81 (2H, br s, N-H), 8.78 (2H, d, *J* = 4.5 Hz, β-pyrrole), 8.78 (2H, d, *J* = 4.5 Hz, β-pyrrole), 8.76 (2H, d, *J* = 4.5 Hz, β-pyrrole), 8.72 (2H, d, *J* = 4.6 Hz, β-pyrrole), 8.43 (2H, d, *J* = 4.6 Hz, β-pyrrole), 8.38 (2H, d, *J* = 4.6 Hz, β-pyrrole), 8.11 (2H, m, H-4), 8.11–8.07 (8H, m, Ph), 8.01 (4H, m, H-d), 7.77–7.66 (12H, m, Ph), 7.63 (2H, m, H-f), 7.46 (4H, m, H-e), 7.43 (4H, m, H-a), 7.38 (2H, d, *J* = 7.4 Hz, H-7), 7.09 (2H, m, H-6), 7.08 (2H, m, H-c), 6.90 (4H, m, H-b), 3.22 (4H, m, CH$_2$-2), 3.05 (4H, m, CH$_2$-1). ^{13}C-NMR (125 MHz, CDCl$_3$, 25 °C) δ = 164.8, 152.0, 150.7, 150.6, 150.2, 149.9, 149.6, 149.5, 149.0, 142.7, 142.6, 142.1, 141.4, 141.1, 140.0, 139.8, 135.3, 134.4, 134.3, 134.2, 133.8, 133.3, 133.0, 132.0, 131.8, 131.7, 131.5, 131.4, 130.9, 128.8, 128.1, 127.43, 127.40, 127.0, 126.7, 126.6, 126.50, 126.46, 126.1, 124.9, 121.9, 121.6, 121.0, 120.23, 120.19, 117.4, 35.3, 31.0. Z-2 UV-Vis (CH$_2$Cl$_2$) λ$_{max}$: 425 nm (ε = 6.0 × 10^5 M^{-1}·cm^{-1}), 552 nm (ε = 5.1 × 10^4 M^{-1}·cm^{-1}). MS (MALDI-MS, dithranol, positive mode) *m/z*: 1669 [M + H]$^+$). HRMS: *m/z* calcd. for C$_{108}$H$_{70}$N$_{10}$O$_2$Zn$_2$, [M + H]$^+$: 1669.4269; found: 1669.4193.

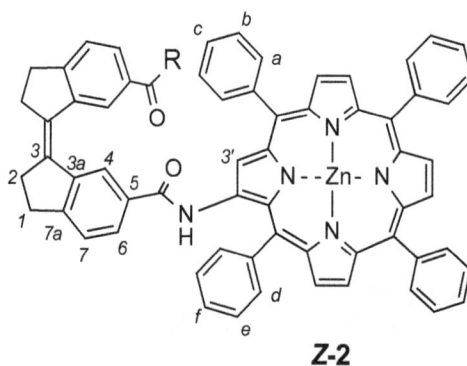

Z-2

4. Conclusions

The Zn-porphyrin tweezers **1**, **2**, and **3**, with differing conformational flexibility, have been investigated. Differences in spacer flexibility affect their possibilities for interporphyrin distance variation, porphyrin rotation, translocation and twisting upon ditopic binding to dinitrogen guest molecules. These differences result in a variation of binding constants, and are further detected via NMR spectra of host–guest complexes. CD spectroscopy offers a further possibility to monitor distortions of the tweezer geometry in the complexes. Flexibility regarding interporphyrin distance variation is the single most important factor that contributes to stronger binding. However, a higher degree of preorganization owing to limited porphyrin rotation as in **3** does not result in higher binding constants when compared to tweezer **1** with the most flexible spacer. Tweezer **2** has the least flexible spacer, hence weaker binding, but due to "built-in" helicity, it shows the highest CD signal amplitudes.

Acknowledgments: Financial support by the Swedish Research Council (A.G., grant nr. 621-2012-3379, J.B., grant 621-2011-4423) is gratefully acknowledged. We thank Bo Ek for assistance with the MALDI-MS and HR-MS measurements. The computations were performed on resources provided by the Swedish National Infrastructure for Computing (SNIC) at the NSC, Linköping University.

Author Contributions: M.B. synthesized part of the compounds, performed binding and CD studies, and participated in writing the manuscript. S.N. synthesized part of the compounds, performed preliminary binding studies, and participated in writing a preliminary manuscript. C.H.A. synthesized part of the compounds. H.H. synthesized the compound for X-ray crystallography. M.E.L. performed the X-ray crystallographic analysis. J.B. performed MALDI-MS and HR-MS analyses. H.G. contributed to conceiving the study and was involved in the UV analyses. A.G. conceived the study, performed conformational analyses, evaluation of the data, and wrote the manuscript. All authors have read and approved the final manuscript.

Conflicts of Interest: The authors declare no conflict of interest.

References

1.   Valderreya, V.; Aragaya, G.; Ballester, P. Porphyrin tweezer receptors: Binding studies, conformational properties and applications. *Coord. Chem. Rev.* **2014**, *258–259*, 137–156. [CrossRef]

2.   Beletskaya, I.; Tyurin, V.S.; Tsivadze, A.Y.; Guilard, R.; Stern, C. Supramolecular chemistry of metalloporphyrins. *Chem. Rev.* **2009**, *109*, 1659–1713. [CrossRef] [PubMed]

3.   Hayashi, S.; Yotsukura, M.; Noji, M.; Takanami, T. Bis(zinc porphyrin) as a CD-sensitive bidentate host molecule: Direct determination of absolute configuration of mono-alcohols. *Chem. Commun.* **2015**, *51*, 11068–11071. [CrossRef] [PubMed]

4.   Brahma, S.; Ikbal, S.A.; Dhamija, A.; Rath, S.P. Highly Enhanced Bisignate Circular Dichroism of Ferrocene-Bridged Zn(II) Bisporphyrin Tweezer with Extended Chiral Substrates due to Well-Matched Host-Guest System. *Inorg. Chem.* **2014**, *53*, 2381–2395. [CrossRef] [PubMed]

5.   Brahma, S.; Ikbal, S.A.; Rath, S.P. Synthesis, Structure, and Properties of a Series of Chiral Tweezer-Diamine Complexes Consisting of an Achiral Zinc(II) Bisporphyrin Host and Chiral Diamine Guest: Induction and Rationalization of Supramolecular Chirality. *Inorg. Chem.* **2014**, *53*, 49–62. [CrossRef] [PubMed]

6.   Ikbal, S.A.; Brahma, S.; Rath, S.P. Transfer and control of molecular chirality in the 1:2 host-guest supramolecular complex consisting of Mg(II)bisporphyrin and chiral diols: The effect of H-bonding on the rationalization of chirality. *Chem. Commun.* **2014**, *50*, 14037–14040. [CrossRef] [PubMed]

7.   Chaudhary, A.; Ikbal, Sk.A.; Brahma, S.; Rath, S.P. Formation of exo–exo, exo–endo and tweezer conformation induced by axial ligand in a Zn(II) bisporphyrin: Synthesis, structure and properties. *Polyhedron* **2013**, *52*, 761–769. [CrossRef]

8.   Tanasova, M.; Borhan, B. Conformational Preference in Bis(porphyrin) Tweezer Complexes: A Versatile Chirality Sensor for α-Chiral Carboxylic Acids. *Eur. J. Org. Chem.* **2012**, *2012*, 3261–3269. [CrossRef]

9.   Petrovic, A.G.; Vantomme, G.; Negrón-Abril, Y.; Lubian, E.; Saielli, G.; Menegazzo, I.; Cordero, R.; Proni, G.; Nakanishi, K.; Carofiglio, T.; *et al.* Bulky Melamine-Based Zn-Porphyrin Tweezer as a CD Probe of Molecular Chirality. *Chirality* **2011**, *23*, 808–819. [CrossRef] [PubMed]

10. Huang, X.; Nakanishi, K.; Berova, N. Porphyrins and metalloporphyrins: Versatile circular dichroic reporter groups for structural studies. *Chirality* **2000**, *12*, 237–255. [CrossRef]

11. Huang, X.; Fujioka, N.; Pescitelli, G.; Koehn, F.E.; Williamson, R.T.; Nakanishi, K.; Berova, N. Absolute Configurational Assignments of Secondary Amines by CD-Sensitive Dimeric Zinc Porphyrin Host. *J. Am. Chem. Soc.* **2002**, *124*, 10320–10335. [CrossRef] [PubMed]

12. Huang, X.; Borhan, B.; Rickman, B.H.; Nakanishi, K.; Berova, N. Zinc Porphyrin Tweezer in Host ± Guest Complexation: Determination of Absolute Configurations of Primary Monoamines by Circular Dichroism. *Chem. Eur. J.* **2000**, *6*, 216–224. [CrossRef]

13. Borovkov, V.V.; Hembury, G.A.; Inoue, Y. Origin, Control, and Application of Supramolecular Chirogenesis in Bisporphyrin-Based Systems. *Acc. Chem. Res.* **2004**, *37*, 449–459. [CrossRef] [PubMed]

14. Borovkov, V.V.; Lintuluoto, J.M.; Hembury, G.A.; Sugiura, M.; Arakawa, R.; Inoue, Y. Supramolecular Chirogenesis in Zinc Porphyrins: Interaction with Bidentate Ligands, Formation of Tweezer Structures, and the Origin of Enhanced Optical Activity. *J. Org. Chem.* **2003**, *68*, 7176–7192. [CrossRef] [PubMed]

15. Hayashi, T.; Nonoguchi, M.; Aya, T.; Ogoshi, H. Molecular recognition of α,ω-diamines by metalloporphyrin dimer. *Tetrahedron Lett.* **1997**, *38*, 1603–1606. [CrossRef]

16. Norrehed, S.; Polavarapu, P.; Yang, W.; Gogoll, A.; Grennberg, H. Conformational restriction of flexible molecules in solution by a semirigid bis-porphyrin molecular tweezer. *Tetrahedron* **2013**, *69*, 7131–7138. [CrossRef]

17. Norrehed, S.; Johansson, H.; Grennberg, H.; Gogoll, A. Improved Stereochemical Analysis of Conformationally Flexible Diamines by Binding to a Bisporphyrin Molecular Clip. *Chem. Eur. J.* **2013**, *19*, 14631–14638. [CrossRef] [PubMed]

18. Tsuge, A.; Kunimune, T.; Ikeda, Y.; Moriguchi, T.; Araki, K. Binding properties of calixarene-based cofacial bisporphyrins. *Chem. Lett.* **2010**, *39*, 1155–1157. [CrossRef]

19. Wahadoszamen, M.; Yamamura, T.; Momotake, A.; Nishimura, Y.; Arai, T. High Binding Affinity of DABCO with Porphyrin in a Porphyrin-cis-Stilbene-Porphyrin Triad. *Heterocycles* **2009**, *79*, 331–337.

20. Fathalla, J.M. Jayawickramarajah. Configurational Isomers of a Stilbene-Linked Bis(porphyrin) Tweezer: Synthesis and Fullerene-Binding Studies. *Eur. J. Org. Chem.* **2009**, 6095–6099. [CrossRef]

21. Tsuge, A.; Ikeda, Y.; Moriguchi, T.; Araki, K. Preparation of cyclophanes having cofacial bisporphyrins and their binding properties. *J. Porphyr. Phthalocyanines* **2012**, *16*, 250–254. [CrossRef]

22. Collman, J.P.; Wagenknecht, P.S.; Hutchinson, J.H. Molecular catalysis for multielectron redox reactions of small molecules: The "cofacial metallodiphorphyrin approach". *Angew. Chem. Int. Ed. Engl.* **1994**, *33*, 1537–1553. [CrossRef]

23. Merkasa, S.; Bouatraa, S.; Reina, R.; Piantanidab, I.; Zinicb, M.; Solladié, N. Pre-organized dinucleosides with pendant porphyrins for the formation of sandwich type complexes with DABCO with high association constants. *J. Porphyr. Phthalocyanines* **2015**, *19*, 535–546. [CrossRef]

24. Solladié, N.; Aziat, F.; Bouatra, S.; Rein, R. Bis-porphyrin tweezers: Rigid or flexible linkers for better adjustment of the cavity to bidentate bases of various size. *J. Porphyr. Phthalocyanines* **2008**, *12*, 1250–1260. [CrossRef]

25. Pescitelli, G.; Gabriel, S.; Wang, Y.; Fleischhauer, J.; Woody, R.W.; Berova, N. Theoretical Analysis of the Porphyrin-Porphyrin Exciton Interaction in Circular Dichroism Spectra of Dimeric Tetraarylporphyrins. *J. Am. Chem. Soc.* **2003**, *125*, 7613–7628. [CrossRef] [PubMed]

26. Brahma, S.; Ikbal, S.A.; Dey, S.; Rath, S.P. Induction of supramolecular chirality in di-zinc(II) bisporphyrin via tweezer formation: Synthesis, structure and rationalization of chirality. *Chem. Commun.* **2012**, *48*, 4070–4072. [CrossRef] [PubMed]

27. Promarak, V.; Burn, P.L. A new synthetic approach to porphyrin-α-diones and a -2,3,12,13-tetraone: Building blocks for laterally conjugated porphyrin arrays. *J. Chem. Soc. Perkin Trans. I* **2001**, 14–20. [CrossRef]

28. Baldwin, J.E.; Crossley, M.J.; DeBernardis, J. Efficient peripheral functionalization of porphyrins. *Tetrahedron* **1982**, *38*, 685–692. [CrossRef]

29. Shine, H.J.; Padilla, A.G.; Wu, S.-M. Ion radicals. 45. Reactions of zinc tetraphenylporphyrin cation radical perchlorate with nucleophiles. *J. Org. Chem.* **1979**, *44*, 4069–4075. [CrossRef]

30. Fanning, J.C.; Mandel, F.S.; Gray, T.L. The reaction of metalloporphyrins with nitrogen dioxide. *Tetrahedron* **1979**, *33*, 1251–1255. [CrossRef]

31. Giraudeau, A.; Callot, H.J.; Jordan, J.; Ezhar, I.; Gross, M. Substituent effects in the electroreduction of porphyrins and metalloporphyrins. *J. Am. Chem. Soc.* **1979**, *101*, 3857–3862. [CrossRef]

32. Crossley, M.J.; King, L.G.; Newsom, I.A.; Sheehan, G.S. Investigation of a "reverse" approach to extended porphyrin systems. Synthesis of a 2,3-diaminoporphyrin and its reactions with α-diones. *J. Chem. Soc. Perkin Trans. 1* **1996**, 2675–2684. [CrossRef]

33. Erdélyi, M.; Gogoll, A. Rapid Homogeneous-Phase Sonogashira Coupling Reactions Using Controlled Microwave Heating. *J. Org. Chem.* **2001**, *66*, 4165–4169. [CrossRef] [PubMed]

34. Li, Q.; Rukavishnikov, A.V.; Petukhov, P.A.; Zaikova, T.O.; Jin, C.; Keana, J.F.W. Nanoscale Tripodal 1,3,5,7-Tetrasubstituted Adamantanes for AFM Applications. *J. Org. Chem.* **2003**, *68*, 4862–4869. [CrossRef] [PubMed]

35. Kosinki, C.; Hirsch, A.; Heineman, F.W.; Hampel, F. An Iterative Approach to cis-Oligodiacetylenes. *Eur. J. Org. Chem.* **2001**, *2001*, 3879–3890. [CrossRef]

36. McMurry, J.E. Carbonyl-coupling reactions using low-valent titanium. *Chem. Rev.* **1989**, *89*, 1513–1524. [CrossRef]

37. Oelgemöller, M.; Frank, R.; Lemmen, P.; Lenoir, D.; Lex, J.; Inoue, Y. Synthesis, structural characterization and photoisomerization of cyclic stilbenes. *Tetrahedron* **2012**, *68*, 4048–4056. [CrossRef]

38. Quick, M.; Berndt, F.; Dobryakov, A.L.; Ioffe, I.N.; Granovsky, A.A.; Knie, C.; Mahrwald, R.; Lenoir, D.; Ernsting, N.P.; Kovalenko, S.A. Photoisomerization Dynamics of Stiff-Stilbene in Solution. *J. Phys. Chem. B* **2014**, *118*, 1389–1402. [CrossRef] [PubMed]

39. Matsumura, M.; Tanatani, A.; Kaneko, T.; Azumaya, I.; Masu, H.; Hashizume, D.; Kagechika, H.; Muranaka, A.; Uchiyama, M. Synthesis of porphyrinylamide and observation of *N*-methylation-induced trans-cis amide conformational alteration. *Tetrahedron* **2013**, *69*, 10927–10932. [CrossRef]

40. Schaefer, W.P.; Abulu, J. An Indanyl Precursor to a Chiral Spiro Compound. *Acta Crystallogr. Sect. C: Cryst. Struct. Commun.* **1995**, *51*, 2364–2366. [CrossRef]

41. Ogawa, K.; Harada, J.; Tomoda, S. Unusually short ethylene bond and large amplitude torsional motion of (*E*)-stilbenes in crystals. X-ray crystallographic study of "stiff" stilbenes. *Acta Crystallogr. Sect. B Struct. Sci.* **1995**, *51*, 240–248. [CrossRef]

42. Jovanovic, J.; Elling, W.; Schurmann, M.; Preut, H.; Spiteller, M. (*Z*)-2,3,2',3'-Tetra-hydro-[1,1']-biindenyl-idene. *Acta Crystallogr. Sect. E: Struct. Rep. Online* **2002**, *58*, o35–o36. [CrossRef]

43. Harada, J.; Ogawa, K.; Tomoda, S. "Stiff" cis-Stilbenes. (*Z*)-6,6'-Dimethyl-1,1'-biindanylidene and (*Z*)-4,4',7,7'-Tetra-methyl-1,1'-biindanylidene. *Acta Crystallogr. Sect. C Cryst. Struct. Commun.* **1995**, *51*, 2125–2127. [CrossRef]

44. Shimasaki, T.; Kato, S.; Shinmyozu, T. Synthesis, Structural, Spectral, and Photoswitchable Properties of cis- and trans-2,2,2',2'-Tetra-methyl-1,1'-indanylindanes. *J. Org. Chem.* **2007**, *72*, 6251–6254. [CrossRef] [PubMed]

45. Cohen, M.D.; Fischer, E. Isosbestic points. *J. Chem. Soc.* **1962**, 3044–3052. [CrossRef]

46. Mauzerall, D. Spectra of Molecular Complexes of Porphyrins in Aqueous Solution. *Biochemistry* **1965**, *4*, 1801–1810. [CrossRef]

47. Solladié-Cavallo, A.; Marsol, C.; Pescitelli, G.; di Bari, L.; Salvadori, P.; Huang, X.; Fujioka, N.; Berova, N.; Cao, X.; Freedman, T.B.; et al. (*R*)-(+)- and (*S*)-(−)-1-(9-Phenanthryl)ethylamine: Assignment of Absolute Configuration by CD Tweezer and VCD Methods, and Difficulties Encountered with the CD Exciton Chirality Method. *Eur. J. Org. Chem.* **2002**, *2002*, 1788–1796. [CrossRef]

48. Hunter, C.A.; Sanders, J.K.M.; Stone, A.J. Exciton coupling in porphyrin dimers. *Chem. Phys.* **1989**, *133*, 395–404. [CrossRef]

49. Hunter, C.A.; Meah, M.N.; Sanders, J.K.M. DABCO-Metalloporphyrin Binding: Ternary Complexes, Host-Guest Chemistry, and the Measurement of T-T Interactions. *J. Am. Chem. Soc.* **1990**, *112*, 5773–5780. [CrossRef]

50. Anderson, H.L.; Hunter, C.A.; Meah, M.N.; Sanders, J.K.M. Thermodynamics of induced-fit binding inside polymacrocyclic porphyrin hosts. *J. Am. Chem. Soc.* **1990**, *112*, 5780–5789. [CrossRef]

51. Borovkov, V.V.; Yamamoto, N.; Lintuluoto, J.M.; Tanaka, T.; Inoue, Y. Supramolecular Chirality Induction in Bis(Zinc Porphyrin) by Amino Acid Derivatives: Rationalization and Applications of the Ligand Bulkiness Effect. *Chirality* **2001**, *13*, 329–335. [CrossRef] [PubMed]

52. Berova, N.; Di Bari, L.; Pescitelli, G. Application of electronic circular dichroism in configurational and conformational analysis of organic compounds. *Chem. Soc. Rev.* **2007**, *36*, 914–931. [CrossRef] [PubMed]

53.   Mohamadi, F.; Richards, N.G.J.; Guida, W.C.; Liskamp, R.; Lipton, M.; Caufield, C.; Chang, G.; Hendrickson, T.; Still, W.C. Macromodel—An integrated software system for modeling organic and bioorganic molecules using molecular mechanics. *J. Comput. Chem.* **1990**, *11*, 440–467. [CrossRef]

54.   Wokaun, A.; Ernst, R.R. Selective detection of multiple quantum transitions in NMR by two-dimensional spectroscopy. *Chem. Phys. Lett.* **1977**, *52*, 407–412. [CrossRef]

55.   Shaka, A.J.; Freeman, R. Simplification of NMR spectra by filtration through multiple-quantum coherence. *J. Magn. Reson.* **1983**, *51*, 169–173. [CrossRef]

56.   Mueller, L.P. E.COSY, a simple alternative to E.COSY. *J. Magn. Reson.* **1987**, *72*, 191–196.

57.   Davis, A.L.; Keeler, J.; Laue, E.D.; Moskau, D. Experiments for recording pure-absorption heteronuclear correlation spectra using pulsed field gradients. *J. Magn. Reson.* **1992**, *98*, 207–216. [CrossRef]

58.   Hurd, R.E.; John, B.K. Gradient-enhanced proton-detected heteronuclear multiple-quantum coherence spectroscopy. *J. Magn. Reson.* **1991**, *91*, 648–653. [CrossRef]

59.   Wagner, R.; Berger, S. Gradient-Selected NOESY—A Fourfold Reduction of the Measurement Time for the NOESY Experiment. *J. Magn. Reson. A* **1996**, *123*, 119–121. [CrossRef] [PubMed]

60.   Bax, A.; Davis, D.G. Practical aspects of two-dimensional transverse NOE spectroscopy. *J. Magn. Reson.* **1985**, *63*, 207–213. [CrossRef]

61.   Braunschweiler, L.; Ernst, R.R. Coherence transfer by isotropic mixing: Application to proton correlation spectroscopy. *J. Magn. Reson.* **1983**, *53*, 521–528. [CrossRef]

62.   Thordarsson, P. Determining association constants from titration experiments in supramolecular chemistry. *Chem. Soc. Rev.* **2011**, *40*, 1305–1323. [CrossRef] [PubMed]

6

Reduced Reactivity of Amines against Nucleophilic Substitution via Reversible Reaction with Carbon Dioxide

Fiaz S. Mohammed and Christopher L. Kitchens *

Academic Editor: Jason P. Hallett

Department of Chemical and Biomolecular Engineering, Clemson University, Clemson, SC 29634, USA;
fiaz.mohammed@chbe.gatech.edu
* Correspondence: ckitche@clemson.edu

Abstract: The reversible reaction of carbon dioxide (CO_2) with primary amines to form alkyl-ammonium carbamates is demonstrated in this work to reduce amine reactivity against nucleophilic substitution reactions with benzophenone and phenyl isocyanate. The reversible formation of carbamates has been recently exploited for a number of unique applications including the formation of reversible ionic liquids and surfactants. For these applications, reduced reactivity of the carbamate is imperative, particularly for applications in reactions and separations. In this work, carbamate formation resulted in a 67% reduction in yield for urea synthesis and 55% reduction for imine synthesis. Furthermore, the amine reactivity can be recovered upon reversal of the carbamate reaction, demonstrating reversibility. The strong nucleophilic properties of amines often require protection/de-protection schemes during bi-functional coupling reactions. This typically requires three separate reaction steps to achieve a single transformation, which is the motivation behind Green Chemistry Principle #8: Reduce Derivatives. Based upon the reduced reactivity, there is potential to employ the reversible carbamate reaction as an alternative method for amine protection in the presence of competing reactions. For the context of this work, CO_2 is envisioned as a green protecting agent to suppress formation of n-phenyl benzophenoneimine and various n-phenyl–n-alky ureas.

Keywords: carbamate; urea; benzophenoneimine; green chemistry

1. Introduction

It has been previously reported that primary and secondary amines react with CO_2 to produce thermally reversible alkylammonium alkylcarbamates as seen in Scheme 1 [1–7]. The CO_2 reacts with one primary amine to form carbamic acid, which then further reacts with free amine to form the carbamate product. This mechanism has been extensively studied in literature with a variety of applications in different fields [8–12]. For example, the reversible amine to carbamate transition exhibits a significant change in the ionic character from non-ionic to ionic upon CO_2 reaction, which has been employed in systems referred to as reversible ionic liquids, reversible surfactants, and switchable solvents [8,13–19]. These systems have included CO_2 reaction with guanidines and amidines reaction with alcohols as two-component systems and primary or secondary amines as single component systems. In these systems, the solvent property changes can be quite appreciable, changing in character from that of hydrophilic methanol to that of hydrophobic chloroform [8,14,15]. The potential implications of such tunable solvent properties are widespread in the realm of reactions and separations, where a change in solvent properties can induce a complete phase separation and facile recovery of products, catalysts, or unreacted reagents. This was initially demonstrated by Jessop *et al.* with the reversible protonation of DBU (1,8-diazabicyclo-[5.4.0]undec-7-ene) in the presence of an alcohol and CO_2 [8]. In the unprotonated state, the DBU/alcohol mixture is considered non-polar and is completely miscible with decane. The addition of CO_2 induces DBU protonation and switch to a polar nature and phase separation from the non-polar decane phase. Specific potential applications that have entertained the use of switchable solvents or reversible ionic liquids have included oil extractions [20] CO_2 capture [17,21] styrene polymerization [15,22], Claisen-Schmidt condensation, Heck reaction [13] Michael addition [13,15] cellulose acetylation [23] and nanomaterial processing [18], to name a few. In each of these applications, amine reactivity can become an impeding issue [13].

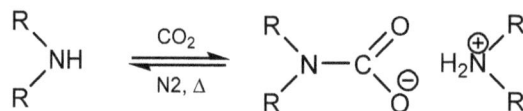

Scheme 1. Reaction scheme of reversible alkylammonium carbamate formation from coupling of carbon dioxide and a secondary amine.

In other work, reversible surfactants use the same chemistry as the switchable solvents or reversible ionic liquids to create and break emulsions or form gels with many potential applications [15,24]. Similarly, the reversible reaction of CO_2 with amines have been demonstrated during the development of ordered laminar materials comprised of amino-terminated silanes [25]. The formation of the hybrid material could not be possible if the absence of the carbamates. Other studies have used the reversible reaction to dramatically dictate the product selectivity during the intramolecular hydroaminomethylation of ethyl methallylic amine by reducing the nucleophilicity of the nitrogen atom [26]. Eckert and coworkers [27] used CO_2 expanded liquids at 30 bar to increase the yield of primary amine synthesis in the hydrogenation reactions of benzonitrile and phenylacetonitrile with $NiCl_2/NaBH_4$ in ethanol. The CO_2 induced carbamate formation prevented side reactions by precipitating the desired product from solution as a carbamate, thus simplifying the purification and increasing the yield. However, the protected carbamate was not used in any further coupling reactions. In each of these applications, CO_2 reaction is beneficial in protecting the amine from undesired product formation.

Protecting groups provide a vital role within the field of organic synthesis, enabling the coupling of various organo-functional groups in the presence of a competing group [28]. This protection also needs to be stable under a broad range of reaction conditions and both moderately and selectively cleavable. Thus, choosing the right type of protective group is of primary significance, especially in the presence of amino containing compounds that tend to be very reactive, basic in nature, and form strong hydrogen bonds. In addition to sulfonamides and amides, carbamates are the most popular and widely used protection mechanism utilized for amino groups. Such protecting groups primarily include carbobenzoxy (CBZ), di-tertiary butoxy carbonyl (Di-t-BOC) [29], and 2-(trimethylsilyl)ethylsulfonyl) (SES) [30] groups as well as other less common groups, such as borane [31]; each having their own set of advantages and disadvantages. The t-BOC and Di-t-BOC protecting groups, for example, are utilized quite frequently in literature due to the stability during catalyzed nucleophilic substitutions and catalytic hydrogenation reactions [32]. However, de-protection requires strong acids, long reaction times and has shown to generate t-butyl cations that then require scavengers to prevent undesirable side reactions [33]. Likewise, regeneration of a CBZ protected amine is achieved by acidolysis, catalytic hydrogenation, or reduction with dissolved metals. In light of these hazards, the use of milder reagents and reaction conditions has been an area of significant potential improvement [34,35], building upon the principles of Green Chemistry. Despite the recent advancements in protection/de-protection mechanisms for amines, these traditional methods are widely lacking in both material and energy efficiency, adding to the cost and/or use of corrosive reagents.

Only recently has the concept of using CO_2 as a simpler, greener approach to amine protection/de-protection arisen [36–38]. Peeters *et al.* demonstrated CO_2 as a reversible amine-protecting agent in selective Michael additions and acylations [36]. Specifically, the reaction of primary amines was inhibited in favor of normally less reactive sulfonamides, cyclic secondary amines, or β-ketoesters for the Michael addition and complete inhibition of benzylamine acylation was achieved, effectively favoring alcohol conversion without significant byproduct. Ethier *et al.* investigated the solvent effect and addition of other bases in addition to DBU on the protection of benzylamines with CO_2 [37]. In each case, protection of the amine was achieved in the presence of a competitive reaction and removal of the CO_2 protecting group was achieved; however, the reactivity of the de-protected amine was not examined.

This work further explores the use of the reversible CO_2 chemistry as a unique method of protecting amines during simple coupling reactions. All reactions were carried out with (1) the non-protected amine; (2) the carbamate analogs formed in solution via reaction with gaseous CO_2; and (3) with the de-protected amine regenerated through a thermal treatment of the carbamate. Significant differences between this work and those of Peeters *et al.* and Ethier *et al.* are the lack of a competing reaction and measurement of reactivity following de-protection. The lack of a competing reaction is significant because it measures the decreased reactivity rather than the competitive reactivity.

2. Results and Discussion

2.1. Synthesis of n-Alkyl, n-Phenyl Urea

Urea formation via amine coupling with isocyanates has been well-studied and characterized [39,40]. The urea synthesis in this work showed 95% yield from propylamine, while other amines achieved comparable yields of 85% from hexylamine, 91% from decylamine, and 90% from octadecylamine (Table 1). Once protected in the carbamate form, the reactivity was significantly reduced for the shorter chained amines with an average decrease of 62%. After the heat treatment, de-protection of the propyl, hexyl and decyl amine was successful based on the increase in urea yield to values comparable to the non-protected amine. This demonstrates the reversibility of the carbamate protection and requirement of no pressurization or separate reaction conditions. The observed lower product yields for the de-protected amines are attributed to incomplete de-protection of the carbamates along with a loss of starting material during de-protection due to the inherent volatilities of the propyl and hexylamine. TGA showed carbamate degradation temperatures higher than their amine conjugates, thus resulting in a loss of starting material during thermal amine regeneration process (Figure S6). Decylamine, with its low volatility and ease of carbamate reversal displayed the best results, with a 93% urea yield from the de-protected amine comparable to that of the pure or un-protected amine (91%). The octadecyl carbamate showed the lowest reduction in yield and the poorest recovery in the de-protected amine reactivity. This was not an expected result as a protection/de-protection cycle had been achieved separately for the octadecyl amine. A significant difference was that the octadecyl-carbamate formation resulted in a white precipitate, however the de-protection regenerated the soluble amine analog. We hypothesize that the long alkyl chains and precipitation inhibited the complete formation of the carbamate and resulted in insufficient protection.

FTIR of the isolated ureas demonstrated high purity through the absence of carbamate or other reaction byproducts for all three scenarios (non-protected, protected and de-protected). In each FTIR spectra (Figure S1), the characteristic carbonyl (C=O) and secondary amine (N-H) absorbance associated with urea formation are consistent within all isolated products while the C-H (2800–3000 cm^{-1}) absorption levels correlate very well to the alkyl chain backbone length. The three states of amine (non-protected, protected and de-protected) all produced an isolated urea product displaying similar absorbance bands (Figure S1). This indicates that the products obtained were spectrally identical and not side products due to residual carbamate presence. The urea purity was further determined by ^{1}H-NMR, displaying chemical shifts characteristic to our desired products (Figures S2 and S3). The main differences in spectra were the chemical shifts at 1.3 ppm attributed to the longer alkyl backbones of hexyl and decylamine, while integration of the peaks further evidenced its structure and purity.

Table 1. Percentage yields of the *n*-phenyl–*n*-alkyl ureas of for pure, protected and de-protected amine.

Amine	*n*	Percentage Yield (%)		
		Non-Protected	**Protected**	**De-Protected**
Propylamine	1	95	28	85
Hexylamine	4	85	20	80
Decylamine	8	91	38	93
Octadecylamine	16	90	69	72

The urea melting temperatures (T_m) were determined by DSC (Table 2). All measured values correlated well with literature values and possessed small melting temperature ranges, indicating high purity [41]. The n-propyl urea exhibited the highest melting point and crystallization temperature despite the shortest carbon chain, and highest volatility. Similarly, the propyl and hexyl carbamates showed an increase in the decomposition onset temperature due to the stronger charged carbamate interactions present. The ureas contain hydrogen bonding donors (NH_2) and hydrogen bonding acceptors (C=O). The shorter alkyl chains contribute less steric hindrance and thus, a higher degree of hydrogen bonding occurs. As the chain length increases, the alkyl chain disrupts the urea-urea hydrogen bonding and order, thus reducing the melting point, as seen with n-hexyl urea (68–69 °C). As the alkyl chain increases, an expected T_m increase from 82 °C for decylamine to 95 °C for stearyl amine is observed, which is due to increases in energy demand for the onset of melting for higher molecular weight urea.

Table 2. *n*-Phenyl, *n*-alkyl urea melting points and crystallization temperatures determine from DSC.

Entry	MP (°C)	Recrystalization Temp. (°C)
n-phenyl, *n*-propyl urea	112–114	89
n-phenyl, *n*-hexyl urea	68–70	47
n-phenyl, *n*-decyl urea	81–83	61
n-phenyl, *n*-stearyl urea	95–97	84

Figure 1 displays a typical DSC curve used to determine T_m for each urea isolated. From the image, we can further conclude urea purity consistency throughout each amine state. As with the FTIR data, the protected and de-protected samples all provide isolated products which are chemically identical.

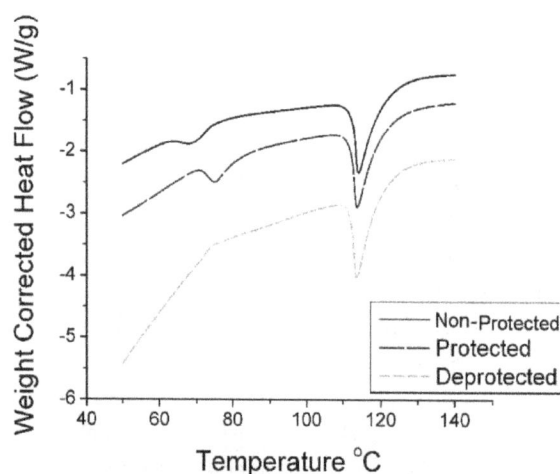

Figure 1. DSC spectra of n-propyl urea isolated from all three amine starting materials. Ramp rate = 5 °C per minute to 140 °C followed by equilibration to 60 °C and repeated heating.

2.2. Synthesis of n-Propyl, Benzophenone(BP) Imine

Titanium(IV) isopropoxide is a low cost mediator in BP imine synthesis [42–46]. According to Figure 2, the initial BP concentration of 0.45 M is slowly reduced to 0.26 M after 3 h and finally 0.11 M after 24 h as the reaction proceeds. Gas chromatography (GC) was used to determine an overall yield of 75% for the non-protected amine while a 25% yield is obtained for the protected and 50% for de-protected propylamine. Once again, a similar trend is seen with the de-protected amines falling short of the original product yield, which is attributed to the incomplete reversal of carbamates or propylamine volatilization during the de-protection process.

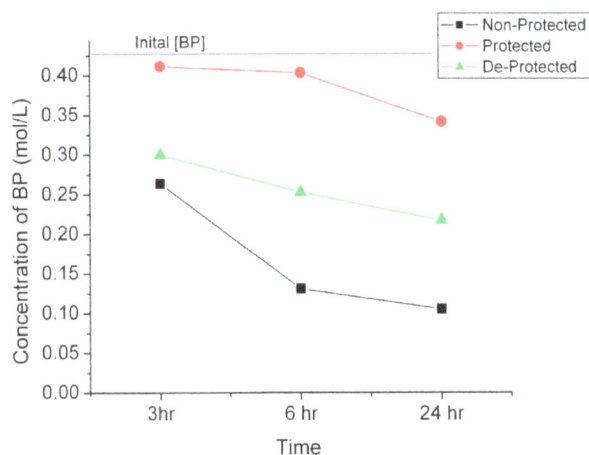

Figure 2. Plot of benzophenone (BP) concentration determined via gas chromatography (GC) for the non-protected, protected and de-protected propylamine reactions at 3, 6 and 24 h.

Although FTIR is not commonly used as a powerful quantitative tool, it was used to confirm the presence of our imine species, as well as, calculate the BP conversion. Figure 3 shows the strongly absorbing C=N shift to 1620 cm^{-1} from the C=O at 1660 cm^{-1} of the original BP [47]. This shift in absorbance demonstrates that the desired imine species was obtained during the BP-propylamine reactions and this was further evidenced via H^{1}NMR and GC-MS.

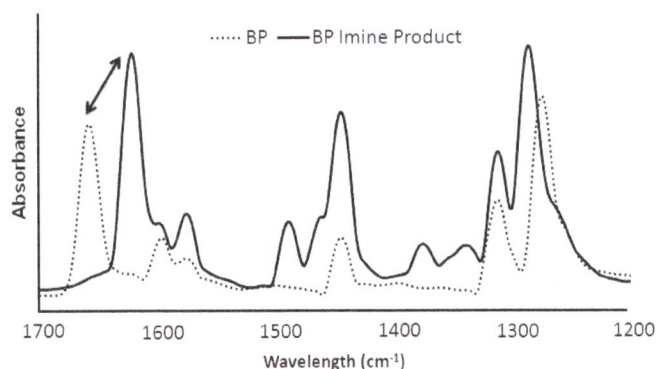

Figure 3. FTIR of pure BP (dashed) compared to isolated BP imine product (solid).

The qualitative analysis was conducted over a period of 12 h, during which time ATR-FTIR spectra of aliquots of the reaction vessels were collected and analyzed to confirm the appearance of the imine product. As the reaction proceeds forward, the concentration of imine in the reaction solution will increase until equilibrium has been reached. By analyzing the absorbance ratio of C=O to C=N, the percentage of imine in the reaction vessel was determined (Table 3). These calculated values correlate very well to the percentage conversion obtained via GC and support our initial hypothesis that CO_2 is a viable tool for the protection and inhibition of free primary amines to undergo coupling reactions.

Table 3. Determined BP conversion from GC analysis compared to BP conversion calculated from IR spectral C=N/C=O absorbance ratio.

Sample	Time (h)	Calculated IR Conversion (%)	Conversion from GC (%)
Non-protected	3	35	33
Non-protected	12	55	65
Protected	3	16	18
Protected	12	26	20 *

* denotes a value that was estimated using Figure S7.

In a separate set of experiments, imine formation was observed to take place without the presence of mechanical stirring (Figure S7). Here, the benzophenone, Ti catalyst and propylamine were combined in 1.5 mL GC vials and monitored *in-situ* with the absence of any a titanium complex intermediates. The percentage conversion correlated very well despite running the reaction in separate vials with maximum conversions identical to those obtained above in Figure 2. The non-protected reaction reaches a 75% equilibrium conversion after 15 h, while the protected propylamine yields a 25% equilibrium conversion after 36 h. The de-protected sample obtains an equilibrium conversion of 50% after 5 h, a phenomenon that has been consistent throughout this work. Confirmed by FTIR, a distinct reduction in CH_2 and CH_3 absorption provides evidence to support the loss of starting amine concentration during de-protection. An alternative explanation for the reduction in imine and urea yield using the de-protected amines is the formation of isocyanate side products during the thermal treatment [48]; however, this theory is not strongly supported.

In the presence of methanol, CO_2 will preferentially react with the alcohol to form an alkylcarbonic acid that then reacts with the amine to form methyl carbamates, as opposed to the alkylammonium carbamates obtained under aprotic conditions [49]. The resulting carbonic structure has a larger energy requirement to cleave the methyl group and release the carbon dioxide. It is hypothesized that the reduced BP conversion for the de-protected reactions could be due to the increased difficulty to reverse the protection as the protic methanol solvent interacts with the carbamate.

Lastly, an explanation into the formation of ureas and benzophenoneimine in spite of CO_2 induced carbamate protection needs to be discussed. For example, the 25% conversion of BP to the imine does not represent a complete protection of the propyl amine. According to Salmi *et al.* [45], the reductive amination of BP proceeds through an imine species with no observable titanium complex intermediates. Hence, the complexation of BP with the Ti is not a cause for the observed reduction in BP concentration, and this conversion is attributed to reactivity of the amine with the BP. Furthermore, control experiments in the absence of propylamine showed negligible loss in BP concentration, further supporting a residual amine reactivity under CO_2 *vs.* complexing of BP to Ti as the source of BP conversion. We conclude that the conversion of starting material under protected conditions is primarily due to the equilibrium which exists between the carbamates and the amine (Scheme 1). As the amine is converted to urea/imine, the equilibrium can shift towards the reactants, resulting in more conversion of the starting material. This theory explains the formation of the urea/imine under the carbamate protection mechanism and an investigation into the kinetics of the carbamate reactivity can also provide useful information on the degree of protection. By fully understanding the relationship that exists between carbamates and the amine-CO_2 system, we can calculate expected yields of the product based on the equilibrium constant of the carbamate formation.

3. Materials and Methods

3.1. General Amine Protection

The required quantity of amine was added to a round-bottom flask before the addition of any other component. A CO_2 atmosphere was then introduced into the closed vessel via a CO_2 gas cylinder equipped with a low pressure regulator for approximately 5 min resulting in the exothermic formation of solid white powders. The reaction solvent was then added with additional CO_2 being bubbled through the solution for a further 3 min. The original volume of the solution was made up with pure solvent after the CO_2 bubbling in order to preserve concentrations. The reactions with protected amines were carried out under a CO_2 atmosphere.

3.2. General Amine De-Protection

Carbamate solutions of the alkylamines were de-protected using indirect heat and thermostat control with stirring. An ice-bath cooled condenser was attached to the top of the reaction vessel while a steady stream of nitrogen was introduced to aid the reversal. This was carried out for 10 min, after which fresh solvent was added to maintain the reaction concentration after nitrogen saturation. This procedure was used in attempt to convert carbamates back to the original amines. The de-protected amines were then employed in the urea/imine synthesis to ensure complete de-protection and no loss of reactivity.

3.3. General Characterization

An aliquot of the reaction mixture (1.5 mL) was taken from each reaction vessel and injected into an Agilent 7695A GC using the Agilent 7683B automatic liquid sampler (Agilent Technologies, Santa Clara, CA, USA). The inlet temperature was set at 300 °C with a pressure of 16 psi, the oven at 80 °C with a heating rate of 20 °C/min to 250 °C. The FID was set at 300 °C. A calibration curve for benzophenone (BP) was created by plotting the integral of the BP peak *vs.* known concentrations of BP (R_t = 14.5 min). Melting points were determined from DSC spectra obtained using a Thermal Analysis SDT Q600 (TA Instruments, New Castle, DE, USA) equipped with alumina pans. A steady heating rate of 5 °C/min was maintained to 350 °C with a nitrogen purge of 100 mL/min. IR spectra were collected on a Nexus 870 spectrometer (Nicolet Instrument Corporation, Madison, WI, USA) with a $4\ cm^{-1}$ resolution using 64 scans. A fixed angle single reflection 60° hemispherical Ge crystal plate, equipped with an ATR pressure clamp, was placed in a sample compartment. The output signal was collected using a deuterated triglycine sulfate (DTGS) room temperature detector. ^{1}H-NMR spectra were obtained on a Bruker 300 MHz in $CDCl_3$ (Bruker Biospin Corporation, Billerica, MA, USA).

3.4. Synthesis of n-Alkyl, n-Phenyl Urea

Exactly 50 µL (0.0545 g, 0.46 mmol) of phenyl isocyanate was added dropwise to a solution of alkylamine (0.92 mmol) in 2 mL of dry $CHCl_3$ at 0 °C. The resulting solution was stirred for 60 min at room temperature and then precipitated into 25 mL of pentane. The product was isolated as a white powder via filtration and washed with several portions of pentane before being dried under vacuum at 50 °C. The products isolated from the alkylcarbamates were alternatively re-dispersed in dry $CHCl_3$ and underwent thermal treatment to remove any residual carbamates. The ureas were then re-crystallized in pentane again and isolated via filtration to determine yield (Scheme 2a).

3.5. Synthesis of n-Propyl, Benzophenone(BP) Imine

Propylamine (5 mmol) was added to 5 mL of dry methanol, followed by 2.5 mmol benzophenone (BP) and 3.3 mmol titanium(IV)-isopropoxide. The solution was stirred under nitrogen at room temperature while aliquots were taken at 3, 6 and 24 h for GC analysis and ATR-FTIR monitoring.

Alternatively, the reaction was separated into 4 GC vials and sampled *in-situ* without stirring. Each GC vial represented a reaction vessel that was used to monitor the reaction kinetics (Scheme 2b).

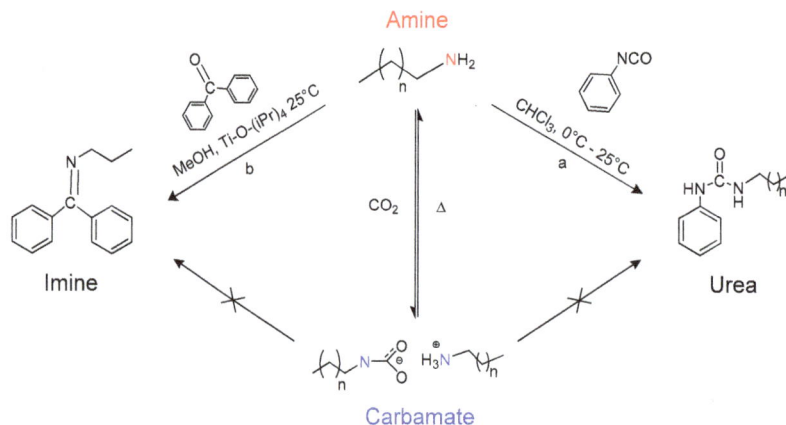

Scheme 2. Reaction scheme showing the pathways for *n*-phenyl, *n*-alkyl urea synthesis in CHCl$_3$ (**a**); and benzophenoneimine synthesis in methanol (**b**).

4. Conclusions

In conclusion, we have shown that the reaction of CO$_2$ with alkyl amines reduced the amine reactivity resulting in reduced urea and imine yields for multiple primary amines. Performing the reaction in the absence of a competing reaction demonstrates a true decrease in reactivity, rather than preferentiality. Reversible carbamate formation via gaseous CO$_2$ alone shows potential as a simple, "green" alternative compared to traditional methods that require harsh chemicals and more energy intensive reaction conditions; adhering to the major strategies of green chemistry which promote reduced solvent and energy use, inherently safer reagents, and reduced number of transformations.

Despite the novelty, the temperature dependence of the reverse reaction does introduce limitations. The ability to efficiently protect and de-protect the amine and the solubility of the carbamates in the reaction media are also potential limitations. In future work, this protection mechanism will be implemented in a competitive reaction where amine functionality is required post protection for a separate coupling reaction. This will test the limits and efficiency of the proposed protection/de-protection technique. Also, this technique will be applied to high pressure systems in an attempt to dictate and shift the carbamate equilibrium to favor the protection. The energy input to maintain higher pressures may increase, but it is offset by the reduction in solvents for conducting the three separate steps in traditional protection methods.

We anticipate that the results from this work will encourage chemists to not only design greener routes for amine protection/de-protection mechanisms, but also to learn about the many advantages that carbamate chemistry has to contribute to safer lab and industrial practices.

Acknowledgments: This work was funded by the American Chemical Society. Authors would like to acknowledge Paul Hernley for his experimental contributions to the study.

Author Contributions: F.M. and C.K. conceived and designed the experiments; F.M. performed the experiments, analyzed the data and drafted the paper. C.K. oversaw the entire research study and coordinated the redaction of the manuscript.

Conflicts of Interest: The authors declare no conflict of interest.

References

1. Jensen, A.; Faurholt, C. Studies on carbamates. V. The carbamates of alpha-alanine and beta-alanine. *Acta Chem. Scand.* **1952**, *6*, 385–394. [CrossRef]

2. Jensen, A.; Jensen, B.J.; Faurholt, C. Studies on carbamates. Vi. The carbamate of glycine. *Acta Chem. Scand.* **1952**, *6*, 395–397. [CrossRef]

3. Olsen, J.; Vejlby, K.; Faurholt, C. Studies on carbamates. Vii. The carbamates of n-propylamine and iso-propylamine. *Acta Chem. Scand.* **1952**, *6*, 398–403. [CrossRef]

4. Werner, E.A. Cxvii.-the constitution of carbamides. Part xi. The mechanism of the synthesis of urea from ammonium carbamate. The preparation of certain mixed tri-substituted carbamates and dithiocarbamates. *J. Chem. Soc. Trans.* **1920**, *117*, 1046–1053.

5. Fichter, F.; Becker, B. Über die bildung ysmmetrisch dialkylierter harnstoffe durch erhitzen der entsprechenden carbaminate. *Ber. Dtsch. Chem. Ges.* **1911**, *44*, 3481–3485. [CrossRef]

6. Murphy, L.J.; Robertson, K.N.; Kemp, R.A.; Tuononen, H.M.; Clyburne, J.A.C. Structurally simple complexes of CO_2. *Chem. Commun.* **2015**, *51*, 3942–3956. [CrossRef] [PubMed]

7. Dong, D.Y.; Yang, L.P.; Hu, W.H. Organic reactions with carbon dioxide. *Prog. Chem.* **2009**, *21*, 1217–1228.

8. Jessop, P.G.; Heldebrant, D.J.; Li, X.; Eckertt, C.A.; Liotta, C.L. Green chemistry: Reversible nonpolar-to-polar solvent. *Nature* **2005**, *436*, 1102–1102. [CrossRef] [PubMed]

9. George, M.; Weiss, R.G. Chemically reversible organogels via "latent" gelators. Aliphatic amines with carbon dioxide and their ammonium carbamates. *Langmuir* **2002**, *18*, 7124–7135. [CrossRef]

10. Xu, H.; Rudkevich, D.M. CO_2 in supramolecular chemistry: Preparation of switchable supramolecular polymers. *Chem. Eur. J.* **2004**, *10*, 5432–5442. [CrossRef] [PubMed]

11. George, M.; Weiss, R.G. Primary alkyl amines as latent gelators and their organogel adducts with neutral triatomic molecules. *Langmuir* **2003**, *19*, 1017–1025. [CrossRef]

12. Mohammed, F.S.; Wuttigul, S.; Kitchens, C.L. Dynamic surface properties of amino-terminated self-assembled monolayers incorporating reversible CO2 chemistry. *Ind. Eng. Chem. Res.* **2011**, *50*, 8034–8041. [CrossRef]

13. Blasucci, V.M.; Hart, R.; Pollet, P.; Liotta, C.L.; Eckert, C.A. Reversible ionic liquids designed for facile separations. *Fluid Phase Equilib.* **2010**, *294*, 1–6. [CrossRef]

14. Pollet, P.; Eckert, C.A.; Liotta, C.L. Switchable solvents. *Chem. Sci.* **2011**, *2*, 609–614. [CrossRef]

15. Jessop, P.G.; Mercer, S.M.; Heldebrant, D.J. CO2-triggered switchable solvents, surfactants, and other materials. *Energy Environ. Sci.* **2012**, *5*, 7240–7253. [CrossRef]

16. Kerton, F. Tunable and switchable solvent systems. In *Alternative Solvents for Green Chemistry*, 2nd ed.; RSC Publishing: London, UK, 2013; pp. 262–284.

17. Switzer, J.R.; Ethier, A.L.; Flack, K.M.; Biddinger, E.J.; Gelbaum, L.; Pollet, P.; Eckert, C.A.; Liotta, C.L. Reversible ionic liquid stabilized carbamic acids: A pathway toward enhanced CO2 capture. *Ind. Eng. Chem. Res.* **2013**, *52*, 13159–13163. [CrossRef]

18. Pollet, P.; Davey, E.A.; Urena-Benavides, E.E.; Eckert, C.A.; Liotta, C.L. Solvents for sustainable chemical processes. *Green Chem.* **2014**, *16*, 1034–1055. [CrossRef]

19. Jessop, P.G. Switchable solvents as media for synthesis and separations. *Aldrichimica Acta* **2015**, *48*, 18–21.

20. Blasucci, V.; Hart, R.; Mestre, V.L.; Hahne, D.J.; Burlager, M.; Huttenhower, H.; Thio, B.J.R.; Pollet, P.; Liotta, C.L.; Eckert, C.A. Single component, reversible ionic liquids for energy applications. *Fuel* **2010**, *89*, 1315–1319. [CrossRef]

21. Shannon, M.S.; Bara, J.E. Reactive and reversible ionic liquids for CO_2 capture and acid gas removal. *Sep. Sci. Technol.* **2012**, *47*, 178–188. [CrossRef]

22. Phan, L.; Chiu, D.; Heldebrant, D.J.; Huttenhower, H.; John, E.; Li, X.W.; Pollet, P.; Wang, R.Y.; Eckert, C.A.; Liotta, C.L.; *et al.* Switchable solvents consisting of amidine/alcohol or guanidine/alcohol mixtures. *Ind. Eng. Chem. Res.* **2008**, *47*, 539–545. [CrossRef]

23. Yang, Y.L.; Xie, H.B.; Liu, E.H. Acylation of cellulose in reversible ionic liquids. *Green Chem.* **2014**, *16*, 3018–3023. [CrossRef]

24. Liu, Y.X.; Jessop, P.G.; Cunningham, M.; Eckert, C.A.; Liotta, C.L. Switchable surfactants. *Science* **2006**, *313*, 958–960. [CrossRef] [PubMed]

25. Alauzun, J.; Besson, E.; Mehdi, A.; Reye, C.; Corriu, R.J.P. Reversible covalent chemistry of CO_2: An opportunity for nano-structured hybrid organic-inorganic materials. *Chem. Mater.* **2008**, *20*, 503–513. [CrossRef]

26. Wittmann, K.; Wisniewski, W.; Mynott, R.; Leitner, W.; Kranemann, C.L.; Rische, T.; Eilbracht, P.; Kluwer, S.; Ernsting, J.M.; Elsevier, C.L. Supercritical carbon dioxide as solvent and temporary protecting group for rhodium-catalyzed hydroaminomethylation. *Chem. Eur. J.* **2001**, *7*, 4584–4589. [CrossRef]

27. Xie, X.; Liotta, C.L.; Eckert, C.A. CO_2-protected amine formation from nitrile and imine hydrogenation in gas-expanded liquids. *Ind. Eng. Chem. Res.* **2004**, *43*, 7907–7911. [CrossRef]

28. Tom, N.J.; Simon, W.M.; Frost, H.N.; Ewing, M. Deprotection of a primary boc group under basic conditions. *Tetrahedron Lett.* **2004**, *45*, 905–906. [CrossRef]

29. Lutz, C.; Lutz, V.; Knochel, P. Enantioselective synthesis of 1,2-, 1,3- and 1,4- aminoalcohols by the addition of dialkylzincs to 1,2-, 1,3- and 1,4- aminoaldehydes. *Tetrahedron* **1998**, *54*, 6385–6402. [CrossRef]

30. Ribiere, P.; Declerck, V.; Martinez, J.; Lamaty, F. 2-(trimethylsilyl)ethanesulfonyl (or ses) group in amine protection and activation. *Chem. Rev.* **2006**, *106*, 2249–2269. [CrossRef] [PubMed]

31. Zajac, M.A. An application of borane as a protecting group for pyridine. *J. Org. Chem.* **2008**, *73*, 6899–6901. [CrossRef] [PubMed]

32. Chankeshwara, S.V.; Chakraborti, A.K. Catalyst-free chemoselective n-tert-butyloxycarbonylation of amines in water. *Org. Lett.* **2006**, *8*, 3259–3262. [CrossRef] [PubMed]

33. Agami, C.; Couty, F. The reactivity of the n-boc protecting group: An underrated feature. *Tetrahedron* **2002**, *58*, 2701–2724. [CrossRef]

34. Heydari, A.; Khaksar, S.; Tajbakhsh, M. 1,1,1,3,3,3-hexafluoroisopropanol: A recyclable organocatalyst for n-boc protection of amines. *Synthesis* **2008**, *2008*, 3126–3130. [CrossRef]

35. Perron, V.R.; Abbott, S.; Moreau, N.; Lee, D.; Penney, C.; Zacharie, B. A method for the selective protection of aromatic amines in the presence of aliphatic amines. *Synthesis* **2009**, *2009*, 283–289.

36. Peeters, A.; Ameloot, R.; de Vos, D.E. Carbon dioxide as a reversible amine-protecting agent in selective michael additions and acylations. *Green Chem.* **2013**, *15*, 1550–1557. [CrossRef]

37. Ethier, A.; Switzer, J.; Rumple, A.; Medina-Ramos, W.; Li, Z.; Fisk, J.; Holden, B.; Gelbaum, L.; Pollet, P.; Eckert, C.; *et al.* The effects of solvent and added bases on the protection of benzylamines with carbon dioxide. *Processes* **2015**, *3*, 497–513. [CrossRef]

38. Bathini, T.; Rawat, V.S.; Bojja, S. In situ protection and deprotection of amines for iron catalyzed oxidative amidation of aldehydes. *Tetrahedron Lett.* **2015**, *56*, 5656–5660. [CrossRef]

39. Perveen, S.; Khan, K.M.; Lodhi, M.A.; Choudhary, M.I.; Voelter, W. Urease and alpha-chymotrypsin inhibitory effects of selected urea derivatives. *Lett. Drug Des. Discov.* **2008**, *5*, 401–405.

40. Zhang, W.; Sita, L.R. Investigation of dynamic intra- and intermolecular processes within a tether-length dependent series of group 4 bimetallic initiators for stereomodulated degenerative transfer living ziegler-natta propene polymerization. *Adv. Synth. Catal.* **2008**, *350*, 439–447. [CrossRef]

41. Izdebski, J.; Pawlak, D. A new convenient method for the synthesis of symmetrical and unsymmetrical n,n'-disubstituted ureas. *Synthesis* **2002**, *1989*, 423–425. [CrossRef]

42. Abdel-Magid, A.; Carson, K.G.; Harris, B.D.; Maryanoff, C.A.; Shah, R.D. Reductive amination of aldehydes and ketones with sodium triacetoxyborohydride. Studies on direct and indirect reductive amination procedures 1. *J. Org. Chem.* **1996**, *61*, 3849–3862. [CrossRef] [PubMed]

43. Kumpaty, H.J.; Bhattacharyya, S.; Rehr, E.W.; Gonzalez, A.M. Selective access to secondary amines by a highly controlled reductive mono-n-alkylation of primary amines. *Synthesis* **2003**, 2206–2210. [CrossRef]

44. Salmi, C.; Letourneux, Y.; Brunel, J.M. Efficient diastereoselective titanium(iv) reductive amination of ketones. *Lett. Org. Chem.* **2006**, *3*, 384–389. [CrossRef]

45. Salmi, C.; Letourneux, Y.; Brunel, J.M. Efficient synthesis of various secondary amines through a titanium(iv) isopropoxide-mediated reductive amination of ketones. *Lett. Org. Chem.* **2006**, *3*, 396–401. [CrossRef]

46. Salmi, C.; Loncle, C.; Letourneux, Y.; Brunel, J.M. Efficient preparation of secondary aminoalcohols through a Ti(iv) reductive amination procedure. Application to the synthesis and antibacterial evaluation of new 3 beta-n-[hydroxyalkyl]aminosteroid derivatives. *Tetrahedron* **2008**, *64*, 4453–4459. [CrossRef]

47. Moretti, I.; Torre, G. A convenient method for the preparation of *n*-alkyl benzophenone imines. *Synthesis* **1970**, *1970*, 141. [CrossRef]

48. Valli, V.L.K.; Alper, H. A simple, convenient, and efficient method for the synthesis of isocyanates from urethanes. *J. Org. Chem.* **1995**, *60*, 257–258. [CrossRef]

49. Balaraman, E.; Gunanathan, C.; Zhang, J.; Shimon, L.J.W.; Milstein, D. Efficient hydrogenation of organic carbonates, carbamates and formates indicates alternative routes to methanol based on CO_2 and CO. *Nat. Chem.* **2011**, *3*, 609–614. [CrossRef] [PubMed]

7

Flavones Isolated from *Scutellariae radix* Suppress *Propionibacterium Acnes*-Induced Cytokine Production *In Vitro* and *In Vivo*

Po-Jung Tsai [1], Wen-Cheng Huang [1], Ming-Chi Hsieh [1], Ping-Jyun Sung [2,3], Yueh-Hsiung Kuo [4,5,*] and Wen-Huey Wu [1,*]

Academic Editor: Derek J. McPhee

[1]   Department of Human Development and Family Studies, National Taiwan Normal University, Taipei 106, Taiwan; pjtsai@ntnu.edu.tw (P.-J.T.); tim810481@yahoo.com.tw (W.-C.H.); lillianhsieh11@gmail.com (M.-C.H.)
[2]   National Museum of Marine Biology and Aquarium, Pingtung 944, Taiwan; pjsung@nmmba.gov.tw
[3]   Graduate Institute of Marine Biology, National Dong Hwa University, Pingtung 944, Taiwan
[4]   Department of Chinese Pharmaceutical Sciences and Chinese Medicine Resources, China Medical University, Taichung 404, Taiwan
[5]   Department of Biotechnology, Asia University, Taichung 413, Taiwan
*   Correspondence: kuoyh@mail.cmu.edu.tw (Y.-H.K.); t10005@ntnu.edu.tw (W.-H.W.)

Abstract: *Scutellariae radix*, the root of *Scutellaria baicalensis*, has long been applied in traditional formulations and modern herbal medications. *Propionibacterium acnes* (*P. acnes*) in follicles can trigger inflammation and lead to the symptom of inflammatory acnes vulgaris. This study was aimed at evaluating the effect of *Scutellariae radix* extract and purified components isolated from it on inflammation induced by *P. acnes in vitro* and *in vivo*. The results showed the ethyl acetate (EA) soluble fraction from the partition of crude ethanolic extract from *Scutellariae radix* inhibited *P. acnes*-induced interleukin IL-8 and IL-1β production in human monocytic THP-1 cells. Seven flavones were isolated from the EA fraction by repeated chromatographies, and identified as 5,7-dihydroxy-6-methoxyflavone (**FL1**, oroxylin), 5,7-dihydroxy-8-methoxyflavone (**FL2**, wogonin), 5-hydroxy-7,8-dimethoxyflavone (**FL3**, 7-O-methylwogonin), 5,6′-dihydroxy-6,7,8,2′-tetramethoxy flavone (**FL4**, skullcapflavone II), 5,7,4′-trihydroxy-8-methoxyflavone (**FL5**), 5,2′,6′-trihydroxy-7,8-dimethoxyflavone (**FL6**, viscidulin II), and 5,7,2′,5′-tetrahydroxy-8,6′-dimethoxyflavone (**FL7**, ganhuangenin). They all significantly suppressed *P. acnes*-induced IL-8 and IL-1β production in THP-1 cells, and **FL2** exerted the strongest effect with half maximal inhibition (IC$_{50}$) values of 8.7 and 4.9 μM, respectively. Concomitant intradermal injection of each of the seven flavones (20 μg) with *P. acnes* effectively attenuated *P. acnes*-induced ear swelling, and decreased the production of IL-6 and tumor necrosis factor-α in ear homogenates. Our results suggested that all the seven flavones can be potential therapeutic agents against *P. acnes*-induced skin inflammation.

Keywords: Chinese herb; *Scutellariae radix*; flavone; anti-inflammation; *Propionibacterium acnes*

1. Introduction

Acne vulgaris is one of the most common skin diseases. The pathogenesis is complex and incompletely understood, but inflammation is believed to be a key component [1]. *Propionibacterium acnes* (*P. acnes*), a Gram-positive anaerobic bacterium species, may play a major role in the initiation of the inflammatory reaction by stimulating the secretion of interleukin (IL)-18, tumor necrosis factor TNFα, IL-8, and IL-12 by monocytic cells, and eventually the development of inflammatory lesions [2,3]. IL-8 is the major inflammatory mediator and a strong chemotactic factor for neutrophils, basophils, and T cells. IL-8 has been implicated in mounting an inflammatory response in acne lesions [4]. In addition, the high levels of IL-1β were observed in human acne lesion, in mouse skin lesion induced by *P. acnes,* and in *P. acnes*-exposed human monocytes [5].

Scutellariae radix, the root of *Scutellaria baicalensis,* has been used as traditional Chinese medicine to treat allergic and inflammatory diseases in Japan and China. It is also often used to treat cardiovascular diseases, respiratory, and gastrointestinal infections [6]. Flavonoids, including baicalin, baicalein, wogonin, and oroxylin-A, have been identified in *Scutellariae radix* [7]. The extracts and the isolated compounds of *Scutellariae radix* are pharmacologically-active and show great potential in the treatment of inflammation, cancers, and virus-related diseases [8].

However, the bioactive components of *Scutellariae radix* have not yet been completely investigated. This study is aimed at exploring the suppressive effects of seven flavones isolated from *Scutellaria radix* on *P. acnes*-induced inflammation *in vitro* and *in vivo,* and their anti-acne potential.

2. Results

2.1. Effects of Scutellariae Radix Extracts on P. acnes-Induced IL-1β and IL-8 Production in THP-Cells

Ethylacetate (EA) fraction was cytotoxic to THP-1 cells when the concentrations applied were higher than 20 μg/mL (Figure 1A). So, the doses of 2.5, 5, and 10 μg/mL were used for the subsequent *in vitro* experiments. Treatment of THP-1 cells with heat-killed *P. acnes* evoked the production of IL-1β and IL-8. EA fraction of *Scutellariae radix* significantly decreased *P. acnes*-induced IL-1β and IL-8 production in a dose-dependent manner (Figure 1B,C).

The butanol fraction also inhibited IL-1β and IL-8 production by *P. acnes*-stimulated THP-1 cells, but less effectively than the EA fraction did. The butanol fraction had half maximal inhibition (IC$_{50}$) values of 80.5 and 135.5 μg/mL for the inhibition of IL-1β and IL-8 secretion, respectively, while the EA fraction had the respective IC$_{50}$ values of 6.6 and 5.6 μg/mL. Therefore, EA fraction was subjected to chromatography on silica gel and further purification by semi-preparative HPLC. Seven known flavones (FL1-7) were found.

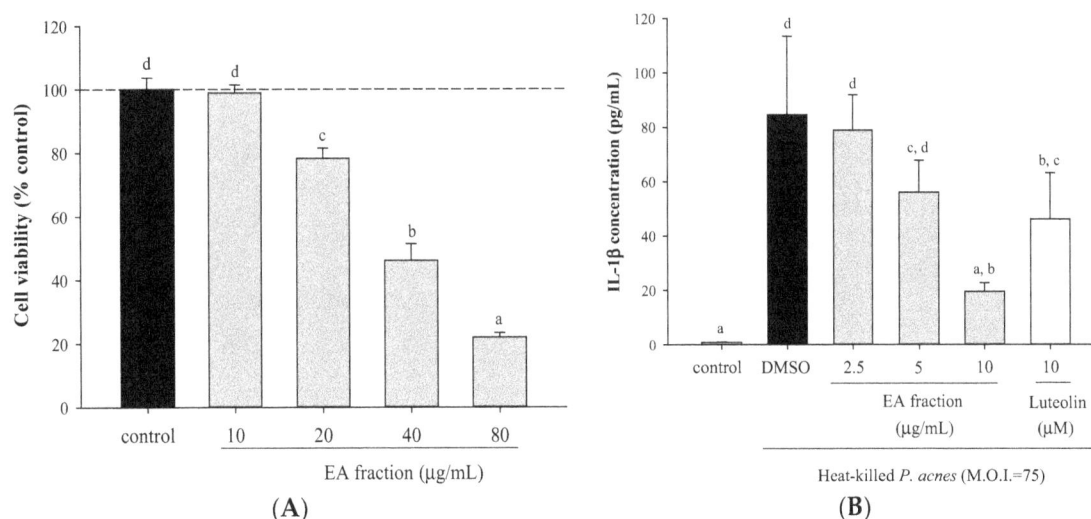

Figure 1. *Cont.*

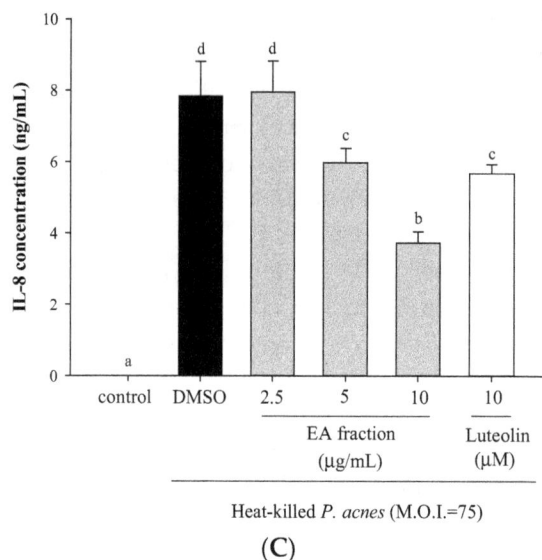

(C)

Figure 1. Effect of ethyl acetate (EA) fraction of ethanolic extract from *Scutellariae radix* on viability of monocytic THP-1 cells, and pro-inflammatory cytokine productions by *P. acnes*-stimulated THP-1 cells. Cell viability (**A**) was determined by MTT assay in cells incubated with vehicle control alone, or the indicated concentrations of EA fraction for 24 h. IL-1β (**B**) and IL-8 (**C**) were determined in cells co-incubated with *P. acnes* (M.O.I. = 75) and the indicated concentrations of samples for 24 h. DMSO (0.1%) was a vehicle control, luteolin was a reference control. A control experiment without *P. acnes* treatment was conducted in parallel. Each column shows the mean ± SD. Values with the same letter are not significantly different as determined by Duncan's multiple range tests.

2.2. Effects of the Seven Flavones Isolated from EA Fraction of Scutellariae Radix on P. acnes-Induced IL-8 and IL-1β Production in Human Monocytic THP-1 Cells

Chemical names, common names, and chemical structures of the seven known flavones isolated from EA fraction of *Scutellariae radix* are shown in Figure 2.

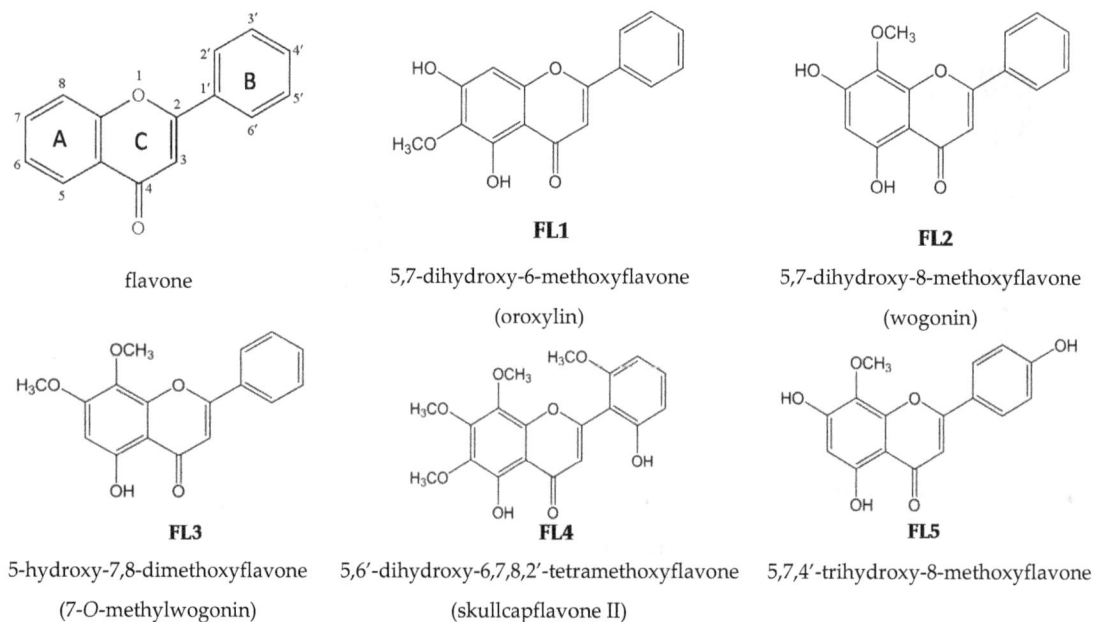

flavone

FL1

5,7-dihydroxy-6-methoxyflavone

(oroxylin)

FL2

5,7-dihydroxy-8-methoxyflavone

(wogonin)

FL3

5-hydroxy-7,8-dimethoxyflavone

(7-O-methylwogonin)

FL4

5,6'-dihydroxy-6,7,8,2'-tetramethoxyflavone

(skullcapflavone II)

FL5

5,7,4'-trihydroxy-8-methoxyflavone

Figure 2. *Cont.*

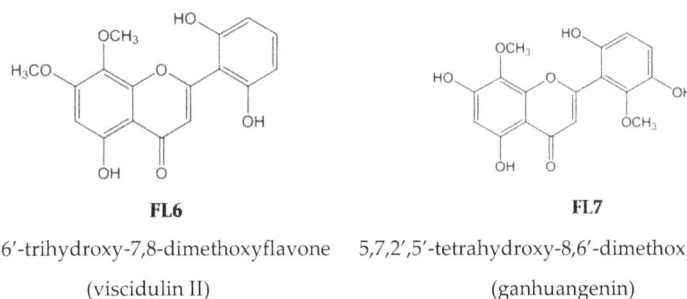

FL6   **FL7**

5,2',6'-trihydroxy-7,8-dimethoxyflavone   5,7,2',5'-tetrahydroxy-8,6'-dimethoxyflavone

(viscidulin II)   (ganhuangenin)

Figure 2. Structures of flavones isolated from *Scutellariae radix*.

The concentrations without apparent cytotoxicity toward THP-1 cells, assayed by MTT, were used for the subsequent experiments (Table 1). Seven flavones, at various concentrations, significantly reduced IL-8 and IL-1β levels (Table 1). The potency of the flavones was expressed as IC_{50} value. IC_{50} values of **FL4** for IL-8, and **FL1** and **FL6** for IL1β could not be precisely determined because the degree of inhibition provided by the highest test concentration was less than 50%. The rank order of potency of these flavones for IL-8 inhibition was **FL2** > **FL5** > **FL1** > (**FL4**) **FL6** > **FL3** > **FL7**; whereas for IL-1β inhibition was **FL2** > **FL4** > **FL5** > (**FL1**) > **FL3** > (**FL6**) **FL7**. Therefore, **FL2** had the most potent inhibitory effect on *P. acnes*-induced IL-1β and IL-8 production *in vitro*.

Table 1. Effects of flavones isolated from *Scutellariae radix* on the cell viability and *P. acnes*-induced cytokine production of THP-1 cells.

Compound	Flavone (μM)	Cell Viability (% of Control)	IL-8 Level (ng/mL)	IC_{50} for IL-8 (μM)	IL-1β Level (ng/mL)	IC_{50} for IL-1β (μM)
Control	0	93.2 ± 4.5	0.2 ± 0.2 **		0.008 ± 0.007 **	
DMSO	0	100.0 ± 2.5	52.4 ± 5.1		2.4 ± 0.2	
FL1	5	101.8 ± 7.8	39.2 ± 3.8 *		1.6 ± 0.2 **	
	10	93.3 ± 4.0	29.3 ± 5.4 **	13.1	1.5 ± 0.2 **	NA (>15)
	15	97.2 ± 3.0	24.1 ± 2.5 **		1.2 ± 0.1 **	
	30	64.5 ± 3.0	ND		ND	
FL2	5	98.6 ± 12.1	55.8 ± 8.0		1.2 ± 0.1 **	
	10	112.0 ± 13.0	18.9 ± 3.5 **	8.7	0.6 ± 0.1 **	4.9
	15	94.7 ± 2.1	5.9 ± 1.3 **		0.3 ± 0.1 **	
	30	88.4 ± 4.0	ND		ND	
FL3	30	99.9 ± 7.1	39.1 ± 12.0 *		1.4 ± 0.1 **	
	60	101.0 ± 7.8	24.7 ± 7.4 **	55.2	1.4 ± 0.1 **	72.8
	120	103.6 ± 4.5	15.8 ± 3.0 **		1.0 ± 0.1 **	
	240	65.6 ± 2.4	ND		ND	
FL4	5	102.8 ± 2.4	56.4 ± 1.9		1.6 ± 0.1 **	
	10	98.8 ± 3.9	39.3 ± 1.6 *	NA (>15)	1.1 ± 0.1 **	9.1
	15	99.6 ± 0.5	29.6 ± 2.2 **		1.0 ± 0.1 **	
	30	87.1 ± 1.8	ND		ND	
FL5	5	103.6 ± 2.0	67.7 ± 10.1 **		2.0 ± 0.1 *	
	10	100.1 ± 2.0	32.4 ± 2.5 **	10.2	1.3 ± 0.2 **	11.3
	15	100.8 ± 2.5	13.5 ± 0.6 **		0.8 ± 0.1 **	
	30	78.2 ± 2.8	ND		ND	
FL6	15	104.3 ± 5.8	41.5 ± 8.2 *		2.0 ± 0.2 **	
	30	101.3 ± 4.6	23.8 ± 1.5 **	26.1	1.6 ± 0.1 **	NA (>60)
	60	100.9 ± 4.2	20.3 ± 3.4 **		1.4 ± 0.1 **	
	120	87.5 ± 3.1	ND		ND	
FL7	60	106.7 ± 6.0	33.5 ± 2.3 **		1.7 ± 0.1 **	
	90	110.9 ± 1.1	28.6 ± 3.0 **	124.3	1.0 ± 0.1 **	84.3
	120	120.7 ± 1.2	26.3 ± 1.2 **		0.8 ± 0.1 **	
	240	89.3 ± 3.2	ND		ND	

IC_{50}, concentrations that provide 50% inhibition; ND, not-determined; NA, not applicable. Control, cells cultured with DMSO (0.1%) alone for 24 h. DMSO, cells cultured with DMSO (as vehicle) and *P. acnes* (M.O.I. = 75) for 24 h. FL, cells cultured with the indicated concentrations of flavones and *P. acnes* (M.O.I. = 75) for 24 h. Cytokine levels were measured using ELISA kits. * $p < 0.05$, ** $p < 0.001$, as compared with DMSO.

2.3. Effects of the Seven Flavones Isolated from Scutellariae Radix on P. acnes-Induced Ear Edema and Cytokine Production in Mouse Ear Homogenates

Mouse ear edema was induced by intradermal injection of *P. acnes* to mouse ears. Concomitant injection of each of the seven flavones with *P. acnes* afforded suppression of *P. acnes*-induced edema as measured by ear thickness, and all the seven flavones had almost equal effects (Figure 3A). All flavones also significantly inhibited IL-6 (Figure 3B) and TNF-α (Figure 3C) production in *P. acnes*-treated mouse ears. For the suppression of IL-1β, except **FL1** and **FL2**, all the other five flavones had significant effect (Figure 3D). Our data indicated that these flavones had protective effects against *P. acnes*-induced skin inflammation.

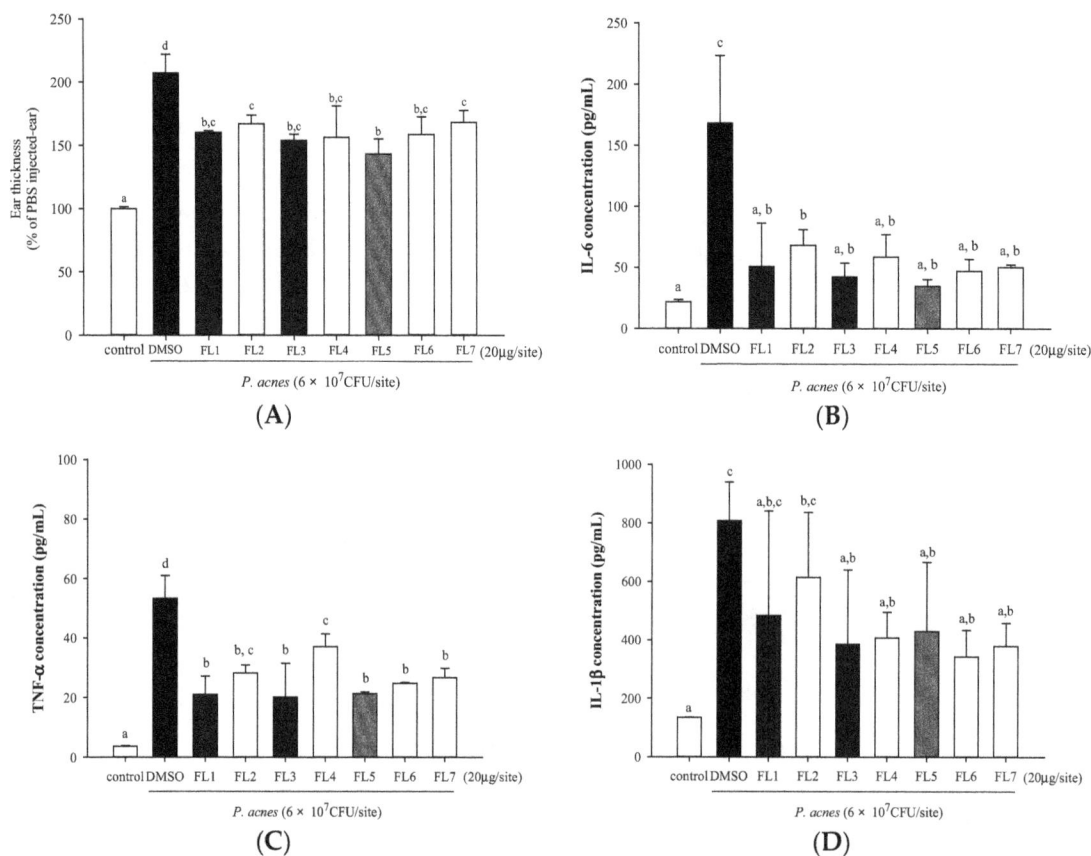

Figure 3. *In vivo* inhibitory effects of seven flavones isolated from *Scutellariae radix* on *P. acnes*-induced skin swelling and pro-inflammatory cytokine levels of mice ears. ICR mice were intradermally injected with PBS (as control), DMSO + *P. acnes* (as vehicle control), or flavones + *P. acnes*. Mouse ear thickness (**A**); and IL-6 (**B**); TNF-α (**C**); and IL-1β (**D**) production in mouse ear homogenates were determined. Values with the same letter are not significantly different as determined by Duncan's multiple range tests.

3. Discussion

The pharmacological effects of the water and organic solvent extracts of *Scutellariae radix* have been extensively studied [8]. There have been over 40 flavonoids isolated from *Scutellariae radix*, with flavones and their glycosides being the most abundant [9]. Seven flavones isolated in this study existed in the forms of aglycones. **FL1** (oroxylin A) and **FL2** (wogonin) are the major flavonoids identified in *Scutellariae radix* [10]. **FL3** to **7** are minor components [8], considerably less attention has been paid to these flavones.

The anti-inflammatory effects of **FL1**, **2**, and **3** have been reported. Oroxylin A (**FL1**) inhibited LPS-induced iNOS and COX-2 gene expression by blocking NF-κB activation *in vitro* [11].

It also suppressed LPS-induced angiogenesis by down-regulation of toll-like receptor TLR-4 and the activity of mitogen-activated protein kinases (MAPK) [12]. Wogonin (**FL2**) had a very potent anti-inflammatory action *in vivo* on several animal models of inflammation, including carrageenan-induced paw edema, adjuvant-induced arthritis, 12-*O*-tetradecanoylphorbol-13-acetate (TPA)-induced skin inflammation and arachidonic acid-induced ear inflammation by oral or topical administration [13–15]. 7-*O*-methylwogonin (**FL3**) significantly inhibited NO and PGE$_2$ release in LPS-stimulated J774A.1 macrophages [16] and effectively suppressed TNF-α, NO and macrophage inflammatory protein (MIP)-2 levels in LPS/IFN-γ-Stimulated RAW 264.7 macrophages [17]. Fewer studies investigated the biological activities of **FL4**, **5**, **6** and **7**. Skullcapflavone II (**FL4**) inhibited ovalbumin (OVA)-induced airway inflammation by reduction of Th2 cytokines, and OVA-specific IgE levels [18]. 5,7,4'-Trihydroxy-8-methoxyflavone (**FL5**) exhibited antiviral activity [19]. Viscidulin II (**FL6**) inhibited activity of adenosine 3',5'-cyclic monophosphate phosphodiesterase [20]. Ganhuangenin (**FL7**) alleviated the type I allergic reaction by inhibiting the release of histamine and LBT4 [21]. The *in vitro* and *in vivo P. acnes*-induced inflammation models utilized here have never been applied to investigate the effect of *Scutellariae radix*. Therefore, our results revealed a new function of the seven flavones isolated from *Scutellariae radix*.

To compare the *in vitro* inhibitory effects of seven flavones on *P. acnes*-induced IL-8 and IL-1β production, IC$_{50}$ values were used. We found wogonin (**FL2**) was the most potent (Table 1). **FL1**, **2**, **4**, and **5** with IC$_{50}$ values smaller than 20 μM for both IL-1β and IL-8 were more potent than **FL3**, **6**, and **7** which had IC$_{50}$ values larger than 50 μM. The two hydroxyl groups at C5 and C7 in **FL1**, **2**, **5** might contribute to their high potency. **FL7** also has hydroxylated C5 and C7, but showed the lowest potency probably because it has too many substituents in B ring. FL3, **4**, **6** with a methoxylated C7, possessing only one hydroxyl group at C5 were assumed to have lower potency. However, **FL4** was an exception, probably due to the extra methoxyl group at C6. The number of methoxyl group may also influence the potency. **FL1**, **2**, **5** have only one methoxyl group, while the less potent flavones, **FL3**, **6**, **7** have two methoxyl groups. So, fewer than two methoxyl groups might be better for potency. However, **FL4** has four methoxyl groups, but still had high potency. Nikaido, *et al.* [20] investigated the structure-activity relationship of flavones from *Scutellariae radix* for cAMP phosphodiesterase inhibition, and found when the flavones had more than five substituents, those with more methoxyl groups were more effective [20]. **FL4** and **FL7** have six, and **FL6** has five substituents. **FL4** has four methoxyl groups, while **FL6** and **7** has two. The higher potency of **FL4** than **FL6** and **7** appeared to be consistent with their hypothesis [20].

To our best knowledge, this is the first study to test the suppression effects of these flavones on *P. acnes*-induced inflammation *in vivo*. Due to the complicated cell populations involved in ear inflammation, and a fixed dose used for all the flavones, the effectiveness of each flavone cannot be completely compared as was ranked *in vitro*. However each of the seven flavones effectively reduced *P. acnes*-induced ear swelling and strongly suppressed the production of TNF-α and IL-6 in mice (Figure 3). IL-6, a pro-inflammatory and chemotactic factor [22], was measured in mice instead of IL-8 that we measured in human THP-1 cells because mice do not produce IL-8 [23].

Luteolin, a positive control in this study, is also a flavone, but without any methoxyl group. Our previous study found it inhibited *P. acnes*-induced pro-inflammatory cytokines expression by inactivating NF-κB through the suppression of mitogen-activated protein kinases (MAPK) phosphorylation [24]. Although the molecular mechanisms by which these seven flavones from *Scutellariae radix* modulate pro-inflammatory cytokine levels are not elucidated, our *in vitro* and *in vivo* results suggest each of the seven flavones has promising therapeutic potential against inflammatory *acne vulgaris*. Our results open a new aspect of the pharmacological role of these flavones.

4. Experimental Section

4.1. Isolation and Structural Elucidation

Scutellariae radix is a Chinese herb and commercially available. We bought it from Sun Ten Pharmaceutical Co. (Taichung, Taiwan). The isolation flowchart of FL1 to 7 from *Scutellariae radix* is shown in Figure 4. Air dried pieces of *Scutellariae radix* (1.7 kg) were extracted twice with ethanol at room temperature (five days each time). The extract was evaporated under reduced pressure using a rotavapor to give brown residue (52.4 g). The residue was suspended in 1.5 L of H_2O and then partitioned with 2 L of ethyl acetate (EA) twice. The water layer was partitioned with n-butanol. The combined EA soluble layer was subjected to chromatography using silica gel and further purification using a high-performance liquid chromatography (HPLC) system (Knauer, Berlin, Germany) with a Phenomenex Luna C18 column (250×10 mm, 5 µm) to furnish seven known flavones (Figure 4). Flavones were identified by comparing their physical and spectral data with literature values as 5,7-dihydroxy-6-methoxyflavone (**FL1**, oroxylin A) [25,26], 5,7-Dihydroxy-8-methoxyflavone (**FL2**, wogonin) [25–27], 5-hydroxy-7,8-dimethoxyflavone (**FL3**, 7-*O*-methylwogonin) [17], 5,6'-dihydroxy-6,7,8,2'-tetramethoxyflavone (**FL4**, skullcapflavone II) [28], 5,7,4'-trihydroxy -8-methoxyflavone (**FL5**) [29,30], 5,2',6'-trihydroxy-7,8-dimethoxyflavone (**FL6**, viscidulin II) [31], and 5,7,2',5'-tetrahydroxy-8,6'-dimethoxyflavone (**FL7**, ganhuangenin) [32] (Figure 2). Nuclear magnetic resonance and infrared data of the seven known compounds are available as Supplementary figures. The purity of flavones was measured by HPLC (Ecom, Prague, Czech Republic) equipped with gradient pumps (Ecom LCP 4100), a UV detector (Ecom LCD 2084) and a LiChrospher® 100 RP-18E (5 µm) HPLC column (125×4 mm i.d., Merck Millipore, Darmstadt, Germany). The mobile phase consisting of a mixture of solvent A (water/methanol, 98:2), and solvent B (methanol/acetic acid, 98:2) was run in the following gradient mode: 0–7 min, from 50% A to 40% A with a flow rate of 0.5 mL/min; 7–12 min, from 40% A to 30% A with a flow rate of 0.3 mL/min; and 12–28 min, from 30% A to 20% A with a flow rate of 0.3 mL/min. The UV detector was at 280 nm. Chromatographic processing was done using the Peak-ABC Chromatography Data Handling System. The purities of **FL1**, **FL2**, **FL3**, **FL4**, **FL5**, **FL6**, and **FL7** were 96.4%, 98.7%, 98.2%, 97.5%, 95.2%, 94.7%, and 98.9%, respectively (Figure 5).

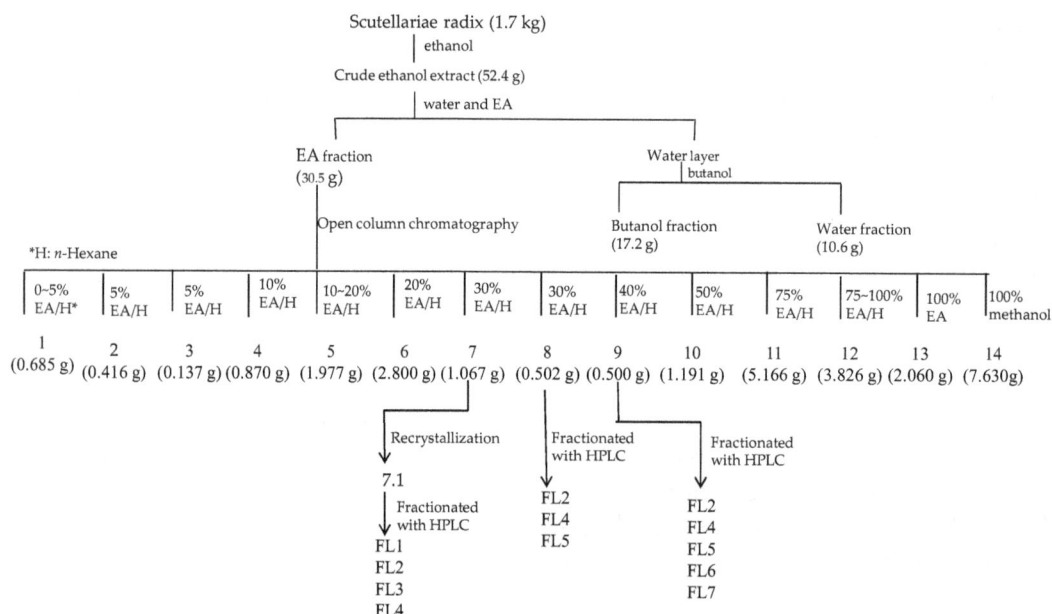

Figure 4. Isolation flowchart of **FL1–7**.

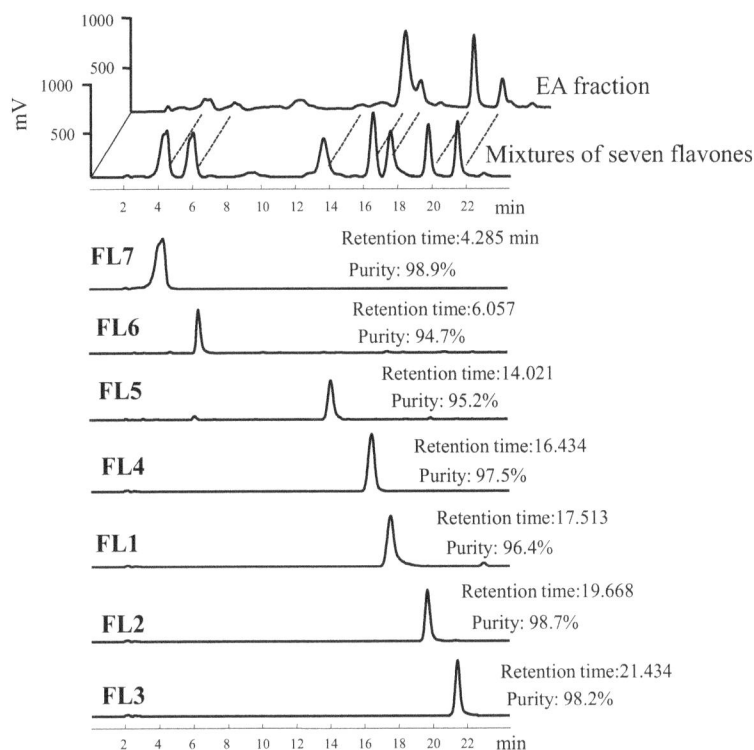

Figure 5. Retention time and purity analysis of **FL1–7** in HPLC.

4.2. Culture of P. acnes and Preparation of Heat-Killed Bacteria

The strain of *P. acnes* (BCRC10723, isolated from facial acne) was obtained from the Bioresource Collection and Research Center (Hsinchu, Taiwan). *P. acnes* was cultured in brain heart infusion (BHI) broth (Difco, Detroit, MI, USA) with 1% glucose in an anaerobic atmosphere using BBL GasPak systems (Becton Dickinson Microbiology Systems, Cockeysville, MD, USA) at 37 °C. A spectrophotometer OD600 (Chrom Tech, Apple Valley, MN, USA) was used to determine the bacterial log phase of growth. The log-phase bacterial culture was harvested, washed thrice with phosphate-buffered saline (PBS) and re-suspended in PBS for induction of mouse ear edema. For the *in vitro* experiments, heat-killed *P. acnes* was prepared. Log-phase bacteria were washed with PBS, incubated at 100 °C for 30 min, re-suspended in RPMI medium (Gibco, Carlsbad, CA, USA) and stored at 4 °C until use.

4.3. Determination of the Viability of THP-1 Cells

The human monocytic THP-1 cell line (BCRC 60430) was obtained from the Bioresource Collection and Research Center (Hsinchu, Taiwan). Cells were maintained in RPMI 1640 supplemented with 10% heat-inactivated fetal bovine serum (FBS, Gibco), penicillin (100 U/mL), and streptomycin (100 μg/mL) at 37 °C in a humidified atmosphere with 5% CO_2. To determine the cytotoxicity of tested samples, a suspension of THP-1 cells (1 × 10^6 cells/mL) was treated with various concentrations of tested samples in 96-well culture plates for 24 h at 37 °C. Whole cell suspension was taken from each well and centrifuged for 4 min at 600 g. Supernatant was removed, and the pellet was incubated with 100 μL of MTT reagent (Sigma-Aldrich) for 3 h at 37 °C. Samples were centrifuged for 2 min at 4500 g, and supernatant was gently removed. Finally, the converted dye from the MTT reagent in cell pellets was solubilized with 500 μL of isopropanol/HCl, and then 200 μL of each sample was transferred in duplicates to a 96-well plate. The absorbance was measured using a Synergy HT multi-detection micro-plate reader (BioTek, Winooski, VT, USA) at 540 nm with 690 nm as the reference wavelength.

4.4. Measurement of Cytokine Production in Human Monocytic THP-1 Cells

THP-1 cells were seeded at 1×10^6 cells/mL in 24-well plates with serum-free medium, and were treated with heat-killed *P. acnes* (7.5×10^7 CFU/mL; multiplicity of infection (M.O.I.) = 75) alone or in combination with different concentrations of tested samples for a 24-h incubation. Cell-free supernatants were collected, and concentrations of IL-1β and IL-8 were analyzed with respective enzyme immunoassay kits (BioLegend, San Diego, CA, USA). The half maximal inhibitory concentration (IC_{50}) values were estimated using a nonlinear regression algorithm (SigmaPlot 12; SPSS Inc. Chicago, IL, USA).

4.5. P. acnes-Induced Ear Edema and Measurement of Cytokine Levels In Vivo

Eight-week-old male ICR mice were purchased from the Animal Center of College of Medicine, National Taiwan University, Taipei, Taiwan. Animal experiments were approved by the Animal Care Committee of the National Taiwan Normal University. Mice were fed with chow diet and water *ad libitum*. To examine the anti-inflammatory effect of flavones *in vivo*, an intradermal injection model was employed [24]. In the preliminary test, 10 μL of flavones (up to 20 μg/site) was injected into mice ears. No noticeable skin irritation occurred (data not shown). Hence, flavones (20 μg/site) were used for the following experiments. Mice were randomly grouped (*n* = 5 per group). *P. acnes* (6×10^7 CFU per 10 μL in PBS) was injected into the left ear of ICR mice. Right ears received an equal amount (10 μL) of PBS. Ten microliters of flavones in 5% DMSO in PBS was injected into the same location of both ears right after *P. acnes* or PBS injection. Twelve hours after bacterial injection, the increase in ear thickness was measured using a micro-caliper (Mitutoyo, Kanagawa, Japan). The increase in ear thickness of the *P. acnes*-injected ear was calculated and expressed as percentage of the PBS-injected control.

4.6. Measurement of Cytokine Levels In Vivo

After thickness had been measured, the ears were excised (*n* = 5 per group). The ear samples were homogenized using a BioMasher III® (Nippi Inc., Tokyo, Japan) in radioimmunoprecipitation (RIPA) buffer (G-Biosciences, St. Louis, MO, USA) supplemented with 1 mM phenylmethylsulfonyl fluoride (PMSF) for 1 min on ice. The homogenates were vortexed and centrifuged at 10,000 *g* for 10 min at 4 °C. The supernatant was reserved and stored at −80 °C for the determinations of TNF-α, IL-1β, and IL-6 levels according to the manufacturer's instruction (BioLegend).

4.7. Statistical Analysis

All data are presented as the mean ± standard deviation (SD). Statistical analyses were performed using the SPSS 19.0 statistical package. The data were evaluated for statistical significance with the one-way ANOVA followed by Duncan's multiple range tests. A *p* value of < 0.05 was considered statistically significant.

Acknowledgments: This work was supported by research grants from the National Taiwan Normal University (Grant No. 100T0700), Taiwan Ministry of Health and Welfare Clinical Trial and Research Center of Excellence (MOHW104-TDU-B-212-113002) and CMU under the Aim for Top University Plan of the Ministry of Education, Taiwan.

Author Contributions: Po-Jung Tsai, Wen-Huey Wu, and Yueh-Hsiung Kuo designed the research and wrote the paper; Ping-Jyun Sung analyzed the spectroscopic data and determined the chemical structures; Po-Jung Tsai and Ming-Chi Hsieh, and Wen-Cheng Huang performed the experimental work.

Conflicts of Interest: The authors declare no conflict of interest.

References

1. Farrar, M.D.; Ingham, E. Acne: Inflammation. *Clin. Dermatol.* **2004**, *22*, 380–384. [CrossRef] [PubMed]

2. Koreck, A.; Pivarcsi, A.; Dobozy, A.; Kemény, L. The role of innate immunity in the pathogenesis of acne. *Dermatology* **2003**, *20*, 96–105. [CrossRef]

3. Kurokawa, I.; Danby, F.W.; Ju, Q.; Wang, X.; Xiang, L.F.; Xia, L.; Chen, W.; Nagy, I.; Picardo, M.; Suh, D.H.; *et al.* New developments in our understanding of acne pathogenesis and treatment. *Exp. Dermatol.* **2009**, *18*, 821–832. [CrossRef] [PubMed]

4. Trivedi, N.R.; Gilliland, K.L.; Zhao, W.; Liu, W.; Thiboutot, D.M. Gene array expression profiling in acne lesions reveals marked upregulation of genes involved in inflammation and matrix remodeling. *J. Investig. Dermatol.* **2006**, *126*, 1071–1079. [CrossRef] [PubMed]

5. Kistowska, M.; Gehrke, S.; Jankovic, D.; Kerl, K.; Fettelschoss, A.; Feldmeyer, L.; Fenini, G.; Kolios, A.; Navarini, A.; Ganceviciene, R.; *et al.* IL-1beta drives inflammatory responses to *Propionibacterium acnes* in vitro and in vivo. *J. Investig. Dermatol.* **2014**, *134*, 677–685. [CrossRef] [PubMed]

6. Shang, X.; He, X.; He, X.; Li, M.; Zhang, R.; Fan, P.; Zhang, Q.; Jia, Z. The genus Scutellaria an ethnopharmacological and phytochemical review. *J. Ethnopharmacol.* **2010**, *128*, 279–313. [CrossRef] [PubMed]

7. Li, H.B.; Jiang, Y.; Chen, F. Separation methods used for *Scutellaria baicalensis* active components. *J. Chromatogr. B* **2004**, *812*, 277–290. [CrossRef]

8. Li, C.; Lin, G.; Zuo, Z. Pharmacological effects and pharmacokinetics properties of Radix Scutellariae and its bioactive flavones. *Biopharm. Drug Dispos.* **2011**, *32*, 427–445. [CrossRef] [PubMed]

9. Wagner, H.; Bauer, R.; Melchart, D.; Xiao, P.G.; Staudinger, A. *Radix Scutellariae*—Huangqin. In *Chromatographic Fingerprint Analysis of Herbal Medicines*; Wagner, H., Bauer, R., Melchart, D., Xiao, P.G., Staudinger, A., Eds.; Springer: Vienna, Austria, 2011; Volume 2, Chapter 63; pp. 755–765.

10. Li, K.L.; Sheu, S.J. Determination of flavonoids and alkaloids in the scute-coptis herb couple by capillary electrophoresis. *Anal. Chim. Acta* **1995**, *313*, 113–120. [CrossRef]

11. Chen, Y.; Yang, L.; Lee, T.J. Oroxylin A inhibition of lipopolysaccharide-induced iNOS and COX-2 gene expression via suppression of nuclear factor-κB activation. *Biochem. Pharmacol.* **2000**, *59*, 1445–1457. [CrossRef]

12. Song, X.; Chen, Y.; Sun, Y.; Lin, B.; Qin, Y.; Hui, H.; Li, Z.; You, Q.; Lu, N.; Guo, Q. Oroxylin A, a classical natural product, shows a novel inhibitory effect on angiogenesis induced by lipopolysaccharide. *Pharmacol. Rep.* **2012**, *64*, 1189–1199. [CrossRef]

13. Kubo, M.; Matsuda, H.; Tanaka, M.; Kimura, Y.; Okuda, H.; Higashino, M.; Tani, T.; Namba, K.; Arichi, S. Studies on Scutellaria radix. VII. Anti-arthritic and anti-inflammatory actions of methanol extract and flavonoid components from Scutellaria radix. *Chem. Pharm. Bull.* **1984**, *32*, 2724–2729. [CrossRef] [PubMed]

14. Yasukawa, K.; Takido, M.; Takeuchi, M.; Nakagawa, S. Effect of chemical constituents from plants on 12-O-tetradecanoylphorbol-13-acetate induced inflammation in mice. *Chem. Pharm. Bull.* **1989**, *37*, 1071–1073. [CrossRef] [PubMed]

15. Chi, Y.S.; Lim, H.; Park, H.; Kim, H.P. Effects of wogonin, a plant flavone from Scutellaria radix, on skin inflammation: *In vivo* regulation of inflammation-associated gene expression. *Biochem. Pharmacol.* **2003**, *66*, 1271–1278. [CrossRef]

16. Chandrasekaran, C.V.; Thiyagarajan, P.; Deepak, H.B.; Agarwal, A. *In vitro* modulation of LPS/calcimycin induced inflammatory and allergic mediators by pure compounds of *Andrographis paniculata* (King of bitters) extract. *Int. Immunopharmacol.* **2011**, *11*, 79–84. [CrossRef] [PubMed]

17. Chao, W.W.; Kuo, Y.H.; Lin, B.F. Anti-inflammatory activity of new compounds from *Andrographis paniculata* by NF-κB transactivation inhibition. *J. Agric. Food Chem.* **2010**, *58*, 2505–2512. [CrossRef] [PubMed]

18. Jang, H.Y.; Ahn, K.S.; Park, M.J.; Kwon, O.K.; Lee, H.K.; Oh, S.R. Skullcapflavone II inhibits ovalbumin-induced airway inflammation in a mouse model of asthma. *Int. Immunopharmacol.* **2012**, *12*, 666–674. [CrossRef] [PubMed]

19. Nagai, T.; Suzuki, Y.; Tomimori, T.; Yamada, H. Antiviral activity of plant flavonoid, 5,7,4′-trihydroxy-8-methoxyflavone, from the roots of *Scutellaria baicalensis* against influenza A (H3N2) and B viruses. *Biol. Pharm. Bull.* **1995**, *18*, 295–299. [CrossRef] [PubMed]

20. Nikaido, T.; Ohmoto, T.; Sankawa, U.; Tomimori, T.; Miyaichi, Y.; Imoto, Y. Inhibition of adenosine 3′,5′-cyclic monophosphate phosphodiesterase by flavonoids. II. *Chem. Pharm. Bull.* **1988**, *36*, 654–661. [CrossRef] [PubMed]

21. Lim, B.O. Effect of ganhuangenin obtained from Scutellaria radix on the chemical mediator production of peritoneal exudate cells and immunoglobulin E level of mesenteric lymph node lymphocytes in Sprague-Dawley rats. *Phytother. Res.* **2002**, *16*, 166–170. [CrossRef] [PubMed]

22. Clahsen, T.; Schaper, F. Interleukin-6 acts in the fashion of a classical chemokine on monocytic cells by inducing integrin activation, cell adhesion, actin polymerization, chemotaxis, and transmigration. *J. Leukoc. Biol.* **2008**, *84*, 1521–1529. [CrossRef] [PubMed]

23. Singer, M.; Sansonetti, P.J. IL-8 is a key chemokine regulating neutrophil recruitment in a new mouse model of *Shigella*-induced colitis. *J. Immunol.* **2004**, *173*, 4197–4206. [CrossRef] [PubMed]

24. Huang, W.C.; Tsai, T.H.; Huang, C.J.; Li, Y.Y.; Chyuan, J.H.; Chuang, L.T.; Tsai, P.J. Inhibitory effects of wild bitter melon leaf extract on *Propionibacterium acnes*-induced skin inflammation in mouse and cytokine production *in vitro*. *Food Funct.* **2015**, *6*, 2550–2560. [CrossRef] [PubMed]

25. Huang, W.H.; Chien, P.Y.; Yang, C.H.; Lee, A.R. Novel synthesis of flavonoids of *Scutellaria baicalensis* Georgi. *Chem. Pharm. Bull.* **2003**, *51*, 339–340. [CrossRef] [PubMed]

26. Lin, Y.L.; Ou, J.C.; Chen, C.F.; Kuo, Y.H. Flavonoids from the roots of *Scutellaria luzonica* Rolfe. *J. Chin. Chem. Soc.* **1991**, *38*, 619–623. [CrossRef]

27. Hua, Y.; Wang, H.Q. Chemical components of *Anaphalis sinica* Hance. *J. Chin. Chem. Soc.* **2004**, *51*, 409–415. [CrossRef]

28. Furukawa, M.; Suzuki, H.; Makino, M.; Ogawa, S.; Iida, T.; Fujimoto, Y. Studies on the constituents of *Lagochilus leiacanthus* (Labiatae). *Chem. Pharm. Bull.* **2011**, *59*, 1535–1540. [CrossRef] [PubMed]

29. Jang, J.; Kim, H.P.; Park, H. Structure and antiinflammatory activity relationships of wogonin derivatives. *Arch. Pharm. Res.* **2005**, *28*, 877–884. [CrossRef] [PubMed]

30. Stevens, J.F.; Wollenweber, E.; Ivancic, M.; Hsu, V.L.; Sundberg, S.; Deinzer, M.L. Leaf surface flavonoids of *Chrysothamnus*. *Phytochemistry* **1999**, *51*, 771–780. [CrossRef]

31. Tanka, T.; Iinuma, M.; Mizuno, M. Synthesis of flavonoids in *Scutellaria* spp. II. Synthesis of 2′,6′-dioxygenated flavones. *Yakugaku Zasshi* **1987**, *107*, 827–829.

32. Iinuma, M.; Tanaka, T.; Mizuno, M. Flavonoids synthesis II. Synthesis of flavones with a 2′,3′,6′-trioxygenated ring B. *Chem. Pharm. Bull.* **1985**, *33*, 4034–4036. [CrossRef]

8

Heterocycles 36. Single-Walled Carbon Nanotubes-Bound *N,N*-Diethyl Ethanolamine as Mild and Efficient Racemisation Agent in the Enzymatic DKR of 2-Arylthiazol-4-yl-alanines

Denisa Leonte [1], László Csaba Bencze [2], Csaba Paizs [2], Monica Ioana Toşa [2], Valentin Zaharia [1,*] and Florin Dan Irimie [2,*]

Academic Editor: Derek J. McPhee

[1] Department of Organic Chemistry, "Iuliu Haţieganu" University of Medicine and Pharmacy, RO-400012 Cluj-Napoca, Victor Babeş 41, Romania; hapau.denisa@umfcluj.ro
[2] Biocatalysis and Biotransformation Research Group, Babeş-Bolyai University, RO-400028 Cluj-Napoca, Arany János 11, Romania; cslbencze@chem.ubbcluj.ro (L.C.B.); paizs@chem.ubbcluj.ro (C.P.); mtosa@chem.ubbcluj.ro (M.I.T.)
* Correspondence: vzaharia@umfcluj.ro (V.Z.); irimie@chem.ubbcluj.ro (F.D.I.)

Abstract: In this paper we describe the chemoenzymatic synthesis of enantiopure L-2-arylthiazol-4-yl alanines starting from their racemic *N*-acetyl derivatives; by combining the lipase-catalysed dynamic kinetic resolution of oxazol-5(4*H*)-ones with a chemical and an enzymatic enantioselective hydrolytic step affording the desired products in good yields (74%–78%) and high enantiopurities (*ee* > 99%). The developed procedure exploits the utility of the single-walled carbon nanotubes-bound diethylaminoethanol as mild and efficient racemisation agent for the dynamic kinetic resolution of the corresponding oxazolones.

Keywords: hydrolases; dynamic kinetic resolution; racemisation agent; L-α-amino acids; thiazole

1. Introduction

Optically-active α-amino acids bearing heterocyclic side chains are of great utility in various fields, not only individually, but especially incorporated in more complex structures, such as peptides and proteins, for the creation of new peptide-based pharmaceutical drug candidates [1,2]. The thiazole core frequently appears in many natural peptides, such as the Bleomycin family (anti-cancer glycopeptide antibiotics) [3], Nocathiacins [4], Aeruginazoles [5], and Thiazomycins [6] (a new class of cyclic thiopeptide antibiotics). The biological potential of this heterocyclic ring system is actually exploited for the design of new thiazole-bearing biologically active compounds, many of them being introduced in therapy. Enantiopure L-α-2-arylthiazole-4-yl alanines constitute chiral synthons with potential applications in drug design, especially when an extended conjugation is beneficial for interaction with pharmacological receptors. For example, the synthesis of new melanotropin analogues incorporating L-α-2-arylthiazole-4-yl alanines has recently been reported [2].

Lipases are often used as biocatalysts for the stereoselective production of variously functionalized optically-active products, due to their ability to transform a wide range of unnatural substrates in a regio- and stereoselective manner, not only in hydrolysis, but also in alcoholysis, aminolysis, or hydrazinolysis reactions using esters, lactones, or lactames as substrates [7].

Despite the success of enzyme-catalysed kinetic resolutions (KR) for the synthesis of a wide range of chiral building blocks, the increasing demand to develop transformations that are not limited by a maximum yield of only 50% drives the development of dynamic kinetic resolution (DKR) processes [8] in which the unreactive enantiomer equilibrates *in situ* under the reaction conditions with the most reactive antipode. Thus, DKR reactions provide the products in theoretical quantitative yields, with high enantiomeric excesses.

The enzymatic DKR of oxazolones was successfully employed for the synthesis of various alanine derivatives [9–14]. The oxazolones, due to the low p*Ka* of the C-4 proton and their inherent reactivity towards lipase-catalysed alcoholysis [9], are excellent substrates for the DKR reaction (Scheme 1). For an efficient DKR, one important requirement is related to the racemisation of the less-reactive enantiomer, which must be rapid under the reaction conditions, and the racemizing agent should not catalyse non-enzymatic secondary reactions, which could decrease the enantiopurity of the desired product.

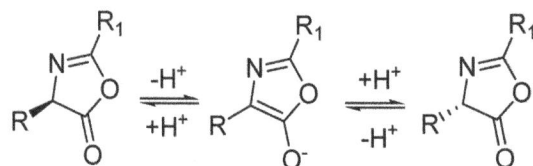

Scheme 1. Base-catalysed racemization of oxazolones [12].

If the spontaneous racemisation is faster than the enzymatic alcoholysis, there is no need to use racemisation agents as in the case of the chemoenzymatic procedures developed for the synthesis of benzofuranyl and benzotiophenyl alanines [13]. However, in case of the recently-reported DKR of phenylfuranyl derivatives, the enzymatic reactions showed higher velocity than the substrate racemisation, forcing the use of triethylamine as racemisation agent, which decreased the enzyme selectivity [14]. In order to alleviate the selectivity decrease caused by the racemisation agent, herein we describe the use of single-walled carbon nanotubes (SWCNT)-bound diethylaminoethanol in the lipase-catalysed dynamic kinetic resolution of the arylthiazole-based oxazolones. The covalent binding of the *N,N*-diethylaminoethanol on carboxy-functionalized SNWCNT$_{COOH}$ was performed using glycerol diglycidyl ether as cross-linker, according to the procedure developed for the immobilization of *Pc*PAL [15] and Lipase B from *Candida antarctica* (CaL-B) [16], the remaining free tertiary amine functionality serving as organic base for the racemization process (Scheme 1). The developed chemoenzymatic procedure for the synthesis of L-2-arylthiazol-4-yl alanines involves two stereoselective enzymatic steps: the lipase-catalysed DKR of the corresponding 4-((2-arylthiazol-4-yl)methyl)-2-methyloxazol-5(4*H*)-ones, followed by Acylase I-mediated hydrolysis.

2. Results and Discussion

2.1. Synthesis of Racemic Substrates

The synthesis of racemic 2-arylthiazol-4-yl alanines *rac*-**6a–d** and their derivatives *rac*-**3-5a–d** is depicted in Scheme 2. 2-Aryl-4-chloromethylthiazoles **1a–d** were synthesized through the Hantzsch condensation of the corresponding thioamides with 1,3-dichloroacetone [17]. 2-Acetamido-3-(2-arylthiazol-4-yl)propanoic acids *rac*-**3a–d** were obtained according to the general malonic ester synthesis [13], starting from the halogenated derivatives **1a–d** through a coupling step with diethylacetamidomalonate, followed by a basic hydrolysis and a decarboxylation reaction.

The racemic esters *rac*-**4a–d** were obtained by treatment of *rac*-**3a–d** with different alcohols (methanol, ethanol, *n*-propanol, *n*-butanol) in the presence of carbonyldiimidazole (CDI).

The cyclisation of *rac*-**3a–d** in the presence of *N,N'*-dicyclohexylcarbodiimide (DCC), in dry dichloromethane, afforded the corresponding oxazol-5(4*H*)-ones *rac*-**5a–d**.

The racemic 2-arylthiazol-4-yl alanines *rac*-**6a–d** were obtained by acidic hydrolysis of the corresponding *N*-acetyl derivatives *rac*-**3a–d**.

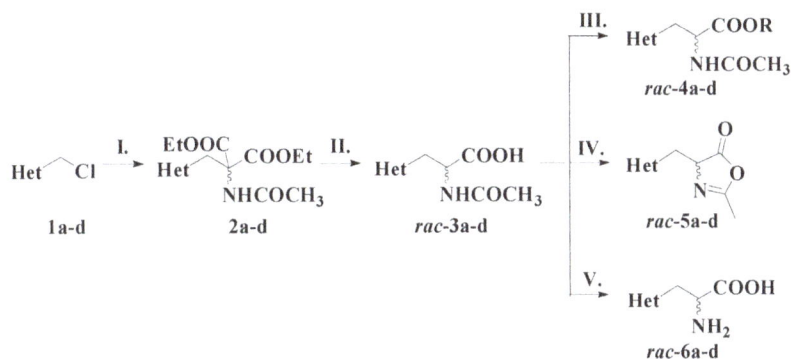

Scheme 2. Synthesis of racemic 2-arylthiazol-4-yl alanines and derivatives. Reagents and conditions: **I.** NaH, CH$_3$CONHCH(COOEt)$_2$/DMF, 60 °C; **II.** (a). Hydrolysis of the ester groups: 10% KOH, reflux, 4 h; (b). Decarboxylation: toluene, reflux, 2 h; **III.** Alcohol (MeOH, EtOH, n-PrOH, n-BuOH), CDI/THF; **IV.** DCC/CH$_2$Cl$_2$; **V.** 18% HCl, reflux, 4 h.

2.2. Chemoenzymatic Synthesis of L-2-Arylthiazol-4-yl Alanines

Racemic 2-acetamido-3-(2-arylthiazol-4-yl)propanoic acids rac-**3a–d** were used as starting materials for the stereoselective chemoenzymatic synthesis of L-2-arylthiazol-4-yl alanines (Scheme 3). The racemic oxazol-5(4H)-ones rac-**5a–d** obtained through cyclisation were used as substrates in the lipase-catalysed DKR process, with various alcohols as nucleophiles. The resulting L-2-acetamido-3-(2-arylthiazol-4-yl)propanoic esters L-**4a–d** were chemically hydrolysed under mild basic conditions ensured by Na$_2$CO$_3$. The obtained L-2-acetamido-3-(2-arylthiazol-4-yl)propanoic acids L-**3a–d** were converted into the corresponding L-2-arylthiazol-4-yl alanines L-**6a–d** by enantioselective hydrolysis of the amide bond, catalysed by Acylase I.

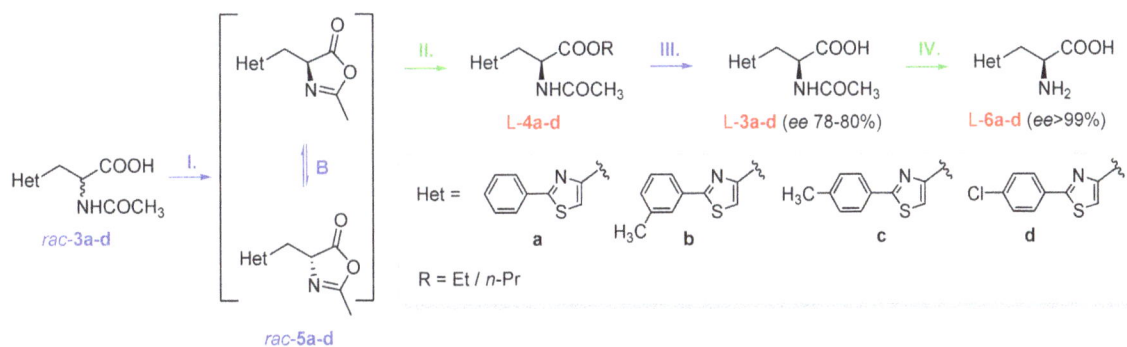

Scheme 3. Stereoselective chemoenzymatic synthesis of L-2-arylthiazol-4-yl alanines and their derivatives (preparative scale). Reagents and conditions: **I.** DCC/CH$_2$Cl$_2$, 0 °C; **II.** CaL-B, ethanol (for dynamic kinetic resolution, DKR, of rac-**5a–c**)/n-Propanol (for DKR of rac-**5d**), acetonitrile; **III.** Na$_2$CO$_3$, H$_2$O, reflux; **IV.** Acylase I, pH 7–8.

In order to investigate the stereoselectivity of the enzymatic processes, first the chiral HPLC separation of rac-**3-6a–d** was established (Figure 1a, see Section 3.1). Further, the enzymatic DKR of oxazol-5(4H)-ones rac-**5a–d** was studied through the screening of the reaction conditions (enzyme, nucleophile, solvent, racemisation catalyst) using the unsubstituted compound rac-**5a** as model substrate in order to obtain the highest enantiopurities. Therefore, we first tested the alcoholysis of rac-**5a** in the presence of various lipases in neat alcohol. Among the tested lipases, only two showed promising results (Table 1): Lipozyme *Mucor miehei* gave poor stereoselectivity (44% *ee*, Table 1, entry 7), while CaL-B (Novozyme 435) showed a higher stereoselectivity (72% *ee*, Table 1, entry 4), using ethanol as nucleophile. Consequently, CaL-B was chosen for further DKR studies.

Table 1. Lipase catalysed kinetic resolution of *rac*-**5a** in ethanol, after 4.5 h reaction time.

Entry	Lipase	c %	ee_p (%)
1	*Candida rugosa* lipase	10	<2
2	Lipase AK "Amano"	92	17 *
3	*Burkholderia cepacia* lipase	14	11 *
4	*Candida antarctica* lipase B	70	72
5	*Candida cylindraceae* lipase	9	<2
6	Lipase F	10	9
7	Lipozyme *Mucor miehei*	95	44

* inverse selectivity.

It is known that the nature of the nucleophile and of the solvent could significantly influence the stereoselectivity of the enzymatic reaction, therefore the CaL-B mediated ring opening of *rac*-**5a** was performed in the presence of various alcohols (methanol, ethanol, *n*-propanol, and *n*-butanol) as nucleophiles (Table 2). Ethanol as nucleophile (52% *ee*, Table 2, entry 2) provided the highest selectivity value, therefore further screening was performed with ethanol as nucleophile (Table 3).

Table 2. The CaL-B mediated ring opening of oxazolone *rac*-**5a** using various alcohols as nucleophiles, after complete conversion of substrates.

Entry	Alcohol	Product	ee %
1	Methanol	L-**4a** methyl ester	8 [a]
2	Ethanol	L-**4a** ethyl ester	52 [b]
3	*n*-Propanol	L-**4a** *n*-propyl ester	47 [a]
4	*n*-Butanol	L-**4a** *n*-butyl ester	35 [a]

[a] after total conversion in 8 h; [b] after total conversion in 24 h.

Table 3. Solvent screening for the enantioselective alcoholysis of oxazolone *rac*-**5a**, with CaL-B and ethanol, after total consumption of the substrate (6 days).

Entry	Solvent	ee_p %
1	1,4-Dioxane	51.3
2	Dichloromethane [1]	-
3	Toluene	58.9
4	Acetonitrile	37.3
5	Tetrahydrofurane	29.2
6	Diethylether	30.0

[1] low conversion (<2%).

The solvent screening showed that higher selectivities and longer reaction times were obtained when compared with the reactions performed in neat ethanol. Furthermore, the observed decrease of the enantiopurities of the produced L-**4a** with increasing conversions (Figure 1b,c) indicates a similar behaviour of the DKR as those reported for the phenylfuran-2-yl derivatives [14], when the racemisation of oxazolone enantiomers was slower than enzymatic alcoholysis. Therefore, the lowered reaction rate of enzymatic alcoholysis from the solvent screening can be beneficial, supported by the increased enantiomeric excess of the product (Table 3, entry 3).

Additionally, in order to increase the racemisation process, the use of organic bases (triethylamine, pryidine, *etc.*) proved successful in several cases [10,14], however their amount in the reaction media must be carefully controlled in order to avoid the decrease in the selectivity and activity of the enzyme [14]. When performing the racemisation in the presence of different low, catalytic amounts (0.5, 0.25, 0.1 eq.) of organic base, the DKR reached total conversions more rapidly (1 day instead of 6 days for DKR process without racemisation catalysts), however the selectivities still remained unsatisfactory (*ee* < 56%, Figure 1d). The blank reaction performed without enzyme proved that even small amounts of weak bases are catalysing the chemical alcoholysis of the oxazolone *rac*-**5a**, resulting in lowered enantiomeric excess of the L-**4a**. In order to avoid the chemical alcoholysis of the substrates and the free entrance of the reactive organic base into the catalytic site of the enzyme, which might also be responsible for the decrease of selectivity, we decided to develop an immobilized racemisation catalyst with large hydrophobic surface area and low diffusional resistance. The developed catalyst would allow for the fast racemisation of the substrate, but would be unable to enter and interact directly with the catalytic site of the enzyme. Thus, catalytic amounts (0.5, 0.25, 0.1 eq.) of diethylaminoethanol covalently bound to carboxy-functionalized single-walled carbon nanotubes (SWCNT$_{COOH}$) (Scheme 4) were tested in the CaL-B catalysed DKR of *rac*-**5a** in various solvents, using ethanol as nucleophile. Accordingly, for the enzymatic DKR of *rac*-**5a**, the enantiomeric excess of the obtained L-**4a** product considerably increased, proving the beneficial effects of the immobilized racemisation catalyst, which is maintaining the substrate in racemic form during the reaction (Figure 1e). The optimal conditions were found to be acetonitrile as solvent (Table 4, entry 4) and ethanol as nucleophile (Table 5, entry 2). Moreover, in the presence of the nanotube-supported base, no chemical ethanolysis was detected in enzyme-less blank experiments, which can be explained by the high affinity of the hydrophobic oxazolones to the surface of the carbon nanotube, which at the same time keeps away the polar nucleophile, avoiding chemical ethanolysis.

In order to increase the enantioselectivity of the DKR process, different ratios between lipase, substrate, racemisation catalyst, and ethanol were tested. Two different amounts of CaL-B were used (5 mg and 10 mg of lipase for 10 mg of substrate), in the presence of different amounts of immobilized racemisation catalyst (3 mg and 6 mg) and ethanol (3 eq. and 6 eq.) in acetonitrile at room temperature. The enantiomeric excesses of the DKR products remained in the interval 70%–80%. The best results (80% *ee*) were obtained with 10 mg CaL-B and 6 mg of racemisation catalyst for 10 mg of substrate, with 3 eq. of ethanol.

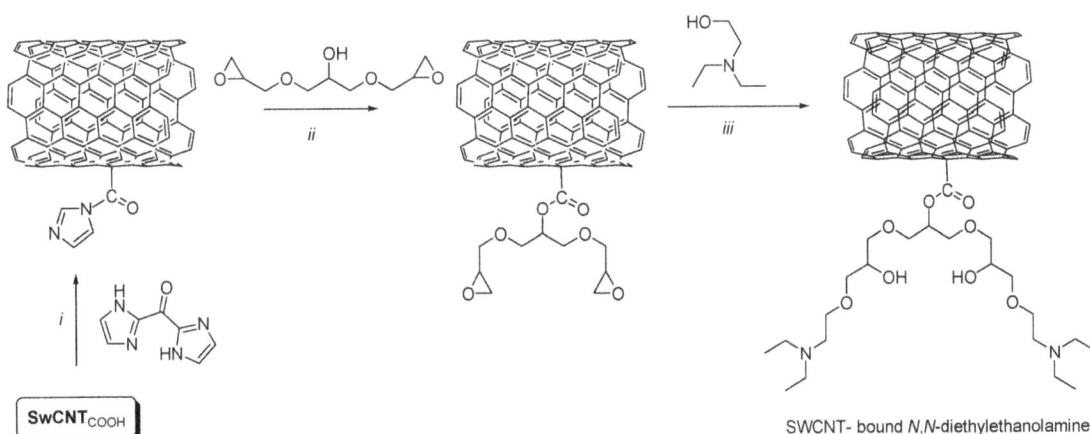

Scheme 4. Immobilisation of *N,N*-diethylaminoethanol on carboxy-functionalized single-walled carbon nanotubes (SWCNT$_{COOH}$). Reaction conditions: **i.** 1,1′-carbonyldiimidazole, anhydrous dichloromethane, r.t.; **ii.** glycerol diglycidyl ether, anhydrous dichloromethane, 24 h, r.t.; **iii.** *N,N*-diethylaminoethanol, anhydrous dichloromethane, 24 h, r.t.

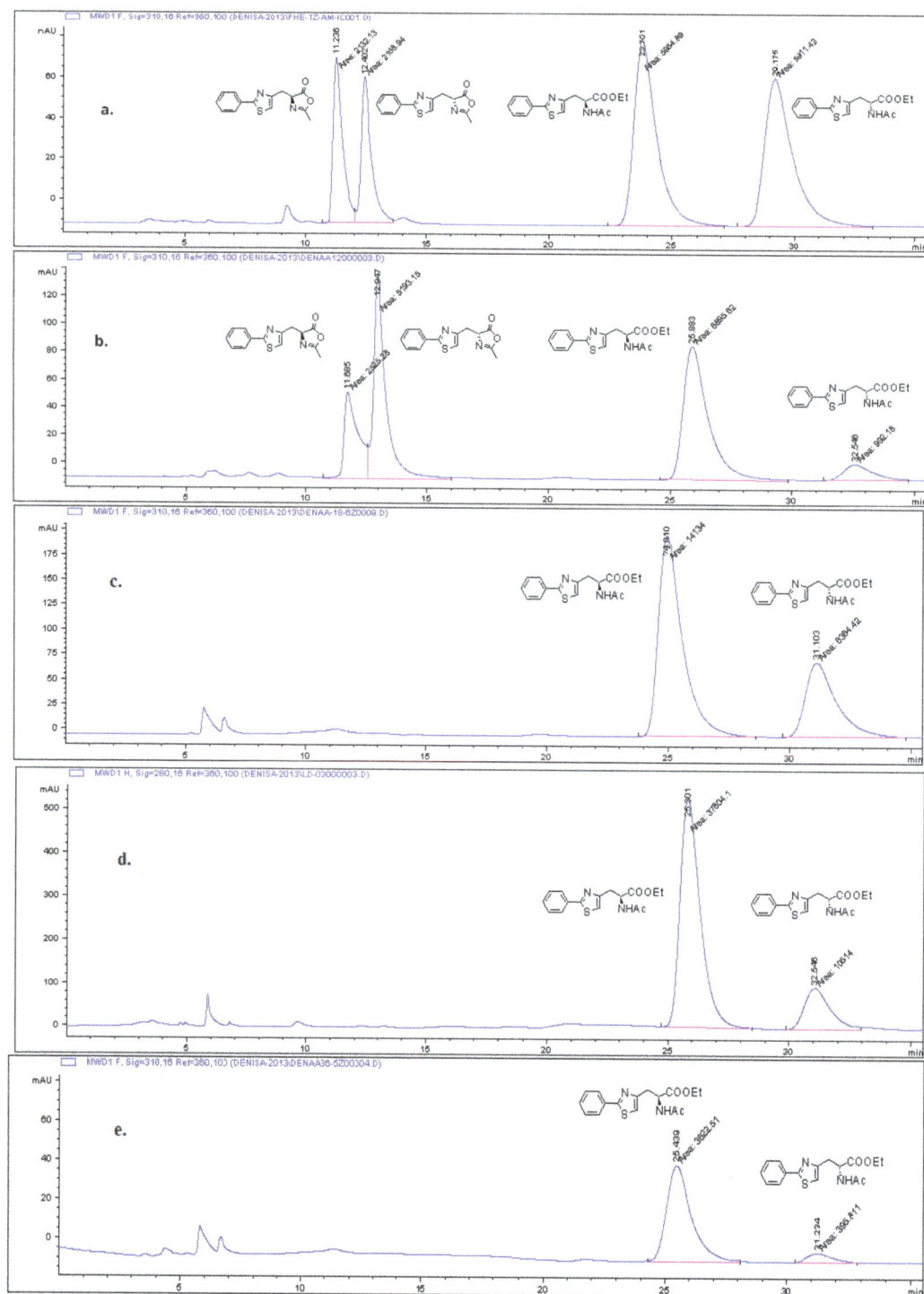

Figure 1. (a) Elution diagram of the mixture of the racemic starting material *rac*-**5a** and racemic product *rac*-**4a** for the lipase-catalysed DKR; (b) Elution diagram of the CaL-B catalysed DKR of oxazolone *rac*-**5a** in acetonitrile, using ethanol as nucleophile, without racemisation catalyst after 24 h and (c) after total consumption of the substrate (6 days); (d) Elution diagram of the CaL-B catalysed DKR of oxazolone *rac*-**5a** with ethanol in acetonitrile, using 0.25 eq. triethylamine as racemisation catalyst after total consumption of the substrate (24 h); (e) Elution diagram of the CaL-B catalysed DKR of racemic oxazolone *rac*-**5a** with ethanol, in acetonitrile, in the presence of the racemisation catalyst *N,N*-diethylaminoethanol immobilized on SWNCT, after total consumption of the substrate (36 h). HPLC method: Chiralpak IC, *n*-hexane: 2-propanol 80:20 *v/v*, UV-detection.

Table 4. Solvent screening for the enantioselective alcoholysis of *rac*-**5a** with CaL-B and ethanol, in the presence of racemisation catalyst (diethylaminoethanol immobilized on SWCNT) (reaction time: 44 h).

Entry	Solvent	c %	ee_p %
1	1,4-Dioxane	91	61
2	Dichloromethane	<2	<2
3	Toluene	>99	61
4	Acetonitrile	>99	80
5	Tetrahydrofurane	60	64
6	Diethylether	>99	30

By performing the DKR of *rac*-**5a** at 30 °C, 40 °C, and 50 °C, an increase of the reaction rate was observed, however no increase of the enantioselectivity was observed; moreover, at 50 °C, the *ee* of the product decreased to 70%.

Having in hand the optimized reaction conditions of the enzymatic DKR of *rac*-**5a**, we decided to extend the same DKR procedure for the other thiazole-based oxazolones *rac*-**5b–d**. Similar good results were obtained when *rac*-**5b,c** were used as substrates, with ethanol (3 eq.) and acetonitrile as solvent (Table 5, entries 5, 6), while in the case of *rac*-**5d**, the obtained lower enantioselectivities forced us to retake the nucleophile screening. Finally, the use of *n*-propanol as nucleophile provided the best results (Table 5, entry 7).

Table 5. CaL-B mediated DKR of *rac*-**5a–d**, in acetonitrile, with different alcohols, in the presence of 0.25 eq. immobilized diethylaminoethanol as racemisation catalyst, at total conversion of the substrate (reaction time: 44 h).

Entry	Substrate	Alcohol	Product	ee_p %
1	*rac*-**5a**	Methanol	L-**4a** methyl ester	71
2	*rac*-**5a**	Ethanol	L-**4a** ethyl ester	80
3	*rac*-**5a**	*n*-Propanol	L-**4a** *n*-propyl ester	76
4	*rac*-**5a**	*n*-Butanol	L-**4a** *n*-butyl ester	58
5	*rac*-**5b**	Ethanol	L-**4b** ethyl ester	78
6	*rac*-**5c**	Ethanol	L-**4c** ethyl ester	78
7	*rac*-**5d**	*n*-Propanol	L-**4d** *n*-propyl ester	80

Using the optimal conditions found for the small scale reactions, the preparative scale enzymatic DKR of oxazolones *rac*-**5a–d** was performed at room temperature, affording the corresponding L-2-acetamido-3-(2-arylthiazol-4-yl)propanoic esters L-**4a–d** in excellent yields (>93%) and moderate enantiomeric excesses (78%–80%, Table 6). The total conversions of the preparative scale DKR processes were achieved after 2 days, in the case of *rac*-**5a–c**, and respectively after 3 days when *rac*-**5d** was used as substrate.

Further, the enantiomerically-enriched DKR products L-**4a–d** were hydrolysed under mild basic conditions, with good yields (>98%) to L-**3a–d**, without altering the *ee* (Table 6, entry 2). In order to increase the *ee* of the final products in enantiopure form, the formed N-acetyl amino acids L-**3a–d** were submitted to the Acylase I-catalysed enantioselective hydrolysis of the amide group (Scheme 2), affording the corresponding amino acids L-**6a–d** with excellent enantiopurity and good global yields (Table 7).

The expected L-configuration of the obtained enantiopure 2-arylthiazole-4-yl alanines **6a–d** was confirmed by measuring their specific rotation, which were consistent with the literature values [18].

Table 6. Yields and *ee* values for each step of the preparative scale chemoenzymatic synthesis of L-2-arylthiazole-4-yl alanines L-**6a–d**.

Entry	Enzymatic/Chemical Step of the Preparative Scale Synthesis	Substrate	Product	Yield %	ee_p %
1	CaL-B catalysed DKR of oxazolones *rac*-**5a–d** with ethanol/*n*-propanol (3 eq.) and immobilized diethylaminoethanol, in acetonitrile	*rac*-**5a**	L-**4a** ethyl ester	95	80
		rac-**5b**	L-**4b** ethyl ester	96	78
		rac-**5c**	L-**4c** ethyl ester	93	78
		rac-**5d**	L-**4d** *n*-propyl ester	95	80
2	Chemical hydrolysis of the *N*-acetyl amino esters L-**4a–d** under mild basic conditions (Na_2CO_3/H_2O)	L-**4a** ethyl ester	L-**3a**	99	80
		L-**4b** ethyl ester	L-**3b**	98	78
		L-**4c** ethyl ester	L-**3c**	99	78
		L-**4d** *n*-propyl ester	L-**3d**	99	80
3	Acylase I-catalysed enantioselective hydrolysis of the *N*-acetyl amino acids L-**3a–d**	L-**3a**	L-**6a**	92	>99
		L-**3b**	L-**6b**	92	>99
		L-**3c**	L-**6c**	91	>99
		L-**3d**	L-**6d**	92	>99

Table 7. Global yields and specific rotations for enantiopure L-2-arylthiazole-4-yl alanines L-**6a–d**, obtained by CaL-B mediated DKR of *rac*-**5a–d** followed by Acylase I-mediated enantioselective hydrolysis of L-**3a–d**.

Entry	Product	Global Yield [a] (%)	*ee* (%)	$[\alpha]_D^{28}$
1	L-**6a**	78	>99	−0.20 [b]
2	L-**6b**	77	>99	−0.26 [b]
3	L-**6c**	74	>99	−0.27 [b]
4	L-**6d**	78	>99	−0.35 [b]

[a] calculated based on the starting material *rac*-**3a–d**; [b] (CH_3COOH, *c* = 5 mg/mL).

3. Experimental Section

3.1. Analytical Methods

The ^{1}H-NMR and ^{13}C-NMR spectra were recorded on a Bruker Avance DPX-300 spectrometer (Bruker, Billerica, MA, USA) operating at 600 and 150 MHz, respectively. Chemical shifts on the δ scale are expressed in ppm values from tetramethylsilane as internal standard. ESI$^+$ MS spectra were recorded on a GC-MS Shimadzu QP 2010 Plus spectrometer (Shimadzu Europa GmbH, Duisburg, Germany) using direct injection, at 30–70 eV.

High performance liquid chromatography analyses were conducted with an Agilent 1200 instrument (Agilent Technologies, Santa Clara, CA, USA), using a Chiralpak IC column (4.6 × 250 mm, Daicel Chiral Technologies Europe, Essex, UK) and a mixture of *n*-hexane and 2-propanol 80:20 (*v/v*) as eluent for the enantiomeric separation of *rac*-**4a–d**, an Astec Chirobiotic V2 column, and a mixture of methanol, acetic acid, triethylamine (TEA) 200:0.15:0.15 (*v/v/v*) as eluent for the enantiomeric separation of *rac*-**3a–d**, and a Chiralpak Zwix(+) column with a mixture of methanol (50 mM formic acid, 25 mM diethylamine (DEA)), acetonitrile, water 49:49:2 (*v/v/v*), for the chiral separation of *rac*-**6a–d**, all at 1 mL/min flow rate. The gradient separation method and the retention times of the enantiomers are shown in Table 8.

Thin layer chromatography (TLC) was carried out using Merck Kieselgel 60F$_{254}$ sheets (Merck, Darmstadt, Germany). Spots were visualized in UV light or by treatment with 5% ethanolic ninhydrin solution and heating of the dried plates. Preparative chromatographic separations were performed using column chromatography on Merck Kieselgel 60 Å (63–200 μm). Optical rotations were determined on a Bellingham-Stanley ADP 220 polarimeter using acetic acid as solvent.

Table 8. The retention times for the enantiomers of *rac*-3-6a–d.

Separation Conditions	RP-HPLC Astec chirobiotic V2 column, eluent: MeOH:CH$_3$COOH:Et$_3$N 200:0.15:0.15 $v/v/v$							
Compound	L-**3a**	D-**3a**	L-**3b**	D-**3b**	L-**3c**	D-**3c**	L-**3d**	D-**3d**
R$_t$ (min)	4.4	6.1	4.5	6.0	4.8	6.4	5.5	7.0
Separation Conditions	HPLC Chiralpak IC, eluent: *n*-hexane:2-propanol 80:20 v/v							
Compound	(*S*)-**5a**	(*R*)-**5a**	(*S*)-**5b**	(*R*)-**5b**	(*S*)-**5c**	(*R*)-**5c**	(*S*)-**5d**	(*R*)-**5d**
R$_t$ (min)	11.2	12.4	11.5	12.6	11.3	12.5	10.2	12.0
Compound	L-**4a**	D-**4a**	L-**4b**	D-**4b**	L-**4c**	D-**4c**	L-**4d**	D-**4d**
R$_t$ (min)	29.9 [a]	36.4 [a]					28.9 [a]	35.5 [a]
	24.4 [b]	30.2 [b]	25.0 [b]	29.2 [b]	28.5 [b]	34.0 [b]	25.0 [b]	30.0 [b]
	20.3 [c]	25.2 [c]					19.0 [c]	23.2 [c]
	19.7 [d]	24.0 [d]					18.6 [d]	22.7 [d]
Separation Conditions	RP-HPLC Chiralpak Zwix(+), eluent: MeOH (50 mM HCOOH, 25 mM diethylamine):acetonitrile:water 49:49:2 $v/v/v$							
Compound	*rac*-**3a**		*rac*-**3b**		*rac*-**3c**		*rac*-**3d**	
R$_t$ (min)	5.9		4.2		5.2		4.1	
Compound	L-**6a**	D-**6a**	L-**6b**	D-**6b**	L-**6c**	D-**6c**	L-**6d**	D-**6d**
R$_t$ (min)	10.9	18.8	9.5	20.2	10.3	21.2	11.3	24.9

[a] methyl ester; [b] ethyl ester; [c] *n*-propyl ester; [d] *n*-butyl ester.

Melting points were determined on open glass capillaries using an Electrothermal IA 9000 digital apparatus.

3.2. Reagents and Solvents

The commercial chemicals and solvents were products of Sigma Aldrich (Sigma Aldrich Chemie Gmbh, Steinheim, Germany) or Fluka (Buchs, Switzerland). All solvents were purified and dried by standard methods as required. Carboxy-functionalized Single walled carbon nanotubes (SWCNT$_{COOH}$) were purchased from Organic Chemicals Co. Ltd. (Chengdu, China). Lipase B from *Candida antarctica* (CaL-B, Novozym 435) was purchased from Novozymes, Bagsvaerdt, Denmark. Lipases from *Candida rugosa* (CrL), *Candida cylindracea* (CcL), *Mucor miehei* (MmL) and Acylase I were purchased from Fluka. Lipases from *Pseudomonas fluorescens* (AK free), *Burkholderia cepacia* (BcL), and lipase F were from Amano, Chipping Norton, UK.

3.3. Chemical Synthesis of Racemic 2-Arylthiazol-4-yl Alanines and Their Derivatives

3.3.1. Synthesis of Racemic 2-Acetamido-3-(2-arylthiazol-4-yl)propanoic Acids *rac*-**3a–d**

A dispersion of 60% NaH in mineral oil (0.84 g, 21 mmol) was suspended in dry *N,N*-dimethylformamide (12 mL) and stirred under argon at room temperature. After 30 min, diethyl acetamidomalonate (4.34 g, 20 mmol) was added, the mixture was stirred for 30 min and cooled, followed by the dropwise addition of the halogenated thiazole derivative **1a–d** (22 mmol) dissolved in dry *N,N*-dimethylformamide (5 mL). The reaction mixture was stirred for 3 h at room temperature, and for the next 4 h at 60 °C. The solution was cooled and poured on a water–ice mixture. The formed precipitate was filtered off, dried, and suspended in an aqueous solution of 10% KOH (4–5 mL). The reaction mixture was refluxed for 4 h, in order to hydrolyse the ester groups. The resulting solution was cooled and the pH was adjusted to 1 with concentrated HCl. The formed precipitate was filtered, dried, suspended in toluene (10 mL), and refluxed for 2 h, until complete decarboxylation. The formed white crystals of 2-acetamido-3-(2-arylthiazol-4-yl)propanoic acids were isolated by filtration and dried.

*2-Acetamido-3-(2-phenylthiazol-4-yl)propanoic acid (rac-**3a**)*: Yield: 64%; white solid; m.p. 175–176 °C; ^{1}H-NMR (600 MHz, DMSO) δ 8.25 (1H, NH), 7.91 (dd, J = 7.9, 1.4 Hz, 2H), 7.52–7.46 (m, 3H), 7.38 (s, 1H), 4.60 (td, J = 8.7, 5.0 Hz, 1H), 3.14 (ddd, J = 23.7, 14.6, 7.0 Hz, 2H), 1.81 (s, 3H). ^{13}C-NMR (151 MHz, DMSO) δ 172.99, 169.29, 166.38, 153.50, 133.11, 130.14, 129.21, 126.06, 116.24, 51.82, 32.89, 22.40; ESI-MS: 291.0800 (calculated: 291.0798, for $C_{14}H_{14}N_2O_3S$ [M + H]$^+$); m/z (%): 313 (24, [M + Na]$^+$), 293 (4.5, [M + 3H]$^+$), 292 (15.2, [M + 2H]$^+$), 291 (100, [M + H]$^+$), 284 (2.8), 279 (1.3), 273 (1.2).

*2-Acetamido-3-(2-m-tolylthiazol-4-yl)propanoic acid (rac-**3b**)*: Yield: 65%; white solid; m.p. 160–161 °C; ^{1}H-NMR (600 MHz, DMSO) δ 7.97 (1H, NH), 7.73 (s, 1H), 7.68 (d, J = 7.7 Hz, 1H), 7.36 (t, J = 7.6 Hz, 1H), 7.30 (s, 1H), 7.26 (d, J = 7.4 Hz, 1H), 4.43 (m, 1H), 3.13 (ddd, J = 23.2, 14.6, 6.6 Hz, 2H), 2.37 (s, 3H), 1.79 (s, 3H). ^{13}C-NMR (151 MHz, DMSO) δ 173.32, 168.80, 165.97, 154.70, 138.51, 133.24, 130.66, 129.08, 126.42, 123.28, 115.31, 52.99, 33.75, 22.61, 20.91; ESI$^+$-MS: 305.0960 (calculated: 305.0954 for $C_{15}H_{16}N_2O_3S$ [M + H]$^+$); m/z (%): 327 (85.3, [M + Na]$^+$), 307 (4.4, [M + 3H]$^+$), 306 (16.5, [M + 2H]$^+$), 305 (100, [M + H]$^+$), 292 (3.0), 291 (21.2), 288 (4.2), 284 (1.7), 263 (30.4), 251 (21.8), 210 (4.8).

*2-Acetamido-3-(2-p-tolylthiazol-4-yl)propanoic acid (rac-**3c**)*: Yield: 67%; white solid; m.p. 181–182 °C; ^{1}H-NMR (600 MHz, DMSO) δ 8.17 (1H, NH), 7.80 (d, J = 8.1 Hz, 2H), 7.31 (s, 1H), 7.30 (d, J = 8.0 Hz, 2H), 4.56 (dd, J = 13.2, 8.3 Hz, 1H), 3.12 (ddd, J = 23.6, 14.6, 6.9 Hz, 2H), 2.35 (s, 3H), 1.80 (s, 3H). ^{13}C-NMR (151 MHz, DMSO) δ 173.03, 169.17, 166.41, 153.52, 139.88, 130.57, 129.72, 125.99, 115.50, 51.98, 33.03, 22.44, 20.93; ESI$^+$-MS: 305.0967 (calculated: 305.0954, for $C_{15}H_{16}N_2O_3S$ [M + H]$^+$); m/z (%): 343 (100, [M + K]$^+$), 327 (25.4, [M + Na]$^+$), 307 (0.8, [M + 3H]$^+$), 306 (3.2, [M + 2H]$^+$), 305 (18.5, [M + H]$^+$), 291 (0.9), 284 (1.8), 263 (1.5), 251 (1.3), 210 (0.2).

*2-Acetamido-3-(2-p-chlorophenylthiazol-4-yl)propanoic acid (rac-**3d**)*: Yield: 64%; white solid; m.p. 199–200 °C; ^{1}H-NMR (600 MHz, DMSO) δ 8.20 (1H, NH), 7.92 (d, J = 7.2 Hz, 2H), 7.55 (d, J = 6.8 Hz, 2H), 7.40 (s, 1H), 4.57 (dd, J = 11.9, 8.9 Hz, 1H), 3.13 (ddd, J = 23.6, 14.6, 7.0 Hz, 2H), 1.80 (s, 3H). ^{13}C-NMR (151 MHz, DMSO) δ 173.03, 169.29, 164.98, 153.89, 134.63, 131.98, 129.28, 127.77, 116.70, 51.99, 33.01, 22.46; ESI$^+$-MS: 325.0408 (calculated: 325.0408, for $C_{14}H_{13}ClN_2O_3S$ [M + H]$^+$); m/z (%): 347 (12.5, [M + Na]$^+$), 328 (11.1, [M + 2H]$^+$, ^{37}Cl), 327 (70.1, [M + H]$^+$, ^{37}Cl), 326 (3.2, [M + 2H]$^+$, ^{35}Cl), 325 (21.4, [M + H]$^+$, ^{35}Cl), 313 (9.5), 305 (32.6), 251 (18.7), 210 (3.7).

3.3.2. Synthesis of Racemic 2-Acetamido-3-(2-arylthiazol-4-yl)propanoic Esters *rac-**4a–d***

To a solution of racemic 2-acetamido-3-(2-arylthiazol-4-yl)propanoic acid *rac-**3a–d*** (0.5 mmol) and carbonyl diimidazole (90 mg, 0.55 mmol) in anhydrous THF (2 mL), ethanol (321 μL, 5.5 mmol) was added. The reaction mixture was stirred at room temperature overnight. The solvent was removed *in vacuo* and the crude product was purified with column chromatography on silica gel using dichloromethane:acetone 9:1 (*v/v*) as eluent. Methyl, *n*-propyl and *n*-butyl 2-acetamido-3-(2-arylthiazol-4-yl)propanoates were obtained by the same procedure, using methanol, propanol or butanol instead of ethanol.

*Ethyl 2-acetamido-3-(2-phenylthiazol-4-yl)propanoate (rac-**4a**)*: Yield: 64%; white solid; m.p. 116–117 °C; ^{1}H-NMR (600 MHz, CDCl$_3$) δ 7.94 (d, J = 4.0 Hz, 2H), 7.45–7.46 (m, 3H), 7.02 (s, 1H), 4.91–4.94 (m, 1H), 4.19 (q, J = 7.1 Hz, 2H), 3.37 (ddd, J = 17.4, 14.4, 3.7 Hz, 2H), 2.04 (s, 3H), 1.24 (t, J = 7.1 Hz, 3H). ^{13}C-NMR (151 MHz, CDCl$_3$) δ 171.32, 170.19, 168.65, 152.31, 133.18, 130.70, 129.25, 126.68, 115.93, 61.67, 52.16, 32.84, 23.42, 14.33; ESI$^+$-MS: 319.1121 (calculated: 319.1111 for $C_{16}H_{18}N_2O_3S$ [M + H]$^+$); m/z (%): 357 (44.3, [M + K]$^+$), 341 (16.5, [M + Na]$^+$), 320 (18.8, [M + 2H]$^+$), 319 (100, [M + H]$^+$), 305 (10), 277 (14).

*Ethyl 2-acetamido-3-(2-m-tolylthiazol-4-yl)propanoate (rac-**4b**)*: Yield: 63%; yellow solid; m.p. 81 °C; ^{1}H-NMR (600 MHz, CDCl3) δ 7.72 (s, 1H), 7.70 (d, J = 7.8 Hz, 1H), 7.32 (t, J = 7.6 Hz, 1H), 7.24 (d, J = 7.5 Hz, 1H), 6.97 (s, 1H), 4.92 (dd, J = 12.9, 5.2 Hz, 1H), 4.18 (q, J = 7.1 Hz, 2H), 3.32 (ddd, J = 46.0, 14.8, 5.2 Hz, 2H), 2.41 (s, 3H), 2.03 (s, 3H), 1.22 (t, J = 7.1 Hz, 3H). ^{13}C-NMR (151 MHz, CDCl3) δ 171.40, 170.05, 168.60, 152.47, 138.88, 133.22, 131.18, 129.03, 127.12, 123.70, 115.61, 61.56, 52.12, 33.03,

23.36, 21.49, 14.27. ESI$^+$-MS: 333.1270 (calculated: 333.1267 for $C_{17}H_{20}N_2O_3S$ [M + H]$^+$); m/z (%): 371 (61.0, [M + K]$^+$), 355 (36.9, [M + Na]$^+$), 334 (19.8, [M + 2H]$^+$), 333 (100.0, [M + H]$^+$), 319 (2.7), 305 (1), 291 (6.2).

Ethyl 2-acetamido-3-(2-p-tolylthiazol-4-yl)propanoate (*rac-4c*): Yield: 65%; white solid; m.p. 98 °C; ^{1}H-NMR (600 MHz, CDCl$_3$) δ 7.78 (d, J = 8.0 Hz, 2H), 7.24 (d, J = 7.9 Hz, 2H), 6.94 (s, 1H), 4.91 (dt, J = 7.5, 5.1 Hz, 1H), 4.18 (q, J = 7.1 Hz, 2H), 3.31 (ddd, J = 50.6, 14.8, 5.0 Hz, 2H), 2.40 (s, 3H), 2.04 (s, 3H), 1.22 (t, J = 7.1 Hz, 3H). ^{13}C-NMR (151 MHz, CDCl$_3$) δ 171.45, 170.07, 168.56, 152.54, 140.60, 130.96, 129.81, 126.42, 115.22, 61.58, 52.15, 33.11, 23.43, 21.56, 14.31; ESI$^+$-MS: 333.1265 (calculated: 333.1267 for $C_{17}H_{20}N_2O_3S$ [M + H]$^+$); m/z (%): 371 (12.1, [M + K]$^+$), 355 (47.6, [M + Na]$^+$), 334 (21.5, [M + 2H]$^+$), 333 (100.0, [M + H]$^+$), 319 (35), 305 (17.5), 291 (5), 259 (12.0), 253 (22.9), 217 (24.9).

n-Propyl 2-acetamido-3-(2-p-chlorophenylthiazol-4-yl)propanoate (*rac-4d*): Yield: 64%; white solid; ^{1}H-NMR (600 MHz, CDCl$_3$) δ 7.86 (d, J = 8.5 Hz, 2H), 7.40 (d, J = 8.5 Hz, 2H), 6.97 (s, 1H), 4.94 (dt, J = 7.7, 5.1 Hz, 1H), 4.08 (t, J = 6.7 Hz, 2H), 2.02 (s, 3H), 1.73–1.65 (m, 2H), 0.89 (t, J = 7.4 Hz, 3H). ^{13}C-NMR (151 MHz, CDCl$_3$) δ 171.55, 170.34, 168.02, 153.77, 148.43, 131.36, 129.35, 127.67, 115.97, 67.24, 52.07, 34.06, 23.40, 21.70, 10.46; ESI$^+$-MS: 367.0879 (calculated: 367.0878 for $C_{17}H_{19}ClN_2O_3S$ [M + H]$^+$); m/z (%): 405 ([M + K]$^+$), 389 ([M + Na]$^+$), 370 (6.6, [M + 2H]$^+$, ^{37}Cl), 369 (37.2, [M + H]$^+$, ^{37}Cl), 368 (19.5, [M + 2H]$^+$, ^{35}Cl), 367 (100, [M + H]$^+$, ^{35}Cl), 333 (10), 319 (1.8), 305 (1.7), 287 (1.5), 244 (1.8).

3.3.3. Synthesis of Racemic 4-((2-Arylthiazol-4-yl)methyl)-2-methyloxazol-5(4*H*)-ones *rac-5a–d*

To a solution of racemic 2-acetamido-3-(2-arylthiazol-4-yl)propanoic acid *rac-3a–d* (1 mmol) in anhydrous dichloromethane (5 mL), a solution of *N,N'*-dicyclohexyl-carbodiimide (247.2 mg, 1.2 mmol) in anhydrous dichloromethane (2 mL) was added dropwise at 0 °C. The reaction mixture was stirred for 1 h at 0 °C. After the completion of the reaction (verified by TLC, eluent dichloromethane:acetone 9:1), the formed precipitate of dicyclohexyl urea was removed by filtration. The solvent was distilled off at reduced pressure, obtaining without further purifications the pure oxazol-5(4*H*)-ones *rac-5a–d*, which were directly used in the enzymatic reactions.

3.3.4. Synthesis of Racemic 2-Arylthiazole-4-yl Alanines *rac-6a–d*

A suspension of racemic 2-acetamido-3-(2-arylthiazol-4-yl)propanoic acid *rac-3a–d* (50 mg) in 18% HCl (6 mL) was refluxed for 4 h. The solvent was removed by distillation at reduced pressure, affording the corresponding 2-arylthiazole-4-yl alanine *rac-6a–d* as hydrochloride salt, which was dried and washed several times with diethyl ether.

2-Amino-3-(2-phenylthiazol-4-yl)propanoic acid (*rac-6a*): Yield: 91%; white powder; m.p. 241–248 °C for the hydrochloride salt, respectively m.p. over 300 °C with decomposition for the free amino acid; ^{1}H-NMR (600 MHz, D$_2$O) δ 7.82–7.79 (m, 2H), 7.62–7.38 (m, 4H), 4.44 (t, J = 6.5 Hz, 1H), 3.45 (ddd, J = 22.5, 15.6, 6.4 Hz, 2H). ^{13}C-NMR (151 MHz, D$_2$O) δ 172.4, 170.98, 170.67, 146.59, 132.00, 129.50, 126.90, 119.69, 52.24, 29.84; ESI$^+$-MS: 249.0699 (calculated: 249.0692 for $C_{12}H_{12}N_2O_2S$ [M + H]$^+$); m/z (%): 263 (100), 249 (2.9, [M + H]$^+$), 203 (1.5).

2-Amino-3-(2-m-tolylthiazol-4-yl)propanoic acid (*rac-6b*): Yield: 88%; white powder; m.p. 225–238 °C for the hydrochloride salt, respectively m.p. 227–242 °C with decomposition for the free amino acid; ^{1}H-NMR (600 MHz, D$_2$O) δ 7.54 (m, 3H), 7.32–7.31 (m, 2H), 4.42 (t, J = 6.6 Hz, 1H), 3.43 (ddd, J = 35.2, 15.6, 6.7 Hz, 2H), 2.28 (s, 3H). ^{13}C-NMR (151 MHz, D$_2$O) δ 173.86, 171.21, 170.58, 139.88, 133.06, 133.02, 129.42, 127.34, 123.93, 119.69, 52.13, 29.60, 20.35; ESI$^+$-MS: 263.0857 (calculated: 263.0849 for $C_{13}H_{14}N_2O_2S$ [M + H]$^+$); m/z (%): 277 (100), 263 (4.6, [M + H]$^+$), 217 (0.8).

2-Amino-3-(2-p-tolylthiazol-4-yl)propanoic acid (*rac-6c*): Yield: 92%; white powder; m.p. 180–190 °C for the hydrochloride salt, respectively m.p. 280–286 °C with decomposition for the free amino acid; ^{1}H-NMR (600 MHz, D$_2$O) δ 7.72 (d, J = 8.1 Hz, 2H), 7.69 (s, 1H), 7.34 (d, J = 8.1 Hz, 2H), 4.45 (t, J = 6.9 Hz, 1H), 3.50 (ddd, J = 22.8, 16.5, 6.9 Hz, 2H), 2.32 (s, 3H). ^{13}C-NMR (151 MHz, D$_2$O) δ 172.29,

170.18, 145.07, 143.56, 130.38, 127.38, 124.49, 120.25, 51.82, 28.80, 20.78; ESI$^+$-MS: 263.0855 (calculated: 263.0849 for $C_{13}H_{14}N_2O_2S$ [M + H]$^+$); m/z (%): 277 (100), 263 (3.0, [M + H]$^+$), 217 (0.9).

2-Amino-3-(2-(4-chlorophenyl)thiazol-4-yl)propanoic acid (*rac-6d*): Yield: 93%; white powder; m.p. 250–260 °C for the hydrochloride salt, respectively m.p. over 300 °C with decomposition for the free amino acid; ^{1}H-NMR (600 MHz, Methanol-d_4) δ 7.99 (d, J = 8.5 Hz, 2H), 7.50 (d, J = 8.5 Hz, 2H), 7.46 (s, 1H), 4.45 (t, J = 5.8 Hz, 1H), 3.46 (ddd, J = 22.9, 15.5, 5.9 Hz, 2H). ^{13}C-NMR (151 MHz, Methanol-d_4) δ 170.97, 169.18, 151.97, 137.46, 133.08, 130.34, 129.08, 118.91, 53.62, 32.26; ESI$^+$-MS: 283.0306 (calculated: 283.0303 for $C_{12}H_{11}ClN_2O_2S$ [M + H]$^+$); m/z (%): 307 (4.1, [M + Na]$^+$, ^{37}Cl), 305 (44.6, [M + Na]$^+$, ^{35}Cl), 286 (4.6, [M + 2H]$^+$, ^{37}Cl), 285 (36.4, [M + H]$^+$, ^{37}Cl), 284 (12.9, [M + 2H]$^+$, ^{35}Cl), 283 (100, [M + H]$^+$, ^{35}Cl), 277 (47.6), 263 (6.6), 256 (4.3).

3.4. Small Scale Enzymatic Reactions

3.4.1. Lipase Screening for the Enzymatic Ring Opening of Oxazol-5(4H)-one rac-5a

To a solution of racemic 4-[(2-phenylthiazol-4-yl)methyl]-2-methyloxazol-5(4H)-one *rac-5a* (6 mg) in different alcohols (methanol, ethanol, *n*-propanol, *n*-butanol) (0.6 mL), different lipases (20 mg) were added. The reaction mixture was shaken at 1200 rpm at room temperature for 4.5 h. Samples were taken from the reaction mixture (20 μL), diluted to 1000 μL with a mixture of *n*-hexane and 2-propanol (4:1 *v/v*), filtered, and analysed by HPLC using a chiralpak IC column and a mixture of *n*-hexane and 2-propanol 80:20 (*v/v*) as eluent.

3.4.2. CaL-B Mediated DKR of Oxazol-5(4H)-one rac-5a with Different Alcohols and in Different Solvents

To a solution of racemic 4-[(2-phenylthiazol-4-yl)methyl]-2-methyloxazol-5(4H)-one *rac-5a* (10 mg) in different solvents (600 μL), CaL-B (10 mg), and ethanol (3 eq.) were added. The enzymatic reactions were performed with and/or without adding 6 mg of racemisation catalyst *N,N*-diethylaminoethanol immobilized on single-walled carbon nanotubes. The enzymatic reactions in acetonitrile were performed using different alcohols (methanol, ethanol, *n*-propanol, and *n*-butanol) (3 eq.). The reaction mixtures were shaken at 1200 rpm at room temperature. Samples were taken from the reaction mixture (50 μL), diluted to 1000 μL with a mixture of *n*-hexane and 2-propanol (4:1 *v/v*), filtered, and analysed by HPLC by the same procedure as described in Section 3.4.1. The enzymatic DKR of racemic oxazolones *rac-5b–d* were performed by a similar procedure, in the presence of the racemisation catalyst.

3.4.3. Enzymatic Hydrolysis of Racemic 2-Acetamido-3-(2-arylthiazol-4-yl)propanoic Acids rac-3a–d

To a suspension of *rac-3a–d* (50 mg) in demineralized water (6 mL), adjusted to pH 8 with a solution of LiOH 1.25 M, a catalytic amount of $CoCl_2 \cdot 6H_2O$ (1 mg) and Acylase I (20 mg) were added. The reaction mixture was stirred at 37 °C. The enzymatic reactions were monitored by TLC using a mixture of *n*-butanol:acetic acid:water 3:1:1 (*v/v/v*). For chiral HPLC analysis, samples were taken from the reaction mixture (100 μL), diluted with Tris-buffer (0.1 mM Tris HCl, pH 8), heated with active charcoal to 90 °C for 20 min, cooled, and filtered before injection. The chiral HPLC analysis was performed using a Chiralpak Zwix(+) column and a mixture of methanol (50 mM formic acid, 25 mM DEA):acetonitrile:water 49:49:2 (*v/v/v*) as eluent.

3.5. Large Scale Chemoenzymatic Preparation of L-2-Arylthiazol-4-yl Alanines L-6a–d

To a solution of racemic 2-acetamido-3-(2-arylthiazol-4-yl)propanoic acid *rac-3a–d* (5 mmol) in anhydrous dichloromethane (20 mL), a solution of *N,N'*-dicyclohexyl carbodiimide (1.24 g, 6 mmol) in anhydrous dichloromethane (8 mL) was added dropwise at 0 °C. The reaction mixture was stirred for 1 h at 0 °C. After the completion of the reaction (total conversion verified by TLC, eluent dichloromethane:acetone 9:1), the formed precipitate of dicyclohexyl urea was removed by filtration. The solvent was distilled off at reduced pressure, at room temperature. The obtained oxazol-5(4H)-one rac-**5a–d** was dissolved in anhydrous acetonitrile (85 mL). To the obtained solution CaL-B (1.4 g), the racemisation catalyst N,N-diethylaminoethanol immobilized on carboxyl single-wall carbon nanotubes (860 mg) and ethanol (3 eq., 15 mmol, 875 µL) in the case of rac-**5a–c**, respectively n-propanol (3 eq., 15 mmol, 1120 µL) in the case of rac-**5d** were added. The reaction mixture was stirred at 1200 rpm at room temperature and was monitored by chiral HPLC as described in Section 3.4.1. The total conversion was achieved after 2 days when rac-**5a–c** were used as substrates, and respectively after 3 days when rac-**5d** was used as substrate. After the completion of the enzymatic alcoholysis of rac-**5a–d**, the enzyme and the catalyst were filtered off and were washed three times with acetonitrile and chloroform, for the complete recovery of the reaction products. The solvent was removed under reduced pressure and the crude product was purified by column chromatography, using a mixture of dichloromethane: acetone 9:1 (v/v) as eluent, affording the 2-acetamido-3-(2-arylthiazol-4-yl)propanoic esters L-**4a–d** in 93%–96% yields and 78%–80% ee (Table 6).

The obtained L-2-acetamido-3-(2-arylthiazol-4-yl)propanoic esters L-**4a–d** were treated with a solution of sodium carbonate (0.53 g, 5 mmol) in water (10 mL). The reaction mixture was gently refluxed for 2 h, followed by acidifying with concentrated HCl (pH 3) and evaporation of water under reduced pressure. The obtained solid was redissolved in water (8 mL) by adjusting the pH to 8, using a solution of LiOH 1.25 M. Further Acylase I (60 mg) and $CoCl_2 \cdot 6H_2O$ (10 mg catalytic amount) were added and the reaction mixture was stirred at 37 °C, keeping the pH 7–8 with LiOH 1.25 M solution. The reaction was monitored by HPLC as described in Section 3.4.3. After the completion of the enzymatic hydrolysis (2 days), the reaction mixture was treated with phosphoric acid 5% until acidic pH (pH 1.5) and the enzyme was removed by centrifugation. The aqueous phase was applied to a DOWEX 50X8 cation exchange resin column. The enantiopure L-2-arylthiazol-4-yl alanines L-**6a–d** eluted with 2M NH_4OH solution.

3.6. Immobilisation of N,N-Diethylaminoethanol on Single-Walled Carbon Nanotubes (SWCNT)

One gram of carboxyl-functionalized single-walled carbon nanotubes ($SWCNT_{COOH}$) were suspended in anhydrous dichloromethane (12 mL) and treated with 1,1′-carbonyldiimidazole (120 mg). The mixture was sonicated for 5 min, and then shaken for 8 h at 1300 rpm. The suspension was filtered under reduced pressure and the precipitate was washed with anhydrous dichloromethane. The obtained filtrate was resuspended in anhydrous dichloromethane (12 mL) treated with glycerol diglycidyl ether (2 mL) and shaken for 24 h. The precipitate was then filtered under reduced pressure and washed several times with water. The obtained derivatized SWCNTs were suspended in anhydrous dichloromethane and treated with N,N-diethylaminoethanol (4 mL of anhydrous dichloromethane and 400 µL of N,N-diethylaminoethanol for 300 mg of derivatized SWCNTs). The mixture was shaken for 24 h and then filtered. The obtained precipitate was washed several times with anhydrous dichloromethane, dried, and used in the enzymatic DKR process.

4. Conclusions

An efficient chemoenzymatic procedure for the synthesis of various enantiopure L-2-arylthiazol-4-yl alanines was developed, based on the DKR of the corresponding oxazolones rac-**5a–d**. The novel SWCNT-immobilized amino functionalities proved to be efficient and mild racemisation agents in the CaL-B catalysed DKR process involving the stereoselective ring opening of oxazol-5(4H)-ones in organic media, not affecting the enzyme selectivity and activity, thus affording the corresponding N- and C- protected L-amino acids (78%–80% ee) with 93%–96% yields. In the next steps, the chemical hydrolysis at the ester function in mild basic conditions followed by an enantioselective hydrolysis of the amide bond, mediated by Acylase I, in aqueous media, yields final products with ee values increased to more than 99%, due to the L-specificity of Acylase I. The developed chemoenzymatic DKR-KR procedure was optimized and successfully applied for the preparative production of enantiopure L-2-arylthiazol-4-yl alanines, with 74%–78% global yields.

Acknowledgments: This work was supported by a grant of the Romanian National Authority for Scientific Research, CNCS-UEFISCDI, project number PN-II-ID-PCE-2011-3-0775.

Author Contributions: Valentin Zaharia and Florin Dan Irimie designed the research. Denisa Leonte and László Csaba Bencze performed the experiments and analyzed the data. Csaba Paizs and Monica Ioana Toşa performed the spectral analysis of the obtained compounds and contributed to the data analysis. All authors contributed to the elaboration of the article and approved the manuscript.

Conflicts of Interests: The authors declare no conflict of interest.

References

1. Cardillo, G.; Gentilucci, L.; Tolomelli, A. Unusual amino acids: synthesis and introduction into naturally occurring peptides and biologically active analogues. *Mini Rev. Med. Chem.* **2006**, *6*, 293–304. [CrossRef] [PubMed]

2. Whitby, L.R.; Boger, D.L. Comprehensive Peptidomimetic Libraries Targeting Protein–Protein Interactions. *Acc. Chem. Res.* **2012**, *45*, 1698–1709. [CrossRef] [PubMed]

3. Mazzei, T. Chemistry and Mechanism of Action of Bleomycin. *Chemioterapia* **1984**, *3*, 316–319. [PubMed]

4. Pucci, M.J.; Bronson, J.J.; Barrett, J.F.; DenBleyker, K.L.; Discotto, L.F.; Fung-Tomc, J.C.; Ueda, Y. Antimicrobial Evaluation of Nocathiacins, a Thiazole Peptide Class of Antibiotics. *Antimicrob. Agents Chemother.* **2004**, *48*, 3697–3701. [CrossRef] [PubMed]

5. Adiv, S.; Ahronov-Nadborny, R.; Carmeli, S. New aeruginazoles, a group of thiazole-containing cyclic peptides from *Microcystis aeruginosa* blooms. *Tetrahedron* **2012**, *68*, 1376–1383. [CrossRef]

6. Zhang, C.; Zink, D.L.; Ushio, M.; Burgess, B.; Onishi, R.; Masurekar, P.; Barrett, J.F.; Singh, S.B. Isolation, structure, and antibacterial activity of thiazomycin A, a potent thiazolyl peptide antibiotic from *Amycolatopsis fastidiosa. Bioorg. Med. Chem.* **2008**, *16*, 8818–8823. [CrossRef] [PubMed]

7. Schmid, A.; Dordick, J.S.; Hauer, B.; Kiener, A.; Wubbolts, M.; Witholt, B. Industrial biocatalysis today and tomorrow. *Nature* **2001**, *409*, 258–268. [CrossRef] [PubMed]

8. De Miranda, A.S.; Miranda, L.S.M.; de Souza, R.O. Lipases: Valuable catalysts for dynamic kinetic resolution. *Biotechnol. Adv.* **2015**, *33*, 372–393. [CrossRef] [PubMed]

9. Bodalo, A.; Gomez, J.L.; Gomez, E.; Bastida, J.; Leon, G.; Maximo, M.F.; Hidalgo, A.M.; Montiel, M.C. Kinetic calculations in the enzymatic resolution of DL-amino acids. *Enzyme Microb. Technol.* **1999**, *24*, 381–387. [CrossRef]

10. Turner, N.; Winterman, J.; McCague, R.; Parratt, J.; Taylor, S. Synthesis of homochiral L-(*S*)-*tert*-Leucine via a lipase catalysed dynamic resolution process. *Tetrahedron Lett.* **1995**, *36*, 1113–1116. [CrossRef]

11. Brown, S.; Parker, M.C.; Turner, N. Dynamic kinetic resolution: Synthesis of optically active α-amino acid derivatives. *Tetrahedron Asymmetry* **2000**, *11*, 1687–1690. [CrossRef]

12. De Jersey, J.; Zerner, B. On the spontaneous and enzyme-catalyzed hydrolysis of saturated oxazolinones. *Biochemistry* **1969**, *8*, 1967–1974. [CrossRef] [PubMed]

13. Podea, P.V.; Toşa, M.I.; Paizs, C.; Irimie, F.D. Chemoenzymatic preparation of enantipure L-benzofuranyl- and L-benzo[*b*]thiophenyl alanines. *Tetrahedron Asymmetry* **2008**, *19*, 500–511. [CrossRef]

14. Bencze, L.C.; Komjáti, B.; Pop, L.A.; Paizs, C.; Irimie, F.D.; Nagy, J.; Poppe, L.; Toşa, M.I. Synthesis of enantipure L-(5-phenylfuran-2-yl)alanines by a sequential multienzyme process. *Tetrahedron Asymmetry* **2015**, *26*, 1095–1101. [CrossRef]

15. Bartha-Vári, J.H.; Tosa, M.I.; Irimie, F.-D.; Weiser, D.; Boros, Z.; Vértessy, B.G.; Paizs, C.; Poppe, L. Immobilization of phenylalanine ammonia-lyase on single-walled carbon nanotubes for stereoselective biotransformation in batch and continuous-flow modes. *ChemCatChem* **2015**, *7*, 1122–1128. [CrossRef]

16. Bencze, L.C.; Bartha-Vári, J.H.; Katona, G.; Toşa, M.I.; Paizs, C.; Irimie, F.D. Nanobioconjugates of *Candida antarctica* lipase B and single-walled carbon nanotubes in biodiesel production. *Bioresour. Technol.* **2016**, *200*, 853–860. [CrossRef] [PubMed]

17. Silberg, A.; Simiti, I.; Mantsch, H. Beiträge zum Studium der Thiazole, I. Über die Herstellung und die Eigenschaften von 2-Aryl-4-halogenmethyl-thiazolen. *Chem. Ber.* **1961**, *94*, 2887–2894. [CrossRef]

18. Burger, K.; Gold, M.; Neuhauser, H.; Rudolph, M.; Höß, E. Synthesis of 3-(Thiazol-4-yl)alanine and 3-(Selenazol-4-yl)alanine derivatives from aspartic acid. *Synthesis* **1992**, *11*, 1145–1150. [CrossRef]

9

Preparative Isolation of Two Prenylated Biflavonoids from the Roots and Rhizomes of *Sinopodophyllum emodi* by Sephadex LH-20 Column and High-Speed Counter-Current Chromatography

Yan-Jun Sun [1,2,*], **Li-Xin Pei** [1,2], **Kai-Bo Wang** [3,4], **Yin-Shi Sun** [5,*], **Jun-Min Wang** [1,2], **Yan-Li Zhang** [1,2], **Mei-Ling Gao** [1,2] and **Bao-Yu Ji** [1,2]

Academic Editor: Derek J. McPhee

[1] Collaborative Innovation Center for Respiratory Disease Diagnosis and Treatment & Chinese Medicine Development of Henan Province, Henan University of Traditional Chinese Medicine, Zhengzhou 450046, Henan, China; xlpjby@sina.com (L.-X.P.); wjmhnzz@163.com (J.-M.W.); zyl2013hnzy@163.com (Y.-L.Z.); gaoxiaomei6266@126.com (M.-L.G.); ys20052@sina.com (B.-Y.J.)

[2] School of Pharmacy, Henan University of Traditional Chinese Medicine, Zhengzhou 450046, Henan, China

[3] Key Laboratory of Structure-Based Drug Design & Discovery, Ministry of Education, Shenyang Pharmaceutical University, Shenyang 110016, Liaoning, China; wangkaibo2014@163.com

[4] School of Traditional Chinese Materia Medica, Shenyang Pharmaceutical University, Shenyang 110016, Liaoning, China

[5] Institute of Special Animal and Plant Sciences, Chinese Academy of Agricultural Sciences, Changchun 130112, Jilin, China

[*] Correspondence: sunyanjun2011@hactcm.edu.cn (Y.-J.S.); sunyinshi2002@163.com (Y.-S.S.)

Abstract: Two prenylated biflavonoids, podoverines B–C, were isolated from the dried roots and rhizomes of *Sinopodophyllum emodi* using a Sephadex LH-20 column (SLHC) and high-speed counter-current chromatography (HSCCC). The 95% ethanol extract was partitioned with ethyl acetate in water. Target compounds from the ethyl acetate fraction were further enriched and purified by the combined application of SLHC and HSCCC. *n*-Hexane–ethyl acetate–methanol–water (3.5:5:3.5:5, *v/v*) was chosen as the two phase solvent system. The flow rate of mobile phase was optimized at 2.0 mL·min^{-1}. Finally, under optimized conditions, 13.8 mg of podoverine B and 16.2 mg of podoverine C were obtained from 200 mg of the enriched sample. The purities of podoverines B and C were 98.62% and 99.05%, respectively, as determined by HPLC. For the first time, podoverins B and C were found in the genus *Sinopodophyllum*. Their structures were determined by spectroscopic methods (HR-ESI-MS, ^{1}H-NMR, ^{13}C-NMR, HSQC, HMBC). Their absolute configurations were elucidated by comparison of their experimental and calculated ECD spectra. The cytotoxic activities were evaluated against MCF-7 and HepG2 cell lines. The separation procedures proved to be practical and economical, especially for trace prenylated biflavonoids from traditional Chinese medicine.

Keywords: *Sinopodophyllum emodi*; prenylated biflavonoid; Sephadex LH-20 column chromatography; high-speed counter-current chromatography; cytotoxic activity

1. Introduction

Sinopodophyllum emodi (Wall.) Ying, which belongs to the family of Berberidaceae, is a herbaceous perennial plant widely distributed in the Southwest of China [1]. Listed in Chinese Pharmacopoeia, the dried fruits are clinically applied to the treatment of amenorrhea, dead fetus, and placental retaining as a traditional Tibetan medicine [2]. Its roots and rhizomes have been used for the treatment of certain cancer, various verrucosis [1], constipation, parasitosis [3], rheumatoid ache [4], and pyogenic infection of skin tissue [5]. Previous chemical investigations on *S. emodi* revealed the presence of bioactive aryltetralin [1,3–8] and tetrahydrofuranoid lignans [9], flavonoids [2,10–12], steroids [13], and phenolics [14]. Although flavonoids have been the research focus of this plant [2,10–12], the biflavonoids, podoverines B, C (Figure 1), were found firstly in the genus *Sinopodophyllum*. As anti-inflammatory agents, podoverines B and C have been isolated from *Podophyllum versipelle* callus cell culture [15], further studies are not reported up to now. Naturally-occurring biflavonoids have exhibited a broad spectrum of biological activities, including anticancer, antibacterial, antifungal, antiviral, anti-inflammatory, analgesic, antioxidant, vasorelaxant, anticlotting, *etc.* [15,16]. As potential therapeutic drugs against cancer, biflavonoids strongly interfere in related pathways of the cancer cell growth and death while have little effect on normal cell proliferation [16]. Therefore, it is critical and urgently needed to develop a rapid and efficient method for the preparative isolation of podoverines B and C from *S. emodi*.

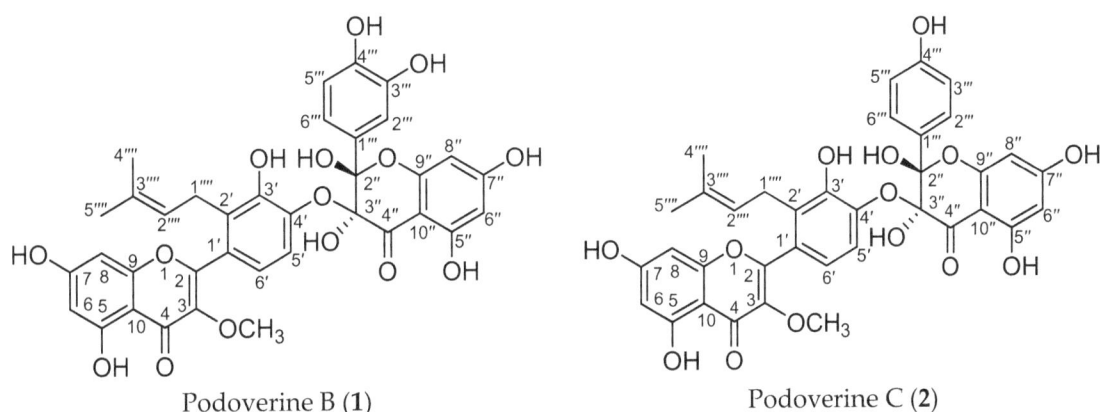

Podoverine B (**1**)   Podoverine C (**2**)

Figure 1. The chemical structures of two prenylated biflavonoids from *S. emodi.*

HSCCC is a support-free liquid-liquid chromatographic technology, based on partitioning of target compounds between two immiscible liquid phases. Some complications arising from solid absorbents can be eliminated, such as irreversible adsorption and denaturation of target compounds [17]. Compared with traditional liquid–solid chromatographic methods, it also has a large number of advantages, including high sample recovery, large loading capacity, low solvent consumption, acceptable efficiency, low cost, and the ease of scaling-up [18,19]. HSCCC has been widely used in the large scale preparative isolation and purification of various kinds of natural products. Although HSCCC has been developed for the purification of normal biflavonoid [20,21], there are no reports on the application of HSCCC for the preparative isolation of prenylated biflavonoids from natural sources. In this work, the HSCCC has been applied in combination with SLHC for the purification of two prenylated biflavonoids (podoverines B and C) from the roots and rhizomes of *S. emodi*. The HSCCC isolation conditions, including two-phase solvent system, flow rate, and revolution speed, were optimized. The chemical structures of the two target compounds were elucidated by HR-ESI-MS, ^{1}H-NMR, ^{13}C-NMR, HSQC, and HMBC. Their absolute configurations were determined by comparison of their experimental and calculated ECD spectra. The cytotoxic activities were also evaluated against MCF-7 and PC-3 cell lines by the MTT method. As far as we know, the HSCCC separation, absolute configurations, and cytotoxic activities of podoverines B and C are now reported for the first time.

2. Results and Discussion

2.1. Enrichment by Sephadex LH-20 Column

The roots and rhizomes of *S. emodi* is particularly rich in aryltetralin lactone lignans and normal flavonoids [22]. Based on the HPLC analysis, it was almost impossible for podoverines B and C to be detected in the ethyl acetate fraction. Furthermore, the polarities and molecular weights of prenylated biflavonoids were similar to those of aryltetralin lactone lignan glycosides. To enrich effectively the target compounds, SLHC chromatography was employed for pre-separation, eluting with a methanol–water gradient. The chromatographic parameters, including mobile phase composition, loading amount, and flow rate, were investigated to produce optimum separation efficiency.

Two kinds of binary solvent systems were examined, including gradient dichloromethane–methanol and methanol–water. When the mixed solvent system of dichloromethane–methanol (v/v, 1:1, 1:1.5, 1:2, 0:1) was used as mobile phase, the target compounds were co-eluted, together with a lot of impurities with similar properties (Figure 2A). The different ratios of isocratic methanol (A)–water (B) (v/v, 10%, 20%, 30%, 40%, 50%, 60%, 70%, 100% A) were also investigated systematically. Brown pigments and lignans, not the target compounds, could be eluted by 10%–50% A. Two target compounds, together with another unknown constituent, mainly existed in 60% A (Figure 2B). Therefore, the target compounds were enriched in a gradient elution mode (50%, 60% A, Figure 2C).

Figure 2. *Cont.*

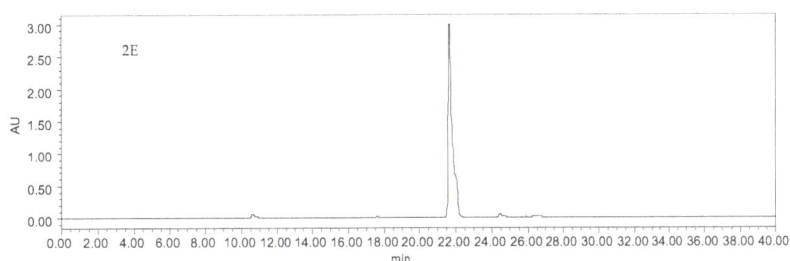

Figure 2. (**A**) HPLC chromatogram of the enriched sample from SLHC, which were isolated with the mixed solvents of dichloromethane–methanol; (**B**) HPLC chromatogram of the enriched sample from SLHC, which were isolated with isocratic 60% methanol; (**C**) HPLC chromatogram of the enriched sample from SLHC, which were isolated with a gradient methanol–water; (**D**) HPLC chromatogram of HSCCC peak fraction 1 in Figure 3; (**E**) HPLC chromatogram of HSCCC peak fraction 2 in Figure 3; Experimental conditions: column, a YMC-Pack ODS A column (5 μm, 250 mm × 4.6 mm); mobile phase, methanol (C) and 0.1% trifluoroacetic acid (D) at the gradient (20%–65% C at 0–20 min, 65%–100% C at 20–40 min); flow rate, 1.0 mL· min^{-1}; detection wavelength, 254 nm; column temperature, 35 °C.

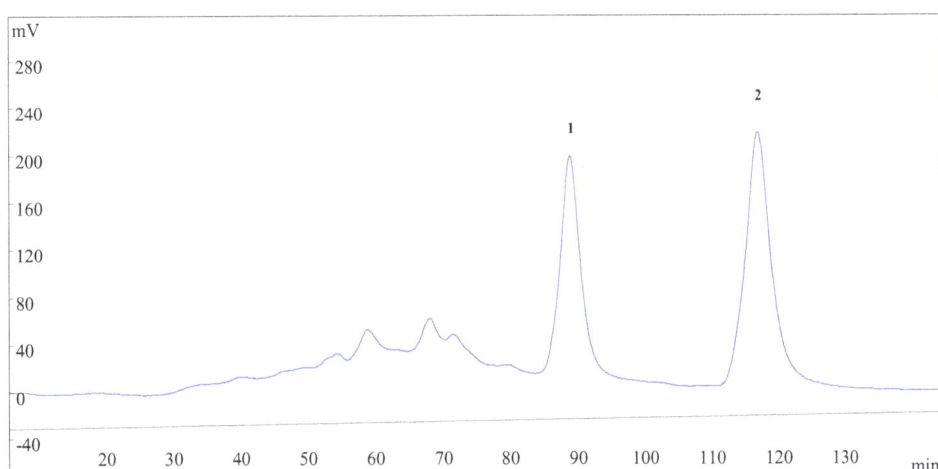

Figure 3. HSCCC chromatogram of the enriched sample from SLHC. Two-phase solvent system: *n*-hexane–ethyl acetate–methanol–water (3.5:5:3.5:5, *v*/*v*); mobile phase: the lower phase; stationary phase: the upper phase; flow rate: 2.0 mL· min^{-1}; revolution speed: 800 rpm; detection wavelength: 254 nm; sample size: 200 mg enriched sample was dissolved in the solvent mixture of *n*-hexane–ethyl acetate–methanol–water (5 mL for each phase).

As sample loading amount exceeded column loading capacity, the separation efficiency on the SLHC decreased obviously. When it was more than 20 mg· g^{-1}, a great deal of lignan glucosides were detected in the eluates containing target compounds. The separation of target compounds was satisfactory with sample loading amount from 10 to 20 mg· g^{-1}. With the decrease of required Sephadex LH-20, the contents of target compounds increased in the eluates. According to these results, the optimal loading amount on the Sephadex LH-20 column was determined as 20 mg· g^{-1}.

For SLHC, flow rate is also an important influencing factor. Generally, the lower the flow rate, the better the resolution. However, the lower the flow rate, the longer the elution time. The effect of flow rate on separation efficiency was further investigated. The contents of impurities in the collected fraction increased with flow rate increasing from 1 to 3 mL· min^{-1}. The separation efficiency gradually increased with the flow rate decreasing from 1 to 0.6 mL· min^{-1}. In order to reduce the total elution time, an ideal flow rate was selected at 1 mL· min^{-1}.

2.2. Selection of Two-Phase Solvent System and Other Conditions of HSCCC

In HSCCC, a suitable two-phase solvent system was critical for successful separation. The two-phase solvent system was selected according to the following requirements [23]: (i) the target compounds should be stable and soluble in the selected system; (ii) the settling time of the two-phase solvent system should be shorter than 30 s; and (iii) the optimal volume ratio of the two-phase solvent system provides an ideal range of the coefficients (K, *i.e.*, usually between 0.5 and 2.5) for the target compounds. Small K values lead to the disappearance of peak resolution, while large K values tend to require a relatively large quantity of organic solvents, long operation time, and produce sample band broadening. For all of the selected two-phase solvent systems, the settling time was less than 20 s. In order to achieve ideal resolution of target compounds, the K values for *n*-hexane–ethyl acetate–methanol–water at different volume ratio were measured systematically (Table 1). Started with *n*-hexane–ethyl acetate–methanol–water (5:5:5:5 *v/v*), two target compounds were eluted close to each other near the solvent front. The distribution capacity of target compounds in the upper layer was adjusted by volume ratio of *n*-hexane–methanol or *n*-hexane–ethyl acetate in mixed solvent system. For the solvent systems consisted of *n*-hexane–ethyl acetate–methanol–water (2:5:2:5), two target compounds were retained in the column for a relatively long time (5 h). In those systems of *n*-hexane–ethyl acetate–methanol–water (4.5:5:4.5:5, and 6:4:5:5), K values were too small. Whereas in those systems of *n*-hexane–ethyl acetate–methanol–water (1:5:1:5), K values were too big. For *n*-hexane–ethyl acetate–methanol–water (4:5:4:5, 3.5:5:3.5:5 and 3:5:3:5), the K values were between 0.5 and 2.5. In contrast, with *n*-hexane–ethyl acetate–methanol–water at a ratio of 3.5:5:3.5:5, the target compounds were separated satisfactorily with high resolution (α = 1.25) in an acceptable run time. Thus, this solvent system was selected for subsequent HSCCC separation.

Table 1. The partition coefficients (K) and separation factors (α) of the target compounds in several solvent systems.

Solvent System	Ratio	K Value		α
		1	2	
n-Hexane–ethyl acetate–methanol–water	6:4:5:5	0.12	0.15	1.25
n-Hexane–ethyl acetate–methanol–water	5:5:5:5	0.18	0.23	1.28
n-Hexane–ethyl acetate–methanol–water	4.5:5:4.5:5	0.29	0.37	1.28
n-Hexane–ethyl acetate–methanol–water	4:5:4:5	0.60	0.69	1.15
n-Hexane–ethyl acetate–methanol–water	3.5:5:3.5:5	0.84	1.05	1.25
n-Hexane–ethyl acetate–methanol–water	3:5:3:5	1.41	1.64	1.16
n-Hexane–ethyl acetate–methanol–water	2:5:2:5	2.56	2.83	1.11
n-Hexane–ethyl acetate–methanol–water	1:5:1:5	5.93	6.14	1.04

The retention of the stationary phase is also one of the most important parameters in HSCCC. The retention of the stationary phase is highly correlated with the flow rate of the mobile phase and the revolution speed of the separation column [21]. Different flow rate (1.6, 1.8, 2.0, 2.2, and 2.5 mL· min^{-1}) of the mobile phase and different revolution speed (600, 700, 800, 900 rpm) were investigated with the selected solvent system. At the flow rate of 2.2 and 2.5 mL· min^{-1}, podoverine C (**2**) was separated satisfactorily, however, podoverine B (**1**) peak always coexisted with some nearby impurities. At the flow rate of 1.6, 1.8, and 2.0 mL· min^{-1}, podoverine B (**1**) was separated from impurities in a higher resolution. The higher the retention of the stationary phase, the better the peak resolution. Low flow rate and high revolution speed increase the retention of the stationary phase [24]. However, with the decrease of the flow rate, the more elution time and more mobile phase will be needed, and the chromatographic peak is widened. High revolution speed may also produce broadening sample bands by violent pulsation [25] and reduce the life of instrument [26]. The retention of the stationary phase was poor at a revolution speed of 600 and 700 rpm, whereas the satisfactory retention was obtained at

800 rpm. Considering these aspects, a flow rate of 2.0 mL· min^{-1} and a revolution speed of 800 rpm were finally selected for HSCCC separation.

Under the optimized conditions, podoverines B (13.8 mg) and C (16.2 mg) were successfully obtained from 200 mg of the enriched sample in one-step elution and less than 150 min. Their purities were 98.62% and 99.05%, respectively. The retention percentage of the stationary phase was 62%.

2.3. Optimization of HPLC Conditions

As shown in Figure 2C, a considerable level of impurities, which showed similar polarities to two target compounds, were detected in the enriched sample. The only structural difference between two target compounds was that podoverine B has one more hydroxyl group at C-3''', so that their properties such as UV absorbance and chromatographic pattern were similar. The contents of podoverines B and C were 25.7% and 33.1%, based on the ratio of peak area, respectively. Optimal HPLC analytical conditions were required to ensure the baseline separation and accurate purity results of the target compounds. Thus, different elution mode, flow rate, column temperature, and detection wavelength were evaluated. For natural products with hydroxyl groups, acid is generally used for reducing the tailing and broadening of peak. To improve separation resolution, trifluoroacetic acid was added into the mobile phase. When isocratic methanol (C)–0.1% trifluoroacetic acid (v/v) (D) (85% C) was used as a mobile phase, the retention time of two target compounds were identical in the HPLC chromatogram. By decreasing methanol ratio in the mobile phase, two target compounds were separated gradually from each other, however, their retention times became longer and nearby impurities were almost hidden by the peak of podoverin B. The screening results indicated the gradient elution of methanol (C)–0.1% trifluoroacetic acid (v/v) (D) (20%–65% C at 0–20 min, 65%–100% C at 20–40 min) gave a satisfactory separation of the target compounds, when the flow rate, column temperature and detection wavelength were at 1.0 mL· min^{-1}, 35 °C and 254 nm.

2.4. Identification of the Separated Peaks

The chemical structures of two prenylated biflavonoids were determined on the basis of spectroscopic evidences (HR-ESI-MS, ^{1}H-NMR, ^{13}C-NMR, HSQC, and HMBC in Supplementary Materials), and their absolute configurations were elucidated by CD analysis. The data were given in Table 2.

Compound **1** was obtained as a yellow amorphous powder and possessed a molecular formula $C_{36}H_{30}O_{15}$ with twenty-two degrees of unsaturation, as revealed from its HR-ESI-MS analysis (m/z 685.1538 [M + H − H$_2$O]$^+$, calcd for $C_{36}H_{29}O_{14}$, 685.1557; m/z 707.1355 [M + Na − H$_2$O]$^+$, calcd for $C_{36}H_{28}O_{14}Na$, 707.1377; m/z 723.1093 [M + K − H$_2$O]$^+$, calcd for $C_{36}H_{28}O_{14}K$, 723.1116). The ^{1}H-NMR spectrum showed four aromatic systems including one 1,2,3,4-tetra-substituted benzene ring δ 7.13 (1H, d, J = 8.4 Hz), 7.01 (1H, d, J = 8.4 Hz), one 1,3,4-tri-substituted benzene ring δ 7.11 (1H, d, J = 2.1 Hz), 6.89 (1H, dd, J = 8.4, 2.1 Hz), 6.68 (1H, d, J = 8.4 Hz), and two 1,2,3,5-tetra-substituted benzene ring at δ 6.34 (1H, d, J = 1.6 Hz), 6.22 (1H, d, J = 1.6 Hz), and δ 5.97 (2H, s); One 3-methyl-2-butenyl δ 5.01 (1H, t, J = 7.0 Hz), 3.28 (2H, d, J = 7.0 Hz), 1.48 (3H, s), and 1.27 (3H, s); One aromatic methoxy group δ 3.67 (3H, s); One chelated phenolic hydroxyl group δ 12.55 (1H, s, 5-OH). The ^{13}C-NMR and HSQC spectrum revealed one 3-methyl-2-butenyl δ 25.5, 121.2, 131.7, 17.3, 25.3; one aromatic methoxy group δ 60.1; one biflavone skeleton including two carbonyl group at δ 187.3, 178.0; four benzene rings, two oxygen-bearing olefinic carbons δ 139.0, 157.5; two di-oxygen-bearing aliphatic quaternary carbons δ 100.1, 90.2. The HMBC correlations of the aromatic protons δ 7.13 (1H, d, J = 8.4 Hz, H-6') with C-2 (δ 157.5), the methylene group protons δ 3.28 (2H, d, J = 7.0 Hz) with C-2' (δ 129.3), and the methoxy group protons δ 3.67 (3H, s) with C-3 (δ 139.0), in combination with one 1,2,3,5-tetra-substituted benzene ring at δ 6.34 (1H, d, J = 1.6 Hz), 6.22 (1H, d, J = 1.6 Hz), indicated that compound **1** contained podoverine A [15] as a subunit. Another subunit was identified as 2,3,3,5,7,3',4'-heptahydroxyflavanone by the HMBC cross peak of the aromatic protons δ 7.11 (1H, d, J = 2.1 Hz, H-2''') and 6.89 (1H, dd, J = 8.4, 2.1 Hz, 6''') with C-2'' (δ 100.1), one 1,2,3,5-penta-substituted benzene ring δ 5.97 (2H, s), and two di-oxygen-bearing aliphatic quaternary carbons δ 100.1, 90.2. A careful comparison of the ^{13}C-NMR spectra of **1** with podoverine A indicated that the subunit of podoverine A was substituted at C-4′, which was confirmed by a chemical shift change from δ 138.3 (C-3′), 142.0 (C-4′), and 115.1 (C-5′) in **1** to δ 144.0 (C-3′), 147.3 (C-4′), and 113.2 (C-5′) in podoverine A. An ether bridge C-4′-O-C-3″ of the two flavonoid subunits was determined by one di-oxygen-bearing aliphatic quaternary carbons δ 90.2 (C-3″). Detailed elucidation on 1D and 2D NMR data led to the construction of the planar structure of **1**. Despite repeated experiments, suitable crystals of **1** for X-ray diffraction were not obtained successfully. The absolute configurations of C-2″ and C-3″ were extrapolated by comparing the experimental and calculated CD spectra, the latter performed by density functional theory. The results showed that experimental and calculated spectra for the 2″*S*, 3″*R*-isomer were in good agreement (Figure 4). Therefore, the absolute configuration at C-2″ and C-3″ were respectively deduced to be *S* and *R*. On the basis of the above evidences and related literature [15], compound **1** was determined as podoverine B.

Table 2. ^{1}H-NMR and ^{13}C-NMR spectroscopic data for compounds **1–2**.

Position	**1** [a]		**2** [a]	
	δ_C	δ_H	δ_C	δ_H
2	157.5		157.4	
3	139.0		139.0	
4	178.0		178.0	
5	161.4		161.3	
6	98.7	6.22 (1H, d, *J* = 1.6)	98.7	6.22 (1H, d, *J* = 2.1)
7	164.3		164.3	
8	93.7	6.34 (1H, d, *J* = 1.6)	93.7	6.34 (1H, d, *J* = 2.1)
9	156.8		156.8	
10	104.7		104.7	
1′	124.7		124.3	
2′	129.3		129.3	
3′	138.3		138.3	
4′	142.0		142.0	
5′	115.1	7.01 (1H, d, *J* = 8.4)	115.1	7.01 (1H, d, *J* = 8.4)
6′	124.1	7.13 (1H, d, *J* = 8.4)	124.1	7.12 (1H, d, *J* = 8.4)
2″	100.1		100.2	
3″	90.2		90.2	
4″	187.3		187.3	
5″	163.0		163.1	
6″	97.1	5.97 (1H, s)	97.1	5.98 (1H, d, *J* = 2.1)
7″	167.8		167.8	
8″	96.2	5.97 (1H, s)	96.1	5.99 (1H, d, *J* = 2.1)
9″	159.1		159.1	
10″	99.8		99.7	
1‴	124.1		124.1	
2‴	114.7	7.11 (1H, d, *J* = 2.1)	129.1	7.44 (1H, dd, *J* = 6.9, 2.1)
3‴	144.6		114.7	6.75 (1H, dd, *J* = 6.9, 2.1)
4‴	146.8		158.6	
5‴	115.6	6.68 (1H, d, *J* = 8.4)	114.7	6.75 (1H, dd, *J* = 6.9, 2.1)
6‴	119.1	6.89 (1H, dd, *J* = 8.4, 2.1)	129.	7.44 (1H, dd, *J* = 6.9, 2.1)
OCH$_3$	60.1	3.67 (3H, s)	60.1	3.66 (3H, s)
1⁗	25.5	3.28 (2H, d, *J* = 7.0)	25.5	3.27 (2H, d, *J* = 6.8)
2⁗	121.2	5.01 (1H, t, *J* = 7.0)	121.2	5.01 (1H, t, *J* = 6.8)
3⁗	131.7		131.6	
4⁗	17.3	1.27 (3H, s)	17.3	1.26 (3H, s)
5⁗	25.3	1.48 (3H, s)	25.2	1.48 (3H, s)
OH		12.55 (1H, s)		12.55 (1H, s)

[a] NMR spectroscopic data were recorded in DMSO-d_6 at 500 MHz (^{1}H-NMR) and 125 MHz (^{13}C-NMR).

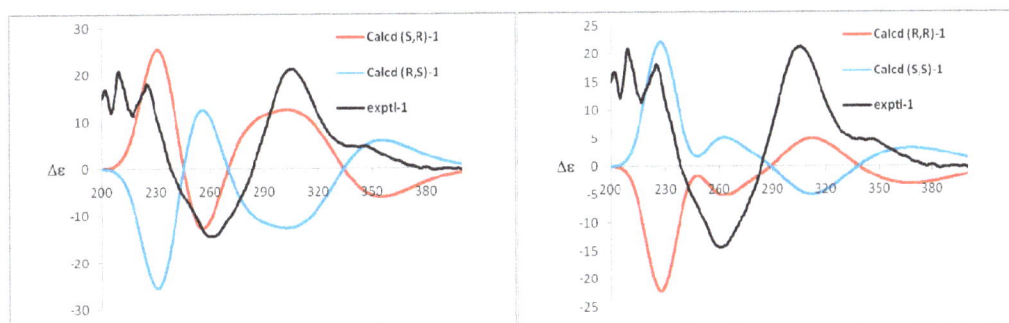

Figure 4. Experimental and calculated ECD spectra of compound **1**.

Compound **2** was obtained as a yellow amorphous powder. Its ^{1}H- and ^{13}C-NMR were quite similar to those of **1**, except that one 1,4-di-substituted benzene ring δ 7.44 (2H, dd, J = 6.9, 2.1 Hz), 6.75 (2H, dd, J = 6.9, 2.1 Hz), were observed instead of one 1,3,4-tri-substituted benzene ring in **1**. Those were further supported by HR-ESI-MS, which gave a molecular formula $C_{36}H_{30}O_{14}$ (m/z 669.1589 [M + H − H$_2$O]$^+$, calcd for $C_{36}H_{29}O_{13}$, 669.1608; m/z 691.1409 [M + Na − H$_2$O]$^+$, calcd for $C_{36}H_{28}O_{13}Na$, 691.1428; m/z 707.1173 [M + K − H$_2$O]$^+$, calcd for $C_{36}H_{28}O_{13}K$, 707.1167), being 16 mass-units less than that of **1**, respectively. The CD spectrum of **2** was identical with **1** which has the absolute configuration (2″S, 3″R). Thus, compound **2** were identified as podoverine C [15].

2.5. The Cytotoxic Activity of Target Compounds

Target compounds were evaluated for their *in vitro* cytotoxic activity against MCF-7 and HepG2 cell lines using MTT assay [8], and IC$_{50}$ values were summarized in Table 3. According to IC$_{50}$ values, compound **1** showed higher cytotoxic activities than **2**, indicating that the hydration at C-3′ resulted in a greater increase of cytotoxic activity.

Table 3. Cytotoxic activity of target compounds (IC$_{50}$, μM).

Compound	MCF-7	HepG2
1	29.8 ± 2.0	41.6 ± 1.9
2	42.6 ± 3.1	67.5 ± 2.6
etoposide	3.17 ± 0.25	0.48 ± 0.03

3. Experimental Section

3.1. Apparatus

The preparative HSCCC experiments were performed on a Model GS-10A high-speed counter-current chromatography (Beijing Institute of New Technology Application, Beijing, China). The instrument was equipped with a PTFE (polytetrafluoroethylene) multilayer coil column (i.d. of the tubing = 1.6 mm, a total capacity = 230 mL) and a manual sample injection valve with a 10 mL loop. The revolution radius was 5 cm, and the β value of the multilayer coil ranged from 0.5 at internal terminal to 0.8 at the external terminal. The maximum revolution speed could be controlled up to 1000 rpm by a speed controller. The HSCCC system was also equipped with BF-2002 CT11 signal collection cell (Chromatography Center of Beifenruili Group Company, Beijing, China), a Model NS-1007A constant-flow pump and a Model 8823B-UV detector (Beijing Institute of New Technology Application, Beijing, China) at 254 nm. The data were collected with HW-2000 chromatography workstation (Qianpu Software Co. Ltd., Shanghai, China).

Analytical HPLC was performed on high-performance liquid chromatography system with a Waters Alliance 2489 separations module equipped with a Waters 2695 UV/visible detector, a quaternary pump, a column temperature control module, and a Waters 717 plus autosampler (Milford,

MA, USA). Empower pro data handling system (Waters Co., Milford, CT, USA) was employed to carry out data acquisition.

The structures of the target compounds were identified by high resolution electrospray ionization mass (HR-ESI-MS) spectrometer (Shimadzu LC-MS 2010, Japan) and nuclear magnetic resonance (NMR) spectrometer (Bruker AM 500, Fällanden, Switzerland). CD spectra were measured on Bio-logic MOS 450 spectropolarimeter (Bio-logic Co., Claix, France). IR spectra were determined on a Nicolet is 10 Microscope Spectrometer (Thermo Scientific, San Jose, CA, USA). UV spectra were recorded on a UV-2401PC apparatus (Shimadzu Corporation, Kyoto, Japan).

3.2. Materials and Reagents

All organic solvents for sample preparation and HSCCC separation were of analytical grade (Fuyu Chemical Reagent Co. Ltd., Tianjin, China). Methanol for HPLC analysis was of chromatographic grade (Siyou Biology Medical Tech Co. Ltd., Tianjin, China), and water was purified by means of a water purifier (18.2 MΩ) (Wanjie Water Treatment Equipment Co. Ltd., Hangzhou, China). The target compounds were enriched by Sephadex LH-20 (Amersham Pharmacia Biotech AB, Uppsala, Sweden). Biological reagents were obtained from Sigma Company (St. Louis, MO, USA). Human heptocellular (HepG2) and breast (MCF-7) cell lines were from Institute of Materia Medica, Chinese Academy of Medical Sciences and Peking Union Medical College (Beijing, China).

The plant materials were collected in Deqin, Yunnan province, People's Republic of China, in September 2014, and were identified as the roots and rhizomes of *S. emodi* (Wall.) Ying according to Chinese Traditional Medicine Dictionary by Professor Cheng-ming Dong (School of Pharmacy, Henan University of Traditional Chinese Medicine).

3.3. Preparation of the Crude Extract

The dried roots and rhizomes of *S. emodi* were ground to powder by a disintegrator. The powders (3.0 kg) were extracted under reflux by 10-fold amounts of 95% ethanol. The extraction procedure was then repeated twice. The extracts were combined together, filtrated with cotton, and then concentrated under reduced pressure to give brown syrup (639 g). This syrup was suspended in 3 L distilled water, and then partitioned with equal volumes of ethyl acetate. After concentration and freeze-drying, the ethyl acetate fraction (25 g) were stored at −10 °C for subsequent experiments.

3.4. Enrichment of the Target Compounds by Sephadex LH-20 Column

The ethyl acetate fraction (6.6 g) was dissolved ultrasonically in methanol (10 mL), and filtered by 0.45 μm microporous membrane. The SLHC was packed as follows: The exit of the chromatographic column (140 cm length × 4 cm i.d.) was plugged with glass wool to retain solids. Sephadex LH-20 (330 g) was swollen with methanol for 4 h. The swollen Sephadex LH-20 slurry was poured into the column in a continuous motion. The column was rinsed with methanol. Before applying the sample, the column was equilibrated with 2 L of 50% methanol, and the level was lowered to the stationary phase. The filtrate containing target compounds was loaded onto the column, and the elution was run with a methanol (A)–water (B) gradient (50%, 60% A, each 2000 mL) at a constant flow rate of 1 mL·min^{-1}. According to TLC results, the eluates containing target compounds were collected and concentrated under reduced pressure. The enrichment procedure was then repeated twice. The enriched sample was stored at −10 °C for the subsequent HSCCC separation.

3.5. Further Purification by HSCCC

3.5.1. Determination of the Partition Coefficient (K) Value

By HPLC, the partition coefficients (*K*) of target compounds were determined as follows: 10 mg of the enriched sample and 2 mL of the each phase of equilibrated two-phase solvent system were added into a 10 mL centrifuge tube. To achieve the thorough equilibration of target compounds between the two phases, the centrifuge tube was then stoppered, vortexed for 1 min, and kept for 30 min at room temperature. The upper and lower phases were separated and evaporated to dryness under N_2 gas. The residue of each phase was re-dissolved in methanol. Then an aliquot of each phase (20 µL) was subjected to HPLC analysis. The K value was expressed as the ratio of the peak area of a given compound in the upper phase divided by that in the lower phase.

3.5.2. Preparation of Two-Phase Solvent System and Sample Solution

Two-phase solvent system for HSCCC was prepared by mixing the corresponding solvents in a separatory funnel at room temperature and thoroughly equilibrated for more than 12 h. Then the lower phase and upper phase were separated shortly and degassed by sonication for 30 min before use. The sample solution was prepared by dissolving 200 mg enriched sample in the solvent mixture of *n*-hexane–ethyl acetate–methanol–water (5 mL for each phase).

3.5.3. HSCCC Separation Procedure

HSCCC separation was performed as follows: the multilayer coiled column was entirely filled with the upper stationary phase of the solvent system. Then, the apparatus was run at a revolution speed of 800 rpm. In the meantime, the lower mobile phase was pumped into the column in a head-to-tail mode at a flow rate of 2.0 mL· min^{-1}. After hydrodynamic equilibrium was established in the column, as indicated by the lower mobile phase front emerging from the tail outlet, about 10 mL of the enriched sample solution was introduced into the column through the injection valve. The eluates from the column outlet were continuously monitored by a UV detector at 254 nm. The fractions during 84–95 min (peak 1) and 112–125 min (peak 2) were collected respectively according to the obtained chromatographic profile (Figure 3). Each collection was evaporated under N_2 gas. The purified compounds were stored at −20 °C before subsequent purity and NMR analyses. After the separation experiment, all solvents in the HSCCC column were ejected with N_2 gas, and the retention of stationary phase was computed. All of the experiments were performed at room temperature (25 °C).

3.5.4. HPLC Analysis and Identification of HSCCC Peaks

The enriched sample from SLHC and each purified HSCCC peak were analyzed on a YMC-Pack ODS A column (5 µm, 250 mm × 4.6 mm) at 35 °C (Figure 2C–E). A methanol (C)–0.1% trifluoroacetic acid (*v/v*) (D) system was used as the mobile phase in gradient elution mode as follows: 20%–65% C at 0–20 min, 65%–100% C at 20–40 min. The flow rate of the mobile phase was 1.0 mL· min^{-1}. The eluates were monitored at 254 nm by a UV-VIS detector. The sample concentration is 0.5 mg· mL^{-1} for the HSCCC peaks, and 1.5 mg· mL^{-1} for the enriched sample from *S. emodi*. All samples were injected with the volume of 20 µL. Based on the peak area normalized to the sum of all observed HPLC peak areas, the purities of the isolated biflavonoids were determined.

Podoverine B (**1**). yellow, amorphous powder; $[\alpha]_D^{25}$ 201.5 (*c* 0.20, MeOH); CD (MeOH) λ_{max} ($\Delta\varepsilon$) 228 (+15.9), 262 (−13.2), 296 (+16.2) nm; UV (MeOH) λ_{max} (log ε) 261 (3.01), 304 (2.92), 338 (1.85) nm; IR (neat) ν_{max} 3301, 2969, 2928, 1637, 1608, 1591, 1507, 1473, 1438, 1357, 1266, 1155, 1087 cm^{-1}; HR-ESI-MS (positive): *m/z* 685.1538 [M + H − H$_2$O]$^+$ (calcd for C$_{36}$H$_{29}$O$_{14}$, 685.1557), *m/z* 707.1355 [M + Na − H$_2$O]$^+$ (calcd for C$_{36}$H$_{28}$O$_{14}$Na, 707.1377), *m/z* 723.1093 [M + K − H$_2$O]$^+$ (calcd for C$_{36}$H$_{28}$O$_{14}$K, 723.1116); NMR data (DMSO-*d$_6$*), see Table 2.

Podoverine C (**2**). yellow, amorphous powder; $[\alpha]_D^{25}$ 204.8 (*c* 0.26, MeOH); CD (MeOH) λ_{max} ($\Delta\varepsilon$) 228 (+14.6), 262 (−12.5), 295 (+15.8) nm; UV (MeOH) λ_{max} (log ε) 260 (0.14), 302 (0.11), 337 (0.09) nm; IR (neat) ν_{max} 3233, 2970, 2929, 1638, 1608, 1591, 1508, 1473, 1440, 1358, 1264, 1156, 1086 cm^{-1}; HR-ESI-MS (positive): *m/z* 669.1589 [M + H − H$_2$O]$^+$ (calcd for C$_{36}$H$_{29}$O$_{13}$, 669.1608), *m/z* 691.1409 [M + Na

$- H_2O]^+$ (calcd for $C_{36}H_{28}O_{13}Na$, 691.1428), m/z 707.1173 $[M + K - H_2O]^+$ (calcd for $C_{36}H_{28}O_{13}K$, 707.1167); NMR data (DMSO-d_6), see Table 2.

3.5.5. Computational Methods

The CONFLEX [27,28] searches based on molecular mechanics with MMFF94S force fields were performed for (SR)-**1** and (RR)-**1**, which gave four stable conformers, respectively. Selected those conformers were further optimized by the density functional theory method at the B3LYP/6-31G (d) level in Gaussian 09 program package [29], which was further checked by frequency calculation and resulted in no imaginary frequencies. The ECD of the conformers of **1** was then calculated by the TDDFT method at the B3LYP/6-31++G (d, p) level with the CPCM model in methanol solution. The calculated ECD spectra for each conformation were combined after Boltzman weighting according to their population contribution using SpecDis 162 software (Revision D.01, Gaussian Inc., Wallingford, CT, USA).

3.5.6. *In Vitro* Cytotoxic Assays

Carcinoma cells were maintained in RPMI-1640 medium containing 10% heat-inactivated fetal bovine serum, penicillin (100 units/mL), and streptomycin (100 ug/mL) under humidified air with 5% CO_2 at 37 °C. Exponentially growing cells were seeded into 96-well tissue culture-treated plates and pre-cultured for 24 h. The tested compounds at various concentrations were added, and the cells were incubated for additional 48 h. The cytotoxic activity was evaluated by MTT assay [8], and the IC_{50} values were obtained from dose-response curves.

4. Conclusions

Sinopodophyllum emodi is a well-known ethnic medicine with a long history. Previous chemical and pharmacological investigations indicated that flavonoids and lignans are mainly responsible for the biological activities of the plant. However, only thirty-six flavonoids had been isolated and identified, and their pharmacological properties were still a neglected field so far [2,10–12,30–32]. For the first time, two prenylated biflavonoids, podoverine B (**1**) and C (**2**), were extracted, isolated and identified from the genus *Sinopodophyllum*, and their cytotoxic activities were evaluated against MCF-7 and HepG2 cell lines. SLHC and HSCCC were developed respectively to enrich and purify two prenylated biflavonoids from *S. emodi*. By the developed method, 13.8 mg of podoverine B and 16.2 mg of podoverine C were obtained with the purities of over 98%. With multiple biological properties, biflavonoids are serving as a rich source of the health products and lead compounds for drug design. However, the biflavonoids generally coexist with other types of natural products or are trace in a complex biological organism. Furthermore, their structures are diverse and complex. For further bioactive investigation or quality control of related traditional Chinese medicine, it is crucial to develop the viable methods for separation and purification of trace natural products with complex structures, especially prenylated biflavonoids. Overall results of our study demonstrated that the combined application of SLHC and HSCCC would be a desirable separation pattern for trace prenylated biflavonoids from *S. emodi*.

Acknowledgments: This work was supported by the National Natural Science Foundation of China (No. 31300284), Basic Science Foundation of Henan University of Traditional Chinese Medicine (No. 2014KYYWF-QN26), Foundation of Henan Province for Excellent Young Teachers of Colleges and Universities (No. 2015GGJS-096), and Doctoral Science Foundation of Henan University of Traditional Chinese Medicine (No. BSJJ2011-13).

Author Contributions: Y.J.S. and Y.S.S. designed research; X.P., K.W., J.W., Y.Z., M.G. and B.J. performed research and analyzed the data; Y.J.S. wrote the paper. All authors read and approved the final manuscript.

Conflicts of Interest: The authors declare no conflict of interest.

References

1. Zhao, C.Q.; Cao, W.; Nagatsu, A.; Ogihara, Y. Three new glycosides from *Sinopodophyllum emodi* (Wall.) Ying. *Chem. Pharm. Bull.* **2001**, *49*, 1474–1476. [CrossRef] [PubMed]

2. Kong, Y.; Xiao, J.J.; Meng, S.C.; Dong, X.M.; Ge, Y.W.; Wang, R.F.; Shang, M.Y.; Cai, S.Q. A new cytotoxic flavonoid from the fruit of *Sinopodophyllum hexandrum*. *Fitoterapia* **2010**, *81*, 367–370. [CrossRef] [PubMed]

3. Zhao, C.Q.; Zhu, Y.Y.; Chen, S.Y.; Ogihara, Y. Lignan glucoside from *Sinopodophyllum emodi* and its cytotoxic activity. *Chin. Chem. Lett.* **2011**, *22*, 181–184. [CrossRef]

4. Yang, X.Z.; Shao, H.; Zhang, L.Q.; Zhou, C.; Xuan, Q.; Yang, C.Y. Present situation of studies on resources of podophyllotoxin. *Chin. Tradit. Herb. Drugs* **2001**, *32*, 1042–1044.

5. Shi, X.L.; Li, X.W.; Liu, J.B.; Zhou, H.Y.; Zhang, H.Q.; Jin, Y.R. Lignan extraction from the roots of *Sinopodophyllum emodi* Wall by matrix solid-phase dispersion. *Chromatographia* **2010**, *72*, 713–717. [CrossRef]

6. Zhao, C.Q.; Huang, J.; Nagatsu, A.; Ogihara, Y. Two new podophyllotoxin glycosides from *Sinopodophyllum emodi* (Wall.) Ying. *Chem. Pharm. Bull.* **2001**, *49*, 773–775. [CrossRef] [PubMed]

7. Zhao, C.Q.; Nagatsu, A.; Hatano, K.; Shirai, N.; Kato, S.; Ogihara, Y. New lignan glycosides from Chinese medicinal plant, *Sinopodophyllum emodi*. *Chem. Pharm. Bull.* **2003**, *51*, 255–261. [CrossRef] [PubMed]

8. Sun, Y.J.; Li, Z.L.; Chen, H.; Liu, X.Q.; Zhou, W.; Hua, H.M. Three new cytotoxic aryltetralin lignans from *Sinopodophyllum emodi*. *Bioorg. Med. Chem. Lett.* **2011**, *21*, 3794–3797. [CrossRef] [PubMed]

9. Sun, Y.J.; Li, Z.L.; Chen, H.; Liu, X.Q.; Zhou, W.; Hua, H.M. Four new cytotoxic tetrahydrofuranoid lignans from *Sinopodophyllum emodi*. *Planta Med.* **2012**, *78*, 480–484. [CrossRef] [PubMed]

10. Sun, Y.J.; Zhou, W.; Chen, H.; Li, Z.L.; Hua, H.M. Isolation and identification of flavonoids from the roots and rhizomes of *Sinopodophyllum emodi*. *J. Shenyang Pharm. Univ.* **2012**, *29*, 185–189.

11. Sun, Y.J.; Sun, Y.S.; Chen, H.; Hao, Z.Y.; Wang, J.M.; Guan, Y.B.; Zhang, Y.L.; Feng, W.S.; Zheng, X.K. Isolation of two new prenylated flavonoids from *Sinopodophyllum emodi* fruit by silica gel column and high-speed counter-current chromatography. *J. Chromatogr. B* **2014**, *969*, 190–198. [CrossRef] [PubMed]

12. Sun, Y.J.; Hao, Z.Y.; Si, J.G.; Wang, Y.; Zhang, Y.L.; Wang, J.M.; Gao, M.L.; Chen, H. Prenylated flavonoids from the fruits of *Sinopodophyllum emodi* and their cytotoxic activities. *RSC Adv.* **2015**, *5*, 82736–82742. [CrossRef]

13. Sun, Y.J.; Li, Z.L.; Chen, H.; Zhou, W.; Hua, H.M. Study on chemical constituents from the roots and rhizomes of *Sinopodophyllum emodi*. *J. Chin. Med. Mat.* **2012**, *35*, 1607–1609.

14. Sun, Y.J.; Zhou, W.; Chen, H.; Li, Z.L.; Hua, H.M. Phenols from roots and rhizomes of *Sinopodophyllum emodi*. *Chin. Tradit. Herb. Drugs* **2012**, *43*, 226–229.

15. Arens, H.; Ulbrich, B.; Fischer, H.; Parnham, M.J.; Römer, A. Novel antiinflammatory flavonoids from *podophyllum versipelle* cell culture. *Planta Med.* **1986**, *52*, 468–473. [CrossRef]

16. Mercader, A.G.; Pomilio, A.B. Naturally-occurring dimers of flavonoids as anticarcinogens. *Anticancer Agents Med. Chem.* **2013**, *13*, 1217–1235. [CrossRef] [PubMed]

17. Sutherland, I.A.; Fisher, D. Role of counter-current chromatography in the modernisation of Chinese herbal medicines. *J. Chromatogr. A* **2009**, *1216*, 740–753. [CrossRef] [PubMed]

18. Sutherland, I.; Hewitson, P.; Ignatova, S. Scale-up of counter-current chromatography: Demonstration of predictable isocratic and quasi-continuous operating modes from the test tube to pilot/process scale. *J. Chromatogr. A* **2009**, *1216*, 8787–8792. [CrossRef] [PubMed]

19. Li, X.; Dong, X.R.; Shuai, M.; Liang, Y.Z. Simultaneous separation and purification of calycosin and formononetin from crude extract of *Astragalus membranaceus* Bge. var. mongholicus (Bge.) using high speed counter current chromatography. *J. Anal. Chem.* **2015**, *70*, 92–97. [CrossRef]

20. Li, S.G.; Zhao, M.F.; Li, Y.X.; Sui, Y.X.; Yao, H.; Huang, L.Y.; Lin, X.H. Preparative isolation of six anti-tumour biflavonoids from *Selaginella doederleinii* Hieron by high-speed counter-current chromatography. *Phytochem. Anal.* **2014**, *25*, 127–133. [CrossRef] [PubMed]

21. Zhang, Y.P.; Shi, S.Y.; Wang, Y.X.; Huang, K.L. Target-guided isolation and purification of antioxidants from *Selaginella sinensis* by offline coupling of DPPH-HPLC and HSCCC experiments. *J. Chromatogr. B* **2011**, *879*, 191–196. [CrossRef] [PubMed]

22. Wang, G.H.; Liu, G.X.; Xu, F.; Shang, M.Y.; Cai, S.Q. Research on chemical fingerprint chromatograms of *Sinopodophyllum hexandrum*. *Chin. J. Chin. Mater. Med.* **2013**, *38*, 3528–3533.

23. Fang, L.; Liu, Y.Q.; Yang, B.; Wang, X.; Huang, L.Q. Separation of alkaloids from herbs using high-speed counter-current chromatography. *J. Sep. Sci.* **2011**, *34*, 2545–2558. [CrossRef] [PubMed]

24. Ito, Y. Golden rules and pitfalls in selecting optimum conditions for high-speed counter-current chromatography. *J. Chromatogr. A* **2005**, *1065*, 145–168. [CrossRef] [PubMed]

25. Jiang, S.J.; Liu, Q.; Xie, Y.X.; Zeng, H.L.; Zhang, L.; Jiang, X.Y.; Chen, X.Q. Separation of five flavonoids from tartary buckwheat (*Fagopyrum tataricum* (L.) Gaertn) grains via off-line two dimensional high-speed counter-current chromatography. *Food Chem.* **2015**, *186*, 153–159. [CrossRef] [PubMed]

26. Fan, Y.Z.; Xiang, H.Y.; Luo, Y.Q.; Song, L.Y.; Liu, Y.; Hou, L.B.; Xie, Y. Preparative separation and purification of three flavonoids from the anti-inflammatory effective fraction of *Smilax china* L. by high-speed counter-current chromatography. *Sep. Sci. Technol.* **2014**, *49*, 2090–2097. [CrossRef]

27. Goto, H.; Osawa, E. Corner flapping: A simple and fast algorithm for exhaustive generation of ring conformations. *J. Am. Chem. Soc.* **1989**, *111*, 8950–8951. [CrossRef]

28. Goto, H.; Osawa, E. An efficient algorithm for searching low-energy conformers of cyclic and acyclic molecules. *J. Chem. Soc. Perkin Trans.* **1993**, *2*, 187–198. [CrossRef]

29. Frisch, M.J.; Trucks, G.W.; Schlegel, H.B.; Scuseria, G.E.; Robb, M.A.; Cheeseman, J.R.; Scalmani, G.; Barone, V.; Mennucci, B.; Petersson, G.A.; *et al. Gaussian 09, Revision D.01*; Gaussian Inc.: Wallingford, CT, USA, 2013.

30. Shang, M.Y.; Wang, Q.H.; Xiao, J.J.; Shang, Y.H.; Kong, Y.; Cai, S.Q. New Isopentenyl Flavone-Like Compounds and Antitumor Application Thereof. C.N. 102382092 A, 21 March 2012.

31. Shang, M.Y.; Kong, Y.; Xiao, J.J.; Ma, X.J.; Ge, Y.W.; Cai, S.Q. Isopentenyl Flavone Compound Extracted from *Sinopodophyllum hexandrum* Fruit, Its Preparation Process and Its Application to Prepare the Medical Preparations for Treating Breast Neoplasm. C.N. 101648934 A, 24 September 2009.

32. Shang, M.Y.; Wang, Q.H.; Xiao, J.J.; Shang, Y.H.; Kong, Y.; Cai, S.Q. Application of Flavonoid Compounds in Treatment of Breast Cancer. C.N. 102335165 A, 15 July 2011.

10

Synthesis, Characterization and Molecular Docking of Novel Bioactive Thiazolyl-Thiazole Derivatives as Promising Cytotoxic Antitumor Drug

Sobhi M. Gomha [1,*], Taher A. Salaheldin [2], Huwaida M. E. Hassaneen [1], Hassan M. Abdel-Aziz [3] and Mohammed A. Khedr [4,5]

Academic Editor: Jean Jacques Vanden Eynde

[1] Department of Chemistry, Faculty of Science, Cairo University, Giza 12613, Egypt; huwaida30@gmail.com
[2] Nanotechnology and Advanced Materials Central Lab, Agricultural Research Center, Giza 12613, Egypt; t1salah@hotmail.com
[3] Chemistry Department, Faculty of Science, Cairo University, Bani Suef Branch, Bani Suef 62514, Egypt; dr_hassan1971@yahoo.com
[4] Department of Pharmaceutical Chemistry, Faculty of Pharmacy, Helwan University, Ein Helwan, Cairo 11795, Egypt; mohammed_abdou0@yahoo.com
[5] Department of Pharmaceutical Sciences, College of Clinical Pharmacy, King Faisal University, P. O. 380, Al-Hasaa 31982, Saudi Arabia
* Correspondence: s.m.gomha@hotmail.com

Abstract: Reactions of ethylidenethiocarbohydrazide with hydrazonoyl halides gave 1,3-thiazole or 1,3,4-thiadiazole derivatives according to the type of hydrazonoyl halides. Treatment of ethylidenethiosemicarbazide with hydrazonoyl halides and dimethylacetylene dicarboxylate (DMAD) afforded the corresponding arylazothiazoles and 1,3-thiazolidin-4-one derivatives, respectively. The structures of the synthesized products were confirmed by IR, ^{1}H-NMR, ^{13}C-NMR and mass spectral techniques. The cytotoxic activity of the selected products against the Hepatic carcinoma cell line (Hepg-2) was determined by MTT assay indicating a concentration dependent cellular growth inhibitory effect, especially for compounds **14c** and **14e**. The dose response curves indicated the IC$_{50}$ (the concentration of test compounds required to kill 50% of cell population) were 0.54 μM and 0.50 μM, respectively. Confocal laser scanning imaging of the treated cells stained by Rhodamin 123 and Acridine orange dyes confirmed that the selected compounds inhibit the mitochondrial lactate dehydrogenase enzymes. The binding mode of the active compounds was interpreted by a molecular docking study. The obtained results revealed promising cytotoxic activity.

Keywords: ethylidenethiocarbohydrazide; ethylidenethiosemicarbazide; hydrazonoyl halides; 1,3-thiazole; 1,3,4-thiadiazole; cytotoxic activity; molecular docking

1. Introduction

Hepatocellular Carcinoma (HCC) is considered the fifth most common cancer type worldwide and the third most common cause of cancer mortality. Globally, over 560,000 people develop liver cancer each year and an almost equal number, 550,000, die of it [1].

At stage I, surgical resection or transplantation is considered a potentially curative modality for HCC; patients with localized unrespectable disease are usually treated with some form of localized therapy. Local therapeutic modalities include targeted chemotherapy through hepatic artery combined with embolization, percutaneous ethanol ablation, radio embolization, radiofrequency ablation, and cryosurgery [2]. Thus, novel approaches for the treatment of unrespectable advanced or metastatic HCC represents a high-unmet medical need. In this work, novel thiazole derivatives were used as a module for management of HCC. Thiazoles are a familiar group of heterocyclic compounds possessing a wide variety of biological activities, and their usefulness as medicines are well established. Thiazole derivatives are reported to exhibit diverse biological activities as antimicrobial [3–5], antioxidant [6], antitubercular [7], and anticonvulsant [8], and anticonvulsant, anticancer [9,10] agents. Moreover, many derivatives of thiazoles are used as selective Cyclooxygenase-2 Inhibitors [11], in addition to their use as 5-HT3 receptor antagonists [12] and as potent and selective acetyl Co-A carboxylase-2 inhibitors [13]. Furthermore, the interesting properties of thiazole derivatives [14,15] in relation to the various changes in the structures of these compounds is worth studying for the synthesis of some less toxic and more potent drugs. Thus, the introduction of other heterocyclic moieties (as the thiazole and thiadiazine ring) should certainly help to fulfill this objective. The aforementioned biological, pharmacological, and industrial importance of these derivatives prompted our interest for the synthesis of some new examples of this class of compounds.

As a part of our research interest towards developing new routes for the synthesis of a variety of heterocyclic systems with promising antitumor activities [9,10,16–22], we report in the present work the synthesis of a new series of thiazolyl-thiazoles as promising cytotoxic antitumor drugs.

2. Results and Discussion

2.1. Chemistry

Ethylidenethiocarbohydrazide (3a) and ethylidenethiosemicarbazide (3b) were prepared via condensation of 5-acetyl-2-amino-4-methylthiazole (1) [23] with thiocarbohydrazide (2a) and thiosemicarbazide (2b) in absolute ethanol in the presence of a catalytic amount of concentrated HCl as depicted in Scheme 1.

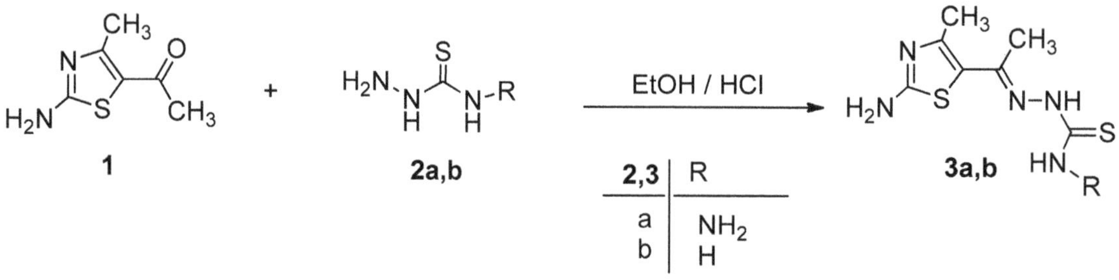

Scheme 1. Synthesis of ethylidenethiocarbohydrazide derivative 3a and ethylidenethiosemicarbazide derivative 3b.

The structure elucidation of the products 3a and 3b were substantiated through spectral data. The mass spectrum of isolated product (m/z 245) was consistent with the expected product 3a. The ^{1}H-NMR spectrum of 3a showed broad singlet bands for NH_2 and NH groups of the thiocarbohydrazide moiety at 4.92, 9.62 and 14.12 ppm, respectively, and the band for NH_2 group of the thiazole ring at 7.42 ppm as well as the two singlet bands for methyl groups at 2.42 and 2.85 ppm.

We commenced our study on the reactivity of ethylidenethiocarbohydrazide 3a toward different types of hydrazonoyl halides 4a–e and 9 to investigate the effect of the presence of carbonyl group on the course of the reaction. Initially, ethylidenethiocarbohydrazide 3a reacted with α-keto-hydrazonoyl halide 4a–e in refluxing ethanol in presence of triethylamine to give dark red color products that proved we are isolated azo-thiazole derivatives 6a–e and not the hydrazo-thiadiazine derivatives 7a–e via elimination of hydrochloric acid and water as shown in Scheme 2. Spectroscopic analyses (IR, MS, ^{1}H- and ^{13}C-NMR) confirmed the structure of 1,3-thiazole derivatives 6a–e and not 7a–e was isolated. The ^{1}H-NMR spectrum showed two broad singlet signal for two NH_2 groups at ~2.5 and ~7.11 ppm.

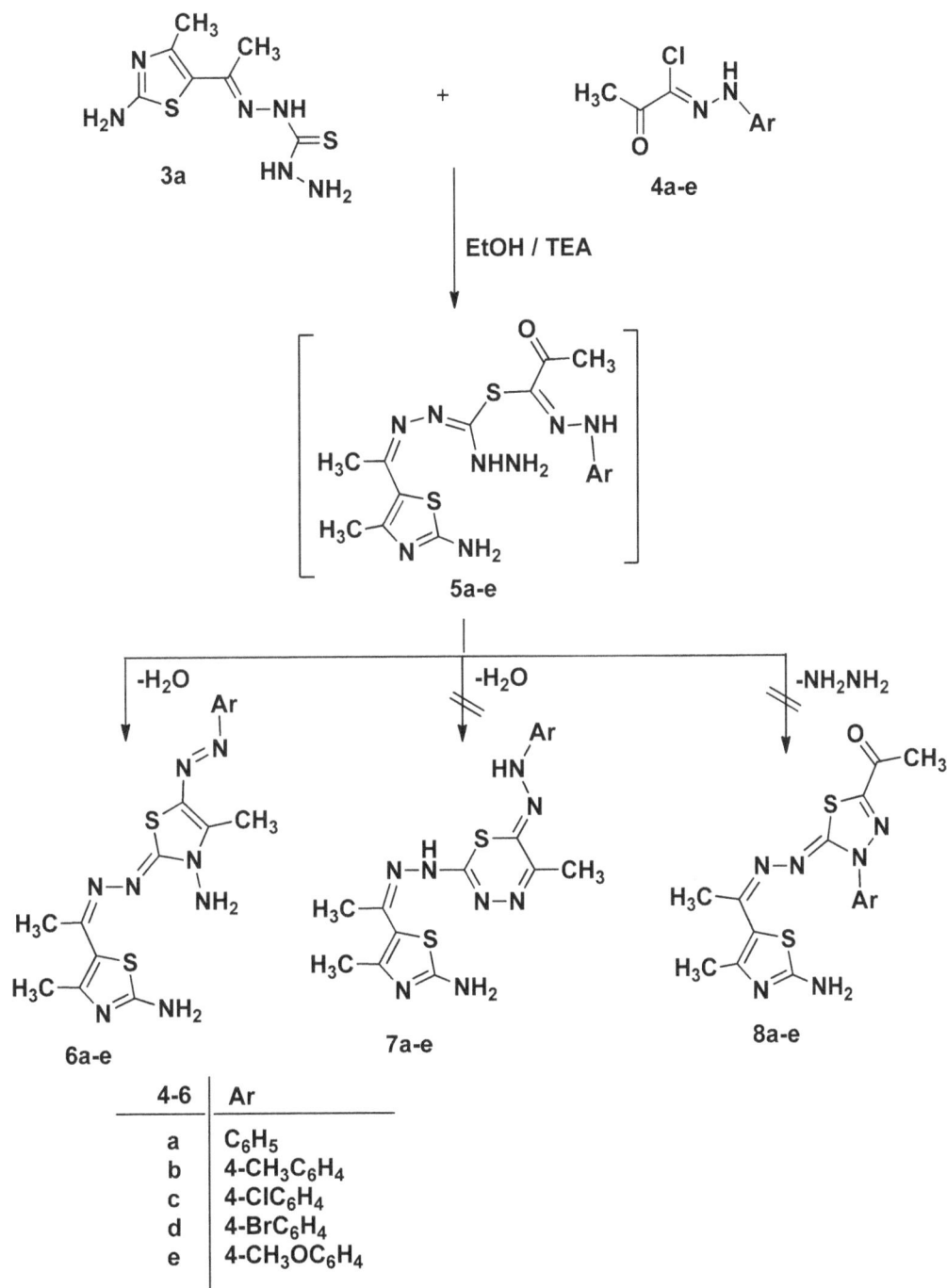

Scheme 2. Synthesis of 1,3-thiazole derivatives **6a–e**.

4-6	Ar
a	C_6H_5
b	$4\text{-}CH_3C_6H_4$
c	$4\text{-}ClC_6H_4$
d	$4\text{-}BrC_6H_4$
e	$4\text{-}CH_3OC_6H_4$

The reaction of N'-phenylbenzohydrazonoyl chloride **9** not containing α-haloketone with **3a** in the same condition gave only one isolable product as examined by thin layer chromatography (TLC) The molecular formula of **11**, $C_{20}H_{18}N_6S_2$, was consistent with elimination of HCl and NH_2NH_2. The structure of **11** was elucidated by elemental analysis and spectroscopic data (see Experimental Section) (Scheme 3). Compound **11** was also obtained via the reaction of 5-acetyl-2-amino-4-methylthiazole (**1**) with 2-hydrazono-3,5-diphenyl-2,3-dihydro-1,3,4-thiadiazole (**12**) [24] in ethanol in presence of drops of acetic acid afford product identical in all respects with that obtained from reaction of the **9** with of ethylidenethiocarbohydrazide **3a** (Scheme 3).

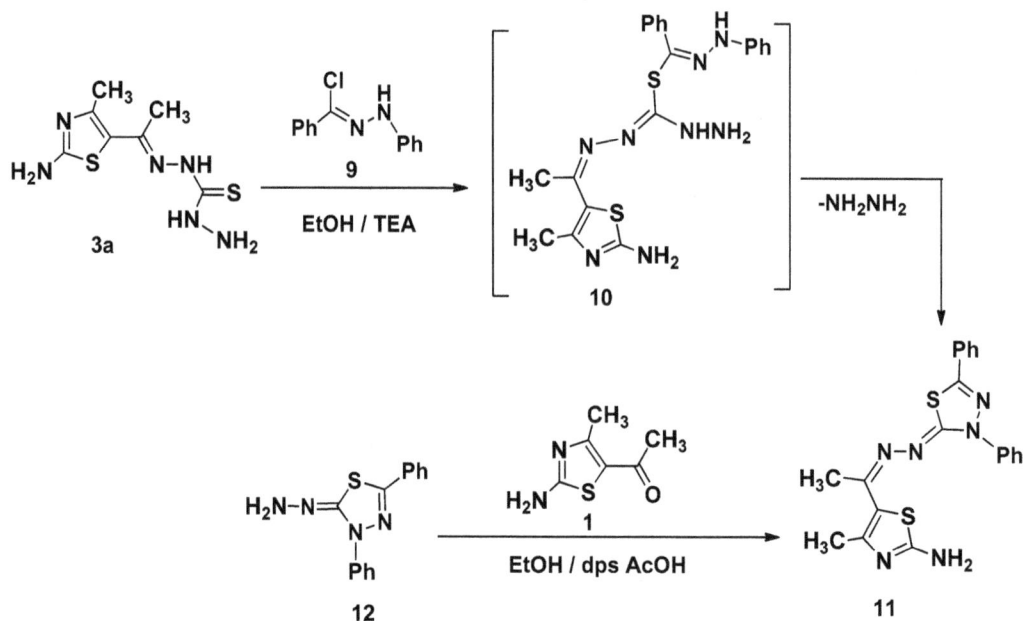

Scheme 3. Synthesis of 1,3,4-thiadiazole **11**.

Next, the reaction of 1-[1-(2-amino-4-methylthiazol-5-yl)ethylidene]thiosemicarbazide **3b** with α-keto-hydrazonoyl halides **4a–f** was performed under similar reaction conditions, and afforded the final product 5-hydrazono-thiazole derivatives **14a–f** via same route of elimination HCl and H_2O (Scheme 4) [25]. The ^{1}H-NMR of **14a** exhibited three singlet signals for methyl groups and a multiplet resonance at δ 6.95–7.35 for aromatic group (see Experimental Section & Figure S3).

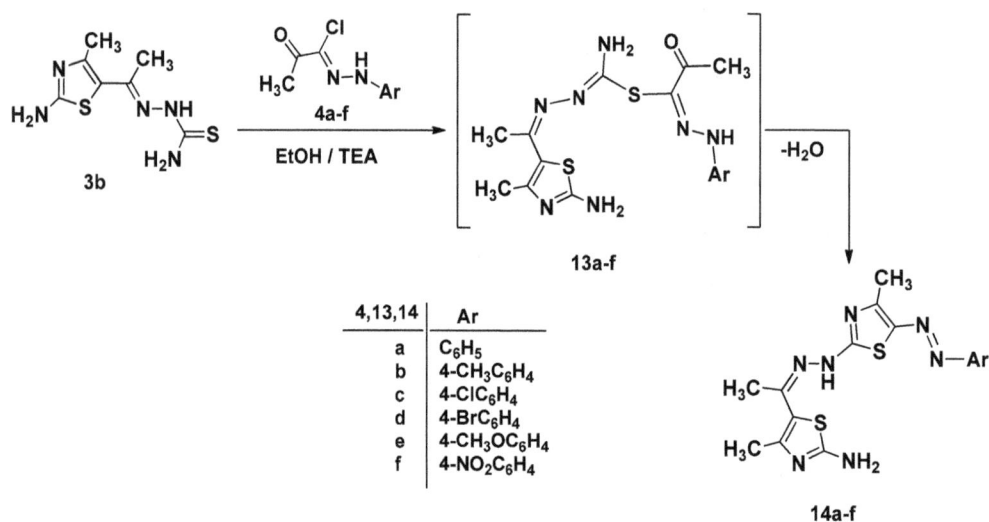

4,13,14	Ar
a	C_6H_5
b	$4-CH_3C_6H_4$
c	$4-ClC_6H_4$
d	$4-BrC_6H_4$
e	$4-CH_3OC_6H_4$
f	$4-NO_2C_6H_4$

Scheme 4. Synthesis of 5-hydrazono-thiazole derivatives **14a–f**.

Finally, refluxing an equimolecular mixture of **3b** and dimethylacetylenedicarboxylate **15** in methanol yielded methyl 4-oxo-thiazolidin-5-ylidene acetate **17** (Scheme 5). The structure of **17** was established on the basis of analytical and spectral data. Thus, the ^{1}H-NMR spectrum showed singlet signal at δ 12.77 ppm (D_2O-exchangeable), assignable to NH group. In addition, the presence of a signal at δ 6.60 assigned to the =C*H*, and singlet at δ 3.29 for ester methyl protons [9,26–28].

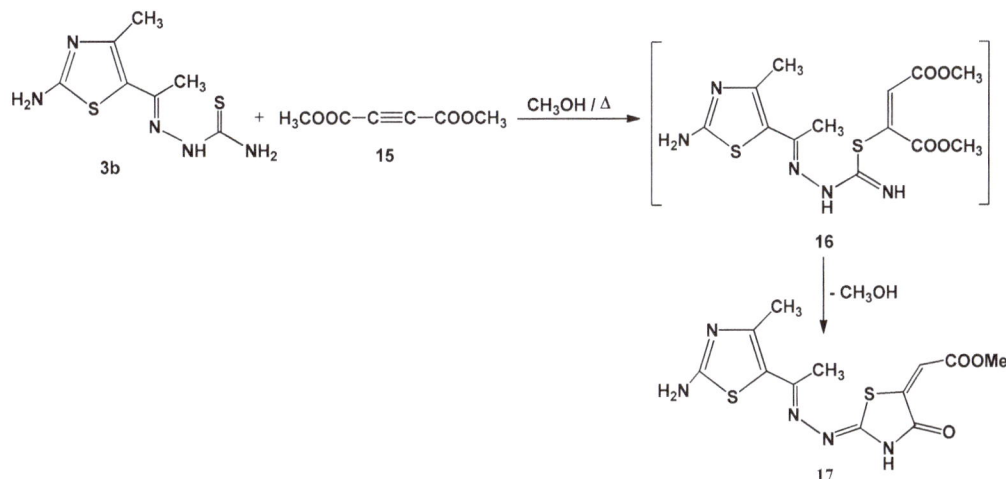

Scheme 5. Synthesis of thiazolidin-4-one **17**.

2.2. Cytotoxiclogical Activity against HEP G2 Cell Line

The cytotoxiclogical activity of synthesized products Group 1 (**6a**, **6b**, **6c**, **6d**, **6e**, **6f**, **11**) and Group 2 (**14a**, **14c**, **14d**, **14e**, **14f**, **17**) were evaluated against HEP G2) using the WST-1 cell proliferation assay as a fast and sensitive quantification of cell proliferation and viability. In brief, the assay is based on the cleavage of the tetrazolium salt WST-1 to formazan by cellular mitochondrial dehydrogenases. Expansion in the number of viable cells results in an increase in the overall activity of the mitochondrial dehydrogenases in the sample. The augmentation in enzyme activity leads to the increase in the amount of formazan dye formed. The formazan dye produced by viable cells can be quantified by a multiwell spectrophotometer (microplate reader) by measuring the absorbance of the dye solution at 450 nm. The viability of the tested compounds in response to their concentration is illustrated in Figures 1 and 2. Data generated were used to plot a dose response curve to determine the concentration of test compounds required to kill 50% of cell population (IC_{50}) was estimated exponentially.

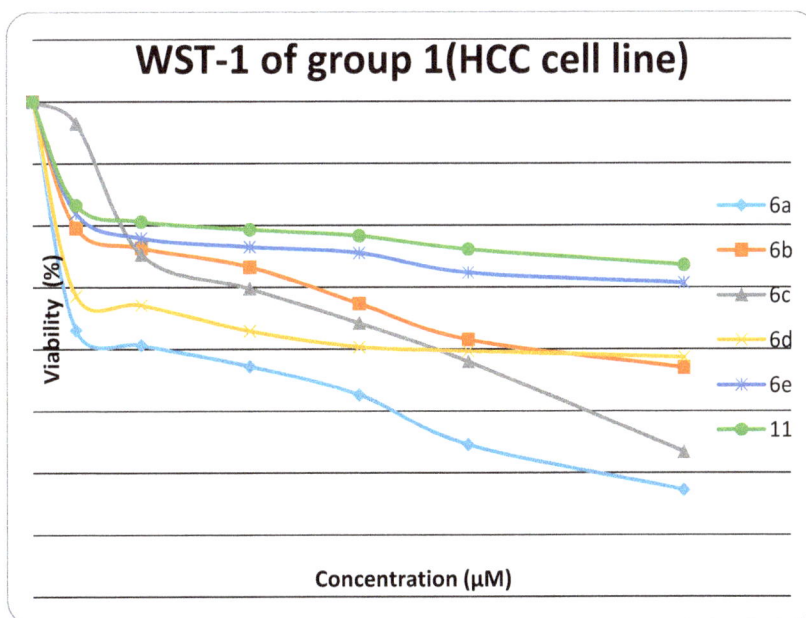

Figure 1. Viability chart of tested Group 1 compounds against HEP G2 cell line.

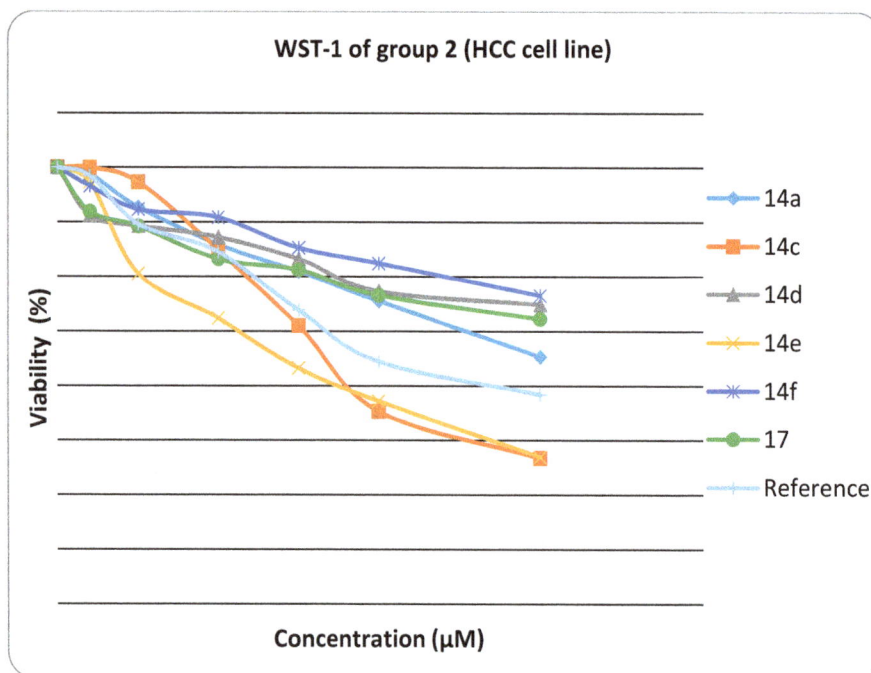

Figure 2. Viability chart of tested Group 2 compounds against HEP G2 cell line.

Cytotoxic activity was expressed as the mean IC_{50} of three independent experiments. The results are represented in Table 1.

Table 1. IC_{50} values of tested compounds $\pm$ standard deviation against HEP G2.

Compound No.	IC_{50} (μM)	Compound No.	IC_{50} (μM)
Doxorubicin	0.68 ± 0.03	**14a**	0.84 ± 0.04
6a	1.00 ± 0.08	**14c**	0.52 ± 0.03
6b	1.49 ± 0.1	**14d**	1.19 ± 0.09
6c	1.04 ± 0.07	**14e**	0.50 ± 0.02
6d	1.73 ± 0.12	**14f**	1.28 ± 0.08
6e	2.17 ± 0.13	**17**	1.07 ± 0.06
11	2.91 ± 0.15		

The results presented in Table 1 showed that:

The *in vitro* inhibitory activities of tested compounds against the hepatocellular carcinoma cell line (HEP G2) have the descending order as follow: **14e > 14c > 14a > 6a > 6c > 17 >14d >14f > 6b > 6d > 6e > 11**. The thiazole rings **6a–e** and **14a–f** have better *in vitro* inhibitory activity than the thiadiazole ring **11**. The thiazolone ring **17** has better *in vitro* inhibitory activity than the thiadiazole ring **11**. The introduction of a amino group on N3 of thiazole ring increases the activity when fixing the substituent at 4-position of phenyl group at position 5 in the thiazole ring (thiazole **14a** is more active than thiazole **6a,** thiazole **14c** is more active than thiazole **6c**, thiazole **14d** is more active than thiazole **6d**, and thiazole **14e** is more active than thiazole **6e**).

The results revealed that compounds **14e, 14c** and **14a** (IC_{50} were 0.5 ± 0.02 μM, 0.52 ± 0.03 μM and 0.84 ± 0.04 μM, respectively) have promising antitumor activity against hepatocellular carcinoma cell line (HEP G2) when compared to doxorubicin as a reference drug (IC_{50} value of doxorubicin = 0.68 ± 0.03 μM), while **6a, 6c, 17, 14d** and **14f** have moderate activity (IC_{50} were 1.04 ± 0.07 μM, 1.07 ± 0.06 μM, 1.19 ± 0.09 μM, and 1.28 ± 0.08 μM, respectively). On the other hand, **6b, 6d, 6e** and **11** have lower inhibitory activity against (HEP G2) ($IC_{50} = 1.49 \pm 0.1$ μM, 1.73 ± 0.12 μM, 2.17 ± 0.13 μM and

2.91 ± 0.15 µM, respectively). The small values of IC_{50} for the selected compounds indicate that, for more anticancer effect, higher concentrations can be used.

The WST-1 assay results revealed a significant decrease in the mitochondrial dehydrogenase activity as a function of the growth rate of the tumor cells but did not explain the mode of action of the compounds. Confocal laser scanning microscopic (CLSM) imaging of HEP G2 cell line stained with acridine orange dye for nucleic acids (green stain) and Rodamine 123 (Orange stain) for inner mitochondrial membrane, where the dehydrogenases located, reflects the activity of mitochondrial dehydrogenase enzyme activity. It was obvious that the activity of dehydrogenases significantly decreased with cells treated with **14e**, **14c** and **14a** compared to the untreated control (Figure 3), while moderate decrease for **6a**, **6c**, **17**, **14d** and **14f** (Figure 4), and mild decrease for the rest of the compounds (Figure 5). The higher amount of orange color reflects the higher activity of mitochondrial dehydrogenase enzyme, which means higher viability and *vice versa*.

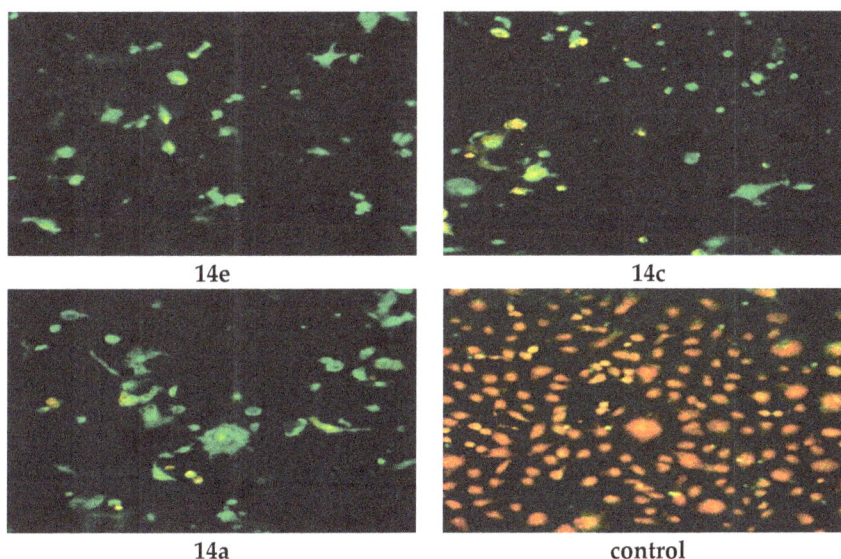

Figure 3. Confocal Laser Scanning Microscopy (CLSM) image of HEP G2 cell line treated with 0.6 µM tested compounds (**14e**, **14c**, **14a** and untreated control). Stained by Acridine orange (green) and Rodamine 123 (Orange).

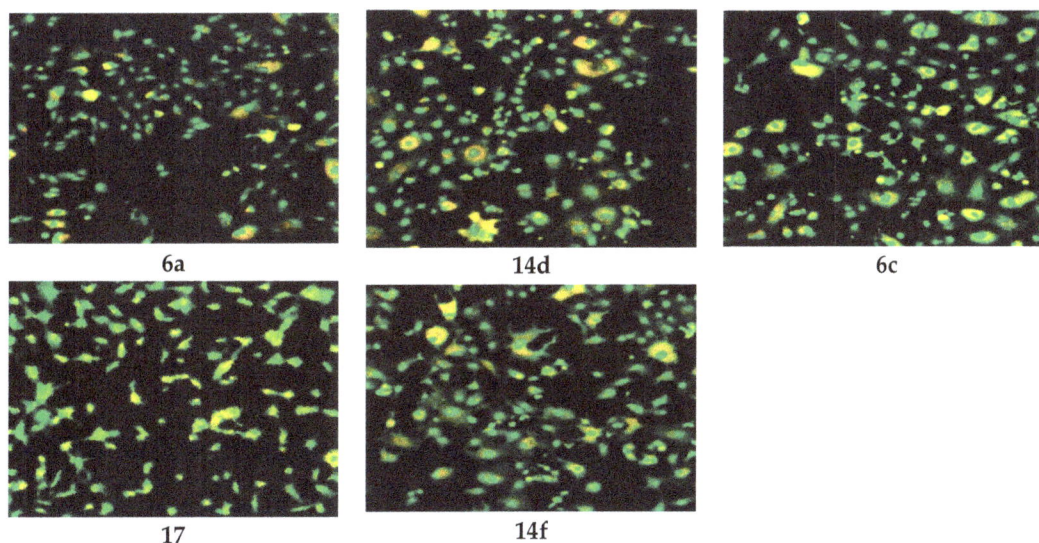

Figure 4. CLSM image of HEP G2 cell line treated with 0.6 µM tested compounds (**6a**, **6c**, **17**, **14d** and **14f**). Stained by Acridine orange (green) and Rodamine 123 (Orange).

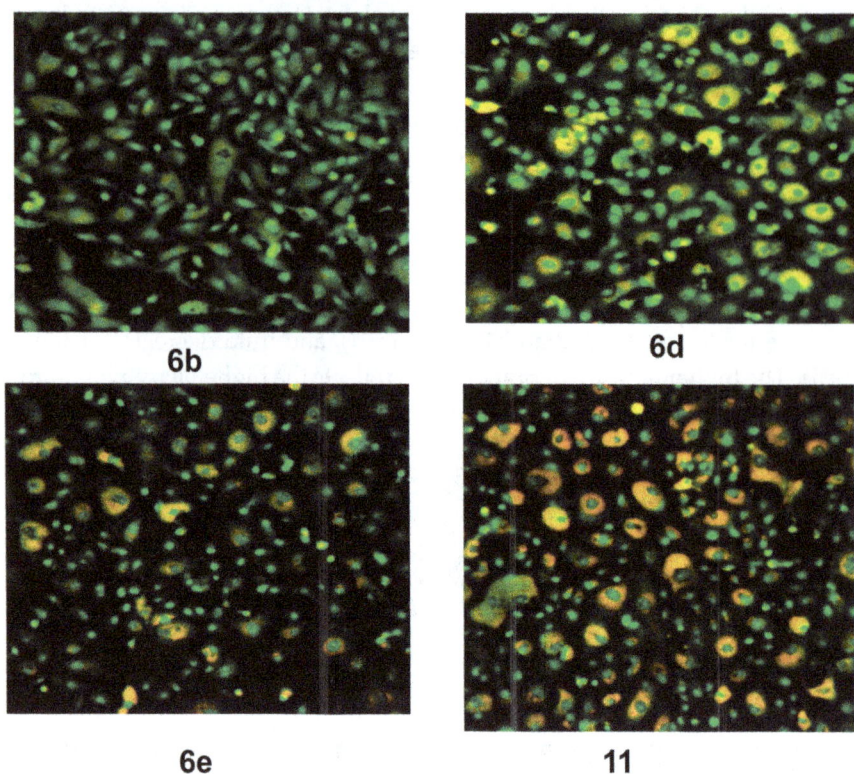

Figure 5. CLSM image of HEP G2 cell line treated with 0.6 µM tested compounds (**6b**, **6d**, **6e** and **11**). Stained by Acridine orange (green) and Rodamine 123 (Orange).

2.3. Molecular Docking Study

In cancer cells, the increased glucose uptake results in high glycolytic activity, which, in turn, causes elevated levels of lactate production [29]. The lactate production is controlled by lactate dehydrogenase-5 (LDH-5), an isoenzyme from the lactate dehydrogenase that is mainly found in the liver cells [30]. It has been reported that LDH-5 has an important role in tumor maintenance in many human cancers like the hepatic cancer [31]. Recently, LDH-5 inhibitors have been reported to show potential anticancer activity [32], which confirms that LDH-5 inhibition is a good target for developing anticancer agents. Molecular docking study was conducted to interpret the LDH inhibitory activity for the synthesized compounds that were confirmed biologically. The docking was performed using Leadit 2.1.5 software to calculate and analyze all parameters that may have a direct relationship to the LDH inhibition (Table 2). A direct correlation between the activity and the binding affinity was observed (Figure 6). For example, compound **14e** showed the best affinity (–24.85 kcal/mol) and the best IC$_{50}$ as well. The entropy ligand conformation score for the highly active group was a neglected value (0.00), which is favored for best fitting.

There is no doubt that all the tested compounds showed promising cytotoxic activity against HEP G2 cell line in low micromolar range ranged from 0.50 µM (compound **14e**) to 2.91 µM (compound **11**). All compounds shared the same 5-(substituted)-ethylidene-hydrazono-2-amine-4-methyl-1,3-thiazole scaffold. This scaffold showed the ability to form some interactions with the main residues in the active site of LDH-5 that was clear in the NH-N= moiety hydrogen bond formation with Arg 99, and the 2-amino group of the 1,3-thizole ring and its hydrogen bond formation with Gly 29, Gly 27, and Thr 95. The compounds were divided into three groups according to their activity: highly active group (**14e**, **14c**, and **14f**), moderately active group (**6a**, **6c**, **17**, **14d**, and **14f**) and low active group (**6b**, **6d**, **6e** and **11**).

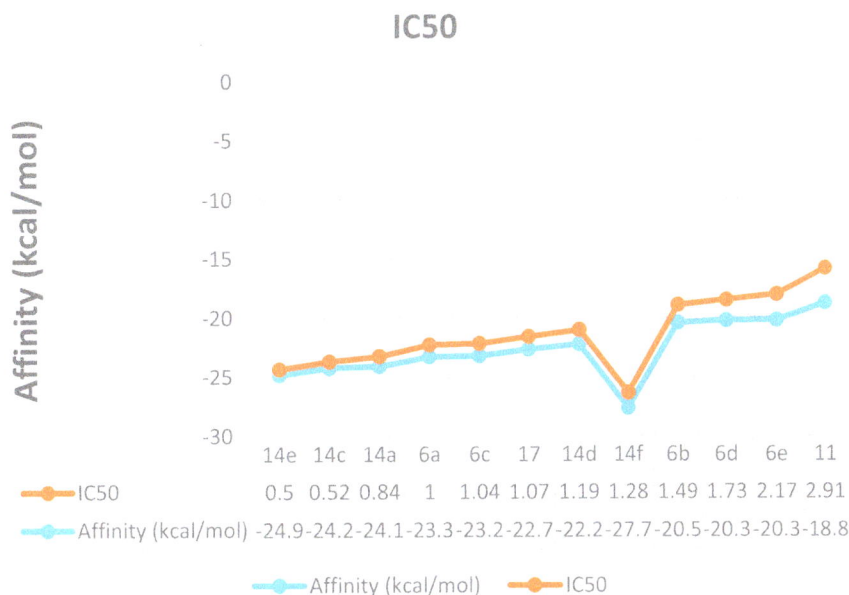

Figure 6. Correlation between the docking affinity and the IC_{50}.

Table 2. Docking Results of the active compounds using Leadit 2.1.5 software (software license was purchased from BioSolveIT GmbH, Germany).

Compounds	Affinity Score kcal/mol	Lipophilic Contribution Score	Clash Score	Ligand Entropy Conformation Score
14e	−24.85	−8.50	4.32	0.00
14c	−24.23	−8.56	4.54	0.00
14a	−24.10	−12.27	4.12	0.00
6a	−23.50	−8.31	5.82	1.40
6c	−23.23	−8.28	5.88	1.40
17	−22.68	−6.36	3.85	1.40
14d	−22.23	−13.86	7.55	0.00
14f	−21.66	−6.91	8.23	0.00
6b	−20.46	−8.52	5.89	1.40
6d	−20.27	−8.31	5.88	1.40
6e	−20.25	−8.39	4.54	2.80
11	−18.80	−9.23	5.11	0.00

For the highly active group, a common binding mode was observed in which the Gly 97, Gly 32, Val 31, Arg 99, Gln 30 and Thr 95 were the most common involved residues. The N=N, C=N, NH_2, and OCH_3 groups were the most important groups for formation of the hydrogen bonds (Figure 7). The oxygen atom of the OCH_3 group in case of **14e** interacted with Gly 97 to form a hydrogen bond. This feature was absent in compound **14c** with the *p*-chloro phenyl and compound **14a** with the non-substituted phenyl ring.

In the moderately active group (**6a**, **6c**, **17**, **14d**, and **14f**), the presence of 3-amino group in the thiazol-2-(3*H*) moiety was important to form a hydrogen bond with Thr 99, Asn 113 and Ala 96. The N=N and C=N groups were involved in the interactions. An overall good fitting and placement was observed for the docked compounds inside the active site where the nicotinamide-adenine dinucleotide was bound. Regarding the low active group (**6b**, **6d**, **6e**, and **11**), the most obvious result was in compound **11** that has a 3-phenyl thiadiazole moiety. This phenyl ring restricts the interactions and the flexibility of the compound resulting in a low binding affinity. The steric hindrance of compound **11** resulted from the 3,5-diphenyl-2,3-dihydro-1,3,4-thiadiazole ring system that was the main reason for bad fitting and low affinity (Figure 8).

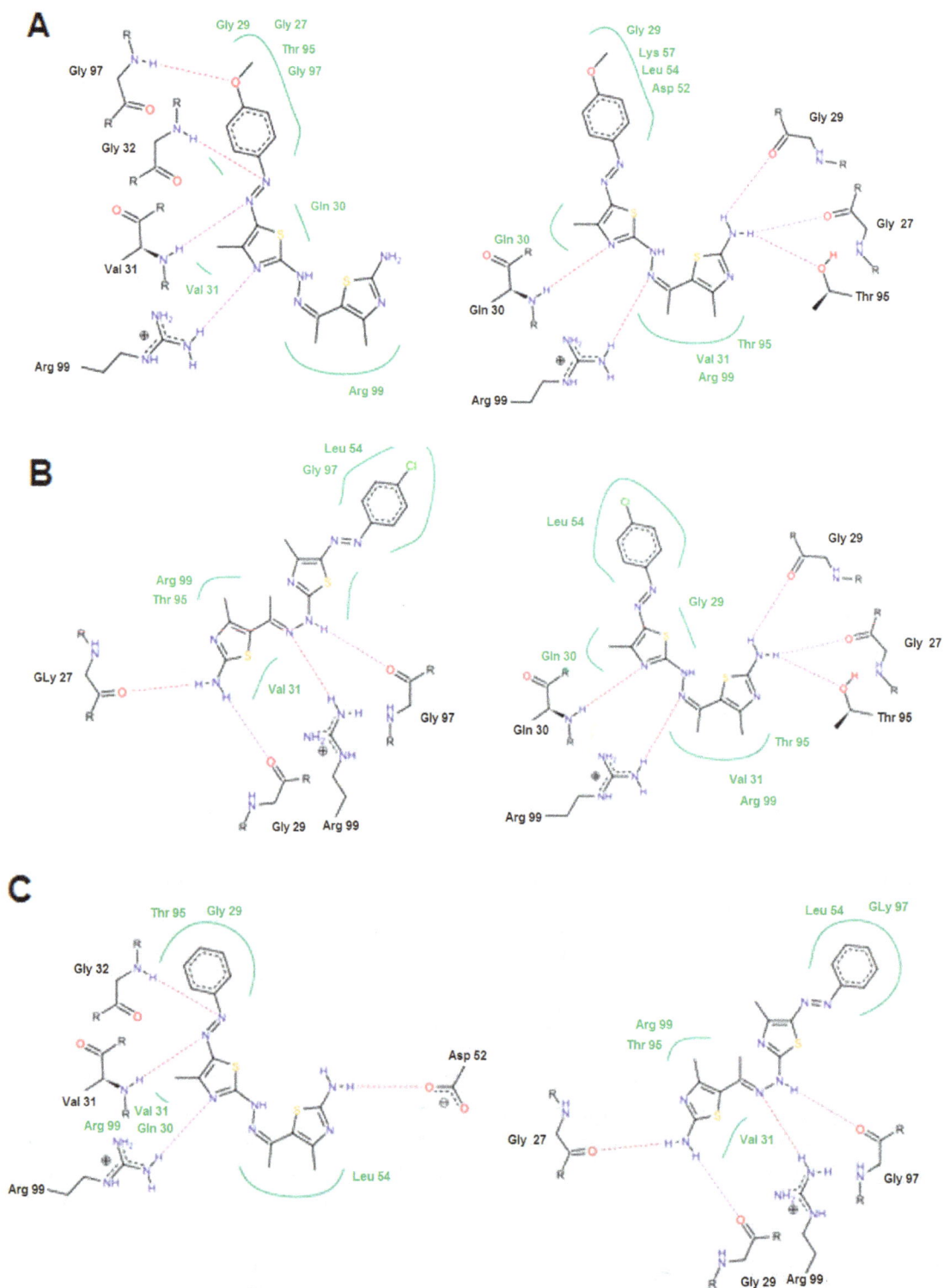

Figure 7. Possible binding modes of compounds (**A**) **14e**; (**B**) **14c**; and (**C**) **14a**.

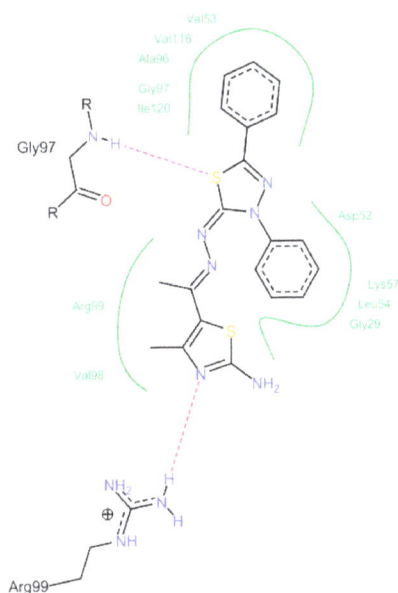

Figure 8. The possible binding mode of compound **11**.

Generally, in the binding site area with mild hydrophobic amino acids, like Ala 96, Gly 97 and Gly 32, the activity was achieved by non-substituted phenyl ring, -OCH$_3$ substitution, and/or chloro substitution that do not affect strongly the electron cloud on the phenyl ring. The main interactions in this area are not hydrophobic but, hydrogen bonding so, any substitution by bromo atom or nitro group affected the electron cloud and the π-π system resulting in shifting in the activity.

3. Experimental Section

3.1. Chemistry

Melting points were measured on an Electrothermal IA 9000 series digital melting point apparatus (Bibby Sci. Lim. Stone, Staffordshire, UK). IR spectra were recorded in potassium bromide discs on Shimadzu FTIR 8101 PC infrared spectrophotometer (Shimadzu, Tokyo, Japan). NMR spectra were recorded on a Varian Mercury VX-300 NMR spectrometer (Varian, Inc., Karlsruhe, Germany) operating at 300 MHz (^{1}H-NMR) or 75 MHz (^{13}C-NMR) and run in deuterated dimethylsulfoxide (DMSO-d_6). Chemical shifts were related to that of the solvent. Mass spectra were recorded on a Shimadzu GCMS-QP1000 EX mass spectrometer at 70 eV (Tokyo, Japan). Elemental analyzes were measured by using a German made Elementar vario LIII CHNS analyzer (GmbH & Co. KG, Hanau, Germany). Antitumor activity was evaluated by the Nanotechnology & Advanced materials central lab, Agricultural Research Center, Giza, Egypt. Hydrazonoyl halides were prepared as previously reported in the respective literature [33,34].

3.1.1. Synthesis of **3a** and **3b**

5-Acetyl-2-amino-4-methylthiazole **1** (7.8 g, 50 mmol) was dissolved in 100 mL of absolute ethanol and stirred with an equimolar quantity of thiocarbohydrazide (**2a**) or thiosemicarbazide (**2b**) for 24 h at room temperature with catalytic amounts of concentrated HCl. The desired products precipitated from reaction mixture were filtered, washed with ethanol and recrystallized from acetic acid to give pure product of compound **3a** and **3b**.

1-[1-(2-Amino-4-methylthiazol-5-yl)ethylidene]thiocarbohydrazide (**3a**). White crystals (84%); mp 234–236 °C; IR (KBr): ν 3436, 3231 (NH$_2$), 3182 (NH), 1631(C=N), 1258 (C=S) cm^{-1}; ^{1}H-NMR (DMSO-d_6): δ 2.42 (s, 3H, CH$_3$), 2.85(s, 3H, CH$_3$-thiazole), 4.92 (s, br, 2H, D$_2$O-exchangeable, NH$_2$, CSNHNH$_2$), 7.42 (s, br, 2H, D$_2$O-exchangeable, NH$_2$-thiazole), 9.64 (s, br, 1H, D$_2$O-exchangeable, NH, CS<u>NH</u>NH2),

14.12 (s, br, 1H, D_2O-exchangeable, NH, C=N-<u>NH</u>-CS); MS m/z (%): 245 (M^+, 19), 244 (M^+, 21), 219 (27), 122 (29), 79 (100). Anal. Calcd. for $C_7H_{12}N_6S_2$ (244.34): C, 34.41; H, 4.95; N, 34.39. Found C, 34.48; H, 4.86; N, 34.24%.

1-[1-(2-Amino-4-methylthiazol-5-yl)ethylidene]thiosemicarbazide (**3b**). White crystals (76%); mp 245–247 °C; IR (KBr): ν 3406, 3275 (NH_2), 3217 (NH), 1639 (C=N), 1242 (C=S) cm^{-1}; ^1H-NMR (DMSO-d_6): δ 2.42 (s, 3H, CH_3), 2.86 (s, 3H, CH_3-thiazole), 4.08 (s, br, D_2O-exchangeable, 2H, NH_2), 7.45(s, br, 2H, D_2O-exchangeable, NH_2-thiazole), 9.78 (s, br, 1H, D_2O-exchangeable, NH); MS m/z (%): 229 (M^+, 6), 212 (7), 128 (19), 42 (100). Anal. Calcd. for $C_7H_{11}N_5S_2$ (229.33): C, 36.66; H, 4.83; N, 30.54. Found C, 36.60; H, 4.69; N, 30.41%.

3.1.2. Synthesis of 5-[1-((3-Amino-4-methyl-5-(substitutedphenyldiazenyl)thiazol-2(3*H*)-ethylidene) hydrazono)]-2-amine-4-methyl-1,3-thiazole (**6a–e**)

A mixture of 1-[1-(2-amino-4-methylthiazol-5-yl)ethylidene]thiocarbohydrazide **3a** (0.244 g, 1 mmol) and appropriate hydrazonoyl halides **4a–e** (1 mmol) in dioxan (30 mL) containing triethylamine (0.1 g, 1 mmol) was refluxed for 6–8 h (monitored by TLC). The formed precipitate after cooling at room temperature was isolated by filtration, washed with methanol, dried and recrystallized from appropriate solvent to give products **6a–e**.

5-[1-((3-Amino-4-methyl-5-(phenyldiazenyl)thiazol-2(3H)-ethylidene)hydrazono)]-2-amine-4-methyl-1,3-thiazol (**6a**). Red solid (67%); mp 192–194 °C; IR (KBr): ν 3406, 3298, 3121 (2NH_2), 1639 (C=N), 1552 (N=N) cm^{-1}; ^1H-NMR (DMSO-d_6): δ 2.42 (s, 3H, CH_3), 2.52 (s, br., 2H, D_2O-exchangeable, NH_2), 2.85 (s, 3H, CH_3), 3.43 (s, 3H, CH_3), 7.15(s, br., 2H, D_2O-exchangeable, NH_2), 7.31–7.73 (m, 5H, Ar-H); ^{13}C-NMR (DMSO-d_6): δ 16.21 (CH_3), 17.10 (CH_3), 19.89 (CH_3), 114.67, 120.32, 121.56, 128.61, 139.54, 143.14 (Ar-C), 154.15, 161.67, 167.45, 169.78, 176.75 (C=N); MS m/z (%): 387 (M^+ + 1, 6), 387 (M^+, 8), 163 (18), 110 (67), 77 (100). Anal. Calcd. for $C_{16}H_{18}N_8S_2$ (386.50): C, 49.72; H, 4.69; N, 28.99. Found C, 49.65; H, 4.60; N, 28.78%.

5-[1-((3-Amino-4-methyl-5-(4-methylphenyldiazenyl)thiazol-2(3H)-ethylidene)hydrazono)]-2-amine-4-methyl-1,3-thiazol (**6b**). Red solid (68%); mp 216–218°C; IR (KBr): ν 3407, 3317, 3113 (2NH_2), 1636 (C=N), 1556 (N=N) cm^{-1}; ^1H-NMR (DMSO-d_6): δ 2.23 (s, 3H, CH_3), 2.42 (s, 3H, CH_3), 2.50 (s, br., 2H, D_2O-exchangeable, NH_2), 2.84(s, 3H, CH_3), 3.47 (s, 3H, CH_3), 7.11 (s, br., 2H, D_2O-exchangeable, NH_2), 7.32–7.75 (m, 4H, Ar-H); MS m/z (%): 402 (M^+ + 1, 11), 401 (M^+ + 1, 18), 400 (M^+, 29), 233 (19), 153 (69), 105 (100). Anal. Calcd. for $C_{17}H_{20}N_8S_2$ (400.52): C, 50.98; H, 5.03; N, 27.98. Found C, 50.68; H, 5.00; N, 27.69%.

5-[1-((3-Amino-5-((4-chlorophenyl)diazenyl)-4-methylthiazol-2(3H)-ethylidene)hydrazono)]-2-amine-4-methyl-1,3-thiazole (**6c**). Red solid (65%); mp 243–245 °C; IR (KBr): ν 3409, 3312, 3151 (2NH_2), 1637 (C=N), 1553 (N=N) cm^{-1}; ^1H-NMR (DMSO-d_6): δ 2.43 (s, 3H, CH_3), 2.52 (s, br., 2H, D_2O-exchangeable, NH_2), 2.86 (s, 3H, CH_3), 3.49 (s, 3H, CH_3), 7.35 (s, br., 2H, D_2O-exchangeable, NH_2), 7.38-7.83 (m, 4H, Ar-H); MS m/z (%): 422 (M^+ + 2, 2), 421 (M^+ + 1, 4), 420 (M^+, 10), 299 (17), 244 (43), 212 (46), 155 (52), 113 (85), 72 (100). Anal. Calcd. for $C_{16}H_{17}ClN_8S_2$ (420.94): C, 45.65; H, 4.07; N, 26.62. Found C, 45.60; H, 4.14; N, 26.38%.

5-[1-((3-Amino-5-(4-bromophenyl)diazenyl)-4-methylthiazol-2(3H)-ethylidene)hydrazono)]-2-amine-4-methyl-1,3-thiazol (**6d**). Red solid (66%); mp 218–219 °C; IR (KBr): ν 3409, 3294, 3113 (2NH_2), 1632 (C=N), 1557 (N=N) cm^{-1}; ^1H-NMR (DMSO-d_6): δ 2.46 (s, 3H, CH_3), 2.51 (s, br., 2H, D_2O-exchangeable, NH_2), 2.85 (s, 3H, CH_3), 3.47 (s, 3H, CH_3), 7.26 (s, br., 2H, D_2O-exchangeable, NH_2), 7.30–7.81 (m, 4H, Ar-H); MS m/z (%): 467 (M^+ + 2, 15), 466 (M^+ + 1, 19), 465 (M^+, 21), 405 (14), 299 (19), 153 (86), 113 (62), 65 (100). Anal. Calcd. for $C_{16}H_{17}BrN_8S_2$ (465.39): C, 41.29; H, 3.68; N, 24.08. Found C, 41.11; H, 3.57; N, 24.01%.

5-[1-((3-Amino-5-(4-methoxyphenyl)diazenyl)-4-methylthiazol-2(3H)-ethylidene)hydrazono)]-2-amine-4-methyl-1,3-thiazol (**6e**). Red solid (68%); mp 212–214 °C; IR (KBr): ν 3403, 3298, 3117 (2NH_2), 1635 (C=N),

1555 (N=N) cm^{-1}; ^{1}H-NMR (DMSO-d_6): δ 2.43 (s, 3H, CH$_3$), 2.50 (s, br., 2H, D$_2$O-exchangeable, NH$_2$), 2.81(s, 3H, CH$_3$), 3.41 (s, 3H, CH$_3$), 3.64 (s, 3H, CH$_3$), 7.12 (s, br., 2H, D$_2$O-exchangeable, NH$_2$), 7.24-7.72 (m, 4H, Ar-H); MS m/z (%): 416 (M$^+$, 4), 322 (13), 273 (12), 182 (61), 124 (100), 109 (99). Anal. Calcd. for C$_{17}$H$_{20}$N$_8$OS$_2$ (416.52): C, 49.02; H, 4.84; N, 26.90. Found C, 48.89; H, 4.76; N, 26.72%.

3.1.3. Synthesis of 2-[(1-(2-Amino-4-methylthiazol-5-yl)ethylidene)hydrazono]-3,5-diphenyl-2,3-dihydro-1,3,4-thiadiazole (11)

A mixture of 1-[1-(2-amino-4-methylthiazol-5-yl)ethylidene]thiocarbohydrazide 3a (0.244 g, 1 mmol) and N'-phenylbenzohydrazonoyl chloride 9 (0.230 g, 1 mmol) in dioxane (20 mL) containing triethylamine (0.1 g, 1 mmol) was refluxed for 6 h. The formed precipitate was isolated by filtration, washed with methanol, dried and recrystallized from DMF to give product 11.

Yellow solid (76%); mp = 240–242 °C; IR (KBr): ν 3407, 3213 (NH$_2$), 1634 (C=N) cm^{-1}; ^{1}H-NMR (DMSO-d_6): δ 2.39 (s, 3H, CH$_3$), 2.54 (s, 3H, CH$_3$), 3.37 (s, br., 2H, D$_2$O-exchangeable, NH$_2$), 7.30–8.14 (m, 10H, Ar-H); MS m/z (%): 407 (M$^+$ + 1, 67), 406 (M$^+$, 52), 361 (59), 304 (64), 233 (91), 172 (64), 138 (67), 90 (79), 63 (100). Anal. Calcd. for C$_{20}$H$_{18}$N$_6$S$_2$ (406.53): C, 59.09; H, 4.46; N, 20.67. Found C, 59.01; H, 4.54; N, 20.45%.

3.1.4. Alternative Synthesis of 11

A mixture of 5-acetyl-2-amino-4-methylthiazole (1) (0.156 g, 1 mmol) and 2-hydrazono-3,5-diphenyl-2,3-dihydro-1,3,4-thiadiazole (12) (0.268 g, 1 mmol) in 10 mL of ethanol with catalytic amounts of glacial acetic acid were refluxed for 4h. The solid precipitated after cooling was filtered, washed with ethanol and recrystallized from acetic acid to give pure 11 which identical in all respects (m.p., mixed m.p. and IR spectra) with those obtained from reaction of 3a with 9 but in 66% yield.

3.1.5. Synthesis of 2-Amino-4-methyl-5-(1-(2-(4-methyl-5-(substitutedphenyldiazenyl)thiazol-2-yl)hydrazono)ethyl)thiazole (14a–f)

A mixture of 1-[1-(2-amino-4-methylthiazol-5-yl)ethylidene]thiosemicarbazide 3b (0.229 g, 1 mmol) and appropriate hydrazonoyl halides 4a–f (1 mmol) in dioxane (30 mL) containing triethylamine (0.1 g, 1 mmol) was refluxed for 6–8 h. (monitored by TLC). The formed precipitate after cooling was isolated by filtration, washed with methanol, dried and recrystallized from appropriate solvent to give products 14a–f.

2-Amino-4-methyl-5-(1-(2-(4-methyl-5-(phenyldiazenyl)thiazol-2-yl)hydrazono)ethyl)thiazole (14a). Red solid (80%); mp 168–169 °C; IR (KBr): ν 3438, 3283 (NH$_2$, NH), 1597 (C=N) cm^{-1}; ^{1}H-NMR (DMSO-d_6): δ 2.43 (s, 3H, CH$_3$), 2.56 (s, 3H, CH$_3$), 3.56 (s, 3H, CH$_3$), 6.95–7.35 (m, 5H, Ar-H), 7.57 (s, 2H, D$_2$O-exchangeable, NH$_2$), 10.51 (s, 1H, D$_2$O-exchangeable, NH); ^{13}C-NMR (DMSO-d_6): δ 16.87 (CH$_3$), 17.29 (CH$_3$), 19.39 (CH$_3$), 114.58, 119.46, 122.31, 129.69, 138.97, 144.14, 153.95 (Ar-C), 161.24, 168.59, 169.31, 176.75 (C=N); MS m/z (%): 372 (M$^+$ + 1, 4), 371 (M$^+$, 21), 218 (29), 153 (100), 78 (90), 42 (79). Anal. Calcd. for C$_{16}$H$_{17}$N$_7$S$_2$ (371.48): C, 51.73; H, 4.61; N, 26.39. Found C, 51.70; H, 4.46; N, 26.27%.

2-Amino-4-methyl-5-(1-(2-(4-methyl-5-(methylphenyldiazenyl)thiazol-2-yl)hydrazono)ethyl)thiazole (14b). Red solid (82%); mp 247–249 °C; IR (KBr): ν 3391, 3278 (NH$_2$, NH), 1632 (C=N) cm^{-1}; ^{1}H-NMR (DMSO-d_6): δ 2.25 (s, 3H, CH$_3$), 2.43 (s, 3H, CH$_3$), 2.53 (s, 3H, CH$_3$), 3.56 (s, 3H, CH$_3$), 7.13 (d, 2H, J = 6.9 Hz, Ar-H), 7.26 (d, 2H, J = 6.9 Hz, Ar-H), 7.55 (s, 2H, D$_2$O-exchangeable, NH$_2$), 10.44 (s, 1H, D$_2$O-exchangeable, NH); ^{13}C-NMR (DMSO-d_6): δ 16.85 (CH$_3$), 17.28 (CH$_3$), 19.39 (CH$_3$), 20.83 (CH$_3$), 114.59, 119.48, 130.12, 131.28, 138.34, 141.83, 153.86 (Ar-C), 161.04, 168.69, 169.26, 176.58 (C=N); MS m/z (%): 386 (M$^+$ + 1, 12), 385 (M$^+$, 36), 232 (53), 153 (85), 113 (100). Anal. Calcd. for C$_{17}$H$_{19}$N$_7$S$_2$ (385.51): C, 52.96; H, 4.97; N, 25.43. Found C, 52.91; H, 4.86; N, 25.33%.

2-Amino-4-methyl-5-(1-(2-(4-methyl-5-(chlorophenyldiazenyl)thiazol-2-yl)hydrazono)ethyl)thiazole (**14c**). Red solid (75%); mp 178–180 °C; IR (KBr): ν 3996, 3283 (NH$_2$, NH), 1632, 1606 (C=N) cm^{-1}; ^{1}H-NMR (DMSO-d_6): δ 2.43 (s, 3H, CH$_3$), 2.57 (s, 3H, CH$_3$), 3.56 (s, 3H, CH$_3$), 7.34 (s, 4H, Ar-H), 7.57 (s, 2H, D$_2$O-exchangeable, NH$_2$), 10.57 (s, 1H, D$_2$O-exchangeable, NH); ^{13}C-NMR (DMSO-d_6): δ 16.87 (CH$_3$), 17.29 (CH$_3$), 19.43 (CH$_3$), 116.04, 119.41, 125.79, 129.55, 139.81, 143.13, 154.26 (Ar-C), 161.53, 168.31, 169.38, 176.77 (C=N); MS m/z (%): 407 (M$^+$ + 1, 5), 406 (M$^+$, 18), 252 (18), 153 (99). 112 (95), 72 (100). Anal. Calcd. for C$_{16}$H$_{16}$ClN$_7$S$_2$ (405.93): C, 47.34; H, 3.97; N, 24.15. Found C, 47.39; H, 3.90; N, 24.06%.

2-Amino-4-methyl-5-(1-(2-(4-methyl-5-(bromophenyldiazenyl)thiazol-2-yl)hydrazono)ethyl)thiazole (**14d**). Red solid (75%); mp 176–178 °C; IR (KBr): ν 3368, 3191 (NH$_2$, NH), 1632 (C=N) cm^{-1}; ^{1}H-NMR (DMSO-d_6): δ 2.42 (s, 3H, CH$_3$), 2.53 (s, 3H, CH$_3$), 3.59 (s, 3H, CH$_3$), 7.28 (d, 2H, J = 4.5 Hz, Ar-H), 7.48(d, 2H, J = 4.5 Hz, Ar-H), 7.58 (s, 2H, D$_2$O-exchangeable, NH$_2$), 10.58 (s, 1H, D$_2$O-exchangeable, NH); ^{13}C-NMR (DMSO-d_6): δ 16.87 (CH$_3$), 17.30 (CH$_3$), 19.43 (CH$_3$), 113.65, 116.48, 119.42, 132.55, 139.91, 143.53, 154.25 (Ar-C), 161.53, 168.29, 169.38, 176.75 (C=N); MS m/z (%): 452 (M$^+$ + 1, 2), 451 (M$^+$, 15), 296 (14), 153 (100), 113 (86), 72 (98). Anal. Calcd. for C$_{16}$H$_{16}$BrN$_7$S$_2$ (450.38): C, 42.67; H, 3.58; N, 21.77. Found C, 42.55; H, 3.67; N, 21.65%.

2-Amino-4-methyl-5-(1-(2-(4-methyl-5-(methoxylphenyldiazenyl)thiazol-2-yl)hydrazono)ethyl)thiazole (**14e**). Red solid (78%); mp 187–189 °C; IR (KBr): ν 3414, 3202 (NH$_2$, NH), 1624 (C=N) cm^{-1}; ^{1}H-NMR (DMSO-d_6): δ 2.43 (s, 3H, CH$_3$), 2.53 (s, 3H, CH$_3$), 3.56 (s, 3H, CH$_3$), 3.72 (s, 3H, OCH$_3$), 6.91–7.30 (m, 4H, Ar-H), 7.87 (s, 2H, D$_2$O-exchangeable, NH$_2$), 10.48 (s, 1H, D$_2$O-exchangeable, NH); ^{13}C-NMR (DMSO-d_6): δ 8.93 (CH$_3$), 16.85 (CH$_3$), 17.19 (CH$_3$), 55.73 (OCH$_3$), 115.07, 115.87, 119.38, 137.60, 137.74, 155.30, 160.59 (Ar-C), 169.08, 169.21, 169.26, 176.60 (C=N); MS m/z (%): 402 (M$^+$ + 1, 2), 401 (M$^+$, 13), 248 (17), 153 (38), 113 (42), 44 (100). Anal. Calcd. for C$_{17}$H$_{19}$N$_7$OS$_2$ (401.51): C, 50.85; H, 4.77; N, 24.42. Found C, 50.82; H, 4.65; N, 24.23%.

2-Amino-4-methyl-5-(1-(2-(4-methyl-5-(nitrophenyldiazenyl)thiazol-2-yl)hydrazono)ethyl)thiazole (**14f**). Red solid (78%); mp 183–185°C; IR (KBr): ν 3375, 3194 (NH$_2$, NH), 1643 (C=N) cm^{-1}; ^{1}H-NMR (DMSO-d_6): δ 2.42 (s, 3H, CH$_3$), 2.54 (s, 3H, CH$_3$), 3.57 (s, 3H, CH$_3$), 6.91–7.38 (m, 4H, Ar-H), 7.59 (s, 2H, D$_2$O-exchangeable, NH$_2$), 10.53 (s, 1H, D$_2$O-exchangeable, NH); ^{13}C-NMR (DMSO-d_6): δ 16.45 (CH$_3$), 17.29 (CH$_3$), 19.76 (CH$_3$), 114.54, 119.53, 122.31, 129.60, 138.91, 144.23, 153.92 (Ar-C), 161.34, 168.55, 169.36, 176.75 (C=N); MS m/z (%): 417 (M$^+$ + 1, 3), 416 (M$^+$, 3), 153 (33), 113 (24), 43 (100). Anal. Calcd. for C$_{16}$H$_{16}$N$_8$O$_2$S$_2$ (416.48): C, 46.14; H, 3.87; N, 26.90. Found C, 46.11; H, 3.76; N, 26.64%.

3.1.6. Synthesis of Methyl 2-(2-((1-(2-amino-4-methylthiazol-5-yl)ethylidene)hydrazono)-4-oxothiazolidin-5-ylidene)acetate (**17**)

To a solution of 1-[1-(2-amino-4-methylthiazol-5-yl)ethylidene]thiosemicarbazide **3b** (0.229 g, 1 mmol) in dry methanol (20 mL) was added dimethylacetylenedicarboxylate **15** (0.142 g, 1 mmol). The solution was refluxed for 2 h. The precipitated product after cooling was filtered, washed with methanol, and recrystallized from ethanol to give product **17**.

Canary yellow solid (75%); mp 352–354 °C; IR (KBr): ν 3306, 3167 (NH$_2$, NH), 1708, 1689 (2C=O), 1609 (C=N) cm^{-1}; ^{1}H-NMR (DMSO-d_6): δ 2.39 (s, 3H, CH$_3$), 2.50 (s, 3H, CH$_3$), 3.76 (s, 3H, OCH$_3$), 6.60 (s, 1H, =C*H*COOCH$_3$), 9.43 (s, 2H, D$_2$O-exchangeable, NH$_2$), 12.88 (s, 1H, D$_2$O-exchangeable, NH); ^{13}C-NMR (DMSO-d_6): δ 15.03 (CH$_3$), 16.01 (CH$_3$), 52.92 (OCH$_3$), 114.64, 118.82, 139.60 143.31 (Ar-C), 157.67, 160.07, 165.97 (C=N), 168.37, 168.37 (C=O); MS m/z (%): 339 (M$^+$, 68), 153 (53), 113 (38), 85 (100), 43 (57). Anal. Calcd. for C$_{12}$H$_{13}$N$_5$O$_3$S$_2$ (339.39): C, 42.47; H, 3.86; N, 20.63. Found C, 42.40; H, 3.89; N, 20.43%.

3.2. Biological Assay

3.2.1. WST-1 Assay

The human hepatocellular carcinoma cell line were cultured and tested at Nanotechnology & Advanced materials central lab, Giza, Egypt. The culture was maintained in DMEM with 10% FBS at 37 °C humidified with 5% CO_2. Various concentrations of the compound being test, as well as doxorubicin as a reference drug (0.0, 0.04, 0.1, 0.2, 0.3, 0.4, and 0.6 µg/mL), were added to the cell monolayer in triplicate wells; then, the individual doses and their cytotoxicity were tested using a standard WST-1 cell proliferation assay as a fast and sensitive quantification of cell proliferation and viability in a 96-well microtiter plate for 24 h, measuring the absorbance of the dye solution at 450 nm [35].

3.2.2. Confocal Laser Scanning Microscopy (CLSM)

The Mode of Potential cytotoxicity action was evaluated using confocal laser scanning microscopic (Carrl Zeiss CLSM 710, diverse net ventures, Boston, MA, USA) imaging of Hep G2 treated cell lines at 0.6 µg/mL concentration of the tested compounds. Cells were plated in 96-multiwill plates (10^4 cells/well) for 24 h before treatment with the tested compound to allow attachment of cells to the wall of the plate. A selected concentration of the compounds being tested (0.6 µM) was added to the cell monolayer in triplicate wells for each individual dose; monolayer cells were incubated with the compounds for 24 h at 37 °C and in atmosphere of 5% CO_2. After 24 h, cells were stained by Rhodamine 132 and Acridine orange stains (Sigma-Aldrich, Boston, MA, USA); after further waiting for five minutes, microscopic examination was done using excitation laser lines at 588 nm and 633 nm by two channel detection.

3.3. Molecular Docking Using Leadit 2.1.5

All compounds were built and saved as Mol2. The crystal structure of human LDH-5B in complex with nicotinamide adenine dinucleotide was downloaded from protein data bank (pdb code = 1T2F). The protein was loaded into Leadit 2.1.5 and the receptor components were chosen by selection of chain A as a main chain. Binding site was defined by choosing NAD^+ as a reference ligand to which all coordinates were computed. Amino acids within radius 6.5 A° were selected in the binding site. All chemical ambiguities of residues were left as default. Ligand binding was driven by enthalpy (classic Triangle matching). For scoring, all default settings were restored. Intra-ligand clashes were computed using clash factor = 0.6. Maximum number of solutions per iteration = 200. Maximum number of solutions per fragmentation = 200. The base placement method was used as a docking strategy.

4. Conclusions

New thiazole derivatives have been synthesized using ethylidenethiosemicarbazide and ethylidenethiocarbohydrazide as starting materials under thermal conditions. Compounds **14e**, **14c** and **14a** may have significant and promising anticancer efficiency for hepatocellular carcinoma with low IC_{50}, 0.5 ± 0.02, 0.52 ± 0.03, and 0.84 ± 0.04 µM, respectively. The cytotoxic effect was due to its inhibitory effect to the inner mitochondrial membrane Lactate dehydrogenase enzyme, which, in turn, decreases cellular activity, including the rate of cell division. The molecular docking study confirmed high binding affinities of –24.85, –24.23, and –24.10 kcal/mol for **14e**, **14c** and **14a**, respectively. A direct correlation between the computed affinity and the IC_{50} was observed.

Acknowledgments: The support from Chemistry Department, Faculty of Science, Cairo University, is gratefully acknowledged.

Author Contributions: S.M.G., H.M.E.H. and H.M.A. designed research, performed research, analyzed the data, wrote and read the paper. T.A.S. designed the pharmacological part. M.A.K. designed the molecular modeling part. S.M.G. and M.A.K. approved the final manuscript.

Conflicts of Interest: The authors declare no conflict of interest.

References

1.   Ahmad, J.; Rabinovitz, M. Etiology and Epidemiology of Hepatocellular Carcinoma. In *Hepatocellular Cancer: Diagnosis and Treatment*; Carr, B.I., Ed.; Humana Press Inc.: Totowa, NJ, USA, 2010; pp. 1–22.

2.   Bruix, J.; Sherman, M.; Llovet, J.M. Clinical management of hepatocellular carcinoma. Conclusions of the Barcelona-200 EASL conference. European Association for the Study of the Liver. *J. Hepatol.* **2001**, *35*, 421–430. [CrossRef]

3.   Karegoudar, P.; Karthikeyan, M.S.; Prasad, D.J.; Mahalinga, M.; Holla, B.S.; Kumari, N.S. Synthesis of some novel 2,4-disubstituted thiazoles as possible antimicrobial agents. *Eur. J. Med. Chem.* **2008**, *43*, 261–267. [CrossRef] [PubMed]

4.   Cukurovali, A.; Yilmaz, I.; Gur, S.; Kazaz, C. Synthesis antibacterial and antifungal activity of some new thiazolylhydrazone derivatives containing 3-substituted cyclobutane ring. *Eur. J. Med. Chem.* **2006**, *41*, 201–207. [CrossRef] [PubMed]

5.   Abdel-Wahab, B.F.; Abdel-Aziz, H.A.; Ahmed, E.M. Synthesis and antimicrobial evaluation of 1-(benzofuran-2-yl)-4-nitro-3-arylbutan-1-ones and 3-(benzofuran-2-yl)-4,5-dihydro-5-aryl-1-[4-(aryl)-1,3-thiazol-2-yl]-1*H*-pyrazoles. *Eur. J. Med. Chem.* **2009**, *44*, 2632–2635. [CrossRef] [PubMed]

6.   Shih, M.H.; Ying, K.F. Syntheses and evaluation of antioxidant activity of sydnonyl substituted thiazolidinone and thiazoline derivatives. *Bioorg. Med. Chem.* **2004**, *12*, 4633–4643. [CrossRef] [PubMed]

7.   Shiradkar, M.; Kumar, G.V.S.; Dasari, V.; Tatikonda, S.; Akula, K.C.; Shah, R. Clubbed triazoles: A novel approach to antitubercular drugs. *Eur. J. Med. Chem.* **2007**, *42*, 807–816. [CrossRef] [PubMed]

8.   Amin, K.M.; Rahman, A.D.E.; Al-Eryani, Y.A. Synthesis and preliminary evaluation of some substituted coumarins as anticonvulsant agents. *Bioorg. Med. Chem.* **2008**, *16*, 5377–5388. [CrossRef] [PubMed]

9.   Gomha, S.M.; Riyadh, S.M.; Abbas, I.M.; Bauomi, M.A. Synthetic utility of ethylidene-thiosemicarbazide: Synthesis and anti-cancer activity of 1,3-thiazines and thiazoles with imidazole moiety. *Heterocycles* **2013**, *87*, 341–356.

10. Gomha, S.M.; Salah, T.A.; Abdelhamid, A.O. Synthesis, characterization and pharmacological evaluation of some novel thiadiazoles and thiazoles incorporating pyrazole moiety as potent anticancer agents. *Monatsh. Chem.* **2015**, *146*, 149–158. [CrossRef]

11. Carter, J.S.; Kramer, S.; Talley, J.J.; Penning, T.; Collins, P.; Graneto, M.J.; Seibert, K.; Koboldt, C.; Masferrer, J.; Zweifel, B. Synthesis and activity of sulfonamide-substituted 4,5-diaryl thiazoles as selective cyclooxygenase-2 inhibitors. *Bioorg. Med. Chem. Lett.* **1999**, *9*, 1171–1174. [CrossRef]

12. Zhu, L.P.; Ye, D.Y.; Tang, Y. Structure-based 3D-QSAR studies on thiazoles as 5-HT3 receptor antagonists. *J. Mol. Model.* **2007**, *13*, 121–131. [CrossRef] [PubMed]

13. Clark, R.F.; Zhang, T.; Wang, X.; Wang, R.; Zhang, X.; Camp, H.S.; Beutel, B.A.; Sham, H.L.; Gu, Y.J. Phenoxythiazole derivatives as potent and selective acetyl Co-A carboxylase-2 inhibitors: Modulation of isozyme selectivity by incorporation of phenyl ring substituent. *Bioorg. Med. Chem. Lett.* **2007**, *17*, 1961–1965. [CrossRef] [PubMed]

14. Andreani, A.; Rambaldi, M.; Mascellani, G.; Rugarli, P. Synthesis and diuretic activity of imidazo [2,1-*b*]thiazole acetohydrazones. *Eur. J. Med. Chem.* **1987**, *22*, 19–22. [CrossRef]

15. Ergenc, N.; Capan, G.; Gunay, N.S.; Ozkirimli, S.; Gungor, M.; Ozbey, S.; Kendi, E. Synthesis and hypnotic activity of new 4-thiazolidinone and 2-thioxo-4,5-imidazolidinedione derivatives. *Arch. Pharm. Pharm. Med. Chem.* **1999**, *332*, 343–347. [CrossRef]

16. Gomha, S.M.; Khalil, K.D. A convenient ultrasound-promoted synthesis and cytotoxic activity of some new thiazole derivatives bearing a coumarin nucleus. *Molecules* **2012**, *17*, 9335–9347. [CrossRef] [PubMed]

17. Gomha, S.M.; Abdel-aziz, H.M. Synthesis and antitumor activity of 1,3,4-thiadiazole derivatives bearing coumarine ring. *Heterocycles* **2015**, *91*, 583–592. [CrossRef]

18. Gomha, S.M.; Ahmed, S.A.; Abdelhamid, A.O. Synthesis and cytotoxicity evaluation of some novel thiazoles, thiadiazoles, and pyrido[2,3-*d*][1,2,4]triazolo[4,3-*a*]pyrimidin-5(1*H*)-one incorporating triazole moiety. *Molecules* **2015**, *20*, 1357–1376. [CrossRef] [PubMed]

19. Abbas, I.M.; Gomha, S.M.; Elneairy, M.A.A.; Elaasser, M.M.; Mabrouk, B.K.A. Fused triazolo[4,3-*a*]pyrimidinones: Synthesis and biological evaluation as antimicrobial and anti-cancer agents. *Turk. J. Chem.* **2015**, *39*, 510–531. [CrossRef]

20. Gomha, S.M.; Khalil, K.D.; El-Zanate, A.M.; Riyadh, S.M. A facile green synthesis and anti-cancer activity of *bis*-arylhydrazononitriles, triazolo[5,1-*c*][1,2,4]triazine, and 1,3,4-thiadiazoline. *Heterocycles* **2013**, *87*, 1109–1120.

21. Gomha, S.M.; Riyadh, S.M.; Mahmmoud, E.A.; Elaasser, M.M. Synthesis, molecular mechanics calculations, and anticancer activities of thiazoles, 1,3-thiazines, and thiazolidine using chitosan-grafted-poly(vinylpyridine) as basic catalyst. *Heterocycles* **2015**, *91*, 1227–1243.

22. Gomha, S.M.; Abbas, I.M.; Elneairy, M.A.A.; Elaasser, M.M.; Mabrouk, B.K.A. Antimicrobial and anticancer evaluation of novel synthetic tetracyclic system through Dimroth rearrangement. *J. Serb. Chem. Soc.* **2015**, *80*, 1251–1264. [CrossRef]

23. Wang, S.; Meades, C.; Wood, G.; Osnowski, A.; Anderson, S.; Yuill, R.; Thomas, M.; Mezna, M.; Jackson, W.; Midgley, C.; *et al.* 2-Anilino-4-(thiazol-5-yl)pyrimidine CDK Inhibitors: Synthesis, SAR analysis, X-Ray Crystallography, and Biological Activity. *J. Med. Chem.* **2004**, *47*, 1662–1675. [CrossRef] [PubMed]

24. Abdelhamid, A.O.; Zohdi, H.F.; Rateb, N.M. Reactions with hydrazonoyl halides XXI: Reinvestigation of the reactions of hydrazonoyl bromides with 1,1-dicyanothioacetanilide. *J. Chem. Res.* **1999**, 184–185. [CrossRef]

25. Gomha, S.M.; Badrey, M.G. A Convenient synthesis of some new thiazole and pyrimidine derivatives incorporating naphthalene moiety. *J. Chem. Res.* **2013**, *2*, 86–90. [CrossRef]

26. Imrich, J.; Tomaščiková, J.; Danihel, I.; Kristian, P.; Böhm, S.; Klika, K.D. Selective formation of 5- or 6-membered rings, 1,3-thiazolidin-4-one *vs.* 1,3-thiazin-4-one, from acridine thiosemicarbazides by the use of ethyne acid esters. *Heterocycles* **2010**, *80*, 489–503.

27. Darehkordi, A.; Saidi, K.; Islami, M.R. Preparation of heterocyclic compounds by reaction of dimethyl and diethyl acetylene dicarboxylate (DMAD, DEAD) with thiosemicarbazone derivatives. *Arkivoc* **2007**, *1*, 180–188.

28. Nami, N.; Hosseinzadeh, M.; Rahimi, E. Synthesis of some 3-substituted 1,2-dihydroindoles. *Phosphorous Sulfur Silicon* **2008**, *183*, 2438–2442. [CrossRef]

29. Augoff, K.; Hryniewicz-Jankowska, A.; Tabola, R. Lactate dehydrogenase-5: An old friend and a new hope in the war on cancer. *Cancer Lett.* **2015**, *358*, 1–7. [CrossRef] [PubMed]

30. Van Eerd, J.P.F.M.; Kreutzer, E.K.J. Mouse anti-malaria PAN pLDH monoclonal antibody. *Klinisch. Chem. Analisten Deel* **1996**, *2*, 138–139.

31. Fujiwara, Y.; Takenaka, K.; Kaliyama, K. The characteristics of hepatocellular carcinoma with high levels of serum lactic dehydrogenase. *Hepatogastroenterology* **1997**, *44*, 820–823. [PubMed]

32. Le, A.; Cooper, C.R.; Gouw, A.M.; Dinavahi, R.; Maitra, A.; Deck, L.M.; Royer, R.E.; Jagt, D.L.V.; Semenza, G.L.; Dan, C.V. Inhibition of lactate dehydrogenase A induces oxidative stress and inhibits tumor progression. *Proc. Nat. Acad. Sci. USA* **2010**, *107*, 2037–2042. [CrossRef] [PubMed]

33. Eweiss, N.F.; Osman, A. Part II new routes to acetylthiadiazolines and alkylazothiazoles. *J. Heterocycl. Chem.* **1980**, *17*, 1713–1717. [CrossRef]

34. Shawali, A.S.; Abdelhamid, A.O. Reaction of dimethylphenacylsulfonium bromide with *N*-nitrosoacetarylamides and reactions of the products with nucleophiles. *Bull. Chem. Soc. Jpn.* **1976**, *49*, 321–327. [CrossRef]

35. Zund, G.; Ye, Q.; Hoerstrup, S.P.; Schoeberlein, A.; Schmid, A.C.; Grunenfelder, J.; Vogt, P.; Turina, M. Tissue engineering in cardiovascular surgery: MTT, a rapid and reliable quantitative method to assess the optimal human cell seeding on polymeric meshes. *Eur. J. Cardio Thorac. Surg.* **1999**, *15*, 519–524. [CrossRef]

11

Synthesis and Cytotoxic Effect of Some Novel 1,2-Dihydropyridin-3-carbonitrile and Nicotinonitrile Derivatives

Eman M. Flefel [1,2], Hebat-Allah S. Abbas [2,3,*], Randa E. Abdel Mageid [2] and Wafaa A. Zaghary [4]

Academic Editors: Derek J. McPhee and Bimal K. Banik

[1] Department of Chemistry, College of Science, Taibah University, Al-Madinah Al-Monawarah 1343, Saudi Arabia; emanmflefel@yahoo.com
[2] Department of Photochemistry, National Research Centre, Dokki, Cairo 12622, Egypt; randaabdelmagid@yahoo.com
[3] Department of Chemistry, College of Science, King Khalid University, Abha 9004, Saudi Arabia
[4] Department of Pharmaceutical Chemistry, College of Pharmacy, Helwan University, Ain Helwan, Cairo 11795, Egypt; wzaghary@yahoo.com
* Correspondence: hebatallah201528@yahoo.com or hsabas@kku.edu.sa or hebanrc@yahoo.com

Abstract: 1-(2,4-Dichlorophenyl)-3-(4-fluorophenyl)propen-1-one (**1**) was prepared and reacted with an active methylene compound (ethyl cyanoacetate) in the presence of ammonium acetate to give the corresponding cyanopyridone **2**. Compound **2** reacted with hydrazine hydrate, malononitrile, ethyl bromoacetate and phosphorous oxychloride to afford compounds **4** and **7–11**, respectively. The 2-chloropyridine derivative **11** reacted with different primary amines, namely benzyl amine, piperonyl amine, 1-phenylethyl amine, and/or the secondary amines 2-methyl-pipridine and morpholine to give the corresponding derivatives **12–15**. Hydrazinolysis of chloropyridine derivative **11** with hydrazine hydrate afforded the corresponding hydrazino derivative **17**. Condensation of compound **17** with ethyl acetoacetate, acetylacetone, isatin and different aldehydes gave the corresponding derivatives **18–21**. Some of newly synthesized compounds were screened for cytotoxic activity against three tumor cell lines. The results indicated that compounds **8** and **16** showed the best results, exhibiting the highest inhibitory effects towards the three tumor cell lines, which were higher than that of the reference doxorubicin and these compounds were non-cytotoxic towards normal cells (IC_{50} values > 100 µg/mL).

Keywords: nicotinonitrile; 4-fluorophenylpyridine; acetohydrazide; chloropyridine; cytotoxicity

1. Introduction

Cancer is the second leading cause of death in both developing and developed countries [1]. The leading forms were lung cancer, colorectal cancer, liver cancer and breast cancer [2,3]. Cancer treatment has been a major research and development effort in academia and the pharmaceutical industry for numerous years [4,5]. Despite the fact that there is a large amount of information available dealing with the clinical aspects of cancer chemotherapy, we felt that there was a clear requirement for an updated treatment from the point of view of medicinal chemistry and drug design [6]. Another major goal for developing new anticancer agents is to overcome cancer resistance to drug treatment, which has made many of the currently available chemotherapeutic agents ineffective [7].

Chalcones, one of the major classes of natural products with widespread occurrence in vegetables, fruits, spices and soy-based foodstuffs, have been reported to possess several biological activities such as antibacterial [8,9], anti-fungal [10,11], anti-inflammatory [12], and anti-tumor activities [13,14]. An important feature of chalcones is their ability to act as an intermediate for the synthesis of biologically active heterocyclic compounds such as pyrimidine and pyridine derivatives [15,16]. The pyridine nucleus is an integral part of anti-inflammatory and anticancer agents [17,18]. On the other hand, cyanopyridone and cyanopyridine derivatives have shown to possess promising antimicrobial [19] antioxidant [20,21], antibiotic [22], antiinflamatory [23,24], analgesic, anticonvulsant [25] and anticancer [26–29] properties. 3-Cyano-2-pyridones are analogous to the alkaloid ricinine, the first known alkaloid containing a cyano group. The anticancer activity of 3-cyano-2-pyridone derivatives is of much interest owing to the different types of biological targets they might interfere with, e.g., PDE3, PIM1 kinase, and survivin (Figure 1) [30].

Motivated by the above recent literature observations and our own previous reports [20,21,31–33], herein some new pyridine derivatives were synthesized, leading to interesting heterocyclic scaffolds that are mostly useful for the creation of varied chemical libraries of drug-like molecules for biological screening.

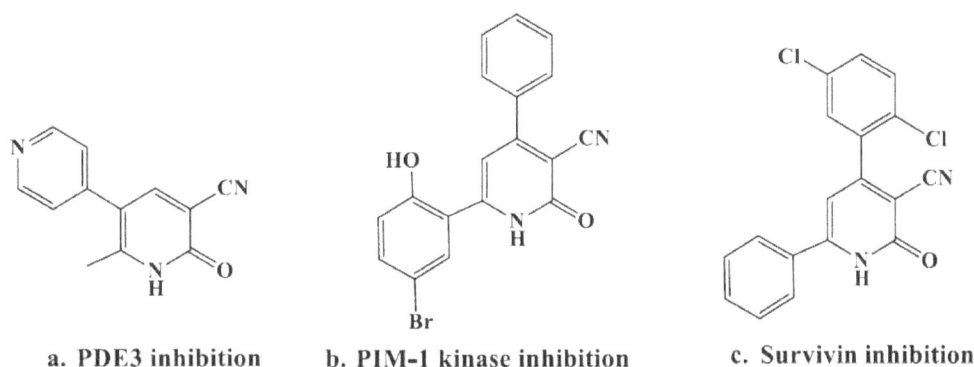

a. PDE3 inhibition   b. PIM-1 kinase inhibition   c. Survivin inhibition

Figure 1. Various 3-cyano-2-oxopyridine derivatives with potential growth inhibitory and/or antiangiogenic actions through PDE3 inhibition (a); PIM-1 kinase inhibition (b); or survivin inhibition (c).

2. Results and Discussion

2.1. Chemistry

The synthesis of the designed target compounds was achieved as outlined in Schemes 1–3. During this investigation, the pyridin-3-carbonitrile starting material **2** was prepared by condensation of the corresponding enone **1** [34] with ethyl cyanoacetate in the presence of excess ammonium acetate (Scheme 1). Compound **2** can also be obtained in high yield through a four-component modified Hantzch reaction, in a one-step synthesis, by refluxing a mixture of 2,4-dichloro-acetophenone, 4-fluorobenzaldehyde, ethyl cyanoacetate and ammonium acetate in *n*-butanol. The structure of pyridin-3-carbonitrile **2** was supported by elemental analysis, IR, (^{1}H, ^{13}C) NMR and mass spectral studies. Its IR spectrum showed absorption bands at 3278, 2219, 1632 cm^{-1} indicating the presence of NH, CN and CO groups, respectively. Its ^{1}H-NMR spectrum displayed a broad D_2O exchangeable singlet at δ 8.10 ppm for the NH proton, while its ^{13}C-NMR spectrum also revealed signals at δ 117.6 and 161.8 ppm for CN and CO moieties, respectively. The mass spectrum showed a molecular ion peak at *m/z* 358 (M$^+$, 98%), which tallies with its molecular formula $C_{18}H_9Cl_2FN_2O$.

Pyridin-3-carbonitrile **2** possesses several reactive sites, *viz.* CN, NH, and CO groups, which can play a great role in the synthesis of heterocyclic derivatives, most of which are interesting from both the chemical and biological point of view. Thus, hydrazinolysis of pyridin-3-carbonitrile **2** with hydrazine hydrate in absolute ethanol for 15 h affords the corresponding pyrazolo[3,4-*b*]pyridin-3-amine derivative **4** through the elimination of a water molecule from the intermediate **3** (Scheme 1). Pyrazolo[3,4-*b*]pyridine derivative **4** was identified by the absence of the cyano and carbonyl groups signals in its IR and the presence of an amino group signal at δ 5.69 ppm and the broad band of the

NH proton at δ 10.05 ppm in its ^{1}H-NMR spectrum. Its mass spectrum showed a molecular ion peak at m/z 372 (M$^+$; 72%), which conforms to its molecular formula $C_{18}H_{11}Cl_2FN_4$.

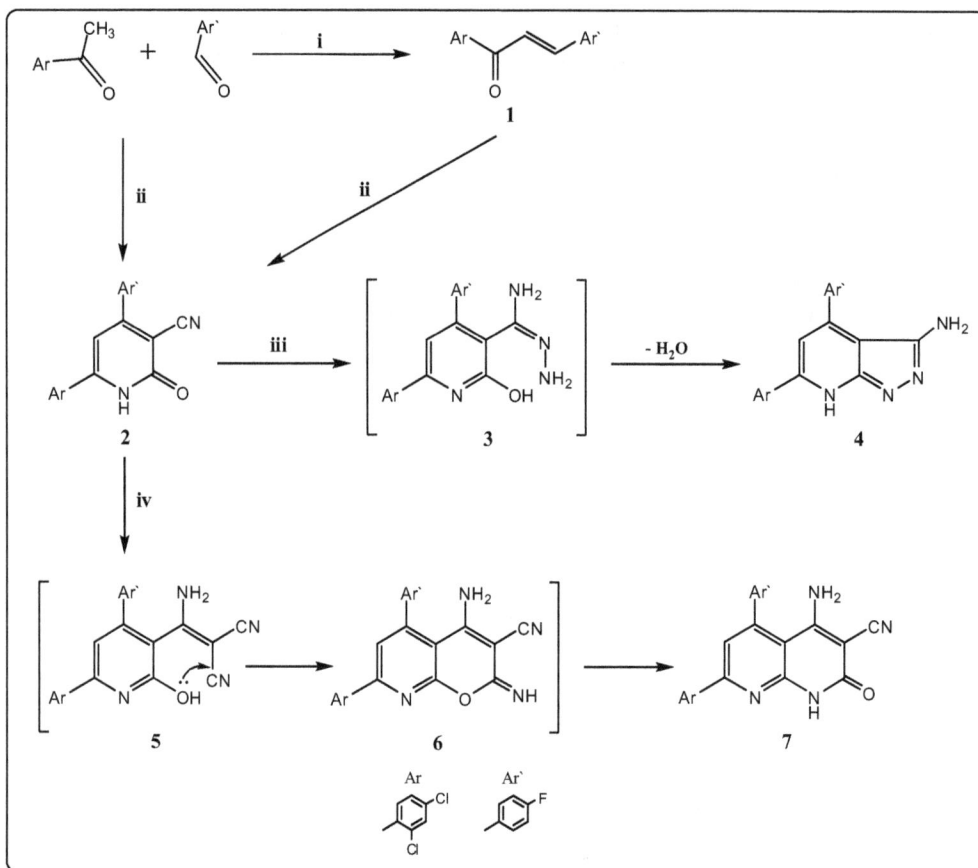

Scheme 1. General methods for the preparation of compounds **2–7**. Reagents and conditions: (**i**) NaOH/EtOH, stirring; (**ii**) ethyl cyanoacetate/CH$_3$COONH$_4$/EtOH, reflux; (**iii**) hydrazine hydrate 98% (1 mL)/EtOH, reflux; and (**iv**) malononitrile/triethylamine (3 mL)/EtOH, reflux.

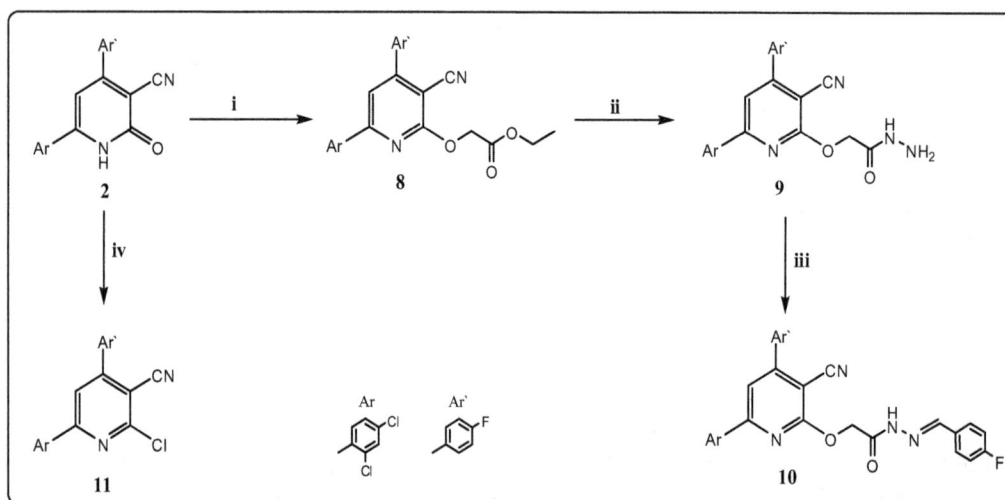

Scheme 2. General methods for the preparation of compounds **8–11**. Reagents and conditions: (**i**) ethyl bromoacetate/anh. K$_2$CO$_3$/dry CH$_3$COCH$_3$, reflux; (**ii**) hydrazine hydrate 98% (2 mL)/EtOH, reflux; (**iii**) 4-flurobenzaldehyde/EtOH, reflux; and (**iv**) phosphorus oxychloride/EtOH, reflux.

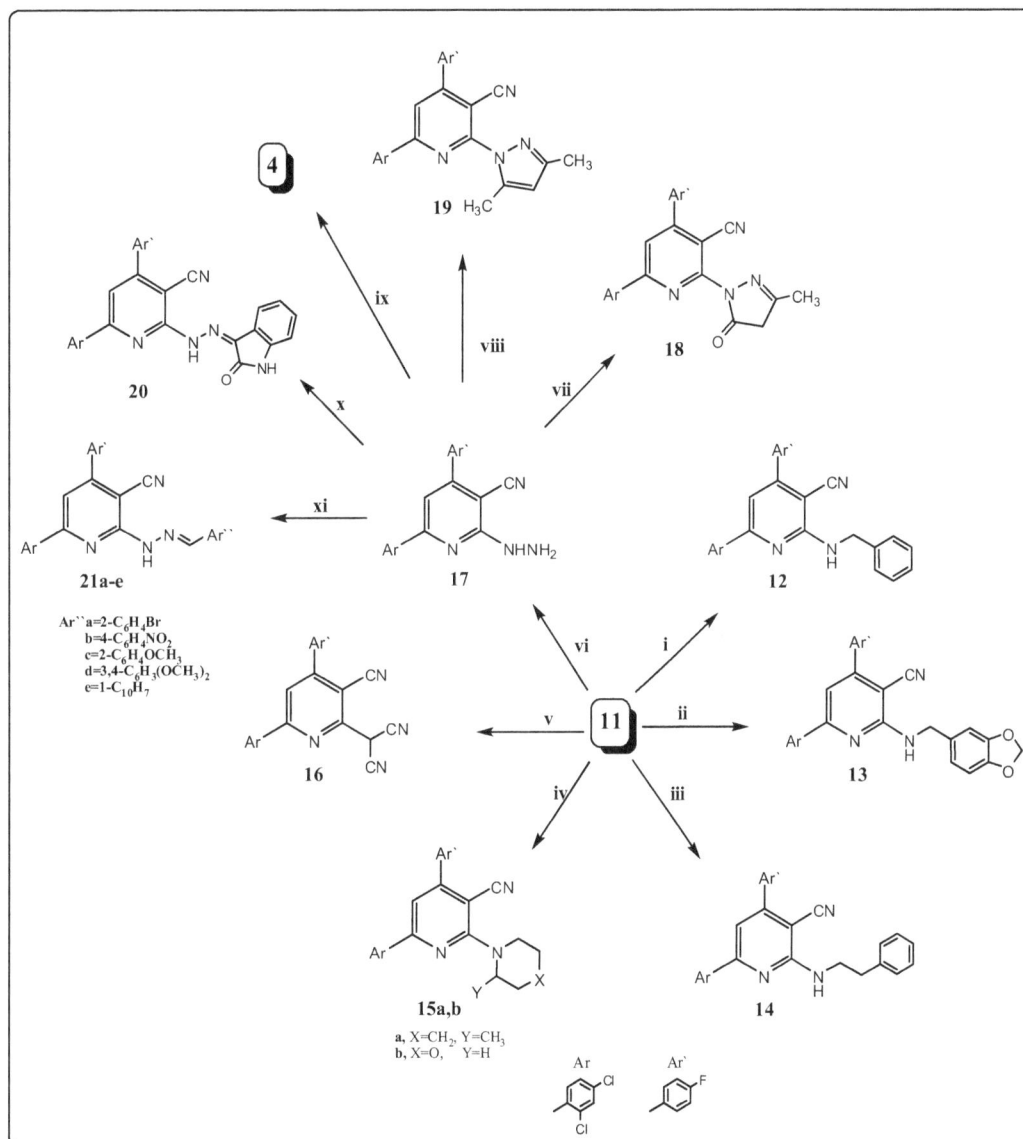

Scheme 3. General methods for the preparation of compounds **12–21a–e**. Reagents and conditions: (**i**) benzylamine/EtOH, reflux; (**ii**) piperonylamine/EtOH, reflux; (**iii**) 1-phenylethylamine/EtOH, reflux; (**iv**) 2-methylpiperidine or morpholine/EtOH, reflux; (**v**) malononitrile/triethylamine (1 mL)/EtOH, reflux; (**vi**) hydrazine hydrate 98% (2 mL)/EtOH, reflux; (**vii**) ethyl acetoacetate/AcOH, reflux; (**viii**) acetylacetone/AcOH, reflux; (**ix**) DMF or AcOH, reflux; (**x**) isatin/3 drops AcOH/EtOH, reflux; and (**xi**) appropriate aromatic aldehyde, namely: 2-bromobenzaldhyde, 4-nitrobenzaldhyde, 2-methoxy-benzaldhyde, 3,4-dimethoxybenzaldhyde and/or1-naphthaldehyde/3 drops AcOH/EtOH, reflux.

Compound **2** was also refluxed with malononitrile to afford 4-amino-7-(2,4-dichlorophenyl)-5-(4-fluorophenyl)-2-oxo-1,2-dihydro-1,8-naphthyridine-3-carbonitrile (**7**) via the intermediates **5** and **6**, as confirmed by elemental analysis, ^{1}H- and ^{13}C-NMR. The R spectrum of compound **7** showed bands at 3312, 3249, 3145 and 1688 cm^{-1} due to NH$_2$, NH and CO groups, respectively; its ^{13}C-NMR spectrum showed signals at δ 118.1 and 168.8 ppm corresponding to CN and CO groups, respectively. Its mass spectrum showed a molecular ion peak at m/z 424 (M$^+$; 92%), which conforms to its molecular formula C$_{21}$H$_{11}$Cl$_2$FN$_4$O.

Moreover, when pyridin-3-carbonitrile **2** was alkylated with ethyl bromoacetate in acetone using anhydrous potassium carbonate as catalyst, the ester derivatives **8** was produced (Scheme 2).

The ^{1}H-NMR spectrum of **8** showed signals at δ 1.15, 4.13 and 5.11 ppm due to the presence of (CH$_3$-ester), (OCH$_2$-ester) and (O–CH$_2$) respectively; and its ^{13}C-NMR exhibited signals at δ 13.5, 43.8, 61.3 and 167.9 ppm due to (CH$_3$), (2CH$_2$) and (CO) groups, respectively.

Ester derivative **8** were condensed with hydrazine hydrate (98%) in ethanol to give 2-[3-cyano-6-(2,4-dichlorophenyl)-4-(4-fluorophenyl)pyridin-2-yloxy]acetohydrazide (**9**) (Scheme 2), confirmed by its IR and NMR spectra. Its IR spectrum showed strong peaks at 3314, 3282 and 3116 cm^{-1} indicating the presence of a –NHNH$_2$ group, and the NMR (^{1}H and ^{13}C) and mass spectra were also in accordance with its structure.

Schiff base **10** can be produced via condensation of acetohydrazide **9** with an aromatic aldehyde, namely 4-flourobenzaldehyde, in ethanol (Scheme 2). The structure of Schiff base **10** was elucidated based on the spectral and analytical data. The IR spectrum revealed the absence of the absorption bands of (NH$_2$) group absorption and its ^{1}H-NMR spectrum showed a singlet at δ 8.10 ppm due to the presence of the (CH=N–) group.

In addition, chlorination of cyanopyridone **2** with phosphorous oxychloride afforded the 2-chloronicotinonitrile derivative **11** (Scheme 2) in good yield, after 8 h. The IR spectrum showed the absence of a characteristic CO group band.

It is known that position 2 in chloronicotinonitrile derivatives shows distinct activities toward nucleophiles, especially nitrogen nucleophiles. Thus, nucleophilic replacement of the chlorine atom of chloronicotinonitrile **11** was performed by refluxing with different primary amines, namely benzyl-amine, piperonylamine, 1-phenylethylamine and/or secondary amines, namely 2-methyl-piperidine and morpholine in boiling ethanol for 6–12 h to afford the corresponding 2-aminopyridine derivatives **12–15a,b**; respectively (Scheme 3). The elemental analysis and spectral data of compounds **12–15a,b** were in agreement with the proposed structures. The ^{1}H-NMR of compound **15b** for example, showed signals at δ 3.31 and 3.72 ppm due to the presence of (2N–CH$_2$) and (2O–CH$_2$) groups, respectively; and its ^{13}C-NMR exhibited signals at δ 47.9, 49.1, 64.10 and 65.9 ppm due to the presence of (2N–CH$_2$) and (2O–CH$_2$), respectively.

Furthermore, nucleophilic displacement was carried out by heating the chloropyridine derivative **11** with malononitrile in ethanol containing a few drops of triethylamine as a catalyst to give 2-[3-cyano-6-(2,4-dichlorophenyl)-4-(4-fluorophenyl)pyridin-2-yl]malononitrile (**16**). The structure of compound **16** was confirmed by its spectral data; the IR spectrum showed the presence of the CN group at 2218, 2225 cm^{-1}. In addition, the NMR (^{1}H and ^{13}C) and mass spectral data were in accordance with its structure. Hydrazinolysis of the chloropyridine derivative **11** was performed by its reaction with excess hydrazine hydrate in refluxing ethanol to give the hydrazino derivative **17** (Scheme 3). The structure of **17** was confirmed by its spectral data. The IR spectrum exhibited the characteristic absorption bands at 3440, 3320, 3150 cm^{-1} indicating the presence of the –NHNH$_2$ group. Its mass spectrum showed a molecular ion peak at m/z 372 (M$^+$; 39%), which conforms to its molecular formula C$_{18}$H$_{11}$Cl$_2$FN$_4$.

The 2-hydrazino-nicotinonitrile **17** is another key compound, which facilitates the synthesis of diverse heterocyclic compounds. Thus, it reacted with different active methylene (β-diketones), namely: ethyl acetoacetate and acetylacetone in glacial acetic acid, and thus the N-pyrazolo derivatives **18** and **19** were produced (Scheme 3). The IR spectrum of compound **19**, for example, showed a characteristic band at 2210 cm^{-1} for the CN group and its ^{1}H-NMR spectrum revealed singlets at δ 2.31, 2.45 and 6.15 ppm due to (2CH$_3$) and the (CH-pyrazole) moieties, respectively. The ^{13}C-NMR data displayed two characteristic signals at δ 18.4, 19.3 and 117.9 ppm for 2CH$_3$ and CN groups, respectively. Also, on heating compound **17** with isatin in ethanol it afforded 6-(2,4-dichlorophenyl)-4-(4-fluorophenyl)-2-(2-(2-oxoindolin-3-ylidene)hydrazinyl)nicotinonitrile (**20**) in good yield (Scheme 3). The structure of compound **20** gave correct elemental analyses values and spectral features.

In addition, to get a new series of Schiff bases expected to be biologically active, heating of 2-hydrazinonicotinonitrile **17** with different aromatic aldehydes, namely 2-bromobenzaldehyde, 4-nitrobenzaldeyde, 2-methoxybenzaldeyde, 3,4-dimethoxybenzaldeyde and/or 1-naphthaldeyde in ethanol gave the corresponding Schiff bases **21a–e**, respectively. The structure of compounds **21a–e** was characterized by the disappearance of the NH_2 group. In addition, the ^{1}H-NMR spectra showed a singlet at around δ 8.31–8.33 due to the presence of the azomethine group (CH=N–). Finally, reaction of 2-hydrazinonicotinonitrile **17** with acetic acid or DMF afforded the corresponding pyrazolo[3,4-*b*]pyridin-3-amine derivative **4** through intramolcular cyclization via the addition of the NH_2 functional group at the CN group.

2.2. In Vitro Anticancer Screening

The *in vitro* cytotoxic activity the newly synthesized compounds against human breast cell line (MCF7), non-small cell lung cancer NCI-H460, CNS cancer SF-268 and WI 38 (normal fibroblast cells) were evaluated using doxorubicin as the reference drug, according to the method reported by Skehan *et al.* [35]. The IC_{50} values of the synthesized compounds compared to the reference drug are shown in Table 1.

Table 1. Cytotoxic activity in (IC_{50}, µg/mL) by the newly synthesized compounds against human cancer cell lines and normal cells.

Comp. No.	IC_{50} (µg/mL)			
	MCF-7	NCI-H460	SF-268	WI 38
4	67.04 ± 6.23 [c]	56.75 ± 8.20 [c]	69.05 ± 9.15 [c]	18.62 ± 1.21
7	36.22 ± 2.14 [c]	74.03 ± 3.65 [c]	62.13 ± 3.61 [c]	22.97 ± 8.2
8	0.02 ± 0.002 [a]	0.01 ± 0.002 [a]	0.02 ± 0.045 [a]	non-cytotoxic
9	2.41 ± 1.24 [a]	2.30 ± 2.86 [a]	0.46 ± 0.06 [a]	62.19 ± 2.02
10	30.58 ± 1.10 [b]	30.67 ± 1.64 [b]	28.18 ± 8.83 [b]	19.80 ± 2.68
13	16.26 ± 1.87 [b]	18.92 ± 1.03 [b]	23.24 ± 4.12 [b]	20.38 ± 4.99
15a	37.07 ± 7.34 [c]	16.37 ± 2.32 [b]	38.94 ± 2.63 [c]	30.62 ± 6.21
16	0.01 ± 0.002 [a]	0.02 ± 0.001 [a]	0.01 ± 0.003 [a]	non-cytotoxic
17	0.61 ± 0.082 [a]	0.86 ± 0.02 [a]	2.19 ± 0.83 [a]	64.11 ± 1.22
18	20.22 ± 2.26 [b]	0.01 ± 0.003 [a]	20.20 ± 3.26 [b]	29.82 ± 4.88
19	75.20 ± 13.86 [c]	62.30 ± 10.35 [c]	10.39 ± 4.19 [a]	50.20 ± 10.22
20	0.66 ± 0.21 [a]	0.90 ± 0.12 [a]	2.34 ± 0.51 [a]	72.45 ± 2.40
21d	66.02 ± 8.25 [c]	44.95 ± 10.46 [c]	32.45 ± 6.04 [b]	non-cytotoxic
DMSO	0	0	0	0
Doxorubicin	0.04 ± 0.008	0.09 ± 0.008	0.09 ± 0.007	non-cytotoxic

MCF-7 (breast adenocarcinoma); NCI-H460 (non-small cell lung cancer); SF-268 (CNS cancer); WI 38 (normal fibroblast cells); Doxorubicin (anticancer positive control); DMSO (solvent, negative control); [a] highly active; [b] moderately active; [c] weakly active.

From the results presented in Table 1 and Figure 2, it is evident that some of the compounds were active against the three human cancer cell lines. Compounds **8** and **16** displayed high cytotoxic activity against the tested cell lines (most of the IC_{50} values ranged from 0.01 ± 0.002 to 0.02 ± 0.001 µg/mL) and these compounds were non-cytotoxic on the normal cells (IC_{50} values > 100 µg/mL) and exhibited better cytotoxicity against most of cancer cell lines than doxorubicin as standard drug. Moreover, compounds **9**, **17** and **20** exhibited high growth inhibitory activity on the various cancer panel cell lines (IC_{50} values ranged from 0.46 ± 0.006 to 2.43 ± 0.51 µg/mL) with weak cytotoxicity on the normal cells (IC_{50} values ranged from 62.19 ± 2.02 to 72.45 ± 2.40 µg/mL). In addition, other compounds showed moderate to weak cytotoxicity against all cancer cell lines (IC_{50} values ranged from 10.39 ± 4.19 to 75.20 ± 13.86 µg/mL) with cytotoxic effects on the human normal cell (IC_{50} values ranged from non-cytotoxic to 50.20 ± 10.22 µg/mL) in comparison with doxorubicin. The resultant data can be analyzed with respect to the chemical structures of the examined compounds; thus it can be noticed that the derivatives **8** and **16** that bear ester or malononitrile side chains on the parent cyanopyrine nucleus showed the highest potency as growth inhibiting agents against the three human cancer cell lines, which might be due to their lipophilicity that allows their accumulation inside tumor tissues inducing growth inhibition effects [36].

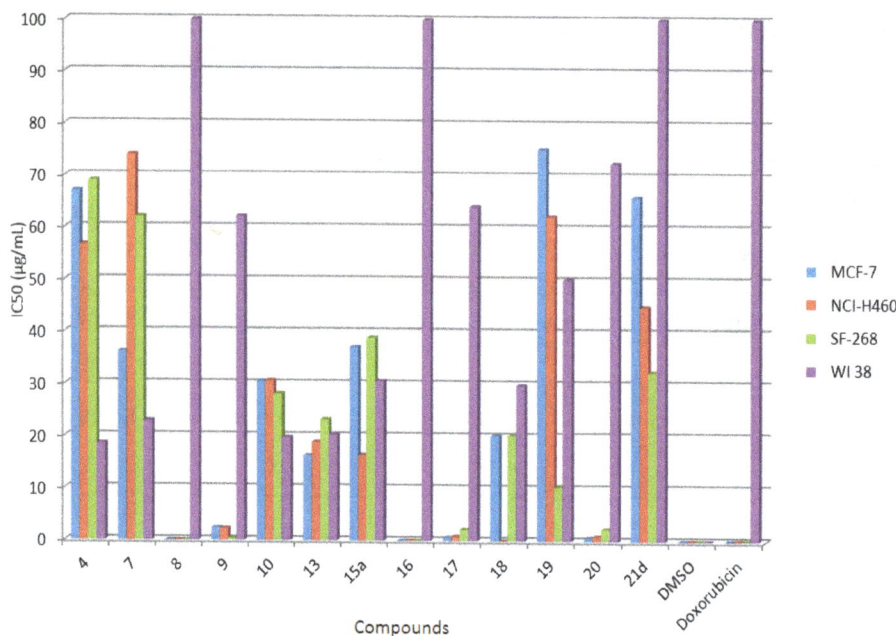

Figure 2. Cytotoxic activity of some newly synthesized compounds against human cancer cell lines and normal cells.

3. Experimental Section

3.1. General Information

All melting points are uncorrected and were determined on a Stuart electric melting point apparatus. The microanalyses were within ±0.4% of the theoretical values and were carried out at the Microanalytical Centre, National Research Centre, Cairo, Egypt. IR spectra (KBr) were recorded on a FT-IR 400D infrared spectrometer (Shizmadu-series, Kyoto, Japan) using the OMNIC program and are reported as frequency of absorption in cm^{-1}. ^{1}H-NMR spectra were recorded on a Bruker (Rheinstetten, Germany) spectrophotometer at 400 MHz using TMS as internal standard and with residual signals of the deuterated solvent δ = 7.26 ppm for $CDCl_3$ and δ 2.51 ppm for DMSO-d_6. ^{13}C-NMR spectra were recorded on the same spectrometer at 100 MHz and referenced to solvent signals δ = 77 ppm for $CDCl_3$ and δ 39.50 ppm for DMSO-d_6. The mass spectra were recorded on a Shimadzu GCMS-QP-1000 EX mass spectrometer (Kyoto, Japan) at 70 eV using the electron ionization technique. Homogeneity of all compounds synthesized was checked by TLC which was performed on Merck 60 (Munich, Germany) ready-to-use silica gel plates to monitor the reactions and test the purity of the new synthesized compounds. The chemical names given for the prepared compounds are according to the IUPAC system.

3.2. Synthetic Procedures

3.2.1. 6-(2,4-Dichlorophenyl)-4-(4-fluorophenyl)-2-oxo-1,2-dihydropyridin-3-carbonitrile (2)

Method A: A mixture of 1-(2,4-dichlorophenyl)-3-(4-fluorophenyl)prop-2-en-1-one (**1**, 2.95 g, 0.01 mol), ethyl cyanoacetate (1.13 g, 0.01 mol) and ammonium acetate (6.16 g, 0.08 mol) in ethanol (40 mL) was refluxed for 10 h. After cooling, the precipitate was filtered, dried and recrystallized from dioxane to give compound **2** (35% yield).

Method B: A mixture of 2,4-dichloroacetophenone (1.88 g, 0.01 mol), 4-fluorobenzaldehyde (1.24 g, 0.01 mol), ethyl cyanoacetate (1.13 g, 0.01 mol) and ammonium acetate (6.16 g, 0.08 mol) in *n*-butanol (20 mL) was refluxed for 3 h, to give yellow crystals that were then filtered, washed with water, dried and recrystallized to give the title compound **2** (85% yield).

Compound **2**: m.p. 276–277 °C; IR (KBr) ν_{max} in cm^{-1}: 3278 (NH), 2219 (CN), 1632 (C=O); ^{1}H-NMR (DMSO-d_6): 6.61 (s, 1H, pyridine H5), 7.41–7.85 (m, 7H, Ar–H), 8.10 (br. s, 1H, NH, D$_2$O exchangeable); ^{13}C-NMR (DMSO-d_6): 117.6 (CN), 119.61, 120.23, 122.00, 122.54, 123.33, 123.59, 133.93, 135.87, 136,12, 142.45, 144.64, 146.21, 146.87, 151,32, 151.98 (16 Ar–C), 161.8 (C=O); MS, *m/z* (%): 358 [M]$^+$ (98), 360 [M + 2]$^+$ (59), 362 (M$^+$ + 4; 11%). Anal. Calcd. for C$_{18}$H$_9$Cl$_2$FN$_2$O (359.18): C, 60.19; H, 2.53; N, 7.80%; found: C, 59.97 ; H, 2.74; N, 7.94%.

3.2.2. 6-(2,4-Dichlorophenyl)-4-(4-fluorophenyl)-7H-pyrazolo[3,4-b]pyridin-3-amine (4)

Method A: A mixture of compound **2** (3.6 g, 0.01 mol) and hydrazine hydrate 98% (1 mL, 0.02 mol) in absolute ethanol (30 mL) was refluxed for 15 h. The reaction mixture was left at room temperature overnight and then poured into ice/cold water to complete precipitation. The product was filtered, dried and recrystallized from benzene to give compound **4** (75% yield).

Method B: Compound **17** (3.7 g, 0.01 mol) in DMF or AcOH (20 mL) was refluxed for 8 h. The reaction mixture was cooled; the solid product that precipitated was filtered, dried and recrystallized from ethanol to give compound **4** (43% yield).

Compound **4**: m.p. 309–311 °C; IR (KBr) ν_{max} in cm^{-1}: 3414, 3335, 3180 (NH$_2$, NH); ^{1}H-NMR (DMSO-d_6): 5.69 (s, 2H, NH$_2$, D$_2$O exchangeable), 7.12–7.97 (m, 8H, Ar–H + pyridine H5), 10.05 (s, H, NH, D$_2$O exchangeable); ^{13}C-NMR (DMSO-d_6): 109.76, 111.54, 112.32, 114.29, 115.21, 115.98, 120.54, 121.87, 123.34, 126.30, 132.64, 138.97, 144.54, 145.92, 148.40 (16 Ar–C), 157.1, 158.9 (2C=N); MS, *m/z* (%): 372 [M]$^+$ (72), 374 [M + 2]$^+$ (40), 376 [M + 4]$^+$ (8). Anal. Calcd. for C$_{18}$H$_{11}$Cl$_2$FN$_4$ (373.21): C, 57.93; H, 2.97; N, 15.01%; found: C, 58.12 ; H, 2.70; N, 15.23%.

3.2.3. 4-Amino-7-(2,4-dichlorophenyl)-5-(4-fluorophenyl)-2-oxo-1,2-dihydro-1,8-naphthyridine-3-carbonitrile (7)

To a solution of compound **2** (3.6 g, 0.01 mol) in absolute ethanol (30 mL), triethylamine (3 mL), and malononitrile (0.7 g, 0.01 mol) were added. The reaction mixture was refluxed for 6 h., then left to cool to room temperature, poured into cold water and neutralized with diluted hydrochloric acid to complete precipitation. The solid obtained was filtered, washed with water, dried and recrystallized from methanol to give compound **7**. Yield 70%; m.p. 188–189 °C; IR (KBr) ν_{max} in cm^{-1}: 3312, 3249, 3145 (NH$_2$, NH), 2223 (CN), 1688 (C=O); ^{1}H-NMR (DMSO-d_6): 5.46 (s, 2H, NH$_2$, D$_2$O exchangeable), 6.89 (s, 1H, pyridine H5), 7.10–7.41 (m, 7H, Ar–H), 8.20 (s, 1H, NH, D$_2$O exchangeable); ^{13}C-NMR (DMSO-d_6): 118.1 (CN), 121.32, 122.87, 123.76, 125.32, 128.43, 132.65, 133.87, 134.12, 138.09, 139.56, 141.54, 147.65, 148.02, 149.53 (18 Ar–C), 158.6 (C=N), 168.9 (C=O); MS, *m/z* (%): 424 [M]$^+$ (92), 426 [M + 2]$^+$ (59), 428 [M + 4]$^+$ (10). Anal. Calcd. for C$_{21}$H$_{11}$Cl$_2$FN$_4$O (425.24): C, 59.31; H, 2.61; N, 13.18%; found: C, 59.12; H, 2.79; N, 13.33%.

3.2.4. Ethyl 2-[3-cyano-6-(2,4-dichlorophenyl)-4-(4-fluorophenyl)pyridin-2-yloxy]acetate (8)

A mixture of compound **2** (3.6 g, 0.01 mol), ethyl bromoacetate (1.2 mL, 0.01 mol) and anhydrous potassium carbonate (2.10 g, 0.015 mol) in dry acetone (50 mL) was refluxed for 24 h. The reaction mixture was cooled and poured onto ice/cold water; the solid that separated out was filtered, dried and recrystallized from dioxane to give compound **8**. Yield 75%; m.p. 119–121 °C; IR (KBr) ν_{max} in cm^{-1}: 2220 (CN), 1755 (C=O ester); ^{1}H-NMR (DMSO-d_6): 1.15 (t, *J* = 7.5 Hz, 3H, CH$_3$-ester), 4.13 (q, *J* = 7.5 Hz, 2H, O-CH$_2$-ester), 5.11 (s, 2H, O-CH$_2$), 7.00–7.40 (m, 8H, Ar–H + pyridine H5); ^{13}C-NMR (DMSO-d_6): 13.5 (CH$_3$), 44.8, 61.3 (2CH$_2$), 118.6 (CN), 123.30, 123.54, 127.10, 127.98, 131.34, 133.54, 137.07, 138.76, 141.99, 142.76, 144.65, 126.87, 148.43 (16 Ar–C), 159.1 (C=N), 167.9 (C=O); MS, *m/z*

(%): 444 [M]$^+$ (61), 446 [M + 2]$^+$ (37), 448 [M + 4]$^+$ (6), 371 [M–COOC$_2$H$_5$]$^+$ (41). Anal. Calcd. for C$_{22}$H$_{15}$Cl$_2$FN$_2$O$_3$ (445.27): C, 59.34; H, 3.40; N, 6.29%; found: C, 59.52 ; H, 3.71; N, 6.05%.

3.2.5. 2-[3-Cyano-6-(2,4-dichlorophenyl)-4-(4-fluorophenyl)pyridin-2-yloxy]acetohydrazide (9)

A mixture of compound **8** (4.5 g, 0.01 mol), hydrazine hydrate 98% (2 mL, 0.04 mol) and absolute ethanol (30 mL) was refluxed for 4 h. The reaction mixture was cooled and the formed solid was filtered, dried and recrystallized from acetic acid to give compound **9**. Yield 76%; m.p. 197–199 °C; IR (KBr) ν_{max} in cm^{-1}: 3314, 3282, 3116 (NH$_2$, NH), 2225 (CN), 1696 (C=O ester); ^{1}H-NMR (DMSO-d_6): 4.92 (s, 2H, CH$_2$), 7.00–7.40 (m, 8H, Ar–H + pyridine H5), 8.2 (s, 1H, NH, D$_2$O exchangeable), 9.8 (s, 2H, NH$_2$, D$_2$O exchangeable); ^{13}C-NMR (DMSO-d_6): 62.4 (CH$_2$), 119.7 (CN), 121.65, 124.05, 124,90, 125.71, 128.23, 129.98, 133.76, 137.54, 142.12, 143.45, 145.86 (16 Ar–C), 158.6 (C=N), 169.4 (C=O); MS, m/z (%): 430 [M]$^+$ (31), 432 [M + 2]$^+$ (20), 434 [M + 4]$^+$ (4), 371 [M–CONHNH$_2$]$^+$ (8). Anal. Calcd. for C$_{20}$H$_{13}$Cl$_2$FN$_4$O$_2$ (431.25): C, 55.70; H, 3.04; N, 12.99%; found: C, 55.52; H, 3.21; N, 13.05%.

3.2.6. 2-[3-Cyano-6-(2,4-dichlorophenyl)-4-(4-fluorophenyl)pyridin-2-yloxy]-N'-(4-fluorobenzylidene)-acetohydrazide (10)

A mixture of compound **9** (4.3 g, 0.01 mol) and 4-fluorobenzaldehyde (1.24 g, 0.01 mol) in ethanol (20 mL) was refluxed for 6 h. The solid formed after cooling was filtered, dried and recrystallized from acetic acid to give compound **10**. Yield 79%; m.p. 219–221 °C ; IR (KBr) ν_{max} in cm^{-1}: 3218 (NH), 2226 (CN), 1665 (C=O); ^{1}H-NMR (DMSO-d_6): 5.1 (s, 2H, O–CH$_2$), 7.32–7.87 (m, 12H, Ar–H + pyridine H5), 8.10 (s, 1H, CH=N (azomethine protone)), 11.08 (s, 1H, NH, D$_2$O exchangeable); ^{13}C-NMR (DMSO-d_6): 64.1 (CH$_2$), 117.5 (CN), 120.98, 121.01, 121.48, 123.40, 125.21, 127.83, 128.03, 128.99, 132.06, 133.56, 137.98, 138.43, 141.32, 142.65, 144.64, 146.10, 149.01, 149.97, 150.16, 150.63 (23 Ar–C), 157.2 (C=N), 159.4 (CH=N); MS, m/z (%): 536 [M]$^+$ (97), 538 [M + 2]$^+$ (64), 540 [M + 4]$^+$ (11). Anal. Calcd. for C$_{27}$H$_{16}$Cl$_2$F$_2$N$_4$O$_2$ (537.34): C, 60.35; H, 3.00; N, 10.43%; found: C, 60.56 ; H, 3.21; N, 10.65%.

3.2.7. 2-Chloro-6-(2,4-dichlorophenyl)-4-(4-fluorophenyl)nicotinonitrile (11)

A mixture of compound **2** (3.6 g, 0.01 mol) and phosphorus oxychloride (4.6 mL, 0.03 mol) was refluxed for 8 h. The reaction mixture was poured into crushed ice and the separated solid was filtered, dried and recrystallized from dioxane to give compound 11. Yield 62%; m.p. 181–182 °C; IR (KBr) ν_{max} in cm^{-1}: 2223 (CN); ^{1}H-NMR (DMSO-d_6): 6.80 (s, 1H, pyridine H5), 7.62–8.01 (m, 7H, Ar–H); ^{13}C-NMR (DMSO-d_6): 119.2 (CN), 122.31, 125.86, 127.90, 128.54, 131.98, 133.32, 138.32, 139.09, 144.89, 145.07, 148.48 (16 Ar–C), 158.7 (C=N); MS, m/z (%): 376 [M]$^+$ (62), 378 [M + 2]$^+$ (60), 380 [M + 4]$^+$ (18); 382 [M + 6]$^+$ (2). Anal. Calcd. for C$_{18}$H$_8$Cl$_3$FN$_2$ (377.63): C, 57.25; H, 2.14; N, 7.42%; found: C, 57.10; H, 2.43; N, 7.65%.

3.2.8. General procedure for the synthesis of 2-(benzylamino)-6-(2,4-dichlorophenyl)-4-(4-fluoro-phenyl)-nicotinonitrile (12), 2-(benzo[d][1,3]dioxol-5-ylmethylamino)-6-(2,4-dichlorophenyl)-4-(4-fluorophenyl)nicotinonitrile (13), 6-(2,4-dichlorophenyl)-4-(4-fluorophenyl)-2-(1-phenylethyl-amino)nicotinonitrile (14), and 6-(2,4-dichlorophenyl)-4-(4-fluorophenyl)-2-(2-substituted-1-yl)nicotinonitriles 15a,b

A mixture of chloropyridine **11** (3.8 g, 0.01 mol) and the appropriate amine, namely benzyl-amine, piperonylamine, 1-phenylethylamine, 2-methylpiperidine and/or morpholine (0.01 mol) in absolute ethanol (30 mL) was refluxed for 6–12 h. The reaction mixture was poured onto ice/cold water, filtered, washed with petroleum ether 60–80 and finally crystallized from ethanol to give the desired derivatives **12–15a,b**, respectively.

2-(Benzylamino)-6-(2,4-dichlorophenyl)-4-(4-fluorophenyl)nicotinonitrile (**12**): Yield 48%; m.p. 158–159 °C; IR (KBr) ν_{max} in cm^{-1}: 3218 (NH), 2220 (CN); ^{1}H-NMR (DMSO-d_6): 4.97 (s, 2H, CH$_2$), 6.98 (br s, 1H, NH, D$_2$O exchangeable), 7.62–8.31 (m, 13H, Ar–H + pyridine H5); ^{13}C-NMR (DMSO-d_6): 51.45 (CH$_2$), 118.7 (CN), 120.01, 120.96, 122.73, 124.65, 128.02, 129.90, 131.45, 133.63, 134.06, 136.81, 139.43, 141.45,

144.09, 146.12, 147.48, 148.18, 149.74, 150.23, 153.42 (22 Ar–C), 159.2 (C=N); MS, m/z (%): 447 $[M]^+$ (30), 449 $[M + 2]^+$ (17), 451 $[M + 4]^+$ (3). Anal. Calcd. for $C_{25}H_{16}Cl_2FN_3$ (448.32): C, 66.98; H, 3.60; N, 9.37%; found: C, 67.10 ; H, 3.40; N, 9.55%.

2-(Benzo[d][1,3]dioxol-5-ylmethylamino)-6-(2,4-dichlorophenyl)-4-(4-fluorophenyl)nicotinonitrile (**13**): Yield 59%; m.p. 149–151 °C; IR (KBr) v_{max} in cm^{-1}: 3229 (NH), 2218 (CN); ^{1}H-NMR (DMSO-d_6): 4.85 (s, 2H, CH$_2$), 6.21 (s, 2H, CH$_2$), 7.02 (br s, 1H, NH, D$_2$O exchangeable), 7.68–8.22 (m, 11H, Ar–H + pyridine H5); ^{13}C-NMR (DMSO-d_6): 49.5, 88.6 (2CH$_2$), 117.2 (CN), 121.72, 122.98, 124.61, 128.03, 131.43, 133.92, 134.98, 137.38, 138.51, 142.26, 143.13, 144.06, 144.97, 148.26, 149.49 (22 Ar–C), 158.7 (C=N); MS, m/z (%): 491 $[M]^+$ (19), 493 $[M + 2]^+$ (13), 495 $[M + 4]^+$ (2). Anal. Calcd. for $C_{26}H_{16}Cl_2FN_3O_2$ (492.33): C, 63.43; H, 3.28; N, 8.53%; found: C, 63.66 ; H, 3.47; N, 8.41%.

6-(2,4-Dichlorophenyl)-4-(4-fluorophenyl)-2-(phenylethylamino)nicotinonitrile (**14**): Yield 32%; m.p. 142–143 °C; IR (KBr) v_{max} in cm^{-1}: 3203 (NH), 2228 (CN); ^{1}H-NMR (DMSO-d_6): 2.93 (m, 2H, CH$_2$), 3.39 (m, 2H, CH$_2$), 7.32 (br s, 1H, NH, D$_2$O exchangeable), 7.76–8.36 (m, 13H, Ar–H + pyridine H5); ^{13}C-NMR (DMSO-d_6): 39.1, 46.9 (2CH$_2$), 119.1 (CN), 122.01, 122.86, 123.84, 124.62, 127.97, 128.12, 131.82, 133.27, 136.46, 139.65, 141.54, 144.87, 146.32, 147.07, 148.86, 149.41 (22 Ar–C), 159.0 (C=N); MS, m/z (%): 461 $[M]^+$ (32), 463 $[M + 2]^+$ (23), 465 $[M + 4]^+$ (3). Anal. Calcd. for $C_{26}H_{18}Cl_2FN_3$ (462.35): C, 67.54; H, 3.92; N, 9.09%; found: C, 67.69 ; H, 3.71; N, 8.89%.

6-(2,4-Dichlorophenyl)-4-(4-fluorophenyl)-2-(2-methylpiperidin-1-yl)nicotinonitrile (**15a**): Yield 47%; m.p. 98–100 °C; IR (KBr) v_{max} in cm^{-1}: 2219 (CN); ^{1}H-NMR (DMSO-d_6): 1.36 (s, 3H, CH$_3$), 1.59–1.73 (m, 6H, 3CH$_2$-piperidine protons), 2.78 (m, 3H, (CH + CH$_2$) piperidine protons), 7.48–8.01 (m, 8H, Ar–H + pyridine H5); ^{13}C-NMR (DMSO-d_6): 18.6 (CH$_3$), 22.4, 25.2, 36.9, 49.3 (4CH$_2$), 56.7 (CH), 119.7 (CN), 120.76, 121.20, 125.85, 128.32, 129.04, 132.85, 133.25, 137.87, 138.13, 139.08, 141.24, 143.79, 143.9 (16 Ar–C), 157.9 (C=N). Anal. Calcd. for $C_{24}H_{20}Cl_2FN_3$ (440.34): C, 65.46; H, 4.58; N, 9.54%; found: C, 65.70 ; H, 4.32; N, 9.39%.

6-(2,4-Dichlorophenyl)-4-(4-fluorophenyl)-2-morpholinonicotinonitrile (**15b**): Yield 32%; m.p. 112–114 °C; IR (KBr) v_{max} in cm^{-1}: 2227 (CN); ^{1}H-NMR (DMSO-d_6): 3.31 (m, 4H, 2N-CH$_2$), 3.72 (m, 4H, 2O-CH$_2$), 7.52–7.78 (m, 8H, Ar–H + pyridine H5); ^{13}C-NMR (DMSO-d_6): 47.9, 49.1 (2N-CH$_2$), 64.10, 65.9 (2O-CH$_2$), 119.1 (CN), 120.46, 122.04, 122.83, 126.23, 127.85, 129.27, 131.47, 133.56, 134.26, 138.93, 139.08, 142.86, 144.75, 146.12, 146.91, 147.15 (16 Ar–C), 158.9 (C=N). Anal. Calcd. for $C_{22}H_{16}Cl_2FN_3O$ (428.29): C, 61.70; H, 3.77; N, 9.81%; found: C, 61.88 ; H, 3.42; N, 9.69%.

3.2.9. 2-[3-Cyano-6-(2,4-dichlorophenyl)-4-(4-fluorophenyl)pyridin-2-yl]malononitrile (**16**)

To a solution of compound **11** (3.8 g, 0.01 mol) in absolute ethanol (30 mL), triethylamine (1 mL), and malononitrile (0.7 g, 0.01 mol) were added. The reaction mixture was refluxed for 6 h, then left to cool to room temperature, poured into cold water and neutralized with diluted hydrochloric acid to complete precipitation. The solid obtained was filtered, washed with water, dried and recrystallized from ethanol to give compound **16**. Yield 39%; m.p. 99–101 °C; IR (KBr) v_{max} in cm^{-1}: 2218, 2225 (CN); ^{1}H-NMR (DMSO-d_6): 5.02 (s, 1H, CH), 7.35–8.13 (m, 8H, Ar–H + pyridine H5); ^{13}C-NMR (DMSO-d_6): 32.8 (CH), 117.8, 119.5 (3CN), 120.76, 122.45, 123.26, 124.01, 124.81, 128.34, 129.27, 132.23, 133.84, 137.37, 139.06, 142.29, 144.52, 145.14, 146.17, 147.91 (16 Ar–C), 159.3 (C=N); MS, m/z (%): 406 $[M]^+$ (96), 408 $[M + 2]^+$ (62), 410 $[M + 4]^+$ (11). Anal. Calcd. for $C_{21}H_9Cl_2FN_4$ (407.23): C, 61.94; H, 2.23; N, 13.76%; found: C, 62.05; H, 2.39; N, 13.63%.

3.2.10. 6-(2,4-Dichlorophenyl)-4-(4-fluorophenyl)-2-hydrazinylnicotinonitrile (**17**)

A mixture of the chloropyridine derivative **11** (3.8 g, 0.01 mol) and hydrazine hydrate (98%, 2 mL, 0.04 mol) in ethanol (20 mL) was stirred under reflux for 6 h. The formed precipitate was filtered, dried and recrystallized from methanol to give the hydrazinyl derivative **17**. Yield 85%; m.p. 223–224 °C; IR (KBr) v_{max} in cm^{-1}: 3440, 3320, 3150 (NH$_2$, NH), 2218 (CN); ^{1}H-NMR (DMSO-d_6): 5.40 (s, 2H,

NH$_2$, D$_2$O exchangeable), 7.39–8.21 (m, 8H, Ar–H + pyridine H5), 9.40 (s, 1H, NH, D$_2$O exchangeable); ^{13}C-NMR (DMSO-d_6): 119.5 (CN), 121.09, 122.58, 123.71, 125.35, 129.34, 132.98, 134.46, 138.47, 139.07, 142.32, 144.35, 145.07, 147.43, 148.54, 149.76, 150.12 (16 Ar–C), 158.9 (C=N); MS, m/z (%): 372 [M]$^+$ (39), 374 [M + 2]$^+$ (21), 346 [M + 4]$^+$ (4). Anal. Calcd. for C$_{18}$H$_{11}$Cl$_2$FN$_4$ (373.21): C, 57.93; H, 2.97; N, 15.01%; found: C, 58.23; H, 2.80; N, 14.89%.

3.2.11. General procedure for the synthesis of 6-(2,4-dichlorophenyl)-4-(4-fluorophenyl)-2-(3-methyl-5-oxo-4,5-dihydro-1H-pyrazol-1-yl)nicotinonitrile (18) and 6-(2,4-dichlorophenyl)-2-(3,5-dimethyl-1H-pyrazol-1-yl)-4-(4-fluorophenyl)nicotinonitrile (19)

A mixture of compound 17 (3.7 g, 0.01 mol) and ethyl acetoacetate or acetylacetone (0.01 mol) in acetic acid (15 mL) was refluxed for 8 h. The solid formed after cooling was filtered, dried and recrystallized from ethanol to give compounds 18, and 19 respectively.

6-(2,4-Dichlorophenyl)-4-(4-fluorophenyl)-2-(3-methyl-5-oxo-4,5-dihydro-1H-pyrazol-1-yl) nicotinonitrile (18): Yield 40%; m.p. 246–248 °C; IR (KBr) ν_{max} in cm^{-1}: 2227 (CN), 1701 (C=O); ^{1}H-NMR (DMSO-d_6): 1.94 (s, 3H, CH$_3$), 2.26 (s, 2H, CH$_2$), 7.52–7.78 (m, 8H, Ar–H + pyridine H5); ^{13}C-NMR (DMSO-d_6): 19.3 (CH$_3$), 41.5 (CH$_2$), 118.3 (CN), 122.62, 123.42, 126.81, 129.17, 131.11, 133.91, 136.23, 137.05, 139.54, 14.13, 142.53, 143.06, 143.94 (16 Ar–C), 158.4, 159.3 (2C=N), 166.5 (C=O); MS, m/z (%): 438 [M]$^+$ (16), 440 [M + 2]$^+$ (11), 442 [M + 4]$^+$ (2). Anal. Calcd. for C$_{22}$H$_{13}$Cl$_2$FN$_4$O (439.24): C, 60.15; H, 2.98; N, 12.75%; found: C, 60.25; H, 3.12; N, 12.54%.

6-(2,4-Dichlorophenyl)-2-(3,5-dimethyl-1H-pyrazol-1-yl)-4-(4-fluorophenyl)nicotinonitrile (19): Yield 32%; m.p. 287–289 °C; IR (KBr) ν_{max} in cm^{-1}: 2210 (CN); ^{1}H-NMR (DMSO-d_6): 2.31 (s, 3H, CH$_3$), 2.45 (s, 3H, CH$_3$), 6.15 (s, 1H, CH-pyrazole), 7.46–8.09 (m, 8H, Ar–H + pyridine H5); ^{13}C-NMR (DMSO-d_6): 18.4, 19.3 (2CH$_3$), 108.4 (CH-pyrazole), 117.9 (CN), 122.94, 123.27, 124.18, 128.97, 129.46, 132.74, 133.24, 136.93, 138.08, 139.60, 143.84, 144.94, 148.23, 148.86 (17 Ar–C), 158.3, 159.5 (2C=N); MS, m/z (%): 436 [M]$^+$ (12), 438 [M + 2]$^+$ (8), 440 [M + 4]$^+$ (1). Anal. Calcd. for C$_{23}$H$_{15}$Cl$_2$FN$_4$ (437.30): C, 63.17; H, 3.46; N, 12.81%; found: C, 62.98; H, 3.17; N, 12.61%.

3.2.12. 6-(2,4-Dichlorophenyl)-4-(4-fluorophenyl)-2-[2-(2-oxoindolin-3-ylidene)hydrazinyl]nicotinonitrile (20)

A mixture of the compound 17 (3.7 g, 1 mmol) and isatin (1.5 g, 1 mmol) in ethanol (25 mL) containing 3 drops of acetic acid was refluxed for 2 h, then left overnight at room temperature. The formed precipitate was filtered, dried and recrystallized from benzene to give 20. Yield 79%; m.p. 268–269 °C; IR (KBr) ν_{max} in cm^{-1}: 3289, 3150 (2NH), 1723 (C=O), 2227 (CN); ^{1}H-NMR (DMSO-d_6): 6.98 (br s, 1H, NH, D$_2$O exchangeable), 7.22–8.24 (m, 12H, Ar–H + pyridine H5), 10.02 (s, 1H, NH, D$_2$O exchangeable); ^{13}C-NMR (DMSO-d_6): 120.01 (CN), 120.42, 121.07, 122.86, 123.07, 124.52, 129.05, 129.96, 131.46, 133.63, 135.64, 137,93, 139.75, 141.25, 143.86, 144.52, 145.28, 147.94, 148.6 149.25 (22 Ar–C), 157.9, 158.3 (2C=N), 167.23 (C=O). MS, m/z (%): 501 [M]$^+$ (85), 503 [M + 2]$^+$ (60), 505 [M + 4]$^+$ (9). Anal. Calcd. for C$_{26}$H$_{14}$Cl$_2$FN$_5$O (502.33): C, 62.17; H, 2.81; N, 13.94%; found: C, 61.97; H, 2.60; N, 14.19%.

3.2.13. General procedure for the synthesis of 6-(2,4-dichlorophenyl)-4-(4-fluorophenyl)-2-[2-(2-substiutedbenzylidene)hydrazinyl]nicotinonitriles (1a–e)

A mixture of compound 17 (3.7 g, 0.01 mol), an appropriate aromatic aldehyde namely 2-bromo-benzaldhyde, 4-nitrobenzaldhyde, 2-methoxybenzaldhyde, 3,4-dimethoxybenzaldhyde and/or 1-naphthaldehyde (0.01 mol) in ethanol (20 mL) containing 3 drops of acetic acid was refluxed for 6–8 h. The precipitate formed after cooling was filtered, dried and recrystallized to give compounds 21a–e, respectively.

2-[2-(2-Bromobenzylidene)hydrazinyl]-6-(2,4-dichlorophenyl)-4-(4-fluorophenyl)nicotinonitrile (21a): Yield 41%; m.p. 226–228 °C; IR (KBr) ν_{max} in cm^{-1}: 3299 (NH), 2217 (CN); ^{1}H-NMR (DMSO-d_6): 7.26–8.19 (m, 12H, Ar–H + pyridine H5), 8.32 (s, 1H, CH=N azomethine proton), 10.13 (s, 1H, NH, D$_2$O exchangeable); ^{13}C-NMR (DMSO-d_6): 118.6 (CN), 121.83, 122.94, 123.10, 123.86, 127.39, 128.93, 129.27, 132.81, 135.23, 139.38, 141.04, 142.50, 144.52, 148.01, 148.87, 149.08, 150.65, 151.31 (22 Ar–C), 158.3 (C=N), 161.3 (CH=N); MS, m/z (%): 538 [M]$^+$ (14), 540 [M + 2]$^+$ (10), 542 [M + 4]$^+$ (1). Anal. Calcd. for $C_{25}H_{14}BrCl_2FN_4$ (540.21): C, 55.58; H, 2.61; N, 10.37%; found: C, 55.68; H, 2.79; N, 10.51%.

6-(2,4-Dichlorophenyl)-4-(4-fluorophenyl)-2-[2-(4-nitrobenzylidene)hydrazinyl]nicotinonitrile (**21b**): Yield 23%; m.p. 281–283 °C; IR (KBr) ν_{max} in cm^{-1}: 3253 (NH), 2223 (CN); ^{1}H-NMR (DMSO-d_6): 7.29–8.14 (m, 12H, Ar–H + pyridine H5), 8.31 (s, 1H, CH=N (azomethine protone)), 10.08 (s, 1H, NH, D$_2$O exchangeable); ^{13}C-NMR (DMSO-d_6): 119.0 (CN), 121.52, 122.61, 123.48, 127.85, 128.03, 129.53, 130.21, 133.25, 134.56, 136.87, 137.04, 139.08, 139.96, 142.19, 143.15, 144.08, 148.60, 149.31 (22 Ar–C), 158.7 (C=N), 160.8 (CH=N); MS, m/z (%): 505 [M]$^+$ (60), 507 [M + 2]$^+$ (39), 509 [M + 4]$^+$ (7). Anal. Calcd. for $C_{25}H_{14}Cl_2FN_5O_2$ (506.32): C, 59.30; H, 2.79; N, 13.83%; found: C, 59.58; H, 2.63; N, 13.51%.

6-(2,4-Dichlorophenyl)-4-(4-fluorophenyl)-2-[2-(2-methoxybenzylidene)hydrazinyl] nicotinonitrile (**21c**): Yield 34%; mp over 300 °C; IR (KBr) ν_{max} in cm^{-1}: 3258 (NH), 2218 (CN); ^{1}H-NMR (DMSO-d_6): 3.39 (s, 3H, OCH$_3$), 7.49–8.09 (m, 12H, Ar–H + pyridine H5), 8.33 (s, 1H, CH=N (azomethine protone)), 10.16 (s, 1H, NH, D$_2$O exchangeable); ^{13}C-NMR (DMSO-d_6): 51.9 (OCH$_3$), 118.7 (CN), 121.09, 122.58, 123.71, 125.35, 129.34, 132.98, 134.46, 137.94, 138.47, 139.07, 142.32, 144.35, 145.07, 147.43, 148.54, 149.76, 150.12, 151.72 (22 Ar–C), 158.1 (C=N), 159.9 (CH=N). Anal. Calcd. for $C_{26}H_{17}Cl_2FN_4O$ (491.34): C, 63.56; H, 3.49; N, 11.40%; found: C, 63.78; H, 3.26; N, 11.57%.

6-(2,4-Dichlorophenyl)-2-[2-(3,4-dimethoxybenzylidene)hydrazinyl]-4-(4-fluorophenyl) nicotinonitrile (**21d**): Yield 31%; m.p. 278–280 °C; IR (KBr) ν_{max} in cm^{-1}: 3294 (NH), 2220 (CN); ^{1}H-NMR (DMSO-d_6): 3.39 (2s, 6H, 2OCH$_3$), 7.36–8.10 (m, 11H, Ar–H + pyridine H5), 8.31 (s, 1H, CH=N (azomethine protone)), 10.23 (s, 1H, NH, D$_2$O exchangeable); ^{13}C-NMR (DMSO-d_6): 55.6, 56.01 (2OCH$_3$), 119.3 (CN), 120.46, 121.08, 122.85, 123.05, 127.30, 128.74, 129.76, 130.64, 133.73, 134.05, 138.12, 138.98, 140.16, 143.54, 144.35, 145.97, 147.46, 148.02, 149.08 (22 Ar–C), 158.7 (C=N), 161.4 (CH=N). Anal. Calcd. for $C_{27}H_{19}Cl_2FN_4O_2$ (521.37): C, 62.20; H, 3.67; N, 10.75%; found: C, 62.38; H, 3.43; N, 10.53%.

6-(2,4-Dichlorophenyl)-4-(4-fluorophenyl)-2-[2-(naphthalen-1-ylmethylene)hydrazinyl] nicotinonitrile (**21e**): Yield 24%; m.p. 296–298 °C; IR (KBr) ν_{max} in cm^{-1}: 3282 (NH), 2213 (CN); ^{1}H-NMR (DMSO-d_6): 7.24–8.26 (m, 15H, Ar–H + pyridine H5), 8.32 (s, 1H, CH=N (azomethine protone)), 10.18 (s, 1H, NH, D$_2$O exchangeable); ^{13}C-NMR (DMSO-d_6): 119.0 (CN), 120.76, 122.45, 123.26, 124.01, 124.81, 128.34, 129.27, 132.23, 133.84, 137.37, 139.06, 142.29, 144.52, 145.14, 146.17, 147.91, 148.96, 149.24, 149,99, 150.65, 151.34, 152.61 (26 Ar–C), 158.3 (C=N), 161.0 (CH=N). Anal. Calcd. for $C_{29}H_{17}Cl_2FN_4$ (511.38): C, 68.11; H, 3.35; N, 10.96%; found: C, 67.96; H, 3.49; N, 11.06%.

3.3. Anticancer Activity

3.3.1. Cell Cultures

The newly synthesized compounds were evaluated *in vitro* against three human cancer cell lines; which are MCF-7 (breast adenocarcinoma), NCI-H460 (non-small cell lung cancer) and SF-268 (CNS cancer), and WI 38 (normal fibroblast cells) were used in this study. MCF-7 was obtained from the European Collection of Cell Cultures (ECACC, Salisbury, UK) but NCI-H460, SF-268 and WI 38 were kindly provided by the National Cancer Institute (NCI, Cairo, Egypt). They grow as monolayers routinely maintained in RPMI-1640 medium supplemented with 5% heat inactivated fetal bovine serum (FBS), 2 mM glutamine and antibiotics (penicillin 100 U/mL, streptomycin 100 µg/mL), at 37 °C in a humidified atmosphere containing 5% CO$_2$. Exponentially growing cells were obtained by plating 1.5 × 105 cells/mL for MCF-7 and SF-268, and 0.75 × 104 cells/mL for NCI-H460 followed by 24 h of incubation. The effect of the vehicle solvent (DMSO) on the growth of these cell lines was evaluated in all experiments by exposing untreated control cells to the maximum concentration (0.5%) of DMSO used in each assay.

3.3.2. Cancer Cell Growth Assay

The effect of compounds on the *in vitro* growth of human tumor cell lines were evaluated according to the procedure adopted by the National Cancer Institute (NCI, Austin, TX, USA) in the *"In vitro Anticancer Drug Discovery Screen"* that uses the protein-binding dye sulforhodamine B (SRB) to assess cell growth [35]. In the assay protocol, all cells were incubated at 37 °C under humidified atmosphere containing 5% CO_2. Briefly, exponentially cells growing in 96-well plates were then exposed for 48 h to five serial concentrations of each compound, starting from a maximum concentration of 150 μg/mL. Following this exposure period, adherent cells were fixed, washed and stained. The bound stain was solubilized and the absorbance was measured at 492 nm in a Power Wave XS plate reader (Bio-Tek Instruments Inc., Winston, NC, USA). For each test compound and cell line, a dose response curve was obtained and the inhibitory concentration of 50% (IC_{50}), corresponding to the concentration of the compounds that inhibited 50% of the net cell growth was calculated as described elsewhere [37]. Doxorubicin was used as a positive control and tested in the same manner.

4. Conclusions

This study focused on the synthesis of a new 1,2-dihydropyridin-3-carbonitrile and nicotinonitrile derivatives as potential anticancer agents. Some of newly synthesized derivatives were examined *in vitro* as cytotoxic agents against three human cancer cell lines. It could be noticed that the ester functionality-bearing derivative **8** and the derivative **16** carrying a malononitrile side chain attached to the parent cyanopyridine nucleus showed the best results, exhibiting the highest inhibitory effects towards the three tumor cell lines, which were higher than that of the reference compound doxorubicin and these compounds were non-cytotoxic towards normal cells (IC_{50} values >100 μg/mL). In addition, compounds **9**, **17** and **20** exhibited high growth inhibitory activity on the various cancer panel cell lines, with weak cytotoxicity on the normal cells.

Author Contributions: The listed authors contributed to this work as described in the following: Eman M. Flefel and Hebat-Allah S. Abbas developed the concept of the work, carried out the synthetic work, interpreted the results and prepared the manuscript, Randa E. Abdel Magid, carried out the synthetic work, interpreted the results and prepared the manuscript and Wafaa A. Zafgary interpreted the results and cooperated in the preparation of the manuscript. All authors read and approved the final manuscript.

Conflicts of Interest: The authors declare no conflict of interest.

References

1.   Zhang, J.Y. Apoptosis-based anticancer drugs. *Nat. Rev. Drug Discov.* **2002**, *1*, 101–102. [CrossRef]
2.   Ali, A.; Fergus, K.; Wright, F.C.; Pritchard, K.I.; Kiss, A.; Warner, E. The impact of a breast cancer diagnosis in young women on their relationship with their mothers. *Breast* **2014**, *23*, 50–55. [CrossRef] [PubMed]
3.   Lam, S.W.; Jimenez, C.R.; Boven, E. Breast cancer classification by proteomic technologies: Current state of knowledge. *Cancer* **2014**, *40*, 129–138. [CrossRef] [PubMed]
4.   Hassan, G.S.; Kadry, H.H.; Abou-Seri, S.M.; Ali, M.M.; Mahmoud, A.E.E. Synthesis and *in vitro* cytotoxic activity of novel pyrazolo[3,4-*d*]pyrimidines and related pyrazole hydrazones toward breast adenocarcinoma MCF-7 cell line. *Bioorg. Med. Chem.* **2011**, *19*, 6808–6817. [CrossRef] [PubMed]
5.   Taher, A.T.; Georgey, H.H.; El-Subbagh, H.I. Novel 1,3,4-heterodiazole analogues: Synthesis and *in vitro* antitumor activity. *Eur. J. Med. Chem.* **2012**, *47*, 445–451. [CrossRef] [PubMed]
6.   Carmen, A.J.; Carlos, M. *Medicinal Chemistry of Anticancer Drugs*, 1st ed.; Elsevier: Amsterdam, The Netherlands, 2008; pp. 1–8.
7.   Borowski, E.; Bontemps-Gracz, M.M.; Piwkowska, A. Strategies for overcoming ABC-transporters-mediated multidrug resistance (MDR) of tumor cells. *Acta Biochim. Pol.* **2005**, *52*, 609–627. [PubMed]
8.   Avila, H.P.; Smania, E.F.; Monache, F.D.; Smania, A. Structure-activity relationship of antibacterial chalcones. *Bioorg. Med. Chem.* **2008**, *16*, 9790–9794. [CrossRef] [PubMed]
9.   Liu, Y.; Sun, X.; Yin, D.; Yuan, F. Syntheses and biological activity of chalcones-imidazole derivatives. *Res. Chem. Intermed.* **2013**, *39*, 1037–1048. [CrossRef]

10. Sortino, M.; Delgado, P.; Juarez, S.; Quiroga, J.; Abonia, R.; Insuasty, B.; Nogueras, M.; Rodero, L.; Garibotto, F.M.; Enriz, R.D.; *et al.* Synthesis and antifungal activity of (Z)-5-arylidenerhodanines. *Bioorg. Med. Chem.* **2007**, *15*, 484–494. [CrossRef] [PubMed]

11. Lopez, S.N.; Castelli, M.V.; Zacchino, S.A.; Dominguez, J.N.; Lobo, G.; Charris-Charris, J.; Cortes, J.C.; Ribas, J.C.; Devia, C.; Rodriguez, A.M.; *et al. In vitro* antifungal evaluation and structure-activity relationships of a new series of chalcone derivatives and synthetic analogues with inhibitory properties against polymers of the fungal cell wall. *Bioorg. Med. Chem.* **2001**, *8*, 1999–2013. [CrossRef]

12. Cheng, J.H.; Hung, C.F.; Yang, S.C.; Wang, J.P.; Won, S.J.; Lin, C.N. Synthesis and cytotoxic, anti-inflammatory, and anti-oxidant activities of 2′,5′-dialkoxylchalcones as cancer chemopreventive agents. *Bioorg. Med. Chem.* **2008**, *16*, 7270–7276. [CrossRef] [PubMed]

13. Katsori, A.M.; Hadjipavlou-Litina, D. Chalcones in cancer: Understanding their role in terms of QSAR. *Curr. Med. Chem.* **2009**, *16*, 1062–1081. [CrossRef] [PubMed]

14. Modzelewska, A.; Pettit, C.; Achanta, G.; Davidson, N.E.; Huang, P.; Khan, S.R. Anticancer activities of novel chalcone and bis-chalcone derivatives. *Bioorg. Med. Chem.* **2006**, *14*, 3491–3495. [CrossRef] [PubMed]

15. Abdelhafez, O.M.; Abdel-Latif, N.A.; Badria, F.A. DNA, Antiviral activities and cytotoxicity of new furochromone and benzofuran derivatives. *Arch. Pharm. Res.* **2011**, *34*, 1623–1632. [CrossRef] [PubMed]

16. Abdel-Latif, N.A. Synthesis and antidepressant activity of some new coumarin derivatives. *Sci. Pharm.* **2005**, *74*, 173–216.

17. Son, J.K.; Zhao, L.X.; Basnet, A.; Thapa, P.; Karki, R.; Na, Y.; Jahng, Y.; Jeong, T.C.; Jeong, B.S.; Lee, C.S.; *et al.* Synthesis of 2,6-diaryl-substituted pyridines and their antitumor activities. *Eur. J. Med. Chem.* **2008**, *43*, 675–682. [CrossRef] [PubMed]

18. Amr, A.G.; Abdulla, M.M. Anti-inflammatory profile of some synthesized heterocyclic pyridone and pyridine derivatives fused with steroidal structure. *Bioorg. Med. Chem.* **2006**, *14*, 4341–4352. [CrossRef] [PubMed]

19. Hammam, A.G.; Abdel Hafez, N.A.; Midura, W.H.; Mikolajczyk, M.Z. Chemistry of seven-membered heterocycles, VI. Synthesis of novel bicyclic heterocyclic compounds as potential anticancer and anti-HIV agents. *Z. Naturforsch.* **2000**, *55*, 417–424. [CrossRef]

20. Kotb, E.R.; Anwar, M.M.; Abbas, H.A.S.; Abd El-Moez, S.I. A concise synthesis and antimicrobial activity of a novel series of naphthylpyridine-3-carbonitrile compounds. *Acta Pol. Pharm. Drug Res.* **2013**, *70*, 667–679.

21. Sayed, H.H.; Morsy, E.M.; Flefel, E.M. Synthesis and reactions of some novel nicotinonitrile, thiazolotriazole, and imidazolotriazole derivatives for antioxidant evaluation. *Synth. Commun.* **2010**, *40*, 1360–1370. [CrossRef]

22. Akira, M.; Aya, N.; Shigeki, I.; Motoki, T.; Kazuo, S. JBIR-54, a new 4-pyridinone derivative isolated from Penicillium daleae Zaleski fE50. *J. Antibiot.* **2009**, *62*, 705–706.

23. Al-Omar, M.A.; Amr, A.E.; A.l-Salahi, R.A. Anti-inflamatory, analgesic, anticonvulsant and antiparkinsonian activities of some pyridine derivatives using 2,6-disubstituted isonicotinic acid hydrazides. *Archiv. Phaem.* **2010**, *343*, 648–656.

24. Martin, C.; Göggel, R.; dal Piaz, V.; Vergelli, C.; Giovannoni, P.; Ernst, M.; Uhlig, S. Airway relaxant and anti-inflammatory properties of a PDE4 inhibitor with low affinity for the high-affinity rolipram binding site. *Naunyn-Schmiedeberg's Arch. Pharmacol.* **2002**, *365*, 284–289. [CrossRef] [PubMed]

25. Amr, A.E.; Sayed, H.H.; Abdulla, M.A. Synthesis and reactions of some new substituted pyridine and pyrimidine derivatives as analgesic, anticonvulsant and antiparkinsonian agents. *Arch. Pharm. Chem. Life Sci.* **2005**, *338*, 433–440. [CrossRef] [PubMed]

26. Al-Abdullah, E.S. Synthesis and anticancer activity of some novel tetralin-6-yl-pyrazoline, 2-thioxopyrimidine, 2-oxopyridine, 2-thioxo-pyridine and 2-iminopyridine derivatives. *Molecules* **2011**, *16*, 3410–3419. [CrossRef] [PubMed]

27. Abo-Ghalia, M.; Abdulla, M.M.Z.; Amr, A.E. Synthesis of some new (N$^\alpha$-dipicolinoyl)-*bis*-L-leucyl-DL-norvalyl linear tetra and cyclic octa bridged peptides as new antiinflammatory agents. *Z. Naturforsch.* **2003**, *58b*, 903–910.

28. Kotb, E.R.; El-Hashash, M.A.; Salama, M.A.; Kalf, H.S.; Abdel Wahed, N.A.M. Synthesis and reactions of some novel nicotinonitrile derivatives for anticancer and antimicrobial evaluation. *Acta Chim. Slov.* **2009**, *56*, 908–919.

29. Kumar, S.; Das, S.; Dey, S.; Maity, P.; Guha, M.; Choubey, V.; Panda, G.; Bandyopadhyay, V. Antiplasmodial activity of [(aryl)arylsulfanylmethyl]pyridine. *Antimicrob. Agents Chemother.* **2008**, *52*, 705–715. [CrossRef] [PubMed]

30. Ghosh, P.S.; Manna, K.; Banik, U.; Das, M.; Sarkar, P. Synthetic strategies and pharmacology of 2-oxo-3-cyanopyridine derivatives: A review. *Int. J Pharm. Pharm. Sci.* **2014**, *6*, 39–42.

31. Abbas, H.-A.S.; El Sayed, W.A.; Fathy, N.M. Synthesis and antitumor activity of new dihydropyridine thioglycosides and their corresponding dehydrogenated forms. *Eur. J. Med. Chem.* **2010**, *45*, 973–982. [CrossRef] [PubMed]

32. Al-Mutairi, M.S.; Al-Abdullah, E.S.; Haiba, M.E.; Khedr, M.A.; Zaghary, W.A. Synthesis, molecular docking and preliminary *in vitro* cytotoxic evaluation of some substituted tetrahydronaphthalene (2′,3′,4′,6′-Tetra-*O*-Acetyl-β-D-Gluco-/Galactopyranosyl) derivatives. *Molecules* **2012**, *17*, 4717–4732. [CrossRef] [PubMed]

33. Kotb, E.R.; Abbas, H.-A.S.; Flefel, E.M.; Sayed, H.H.; Abdel Wahed, N.A.M. Utility of hantzsch ester in synthesis of some 3,5-bisdihydropyridine derivatives and studying their biological evaluation. *J. Heterocycl. Chem.* **2015**, *52*, 1531–1539. [CrossRef]

34. Sayed, H.H.; Flefel, E.M.; Abd El-Fatah, A.M.; El-Sofany, W.I. Focus on the synthesis and reactions of some new pyridine carbonitrile derivatives as antimicrobial and antioxidant agents. *Egypt J. Chem.* **2010**, *53*, 17–35.

35. Skehan, P.; Storeng, R.; Scudiero, D.; Monks, A.; McMahon, J.; Vistica, D.; Warren, J.T.; Bokesch, H.; Kenne, S.; Boyd, M.R. New colorimetric cytotoxicity assay for anticancer-drug screening. *J. Natl. Cancer Inst.* **1990**, *82*, 1107–1112. [CrossRef] [PubMed]

36. Lee, P.; Zhang, R.; Li, V.; Liu, X.; Sun, R.W.Y.; Che, C.M.; Wong, K.K.Y. Enhancement of anticancer efficacy using modified lipophilic nanoparticle drug encapsulation. *Int. J. Nanomed.* **2012**, *7*, 731–737.

37. Monks, A.; Scudiero, D.; Skehan, P.; Shoemaker, R.; Paul, K.; Vistica, D.; Hose, C.; Langley, J.; Cronise, P.; Vaigro-Wolff, A. Feasibility of a high-flux anticancer drug screen using a diverse panel of cultured human tumor cell lines. *J. Natl. Cancer Inst.* **1991**, *83*, 757–766. [CrossRef] [PubMed]

12

New Alcamide and Anti-oxidant Activity of *Pilosocereus gounellei* A. Weber ex K. Schum. Bly. ex Rowl. (Cactaceae)

Jéssica K. S. Maciel [1,†], Otemberg S. Chaves [2,†], Severino G. Brito Filho [2,†], Yanna C. F. Teles [1,†], Marianne G. Fernandes [2,†], Temilce S. Assis [1,3,†], Pedro Dantas Fernandes [4,†], Alberício Pereira de Andrade [4,†], Leonardo P. Felix [4,†], Tania M. S. Silva [5,†], Nathalia S. M. Ramos [5,†], Girliane R. Silva [5,†] and Maria de Fátima Vanderlei de Souza [1,2,*]

Academic Editors: Derek J. McPhee and John A. Beutler

[1]   Post-Graduation Program in Development and Technological Innovation in Medicines, Health Science Center, Federal University of Paraiba, Campus I, João Pessoa, PB, 58051-900, Brazil; jksmaciel@hotmail.com (J.K.S.M.); yannateles@gmail.com (Y.C.F.T.); temilce@gmail.com (T.S.A.)
[2]   Post-Graduation Program in Bioactive Natural and Synthetic Products, Health Science Center, Federal University of Paraiba, Campus I, João Pessoa, PB, 58051-900 Brazil; otemberg_sc@yahoo.com.br (O.S.C.); severinogfilho@yahoo.com.br (S.G.B.F.); marianne.guedes@gmail.com (M.G.F.)
[3]   Health Science Centre, Physiology and Pathology Department, Federal University of Paraiba, Campus I, Cidade Universitária—João Pessoa, PB, 58059-900, Brazil
[4]   Department of Agroecology and Agriculture, Center of Agricultural and Environmental Sciences, University of Paraiba State, 351 Baraúnas Street, Campina Grande, PB, 58429-500, Brazil; pedrodantasfernandes@gmail.com (P.D.F.); albericio3@gmail.com (A.P.A.); lpfelix@hotmail.com (L.P.F.)
[5]   Postgraduate Program in Development and Technological Innovation in Medicines, Department of Molecular Sciences, Rural Federal University of Pernambuco, Campus Dois Irmãos, Recife, PE, 52171-900, Brazil; sarmentosilva@gmail.com (T.M.S.S.); quinathi@gmail.com (N.S.M.R.); girlianeregina@gmail.com (G.R.S.)
*   Correspondence: mfvanderlei@ltf.ufpb.br
†   These authors contributed equally to this work.

Abstract: The Cactaceae family is composed by 124 genera and about 1438 species. *Pilosocereus gounellei*, popularly known in Brazil as xique-xique, is used in folk medicine to treat prostate inflammation, gastrointestinal and urinary diseases. The pioneering phytochemical study of *P. gounellei* was performed using column chromatography and HPLC, resulting in the isolation of 10 substances: pinostrobin (**1**), β-sitosterol (**2**), a mixture of sitosterol 3-O-β-D-glucopyranoside/stigmasterol 3-O-β-D-glucopyranoside (**3a/3b**), 13^2-hydroxyphaeophytin a (**4**), phaeophytin a (**5**), a mixture of β-sitosterol and stigmasterol (**6a/6b**), kaempferol (**7**), quercetin (**8**), 7'-ethoxy-*trans*-feruloyltyramine (mariannein, **9**) and *trans*-feruloyl tyramine (**10**). Compound **9** is reported for the first time in the literature. The structural characterization of the compounds was performed by analyses of 1-D and 2-D NMR data. In addition, a phenolic and flavonol total content assay was carried out, and the anti-oxidant potential of *P. gounellei* was demonstrated.

Keywords: *Pilosocereus gounellei*; 7'-ethoxy-*trans*-feruloyltyramine; anti-oxidant activity

1. Introduction

Cactaceae is a family belonging to the order Caryophyllales with 124 genera and about 1438 species [1] distributed throughout American territory in tropical and temperate dry regions with wide occurrence in Mexico and Brazil [2]. The family is remarkable due to the evolution of several adaptations for aridity, its species showing outstanding diversity of growth forms [3].

Caatinga is a dry eco-region located in northeastern Brazil considered the third diversity center of Cactaceae species, where the family plays an important role, mainly due to the use of its species in human and animal foods [4,5]. Furthermore, many species are used in traditional medicine, for example *Nopalea cochenillifera* is used as an anti-inflammatory, diuretic and hypoglycaemic [6] agent and *Cereus jamacaru* is used to treat ulcers and bronchitis [7].

Previous phytochemical studies on Cactaceae species have reported the presence of flavonoids, such as quercetin, rutin and kaempferol, as well as the anti-oxidant activity of several species, e.g., *Opuntia monocantha*, *Opuntia ficus-indica*, *Cereus jamacaru*, *Pilosocereus pachycladus* and *Pilosocereus arrabidae* [8,9]. In plants, this activity is related to the presence of phenolic molecules such as flavonoids, tannins, phenolic acids, anthocyanins and others [10]. The anti-oxidant activity of phenolics is justified by the presence of conjugated double bonds, which provide resonance, and the hydroxyl groups on an aromatic ring [11], which can chelate metals, eliminate free radicals by donating hydrogen, oxidize and induce enzymes to interact and stabilize free radicals [12,13].

The occurrence of nitrogen compounds such as β-phenethylamines, tetrahydroisoquinolines and their derivatives in this family is also well reported [14–16]. Three alkaloids have been isolated from the *Neobuxbaumia* genus: salsolidine, carnegine and anhalidine [17].

Pilosocereus gounellei A. Weber ex K. Schum. Bly. ex Rowl, popularly known as xique-xique, is an endemic species from the Caatinga region and its cladodes are used by local people as food, cooked or baked to produce flour, cakes and pastries [18,19]. In folk medicine, its roots are used to treat urinary infections and prostate inflammation. Its cladodes are used to treat gastritis [5,20,21].

This work describes the first isolation of secondary metabolites of *P. gounellei* and the evaluation of the anti-oxidant activity of its extracts and methanol fraction. Anti-oxidant activity of Cactaceae species has been reported previously, justifying our investigation of *P. gounellei*'s anti-oxidant potential [8,9].

2. Results and Discussion

2.1. Identification of Isolated Compounds

Chromatographic procedures led to the isolation of 10 compounds from *P. gounellei* (Figures 1 and 2). The isolated compounds were identified by analysis of their 1-D and 2-D NMR data and comparisons with the literature. Compound **1** was isolated as colourless crystals soluble in chloroform and it was identified as the flavonoid pinostrobin [22]. Compounds **2**, **3a/3b** and **6a/6b** were identified as steroids: β-sitosterol, the mixture of sitosterol 3-*O*-β-D-glucopyranoside/stigmasterol, 3-*O*-β-D-glucopyranoside [23] and sitosterol/stigmasterol [24], respectively. These phytosteroids are widespread in plants, being important components of vegetable cell walls and membranes. In addition, they are known as anti-inflammatory agents and precursors of vitamin D [25].

Compounds **4** and **5** were isolated as dark green amorphous solids. The ^{1}H- and ^{13}C-NMR indicated that the compounds are chlorophyll derivatives. When compared with the literature data, they were identified as 13^2-hydroxyphaeophytin a (**4**) [26] and phaeophytin a (**5**) [27]. These porphyrinic compounds are derived from chlorophyll a and are widely present in the vegetable kingdom [28]. The substances **7** and **8** were isolated as a yellow amorphous powder and were identified as the flavonoids kaempferol (**7**) [29] and quercetin (**8**) [30]. Flavonoids have great relevance in the pharmaceutical field, displaying anti-oxidant, anti-inflammatory and antimicrobial activities [31], and have been previously reported from several species of the genera *Opuntia* and *Pilosocereus* [8].

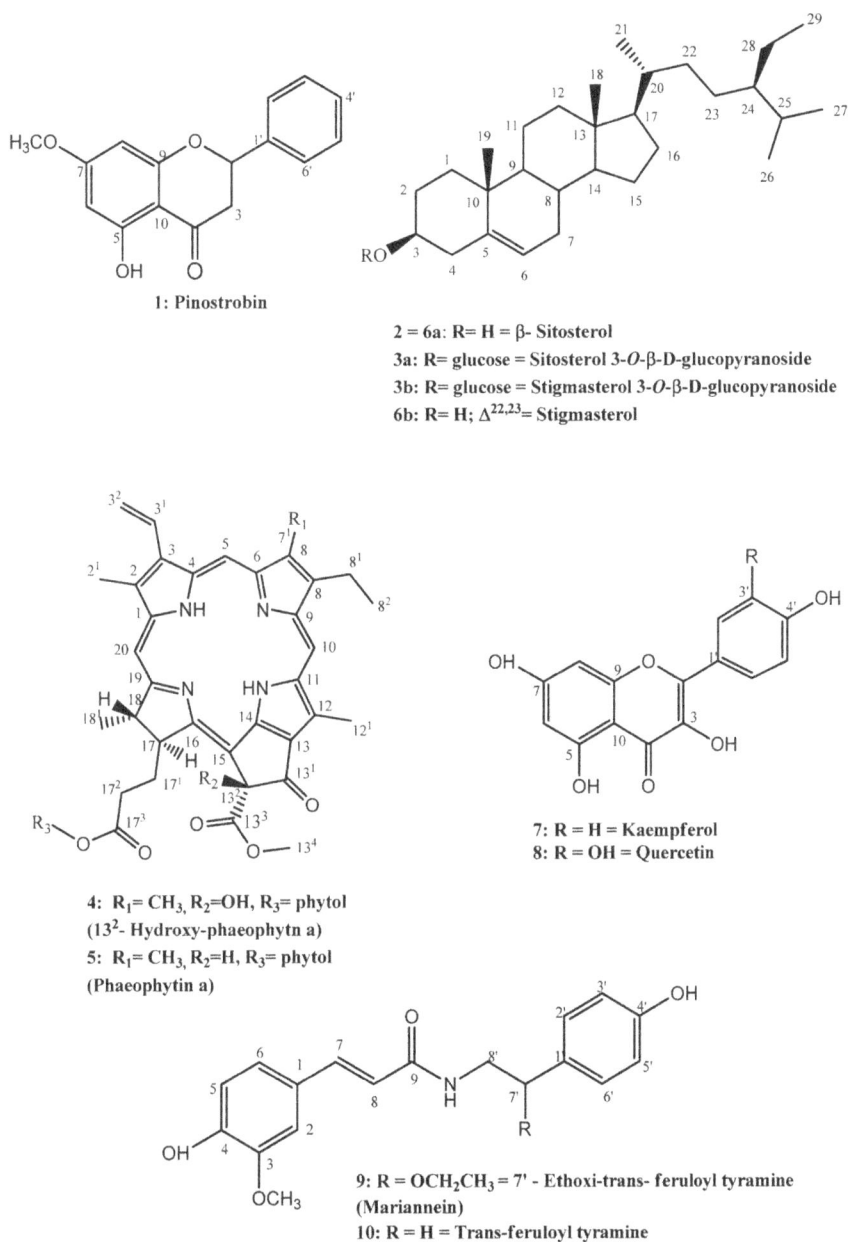

Figure 1. Compounds isolated from *Pilosocereus gounellei*.

The ¹H-NMR spectrum of compound **9** (Figure 2) showed two doublets (δ_H 7.11 and δ_H 6.75) integrating for two protons each, suggesting the presence of a *para*-substituted ring. Two doublets in δ_H 6.79 (H-5), coupling *ortho*, and δ_H 7.11 (H-2), coupling *meta* with H-6 and a double doublet at δ_H 6.98 (H-6) coupling *ortho* and *meta* with H-5 and H-2, respectively, indicated the presence of an AMX ring system. The presence of a methoxy group in one aromatic ring was shown by a singlet at δ_H 3.79 (3H). The presence of the coumaroyl and tyramine units was suggested by the following signals: a pair of doublets at δ_H 6.52 (H-8) and δ_H 7.31 (H-7), both with J = 16 Hz, characteristic of *trans* olefin protons; an interesting triplet at δ_H 8.00, referring to a proton bonded to N [32]; two double doublets at δ_H 3.29 (H-8a') and 3.25 (H-8b') referring to the methylene carbon adjacent to N and one triplet on δ_H 4.27 (H-7'), attributed to a proton on the oximethinic carbon [33]. Other relevant signals in the ¹H-NMR spectra of **9** were two signals at δ_H 3.27 (q) and δ_H 1.07 (t) attributed to an ethoxy group attached to an oximethinic carbon whose proton was found at δ_H 4.27 (t) [33].

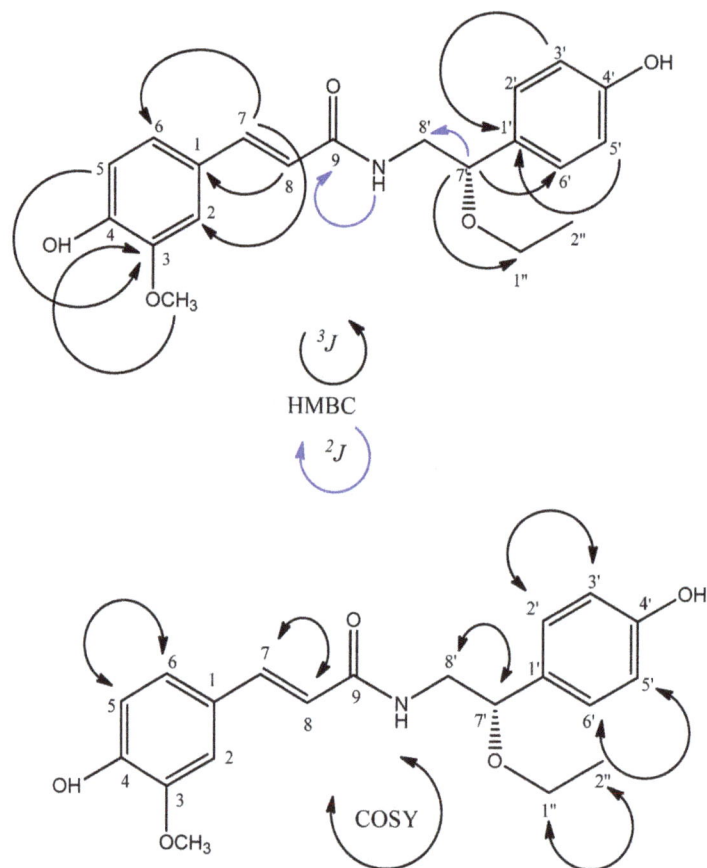

Figure 2. Compound **9** HMBC (^3J and ^2J) and COSY correlations.

The ^{13}C-NMR spectrum showed signals for 20 carbons, supporting the information provided by the ^{1}H-NMR spectrum about the presence of the coumaroyl and tyramine portions of compound **9**. The coumaroyl portion was defined by the signals at δ_C 165.8 (amide α,β-unsaturated carbonyl) and two signals at δ_C 139.46 and δ_C 119.01 (the α,β-unsaturated carbons C-7 and C-8). The tyramine portion was confirmed by the signals at δ_C 45.61 (CH$_2$, C-8′) and an oximethinic carbon δ_C 79.6 (CH, C-7′), indicating the position the ethoxy group is attached to [32]. The presence of one ethoxy group in compound **9** was reinforced by the signals at δ_C 63.51 (CH$_2$, C-1″) and δ_C 15.42 (CH$_3$, C-2″). The ^{13}C-NMR spectrum confirmed the presence of two aromatic rings, with δ_C 128.07 (C-2′/6′), δ_C 115.34 (C-3′/5′) of one AA′BB′ system and δ_C 110.9 (C-2), δ_C 115.83 (C-5) and δ_C 121.82 (C-6) of the AMX system, showing one methoxy group at δ_C 55.75 (H$_3$CO–C-3) [32,34].

The absence of a correlation for proton δ_H 8.00 (t, N-H) with any carbon in the HMQC spectrum, and the correlation shown by the proton at δ_H 8.00 (t, 1H) and the carbon at δ_C 165.8 (^2J) in the HMBC spectrum demonstrated that the compound possesses an amide carbonyl [32–34], being a *trans*-feruloyl derivative [32–34]. The HMBC spectrum showed correlations (^3J) that confirmed the presence of the coumaroyl moiety: H-7 with C-2, C-6 and C-9; H-8 with C-1; H-5 with C-3 and C-1. Other correlations in the HMBC spectrum suggest the presence of the ethoxy group and also the *trans*-feruloyl tyramine portion: H-1″ with C-7′ and H-8′ with C-9 and C-1′. These data allowed identification of compound **9** as 7′-ethoxy-*trans*-feruloyl- tyramine, reported herein for the first time. The COSY spectrum supported the proposed structure by showing correlations between hydrogen (N-H) δ_H 8.00 (t) with H-8′; H-8′ and H-7′; H-1″ and H-2″ (Table 1). The optical rotation of **9** was found to be −10° (0.01; MeOH) establishing the S(−) absolute configuration at the C-7 chiral center [33].

Table 1. ^{13}C-NMR (100 MHz), ^{1}H-NMR (400 MHz), HMQC and HMBC data of 7'-ethoxy-*trans*-feruloyltyramine (**9**) (DMSO-d_6, δ).

Position	HMQC		HMBC	
	δ_C, Type	δ_H (J in Hz)	^3J (H C)	^2J (H C)
1	126.62 C	-		
2	110.9 C-H	7.11, d (1.2)	4, 6, 7	3
3	148.03 C	-		
4	148.46 C	-		
5	115.83 C-H	6.79, dd (8.0, 0.72)	1, 3	4, 6
6	121.82 C-H	6.98, dd (8.0, 1.2)	2, 4, 7	5
7	139.46 C-H	7.31, d (16.0)	2, 6 ,9	1, 8
8	119.01 C-H	6.52, d (16.0)	1	9
9	165.87 C	-		
1'	130.78 C	-		
2'/6'	128.07 C-H	7.12, dd (8.0, 1.2)	2', 6', 4', 7'	3', 5'
3'/5'	115.34 C-H	6.75, dd (8.0, 1.2)	1', 3', 5'	4'
4'	157.08 C	-		
7'	79.64 C-H	4.27, t (5.0)	1", 2', 6'	8'
8'	45.61 C-H$_2$	3.29 dd, (13.0, 5.9) 3.25, dd (13.0, 5.9)	1'	7'
1"	63.51 C-H$_2$	3.27, q (6.9)		2"
2"	15.42 C-H$_3$	1.07, t (6.9)		1"
OMe-3	55.56 C-H$_3$	3.79, s	3	
N-H	-	8.0, t (5.0)		9

The molecular formula of new compound **9** was established as $C_{20}H_{23}NO_5$ from a pseudomolecular ion peak at m/z 358.1706 $[M + H]^+$ (calcd. $C_{20}H_{24}NO_5$, 358.1648) indicating the formula $C_{20}H_{23}NO_5$ for the new compound. The corresponding loss of the ethoxy radical fragment was seen in EIMS $[M + H]^-$, FTMS as 312.17 ($C_{18}H_{18}NO_4$) and the loss of ethanol was seen in EIMS $[M + H]^-$ as 310.17, confirming the proposed structure of compound **9**. The hypotheses that compound **9** is an artefact was eliminated based on previous studies that described the isolation of *trans*-feruloyltyramine and its derivatives by several different extraction methods [33,35]. Liang *et al.* performed the extraction of eight nitrogenated substances from *Portulaca oleracea* using microwave irradiation and different solvents such as dichloromethane, ethyl acetate, methanol, ethanol, ethanol 70%, ethanol 30% and water. The eight isolated substances, including *N*-feruloylnormetanephrine and the *N-trans*-feruloyltyramine, were extracted with all the tested solvents, showing that those solvents did not promote the formation of different radicals or artefacts [35].

Compound **10** was obtained as a pale amorphous solid and its spectra showed a structure similar to compound **9**, differing in the absence of signals related to the ethoxy group. Comparisons of the spectral data of compound **10** with compound **9** and literature data allowed identification of compound **10** as *trans*-feruloyltyramine, previously isolated from other plant species [32–34]. Compound **10** was shown to possess action against weeds, improvement of seed germination [34] and anti-inflammatory activity by inhibiting COX enzymes [36].

2.2. Total Phenolic, Total Flavone and Flavonol Contents and Anti-Oxidant Activity of Extracts and Methanolic Fraction from P. gounellei

The values of total phenolic, flavone and flavonol contents and anti-oxidant activity (DPPH and ABTS) of ethanol extracts of stems, roots, flowers, fruits and methanol fraction are shown in Table 2. The fruit extract showed the best antiradical activity in the ABTS test (IC_{50} = 10.4 ± 0.24) and the flower extract showed the lowest anti-oxidant activity (IC_{50} = 76.9 ± 0.61); the cladode extract showed a slightly higher activity (IC_{50} = 62.4 ± 0.44) than the flower extract. The scavenging activity of free radicals of the roots extract showed an activity (IC_{50} = 41.6 ± 1.06) similar to the methanol fraction (IC_{50} = 40.9 ± 0.69). Thus, we can classify the activity of extracts and methanol fraction of *P. gounellei*

(*Pg*) as follows: *Pg* fruits > MeOH fraction = *Pg* roots > *Pg* cladodes > *Pg* flowers. In the DPPH test, the fruit extract showed the best anti-oxidant activity (IC_{50} = 11.3 ± 0.12), followed by the root extract (IC_{50} = 102.1 ± 1.49). The methanol fraction (IC_{50} = 130.1 ± 3.02) showed similar activity to the cladode extract (IC_{50} = 136.0 ± 3.48) and the least potent was the flower extract (IC_{50} = 194.3 ± 2.33).

Table 2. Total phenolic, total flavones and flavonols contents and DPPH and ABTS free radical scavenging activity.

Samples	Total Phenolics Contents (mg GAE/g ± SEM) [a]	Total Flavones and Flavonols Contents (mg QE/g ± SEM) [b]	Free-Radical Scavenging Activity (IC_{50}) [c]	
			DPPH (μg/mL ± SEM)	ABTS$^{.+}$ (μg/mL ± SEM)
Pg Cladodes	45.1 ± 2.50 [1]	12.625 ± 1.08 [1]	136.0 ± 3.48 [1]	62.4 ± 0.44 [1]
Pg Methanol Fraction	59.0 ± 2.39 [2]	13.667 ± 0.833 [1]	130.1 ± 3.02 [1]	40.9 ± 0.69 [2]
Pg Roots	61.1 ± 1.03 [2]	4.920 ± 0.550 [2]	102.1 ± 1.49 [2]	41.6 ± 1.06 [2]
Pg Flower	43.5 ± 2.16 [1]	8.460 ± 0.550 [3]	194.3 ± 2.33 [3]	76.9 ± 0.61 [3]
Pg Fruits	127.9 ± 1.67 [3]	2.417 ± 0.417 [4]	11.3 ± 0.12 [4]	10.4 ± 0.24 [4]
Ascorbic acid	-	-	3.6 ± 0.06 [5]	-
Trolox	-	-	-	5.0 ± 0.25 [5]

All values are mean ± S.E.M (*n* = 3); [a] GAE = Gallic acid equivalent per gram of sample; [b] QE = Equivalent quercetin per gram of sample; [c] Value defined as the concentration of sample that scavenged 50% of the DPPH or ABTS$^{.+}$. There are no significant differences among values marked with the same superscript numbers in individual columns.

A correlation has been shown between anti-oxidant activity and total phenolic content in natural products, especially between extracts with the two highest values. The fruit extract showed a greater amount of phenolics (127.9 ± 1.67 mg GAE/g) and also showed the greater anti-oxidant activity in DPPH (IC_{50} = 11.3 ± 0.12) and ABTS (IC_{50} = 10.4 ± 0.24) test. The flower extract presented the lowest anti-oxidant activity in DPPH (IC_{50} = 194.3 ± 2.33) and ABTS (IC_{50} = 76.9 ± 0.61) tests as well as the lowest phenolic content (43.5 ± 2.16). When analysing the other extracts and methanol phase of the plant, a sequence of linear correlations between total phenolic content and radical scavenging activity was not observed.

Among all tested samples, only the fruit extract has sufficient anti-oxidant activity to be considered for nutraceutical use. According to the anti-oxidant activity index (AAI), all extracts and phases of *P. gounellei* showed an index lower than 0.5, thus presenting poor activity. Surprisingly, the fruit extract presented an AAI value of 2.01, which is considered as very strong.

Many studies have reported the relationship between total phenolics assay results and anti-oxidant activity [37]; our study confirms these findings. The relationships between the total phenolic content and the antiradical activity of DPPH ($1/EC_{50}$), and antiradical activity of ABTS$^{.+}$ ($1/EC_{50}$) are shown in Figure 3. The Pearson correlation coefficients (*r*) of these plots were approximately 0.943 for the DPPH assay and 0.944 for the ABTS assay. This result suggests that 94% of the anti-oxidant capacity of extracts and methanol fraction from *P. gounellei* is due to the contribution of phenolic compounds. It is interesting to mention that there was no inverse correlation between the total phenolic content and the anti-oxidant activity by the DPPH and ABTS methods and flavones/flavonols content, at least when comparing the extract with the highest anti-oxidant activity, the extract from the fruits. Therefore, it is possible that the anti-oxidant activity may be attributed to other phenolic compounds than flavonols and/or flavones.

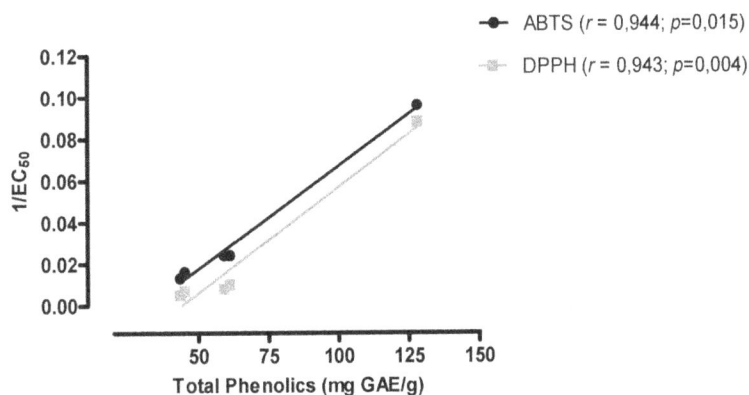

Figure 3. Correlation among the total phenolic content and the antiradical activity DPPH and antiradical activity ABTS$^{.+}$ of ethanol extracts and methanol fraction of *P. gounellei*.

3. Experimental Section

3.1. General Procedures

Column chromatography separations (CC) were performed on glass columns packed with silica gel 60 (Merck, Nottingham, UK) 7734 (0.063–0.2 mm particles, 70–230 mesh), flash silica (0.04–0.0063 mm particles, 230–400 mesh), Amberlite XAD or Sephadex LH-20. Thin layer chromatography (TLC) was performed on silica gel PF254 plates (Merck, Nottingham, UK). For HPLC experiments, a high-performance liquid chromatograph (HPLC–DAD, equipped with a binary pump (LC-6AD), and diode array detector SPD-M20A (Shimadzu, Kyoto, Japan) and Rheodyne injector (7125) with a 20 µL loop for the analytical column and 500 µL for the semipreparative column were used. An analytical Luna C-18 100A column (250 mm × 4.6 mm × 5 µm, Phenomenex, Torrance, CA, USA), semipreparative Luna C-18 column (250 mm × 21.2 mm × 5 µm, Phenomenex) and Millipore filter membranes (0.45 µm Supelco®, Bellefonte, PA, USA) were used.

The solvents used in the chromatographic procedures were p.a. grade: *n*-hexane, dichloromethane, chloroform and ethyl acetate. Methanol HPLC grade (Tedia®, Rio de Janeiro, Brazil). Water was obtained from a Millipore® MilliQ system (Millipore, São Paulo, Brazil).

To evaluate the total phenols, total flavones and flavonols content and anti-oxidant activity, the following reagents were used: Folin-Ciocalteu, gallic acid, Trolox (6-hydroxy-2,5,7,8-tetramethylchroman- 2-carboxylic acid 97%), ABTS (2,2'-azinobis-(3-ethylbenzothiazoline-6-sulfonic acid) diammonium salt 98%) (Sigma-Aldrich, Sternheim, Germany), ethanol, methanol, sodium carbonate, potassium persulfate, aluminium chloride (Vetec, Rio de Janeiro, Brazil) and Milli-Q water. Samples were solubilized in an ultrasonic bath, 3.5 L (Unic 1600A, Unique, São Leopoldo, Brazil). Readings were performed on Asys HiTech UVM 340 apparatus (São Paulo, Brazil).

Isolated compounds were identified using 1-D and 2-D NMR analysis acquired on the following spectrometers: Varian Oxford (200 MHz), Varian (500 MHz) (Varian, Palo Alto, CA, USA) and Avance III (Bruker, Coventry, UK) using deuterated solvents. The high-resolution mass spectra were obtained using LC-HRMS analysis performed on an Accela 600 HPLC system combined with an Exactive (Orbitrap) mass spectrometer from Thermo Fisher Scientific (Bremen, Germany). EIMS was obtained with a Shimadzu QP-2000 spectrometer (Kyoto, Japan). The $[\alpha]_D^{25}$ 25 °C was determined using a MCP 200 polarimeter (Anton Paar, Saint Laurent, QC, Canada).

3.2. Botanical Material

Pilosocereus gounellei was collected in Boa Vista City-PB (Brazil) in November 2010. The plant was identified by Prof. Dr. Leonardo Person Felix (CCA/UFPB) and a voucher specimen (15437) was deposited in the Herbarium Prof. Jaime Coelho Morais of the Agricultural Sciences Center (CCA/UFPB).

3.3. Extraction and Isolation

The botanical material (cladodes) was dried in an oven with circulating air at 40 °C and ground using a mechanical mill, yielding 5.18 kg of powder, which was macerated with 10 L of EtOH at room temperature, for 72 h. The extraction solution was concentrated in a rotary evaporator at 40 °C, yielding 237.13 g of crude ethanolic extract (CCEE). CCEE (10 g) was submitted to vacuum liquid chromatography (VLC) using silica gel and eluted with hexane (Hex), chloroform (CHCl$_3$) and methanol (MeOH) to obtain the corresponding fractions. The chloroform fraction (8.0 g) was chromatographed in a silica column using solvents of increasing polarity: Hex, dichloromethane (CH$_2$Cl$_2$) and MeOH. The fractions Hex–CH$_2$Cl$_2$ (3:7), CH$_2$Cl$_2$ and CH$_2$Cl$_2$–MeOH (9:1) were combined and chromatographed using the same method yielding 188 fractions that were analysed using TLC. The fractions 35/38 (8.00 mg), after recrystallization, gave pure colourless crystals of compound 1. The fractions 54/59 (3.51 g) contained compound 2 and the fractions 146/150 (2.72 g) afforded a white powdery precipitate of compound 3. CCEE (5.00 g) was dissolved in CHCl$_3$:H$_2$O (1:1) and separated using a separating funnel, yielding an aqueous fraction and a chloroform fraction. A portion of the chloroform fraction (2.0 g) was subjected to successive column chromatography on silica gel, following the methodology previously described, resulting in the isolation of compound 4 (20.50 mg).

CCEE (131.84 g) was submitted to column chromatography using Amberlite XAD as the stationary phase and eluted with H$_2$O, MeOH, Hex, ethyl acetate (EtOAc) and acetone. The hexane fraction from XAD was chromatographed using VLC on silica gel 60 with hexane, CH$_2$Cl$_2$ and MeOH. The hexane fraction was chromatographed on flash silica column yielding the subfractions 28/33 (92.30 mg) that were purified using preparative TLC, eluted with Hex–EtOAc (80:20), to give compound 5 (10.20 mg).

The Hex–CH$_2$Cl$_2$ (1:1) fraction was chromatographed on silica gel resulting in fractions 13/15 (white crystals), corresponding to 6 (3.51 g). The methanolic fraction from XAD (9.00 g) was chromatographed on Sephadex eluted with MeOH–CH$_2$Cl$_2$ (7:3) resulting in 33 fractions. Fraction 12 yielded compound 7 (15.00 mg) and fraction 14 yielded compound 8 (12.00 mg), both as yellow powders.

The roots were dried in an oven at 40 °C and ground in a mechanical mill, yielding 1.37 kg of powder that was macerated with 10 L of EtOH at room temperature, for 72 h. The obtained solution was concentrated in a rotary evaporator at 40 °C, resulting in 27.00 g of root crude ethanol extract (RCEE).

To isolate nitrogen-containing compounds, the method described by Souza and Silva [38] was used. The acidified chloroform fraction (ACF, 3.38 g) was chromatographed on a silica column eluted with hexane, EtOAc and MeOH, yielding 159 fractions. The combined fractions 88/114 (580.80 mg) were analysed by HPLC-DAD using a semipreparative column at room temperature. As the mobile phase, Milli-Q water and MeOH were used gradient-wise, and the concentration of the MeOH was increased from 50 to 100% in a 20-min run. The chromatogram showed three peaks, and the one at higher retention time was found to be the major component; thus, it was isolated and identified as compound 9 (8.00 mg).

RCEE (10.00 g) was solubilized in H$_2$O and yielded 5.00 g of precipitate. A sample (2.50 g) of it was dissolved in MeOH–CHCl$_3$ (1:1) and submitted to filtration on Sephadex LH-20 with MeOH–CHCl$_3$ (1:1), resulting in 10 fractions. The subfraction 9/10 (10.00 mg) was proved to be pure by TLC and identified as 10. The remaining 2.50 g of the precipitate was chromatographed on silica gel 60 column and eluted with hexane, EtOAc and MeOH, yielding the pure fractions 5/7 which corresponded to compound 3.

3.4. Spectral Data

Pinostrobin (5-hydroxy-7-methoxy-flavanone, **1**): colorless crystals, molecular formula ($C_{16}H_{14}O_2$) ^{1}H-NMR (500 MHz, $CDCl_3$): 11.99 (1 H, s, 5-OH), 7.45 (5 H, m, H-2', H-3', H-4' e H-5'), 6.06 (1H, d, J = 2.5 Hz, H-6), 6.05 (1H, d, J = 2.5, H-8), 5.41 (1H, dd, J = 13.00 and 3.00 Hz, H-2), 3.79 (3H, s, 7-OMe), 3.07 (1H, dd,17.00Hz e 13.00, H-3), 2.82 (1H, dd, 17.00Hz, 3.00 Hz, H-3). ^{13}C-NMR (125 MHz, $CDCl_3$): 79.20 (C-2), 43.41 (C-3), 195.73 (C-4), 162.79 (C-5), 95.20 (C-6), 167.80 (C-7), 94.30 (C-8), 164.17 (C-9), 138.39 (C-1'), 126.19 (C2'/C-6'), 128.9 (C-3', C-4'/C-5'), 55.69 (OCH_3).

β-Sitosterol (**2**): ^{1}H-NMR and ^{13}C-NMR data were consistent with the literature [24].

Sitosterol 3-O-β-D-glucopyranoside/stigmasterol 3-O-β-D-glucopyranoside (**3**): ^{1}H-NMR and ^{13}C-NMR data were consistent with the literature [24].

13²-Hydroxy-(13²-S)-pheophytin a (**4**): amorphous green powder ^{1}H-NMR and ^{13}C-NMR data were consistent with the literature [26].

Phaeophytin a (**5**): morphous green powder, ^{1}H-NMR and ^{13}C-NMR data were consistent with the literature [27].

β-Sitosterol and stigmasterol (**6**): ^{1}H-NMR and ^{13}C-NMR data were consistent with the literature [24].

Kaempferol (3,4',5,7-Tetrahydroxyflavone, **7**): yellow amorphous powder, ^{1}H-NMR and ^{13}C-NMR data were consistent with the literature [29].

Quercetin (3,3',4',5,7-Pentahydroxyflavone, **8**): yellow amorphous powder, ^{1}H-NMR and ^{13}C-NMR data were consistent with the literature [30].

7'-Ethoxy-trans-feruloyltyramine (**9**): a pale amorphous solid, molecular formula ($C_{20}H_{23}NO_5$), ^{1}H-NMR (δ, 400 MHz, DMSO-d_6) and ^{13}C-NMR (δ, 100 MHz, DMSO-d_6): see Table 1.

trans-Feruloyltyramine (**10**): pale amorphous solid, molecular formula ($C_{18}H_{19}NO_4$), ^{1}H-NMR (δ, 400 MHz, DMSO-d_6), 7.11 (1H, sl, H-2), 6.78 (1H, d, J = 8.0Hz, H-5), 6.98 (1H, d, J = 8.0 Hz, H-7), 7.30 (1H, d, J = 16Hz, H-7), 6.43, (1H, d, J = 15.8 Hz, H-8), 7.00 (2H, d, J = 8.0, H-2'/H-6'), 6.68 (2H, d, J = 8.0 Hz, H-3'/H-5'), 2.64 (2H, t, H-7'), 3.33 (2H, m, H-8'), 3.79 (3H, s, OCH_3), 8.01 (1H, t, NH). ^{13}C-NMR (δ, 100 MHz, DMSO-d_6) 126.93 (C-1), 111.12 (C-2), 148.37 (C-3), 148.75 (C-4), 116.18 (C-5), 122,09 (C-6), 139.56 (C-7), 119.53 (C-8), 128.19 (C-1'), 130.02 (C-2'/C-6'), 115.67 (C-3'/C-5'), 156.18 (C-4'), 34.27 (C-7'), 41.27 (C-8'), 56.07 (OCH_3).

3.5. Total Phenol Content Assay

The total phenolics content was evaluated by the method of Gulcin *et al.*, [39] with some modifications, using the Folin-Ciocalteu reagent and gallic acid as a positive control. Samples of cladodes, roots, flowers and fruit and the methanolic fraction from *P. gounellei*, from a stock solution of 5 mg/mL, solubilized in EtOH, were transferred to a 1.0 mL Eppendorf tube by adding 20.0 μL of Folin-Ciocalteu reagent, stirring for 1 min. Then Na_2CO_3 (60.0 μL, 15%) was added to the mixture and stirred for 30 s. Finally, distilled water (900 μL) was added to give a final concentration of 100 μg/mL. After 2 h, the absorbance of the samples was measured at 760 nm. The concentration of the phenolic compounds was determined as equivalent milligram of gallic acid per gram of sample (mg GAE/g), from the calibration curve constructed with gallic acid standard (2.5 to 15.0 μg/mL), considering the average standard error (SEM).

3.6. Total Flavones and Flavonols Content Assay

The flavones and total flavonols content were determined adapting the methodology described by Mihai *et al.* [40]. Stock solutions (1.0 mg/mL) of extracts from cladodes, roots, flowers and fruits and methanolic fraction from *P. gounellei* were prepared. Each sample solution (400 μL) and methanolic solution of aluminium chloride (200 μL, 2%) were added in a volumetric flask. The final volume was adjusted with the same solvent to 10 mL. Reaction occurred for 30 min in the dark. The reading was performed at a wavelength of 425 nm. The analysis was evaluated in triplicate and the total flavones and flavonols content was determined from the calibration curve constructed with straight line equation of quercetin solutions (1.0 to 40.0 μg/mL) and expressed in equivalent milligrams of quercetin by gram of extract (mg QE/g), considering the average standard error (SEM).

3.7. DPPH· Radical Scavenging Activity Assay

The anti-oxidant activity of ethanolic extracts of the cladodes, roots, flowers and methanol fraction from *P. gounellei* was performed against the free radical DPPH following the methodology of Silva *et al.* [41]. Stock solutions were prepared from the extracts and methanol fraction at several concentrations (0.10 to 5.0 mg/mL). Through preliminary analysis, appropriate quantities of stock solutions of the samples and 450 μL of the solution of DPPH· (23.6 mg/mL in EtOH) were transferred to 0.5 mL Eppendorf tubes and the volume was completed with EtOH, following homogenization. Samples were sonicated for 30 min and the amount of DPPH· was recorded on a UV-vis device at a wavelength of 517 nm in a 96-well plate. Ascorbic acid was used as a positive control and all concentrations were tested in triplicate.

The percentage scavenging activity (% SA) was calculated from the equation:

$$\% \, \text{SA} = 100 \times \frac{\left(Abs_{\text{control}} - Abs_{\text{sample}} \right)}{Abs_{\text{control}}} \tag{1}$$

where Abs_{control} is the absorbance of the control containing only the ethanol solution of DPPH, and Abs_{sample} is the absorbance of the radical in the presence of the sample or standard ascorbic acid. The anti-oxidant activity index (AAI) was calculated according to Scherer and Godoy [42] as follows:

$$\text{AAI} = \frac{\text{final concentration of DPPH} \left(\mu\text{g·mL}^{-1} \right)}{\text{IC}_{50} \left(\mu\text{g·mL}^{-1} \right)} \tag{2}$$

AAI values below 0.5 indicate low anti-oxidant activity, values between 0.5 and 1.0 indicate moderate activity, values between 1.0 and 2.0 indicate strong activity and AAI values above 2.0 indicate very strong anti-oxidant activity.

3.8. Determination of Anti-oxidant Activity against the Radical Cation ABTS·+

The determination of anti-oxidant activity from extracts of cladodes, roots, flowers, fruits and methanolic fraction from *P. gounellei* against the radical cation ABTS·+ was carried out following the methodology described by Re [43] using Trolox as the standard compound. The starting concentrations of the solutions of the samples were 0.1–1.0 mg/mL, with the addition of 450 μL of the radical ABTS·+ solution to give final concentrations of 2.5–100.0 μg/mL samples. Samples were protected from light and sonicated for 6 min. Absorbance of the samples and the positive control were measured at a wavelength of 734 nm using a microplate of 96 wells. Each concentration was tested in triplicate. The percentage of free radical scavenging activity of ABTS·+ was calculated by the equation:

$$\% \, \text{SA} = 100 \times \frac{\left(Abs_{\text{control}} - Abs_{\text{sample}} \right)}{Abs_{\text{control}}} \tag{3}$$

where $Abs_{control}$ is the absorbance of the control containing only the ethanol solution of $ABTS^+$ and Abs_{sample} is the absorbance of the radical in the presence of the sample or standard ascorbic acid.

The antiradical efficiency was established using linear regression analysis and the 95% confidence interval ($p < 0.05$) obtained using the statistical program GraphPad Prism 5.0. The results were expressed as $EC_{50} \pm SEM$ (sample concentration required to eliminate 50% of the DPPH radicals available, plus or minus the SEM).

3.9. Statistical Analyses

The results are expressed as the mean $\pm$ standard error of the mean (SEM). Analysis of variance (ANOVA one-way and Tukey's *post hoc* test) were used to evaluate the differences of the means between groups. The antiradical efficiency was established using linear regression analysis. Pearson correlation coefficients (r) were used to express correlations and confidence interval of 95% ($p < 0.05$) obtained using the statistical program GraphPad Prism 5.0 (GraphPad Software Inc., San Diego, CA, USA). The results were expressed by sample concentration required to eliminate 50% of the DPPH radicals available, plus or minus the SEM ($EC_{50} \pm SEM$).

4. Conclusions

The phytochemical study of *Pilosocereus gounellei* led to the isolation and identification of 10 compounds: pinostrobin, β-sitosterol, a mixture of β-sitosterol/stigmasterol, 13^2-hydroxy-phaeophytin a, phaeophytin a, sitosterol 3-*O*-β-D-glucopyranoside/stigmasterol 3-*O*-β-D-gluco-pyranoside, kaempferol, quercetin, the new substance 7'-ethoxy-*trans*-feruloyltyramine and *trans*-feruloyltyramine. The evaluation of anti-oxidant activity from *P. gounellei* demonstrated that the fruit ethanol extract possesses excellent anti-oxidant activity, mainly because of the presence of phenolic compounds reported in the genus and the Cactaceae family.

Acknowledgments: The authors are grateful to INSA/MCT/CNPq (No. 562730/2010-9), CAPES and FACEPE (PRONEM APQ-0741-1.06/14) for financial support and to UFPB for technical assistance.

Author Contributions: Jéssica K. S. Maciel, Otemberg S. Chaves, Severino G. Brito Filho, Yanna C. F. Teles, Marianne G. Fernandes (in memoriam) and Maria de Fátima Vanderlei de Souza carried out the isolation and identification of compounds. Pedro Dantas Fernandes, Albericio Pereira de Andrade and Leonardo P. Felix collected and identified the plant. Temilce S. Assis carried out the antioxidant assay. Tania M. S. Silva, Nathalia S. M. Ramos and Girliane R. Silva carried out liquid cromatography and DCR.

Conflicts of Interest: The authors declare no conflict of interest.

References

1.   Hunt, D.; Taylor, N.P.; Charles, G. *The New Cactus Lexicon: Descriptions and Illustrations of the Cactus Family*; DH Books: Milborne Port, UK, 2006.

2.   Wallace, R.S. Molecular systematic study of the cactaceae: Using chloroplast DNA variation to elucidate cactus phylogeny. *Bradleya* **1995**, *13*, 1–12.

3.   Hernández-Hernández, T.; Hernández, H.M.; De-Nova, J.A.; Puente, R.; Eguiarte, L.E.; Magallón, S. Phylogenetic relationships and evolution of growth form in Cactaceae (Caryophyllales, Eudicotyledoneae). *Am. J. Bot.* **2011**, *98*, 44–61. [CrossRef] [PubMed]

4.   Ribeiro, E.M.S.; Meiado, M.V.; Leal, I.R. The role of clonal and sexual spread in cacti species dominance at the Brazilian Caatinga. *Gaia* **2015**, *9*, 27–33.

5.   Lucena, C.M.; Costa, G.M.; Sousa, R.F.; Carvalho, T.K.N.; Marreiros, N.A.; Alves, C.A.B.; Pereira, D.D.; Lucena, R.F.P. Conhecimento local sobre cactáceas em comunidades rurais na mesorregião do sertão da Paraíba (Nordeste, Brasil). *Biotemas* **2012**, *25*, 281–291. [CrossRef]

6.   Necchi, R.M.M.; Alves, I.A.; Alves, S.H.; Manfron, M.P. *In vitro* antimicrobial activity, total polyphenols and flavonoids contents of *Nopalea cochenillifera* (L.) Salm-Dyck (Cactaceae). *J. Res. Pharm.* **2012**, *2*, 1–7.

7.   Davet, A.; Virtuoso, S.; Dias, J.F.G.; Miguel, M.D.; Oliveira, A.B.; Miguel, O.G. Atividade antibacteriana de *Cereus jamacaru* DC, Cactaceae. *Rev. Bras. Farmacogn.* **2009**, *19*, 561–564. [CrossRef]

8. Gonçalves, A.S.M.; Peixe, R.G.; Sato, A.; Muzitano, M.F.; de Souza, R.O.M.A.; de Barros Machado, T.; Amaral, A.C.F.; Moura, M.R.L.; Simas, N.K.; Leal, I.C.R. *Pilosocereus arrabidae* (Byles & Rowley) of the Grumari Sandbank, RJ, Brazil: Physical, chemical characterizations and antioxidant activities correlated to detection of flavonoids. *Food Res. Int.* **2015**, *70*, 110–117.

9. Nasciemnto, U.T.; Moura, N.P.; Vasconcelos, M.A.S.; Maciel, I.S.M.; Albuquerque, U.P. Chemical characterization of native wild plants of dry seasonal forest of the semi-arid region of northeastern Brazil. *Food Res. Int.* **2011**, *44*, 2112–2119. [CrossRef]

10. Neill, S.O.; Gould, K.S.; Kilmartin, P.A.; Mitchell, K.A.; Markham, K.R. Antioxidant activities of red *versus* green leaves in Elatostema rugosum. *Plant Cell Environ.* **2002**, *25*, 539–547. [CrossRef]

11. Pascoal, A.; Rodrigues, S.; Teixeira, A.; Feás, X.; Estevinho, L.M. Biological activities of commercial bee pollens: Antimicrobial, antimutagenic, antioxidant and anti-inflammatory. *Food chem. Toxicol.* **2014**, *63*, 233–239. [CrossRef] [PubMed]

12. Bouhlel, I.; Limem, I.; Skandrani, I.; Nefatti, A.; Ghedira, K.; Dijoux-Franca, M.G.; Leila, C.G. Assessment of isorhamnetin 3-*O*-neohesperidoside from Acacia salicina: Protective effects toward oxidation damage and genotoxicity induced by aflatoxin B1 and nifuroxazide. *J. Appl. Toxicol.* **2010**, *30*, 551–558. [CrossRef] [PubMed]

13. Bouhlel, I.; Skandrani, I.; Nefatti, A.; Valenti, K.; Ghedira, K.; Mariotte, A.M.; Hininger-favier, I.; Laporte, F.; Dijoux-Franca, M.G.; Chekir-Ghedira, L. Antigenotoxic and antioxidant activities of isorhamnetin 3-*O*-neohesperidoside from *Acacia salicina. Drug Chem. Toxicol.* **2010**, *32*, 258–267. [CrossRef] [PubMed]

14. Starha, R.; Chybidziurova, A.; Lacny, Z. Alkaloids of the genus Turbinicarpus (Cactaceae). *Biochem. Syst. Ecol.* **1999**, *27*, 839–841. [CrossRef]

15. Ordaz, C.; Ferrigni, N.R.; Mclaughlin, J.L. Dehydroheliamine, A trace alkaloid from the saguaro, *Carneglea gigantea* (Cactaceae). *Phitochemistry* **1983**, *22*, 2101–2102. [CrossRef]

16. Starha, R. Alkaloids from the Cactus Genus Gymnocalycium (Cactaceae)-II. *Biochem. Syst. Ecol.* **1997**, *25*, 363–364. [CrossRef]

17. Ortiz, C.M.F.; Dávila, P.; Portilla, L.B.H. Alkaloids from *Neobuxbaumia species* (Cactaceae). *Biochem. Syst. Ecol.* **2003**, *31*, 581–585. [CrossRef]

18. Almeida, C.A.; Figueirêdo, R.M.F.; Queiroz, A.J.M.; Oliveira, F.M.N. Características físicas e químicas da polpa de xiquexique. *Rev. Ciênc. Agron.* **2007**, *38*, 440–443.

19. Lucena, C.M.; Lucena, R.F.P.; Costa, G.M.; Carvalho, T.K.N.; Costa, G.G.S.; Alves, R.R.N.; Pereira, D.D.; Ribeiro, J.E.S.; Alves, C.A.B.; Quirino, C.G.M.; *et al.* Use and knowledge of Cactaceae in Northeastern Brazil. *J. Ethnobiol. Ethnomed.* **2013**, *9*, 62. [CrossRef] [PubMed]

20. Roque, A.A.; Rocha, R.M.; Loiola, M.I.B. Uso e diversidade de plantas medicinais da Caatinga na comunidade rural de Laginhas, município de Caicó, Rio Grande do Norte (nordeste do Brasil). *Rev. Bras. Plantas Med.* **2010**, *12*, 31–42. [CrossRef]

21. Agra, M.F.; Silva, K.N.; Basílio, I.J.L.D.; Freitas, P.F.; Barbosa-Filho, J.M. Survey of medicinal plants used in the region Northeast of Brazil. *Rev. Bras. Farmacogn.* **2008**, *18*, 472–508. [CrossRef]

22. Murillo, M.C.A.; Suarez, L.E.C.; Salamanca, J.A.C. Actividad insecticida sobre Spodoptera frugiperda (*Lepidóptera: Noctuidae*) de los compuestos aislados de la parte aérea de *Piper septuplinervium* (Miq.) C. DC. y las inflorescencias de *Piper subtomentosum* Trel. & Yunck. (Piperaceae). *Quim. Nova* **2014**, *37*, 442–446.

23. Mclaughlin, J.L.; Rogers, L.L.; Anderson, J.E. The use of biological assys to evaluate botanicals. *Drug. Inf. J.* **1998**, *32*, 513–524.

24. Kojima, H.; Sato, N.; Hatano, A.; Ogura, H. Sterol glucosides from Prunellavulgaris. *Phytochemistry* **1990**, *29*, 2351–2355. [CrossRef]

25. Dannhardt, G.; Kiefer, W. Cyclooxygenase inhibitors-current status and future prospects. *Eur. J. Med. Chem.* **2001**, *36*, 109–126. [CrossRef]

26. Teles, Y.C.F.; Gomes, R.A.; Oliveira, M.S.; Lucena, K.L.; Nascimento, J.S.; Agra, M.F.; Igoli, J.O.; Gray, A.I.; Souza, M.F.V. Phytochemical investigation of *Wissadula periplocifolia* (L.) C. Presl and evaluation of its antibacterial activity. *Quim. Nova* **2014**, *37*, 1491–1495.

27. Chaves, O.S.; Gomes, R.A.; Tomaz, A.C.; Fernandes, M.G.; Graças Mendes, L.J.R.; Agra, M.F.; Braga, V.A.; Souza, M.F.V. Secondary Metabolites from *Sida rhombifolia* L. (Malvaceae) and the Vasorelaxant Activity of Cryptolepinone. *Molecules* **2013**, *18*, 2769–2777. [CrossRef] [PubMed]

28. Brito-Filho, S.G. *Feofitinas e Esteróides glicosilados de Turnera subulata Sm.(TURNERACEAE)*; Dissertação (Mestrado) Programa de Pós-Graduação em Produtos Sintéticos e Bioativos, Universidade Federal da Paraíba: João Pessoa, Brazil, 2011; p. 90.

29. Costa, D.A.; Matias, W.N.; Lima, I.O.; Xavier, A.L.; Costa, V.B.M.; Diniz, M.F.F.M.; Agra, M.F.; Batista, L.M.; Souza, M.F.V. First secondary metabolites from *Herissantia. crispa* L (Brizicky) and the toxicity activity against Artemia. salina Leach. *Quim. Nova* **2009**, *32*, 48–50. [CrossRef]

30. Gomes, R.A.; Maciel, J.K.S.; Agra, M.F.; Souza, M.F.V.; Falcão-Silva, V.S.; Siqueira-Junior, J.P. Phenolic compounds from *Sidastrum micranthum* (A. St.-Hil.) fryxell and evaluation of acacetin and 7,4'-Di-O-methylisoscutellarein as motulator of bacterial drug resistence. *Quim. Nova* **2011**, *34*, 1385–1388. [CrossRef]

31. Williams, R.J.; Spencer, J.R.; Rice-Evans, C. Flavonoids: Antioxidants or signaling molecules? *Free Radic. Biol. Med.* **2004**, *36*, 838–849. [CrossRef] [PubMed]

32. Efdi, M.; Ohguchi, Y.A.; Akao, Y.; Nozawa, Y.; Korestsu, M.; Ishihara, H.M. *N*-TransFeruloyltyramine as a Melain Biosynthesis Inhibitor. *Biol. Pharm. Bull.* **2007**, *30*, 1972–1974. [CrossRef] [PubMed]

33. Greca, M.D.; Previtera, L.; Purcaro, R.; Zarrelli, A. Cinnamic acid amides and lignanamides from *Aptenia cordifolia*. *Tetrahedron* **2006**, *62*, 2877–2882.

34. Cavalcante, J.M.S.; Nogueira, T.B.S.S.; Tomaz, A.C.A.; Silva, D.A.; Agra, M.F.; Souza, M.F.V.; Carvalho, P.R.C.; Ramos, S.R.; Nascimento, S.C.; Gonçalves-Silva, T. Steroidal and phenolic compounds from *Sidastrum paniculatum* Fryxell and evaluation of cytotoxic and anti-inflammatory activities. *Quim. Nova* **2010**, *33*, 846–849. [CrossRef]

35. Liang, X.; Tian, J.; Li, L.; Gao, J.; Zhang, Q.; Gao, P.; Song, S. Rapid determination of eight bioactive alkaloids in Portulacaoleracea L. by the optimal microwave extraction combined with positive-negative on version multiple reaction monitor (+/−MRM) technology. *Talanta* **2014**, *120*, 167–172. [CrossRef] [PubMed]

36. Park, J.B. Isolation and Characterization of *N*-Feruloyltyramine as the *P*-selectine Expression Supressor from Garlic (*Allium sativum*). *J. Agric. Food Chem.* **2009**, *57*, 8868–8872. [CrossRef] [PubMed]

37. Oliveira, M.F.; Pinheiro, L.S.; Pereira, C.K.S.; Matias, W.N.; Gomes, R.A.; Chaves, O.S.; Souza, M.F.V.; Almeida, R.N.; Assis, T.S. Total phenolic content and antioxidant activity the some Malvaceae species. *Antioxidants* **2012**, *1*, 33–43. [CrossRef]

38. Silva, D.A.; Silva, T.M.S.; Claudio, A.; Costa, D.A.; Cavalcante, J.M.S.; Matias, W.N.; Braz Filho, R.; Souza, M.F.V. Constituintes químicos e atividade antioxidante de *Sida galheirensis*. *Quim. Nova* **2006**, *29*, 1250–1254. [CrossRef]

39. Gulcin, I.; Sat, I.G.; Beydemir, S.; Elmastas, M.; Kufrevioglu, O.I. Comparison of antioxidant activity of clove (Eugenia caryophylata Thunb) buds and lavender (*Lavandula stoechas* L.). *Food Chem.* **2004**, *87*, 393–400. [CrossRef]

40. Mihal, C.M.; Mărghitas, L.A.; Dezmirean, D.S.; Chirilă, F.; Moritz, R.F.; Schlüns, H. Interactions among flavonoids of propolis affect antibacterial activity against the honeybee pathogen *Paenibacillus* larvae. *J. Invertebr. Pathol.* **2012**, *110*, 68–72. [CrossRef] [PubMed]

41. Silva, E.M.S.; Freitas, B.M.; Santos, F.A.R. Chemical Composition and Free Radical Scavenging Activity of Pollen loads from Stingless bee *Melipona subnitida* Ducke. *J. Food Compost. Anal.* **2006**, *19*, 507–511. [CrossRef]

42. Scherer, R.; Godoy, H.T. Antioxidant activity index (AAI) by 2,2-diphenyl-1-picrylhydrazyl method. *Food Chem.* **2009**, *112*, 654–658. [CrossRef]

43. Re, R.; Pelegrini, N.; Proteggente, A.; Pannala, A.; Yang, M.; Riceevans, C. Antioxidant activity applying an improved ABTS radical cátion decolorization assay. *Free Radic. Biol. Med.* **1999**, *26*, 1231–1237. [CrossRef]

13

Solubilization Behavior of Polyene Antibiotics in Nanomicellar System: Insights from Molecular Dynamics Simulation of the Amphotericin B and Nystatin Interactions with Polysorbate 80

Meysam Mobasheri [1,*], Hossein Attar [1,2], Seyed Mehdi Rezayat Sorkhabadi [3,4], Ali Khamesipour [5] and Mahmoud Reza Jaafari [6,*]

Academic Editor: James W. Gauld

[1] Department of Chemical Engineering, Science and Research Branch, Islamic Azad University, Tehran 1477893855, Iran; attar.h@srbiau.ac.ir

[2] Tofigh Daru Research and Engineering Company (TODACO), Tehran 1397116359, Iran

[3] Department of Medical Nanotechnology, School of Advanced Technologies in Medicine, Tehran University of Medical Sciences, Tehran 1417755469, Iran; rezayat@sina.tums.ac.ir

[4] Department of Toxicology and Pharmacology, Pharmaceutical Sciences Branch, Islamic Azad University, Tehran 193956466, Iran

[5] Center for Research and Training in Skin Diseases and Leprosy, Tehran University of Medical Sciences, Tehran 1416613675, Iran; khamesipour@tums.ac.ir

[6] Biotechnology Research Center, Nanotechnology Research Center, School of Pharmacy, Mashhad University of Medical Sciences, P. O. Box: 91775-1365, Mashhad 917751365, Iran

* Correspondence: m.mobasheri@srbiau.ac.ir (M.M.); jafarimr@mums.ac.ir (M.R.J.)

Abstract: Amphotericin B (AmB) and Nystatin (Nys) are the drugs of choice for treatment of systemic and superficial mycotic infections, respectively, with their full clinical potential unrealized due to the lack of high therapeutic index formulations for their solubilized delivery. In the present study, using a coarse-grained (CG) molecular dynamics (MD) simulation approach, we investigated the interaction of AmB and Nys with Polysorbate 80 (P80) to gain insight into the behavior of these polyene antibiotics (PAs) in nanomicellar solution and derive potential implications for their formulation development. While the encapsulation process was predominantly governed by hydrophobic forces, the dynamics, hydration, localization, orientation, and solvation of PAs in the micelle were largely controlled by hydrophilic interactions. Simulation results rationalized the experimentally observed capability of P80 in solubilizing PAs by indicating (i) the dominant kinetics of drugs encapsulation over self-association; (ii) significantly lower hydration of the drugs at encapsulated state compared with aggregated state; (iii) monomeric solubilization of the drugs; (iv) contribution of drug-micelle interactions to the solubilization; (v) suppressed diffusivity of the encapsulated drugs; (vi) high loading capacity of the micelle; and (vii) the structural robustness of the micelle against drug loading. Supported from the experimental data, our simulations determined the preferred location of PAs to be the core-shell interface at the relatively shallow depth of 75% of micelle radius. Deeper penetration of PAs was impeded by the synergistic effects of (i) limited diffusion of water; and (ii) perpendicular orientation of these drug molecules with respect to the micelle radius. PAs were solvated almost exclusively in the aqueous poly-oxyethylene (POE) medium due to the distance-related lack of interaction with the core, explaining the documented insensitivity of Nys solubilization to drug-core compatibility in detergent micelles. Based on the obtained results, the dearth of water at interior sites of micelle and the large lateral occupation space of PAs lead to shallow insertion, broad radial distribution, and lack of core interactions of the amphiphilic drugs. Hence, controlled promotion of micelle permeability and optimization of chain crowding in palisade layer may help to achieve more efficient solubilization of the PAs.

Keywords: polyene antibiotics; Amphotericin B; Nystatin; Polysorbate 80; solubilization; molecular dynamics simulation; drug formulation; drug delivery systems; nanomedicine

1. Introduction

The therapeutic index (TI) of many surface active drugs is limited due to their tendency to self-associate in aqueous medium, resulting in low solubility, limited bioavailability, and potential severe side effects of these pharmaceuticals [1,2]. The clinical advantages of overcoming this challenge have provoked a large body of research into the development of formulations with improved pharmaceutical properties [3]. Among the most surface active drugs with an extending therapeutic use is the Amphotericin B (AmB). With a long history of clinical application, this polyene antibiotic (PA) remains the "gold standard" therapy against systemic mycotic infection, among all reasons, due to its broad-spectrum fungicidal activity, superior pharmacokinetic and pharmacodynamic profile compared to the related agents, and low rate of resistance [4,5]. Featured as well with leishmanicidal activity, AmB is also increasingly used as the drug of choice for treatment of visceral leishmaniasis in the cases of resistance against antimonials [6–8]. In spite of this striking therapeutic potency, the full potential efficacy of this drug is not realized in clinical practice due to its diverse dose-dependent side effects ranging from irreversible renal failure to central nervous system and liver damage to infusion-related reactions such as chills, fever, hypotension, dyspnea, rigors, arthralgias, nausea, vomiting and headaches [9]. Another amphiphilic PA with close structural similarity with AmB [10] is Nystatin (Nys), a potent antimycotic agent widely used for treatment of superficial mycoses [11,12]. While Nys has shown broad-spectrum fungicidal activity [11,13,14], and effectiveness against some fungal infections resistant to AmB [13,14], its potential for systemic antifungal therapy has remained untapped because of high toxicity of the drug when administered intravenously [11,12,15,16].

Extensive evidence has linked the toxicity of AmB and Nys to their presence in the aggregated state [17–20]. The mechanism of action of these PAs is based on the ability of their monomers to selectively promote permeability of ergostrol-rich fungal membranes rather than cholesterol-containing mammalian membranes, leading to leakage of ions and small molecules essential for fungal cell life, and ultimately lysis and death of the pathogen [21–24]. At the aggregated state, however, these drugs lose their selective recognition of the membranes and attack the membranes of both fungal and mammalian cells, causing cytotoxicity [17–26]. The solubilized delivery of PAs at monomeric state, hence, has been proposed as a possible pathway to their higher TI [9,27–29].

A powerful biomimetic strategy for enhancing the solubility of amphiphilic drugs is to incorporate them in the micellar assemblies of surface-active agents (surfactants) [2,30,31]. Owing to their particular structure which limits presence of water in the internal sites, micelles provide an energetically more favorable environment for residence of amphiphilic drugs compared with the bulk aqueous solution [32]. The advantages associated with micellar delivery systems, including ease of development, affordable costs, well-studied mechanisms of delivery, and the large parameter space for engineering specific properties, retains them as attractive vehicles for solubilized drug delivery. The Fungizone®, the major commercially available formulation of AmB produced and marketed by Bristol-Myers Squibb [28], is a dispersion of AmB in micellar solution of deoxycholate [1,33,34]. There have also been attempts to develop alternative micellar/vesicular formulations of this antibiotic using various excipients [1,35]. Similar efforts have been made for solubilized delivery of Nys, exemplified by liposomal [15], polymeric micellar [29], and more recently surfactant-based niosomal [36,37] delivery systems, all of which aimed at enabling parenteral administration of the drug. Although these formulations have proven great promise for improving the solubility and alleviating the toxicity of PAs, their widespread use in clinical practice is subject to more breakthrough advances in the field. In this respect, rational design of novel delivery systems inspired by the detailed knowledge of drug-excipient interactions can promote the efficiency of the research.

Rational development of pharmaceutical formulations is being increasingly aided by *in silico* methods [38–41]. Specifically, molecular dynamics (MD) simulation has proven powerful in providing molecular-level insight into drug-expedient interactions, particularly in those aspects not able to be readily explored by routine experimental techniques [39,41,42]. There are intriguing success stories of MD simulations use in drug formulation research, having provided insights into the location and mode of distribution of drug molecules in the vehicles [42], morphology of drug-loaded particles [43], aggregation behavior and mechanism of toxicity of drugs in presence of excipients [44], mechanisms, driving forces, and loading efficiency of drug encapsulation in carriers [45,46], and efficiency of particular formulations for solubilized drug delivery [46,47]. These studies have attracted attention to the MD simulations as a promising approach to optimal engineering of drug delivery systems [38,39,41,47].

In relation to PAs, MD simulations have been extensively used in the conformational analysis of the AmB in aqueous [48–50] and lipidic [51,52] media, characterizing the molecular aspects of AmB-biomembranes interactions [50,53–57], elucidating the mechanism of action of the drug [58,59], and understanding the nature of relationship between its molecular organization and selective toxicity [58,60–62]. Despite, however, MD simulations are not applied thus far in exploring the factors contributing to the solubilization of PAs, in contrast for instance to anticancer drugs such as paclitaxel [46,63]. AmB and Nys are structurally characterized by a rectangular glycosylated lactone ring, with a hydrophobic conjugated polyene chain on one edge, a hydrophilic polyhydroxyl chain on the opposite edge, a polar head comprised of carboxylate anion linked to the main ring, and an ammonium cation attached to the mycosamine moiety [48,64]. This structure renders these drug molecules as both amphiphilic and amphoteric, leading to their poor solubility, limited permeability/bioavailability and complex aggregation behavior. The same molecular characteristics also underlie the difficulties associated with developing formulations of AmB and Nys with desired profile [1,9]. This situation indicates a rationale for the use of *in silico* methods to further explore the molecular peculiarities of AmB and Nys in solubilized state, and identify factors influencing their solubilization efficiency.

Following our previous work on developing liposomal preparation of AmB [65], in this study, we sought to explore the utility of MD simulation in providing information about the of PA-carrier interactions. Our aim is to gain an understanding of the behavior of PAs in micellar systems and derive potential implications for future formulation developments. We selected Polysorbate 80 (P80) as the model surfactant. P80 is a non-toxic non-ionic detergent widely used for solubilizing and emulsifying hydrophobic pharmaceuticals, cosmetics, and food additives [30–32]. The P80 micelles have already been shown to be capable of remarkably enhancing Nys solubility in the aqueous medium [27]. P80 has also been used as a surfactant/co-surfactant in various formulations of AmB, such as microemulsions [33,66], nanoemulsions [67,68], and SEDDS [69,70]. Therefore, the MD simulation study of interactions of PAs with P80 may also provide insight for improving the efficiency of the currently available formulations of these drugs.

Among the barriers to the direct use of MD simulation to drug formulation practice are the associated computational demand and time-inefficiency, which may offset its potential advantages such as reduced experimental costs and labor. The extent of the challenge, however, is increasingly narrowing by the advent of efficient molecular modeling techniques such as coarse-grained (CG) modeling, allowing for less computational cost while maintaining certain levels of accuracy. In the CG modeling, the individual atoms are grouped together according to certain criteria to form CG representative interaction sites (beads), and the MD calculations are then performed between these interaction sites rather than between the individual atoms. This technique results in remarkable improvement of computational efficiency, thereby possibility of monitoring molecular interactions in a broad range of time and scale, which is essential for understanding many time-depending molecular-level phenomena. One of the standardized approaches to CG MD simulations is the MARTINI method developed and extended by Marrink *et al.* [71,72]. This method enables regular mapping of the atomic structure of a particular molecule to a CG molecular model followed by systematic model parameterization. MARTINI model has proven accurate in reproducing structural, dynamic, and thermodynamic properties for a broad range of systems and state points in accord with experimental data [73]. In this research, we used MARTINI approach to study the molecular dynamics of the interactions between PAs and P80. The study focuses on the aggregation of the drugs and the surfactant in aqueous medium, encapsulation behaviors of PAs, the intermolecular interactions involved in the drug encapsulation processes, the structural properties of P80-PA nanomicelles, and the stability of drug-surfactant mixed micellar system. The implications of the results for drug formulation development are discussed.

2. Results and Discussion

2.1. Simulation of P80 Micelle

2.1.1. Simulation of Single P80 Molecule in Aqueous Medium

For characterization of P80-PA interactions is the aqueous medium, it is first essential to gain a description of single and multiple molecular behaviors of each substance, separately. Simulation of a single P80 molecule was carried out by solvating it in 3500 CG water particles (0.5 w%), and running the production for 150 ns (Table 1, Simulation 1). Assuming that the mode of P80 hydration would influence its interaction with PAs, this simulation was carried out to characterize structuring of water around the surfactant molecule. As the radial distribution function (RDF) against water and hydration numbers of P80 moieties (Supplementary Material 1, Figure S1) indicates the hydration pattern represents the expected contrast in the density of water around the hydrophilic and hydrophobic domains.

2.1.2. Formation and Characteristics of P80 Micelle

We studied the self-assembly and properties of P80 micelle in the aqueous medium. A system of 10 w% P80 in water was configured by randomly distributing 60 P80 molecules in 9817 CG water particles, and the production simulation was run for 600 ns (Table 1, Simulation 2). The selected trajectory snapshots of P80 self-assembly are illustrated in Figure 1, and the corresponding animation is provided in Supplementary Materials 2, Video S1. As can be seen, the randomly dispersed surfactant molecules ($t = 0$) self-assemble to two distinct clusters within 0.1 ns. The individual clusters become more packed during the next 0.3 ns. The clusters are then fused to each other ($t = 1.0$ ns) to minimize water-hydrocarbon contact, leading to the formation of a micelle containing all individual surfactant molecules ($P80_{60}$). The plot of energy evolution (Supplementary Materials 1, Figure S2a) shows that the micelle reaches equilibrium state within around 3.0 ns, and remains energetically stable until the end of simulation.

Table 1. Summary of the study simulations.

Study Phase	Simulation	No.	Time (ns)	No. of Molecule(s)				Concentration (w%)			Ratio
				P80	AmB	Nys	Water Particles	P80	AmB	Nys	PA-to-P80
1: Simulation of P80	Single P80 molecule	1	150	1	-	-	3500	0.5	-	-	-
		2	600	60	-	-	9817	10	-	-	-
		3	600	100	-	-	16,361	10	-	-	-
		4	600	150	-	-	24,542	10	-	-	-
		5	600	60	-	-	4363	20	-	-	-
	Multiple P80 molecules	6	600	60	-	-	2545	30	-	-	-
		7	600	60	-	-	1636	40	-	-	-
		8	600	200	-	-	32,723	10	-	-	-
		9	600	200	-	-	14,544	20	-	-	-
		10	600	200	-	-	8484	30	-	-	-

Table 1. *Cont.*

Study Phase	Simulation	No.	Time (ns)	No. of Molecule(s)				Concentration (w%)			Ratio
				P80	AmB	Nys	Water Particles	P80	AmB	Nys	PA-to-P80
2: Simulation of PAs	Single AmB	11	150	-	1		410		3	-	
	Multiple AmB	12	150	-	9		11024	-	1	-	-
	Single Nys	13	150	-	-	1	410		-	3	
	Multiple Nys	14	150	-	-	9	11024	-	-	1	-
3: Simulation of P80-PA systems	Single AmB and single P80	15	150	1	1	-	410	4.1	2.9	-	0.7
	Single AmB encapsulation	16	150	60	1	-	9817	10	0.1	-	0.01
	Multiple AmB encapsulation	17	600	60	9	-	9817	10	1	-	0.1
		18	600	60	17	-	9600	10	2	-	0.2
		19	600	60	26	-	9485	10	3	-	0.3
		20	600	60	34	-	9382	10	4	-	0.4
		21	600	60	43	-	9817	10	5	-	0.5
		22	600	100	9	-	16,153	10	0.7	-	0.06
		23	600	150	9	-	24,426	10	0.4	-	0.04
		24	600	200	29	-	32,351	10	1	-	0.1
		25	600	200	14	-	14,364	20	1	-	0.05
		26	600	200	9	-	8368	30	1	-	0.03
	Single Nys and single P80	27	150	1	-	1	410	4.1	-	2.9	0.7
	Single Nys encapsulation	28	150	60	-	1	9817	10	-	0.1	0.01
	Multiple Nys encapsulation	29	600	60	-	9	9817	10	-	1	0.1
		30	600	60	-	17	9600	10	-	2	0.2
		31	600	60	-	26	9485	10	-	3	0.3
		32	600	60	-	34	9382	10	-	4	0.4
		33	600	60	-	43	9817	10	-	5	0.5
		34	600	100	-	9	16,153	10	-	0.7	0.06
		35	600	150	-	9	24,426	10	-	0.4	0.04
		36	600	200	-	29	32,351	10	-	1	0.1
		37	600	200	-	14	14,364	20	-	1	0.5
		38	600	200	-	9	8368	30	-	1	0.3

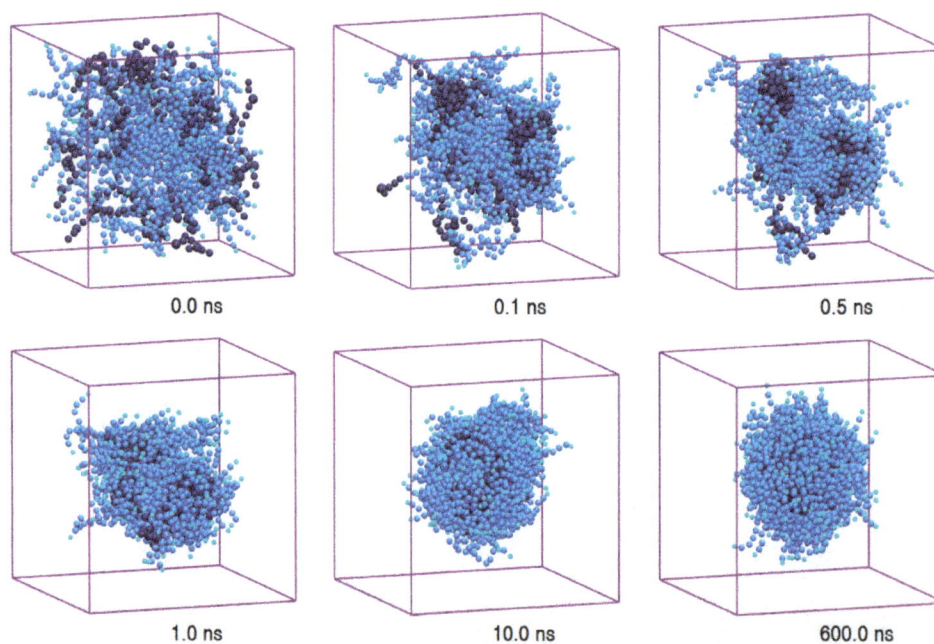

Figure 1. Trajectory snapshots of micellation of Polysorbate 80 (P80) in the aqueous medium at different simulation times (water particles have been removed for clarity).

The time evolution of radius of gyration (R_g) is plotted in Supplementary Materials 1, Figure S2b. The average R_g during the last 500 ns of the simulation was calculated to be 27.5 Å, which shows excellent match with the experimentally determined value of 27.2 Å [74], being obtained more accurately compared with the corresponding value in the CG simulation study of Amani et al. (26.2 Å) [75]. The R_g of the micelle core was 16.9 Å. Using the formula $R_m = \sqrt{\frac{5}{3}} R_g$ [76], these values give an effective micelle radius of 35.5 Å, with the micelle core radius (R_c) of 21.8 Å and the shell thickness (ST) of 13.7 Å. The shape of the P80 micelle was estimated based on the principal moments of inertia (PMI). The ratio of the average principal axes of inertia (PAI) was calculated to be 1.31:1.14:1.00 over the last 500 ns of the simulation. The eccentricity (e) of micelle was determined using the formula $e = \sqrt{1 - \frac{c^2}{a^2}}$, where c and a are the shortest and the longest computed semi-axes, respectively. The eccentricity value was obtained 0.65 which along with PAI ratios reveals the semi-spherical (ellipsoidal) shape of the micelle. The SASA of micelle and the hydrophilic % were estimated to be 556.1 nm^2 and 82%, respectively. Figure 2 displays the radial density functions for head and water from the micelle COM. As seen, the head group peaks around 26 Å and the water penetrates up to 17 Å from the COM.

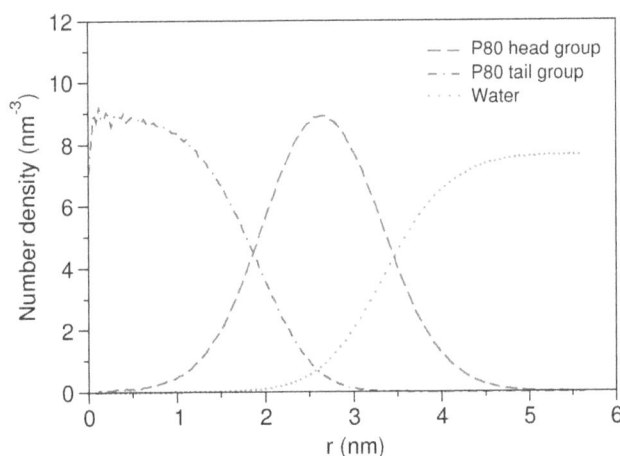

Figure 2. Density distribution of water and head and tail of P80 from micelle center of mass.

The micelle parameters including R_g, eccentricity, SASA, hydrophilic %, and head peak and water penetration radii are in good agreement with the corresponding values reported from atomic-scale simulation of the Tween 80 micelle of canonical structure with equal length of POE head branches at N_{ag} of 60 (27.4 Å, 0.63, 540 nm^2, 87.6%, 22 Å, 16 Å, respectively) [77], demonstrating the comparable accuracy of MARTINI model with atomic-scale simulation in representing our surfactant micellar system. Following the validation of model, we characterized some more properties of the P80 micelle, as described in details Supplementary Materials 1. In short, the micelle structure shows little dependence on N_{ag}, micelle hydrocarbon moieties are poorly hydrated (Supplementary Materials 1, Figure S3) consistent with observation from the density profile, and the calculated lateral diffusivity of the micelle (2.3 × 10^{-7} cm^2/s^{-1}), approaches well to the experimentally reported value of 3.0 × 10^{-7} cm^2/s^{-1} [78].

2.1.3. Effect of Surfactant Concentration on Micelle Size and Morphology

Experimental and in silico studies have shown that the surfactant micellar assemblies typically undergo size and morphological transition in response to the altered concentration [79–81]. The geometrical properties of surfactant carriers may impact their drug loading capacity and delivery efficiency [63,82]. We investigated the potential effect of surfactant concentration on the structural properties the P80 micelle, and the effect of system size and initial configuration on the aggregation behavior of the surfactant. Briefly, a sphere-to-cylinder shape transition (Supplementary Materials 1, Figure S4) and an increased size (Supplementary Materials 1, Figure S5) of the micelles with concentration augment were observed. These observations are consistent with the concentration-dependent structural transition of the non-ionic surfactants reported in previous experimental and *in silico* studies [79,80,83], supporting the validity of our modeling. In addition, the size of the largest micelle formed at various system sizes and initial configurations ranged around the value of 60 verifying both the assumed aggregation number of P80 and the robustness of the simulated micellar system against system size, at least within the simulation timeframe. More detailed description of the relationship between structural properties of P80 micelle and the concentration and the effect of system size on simulation results is provided in Supplementary Materials 1.

2.2. Simulation of AmB and Nys

2.2.1. Simulation of a Single PA Molecule in Water

To characterize hydration and dynamics of PAs in water, a single molecule of AmB and Nys was simulated in presence of 400 CG water particle (3 w%) separately, for 150-ns (Table 1, Simulations 11 and 13, respectively). Figure 3a,b present the RDFs for various moieties of the AmB and Nys against water, respectively. As seen, each of the PA moieties is surrounded by a hydration shell with different degrees of water structuring. The average number of water particles within the hydration shell around AmB (Nys) moieties was calculated to be 8.02 (8.13) for the carboxyl group, 7.69 (7.75) for the amine group, 4.52 (4.28) for the hydroxyl group, and 4.22 (4.22) for the polyene group, which is compliant with the expected order. The higher hydration intensity of the charged moieties (carboxyl and amine groups) reflects their comparatively strong tendency towards the polar solvent. However, one would expect to see a larger difference in hydration number between the hydroxyl and polyene groups given their hydrophilic and hydrophobic nature, respectively, which is not captured by the coarse-grained modeling, in contrast to the atomic-scale simulation [48]. The average distance between the polyol and polyene chains in water for both drug molecules was calculated to be 4.73 ± 0.07 Å, which is nearly equal to the van der Waals radius of the CG water particle. The average distance between the most distant sites on the lactone ring (beads 8 and 15) was calculated to be 14.9 ± 0.13 Å for both drug molecules.

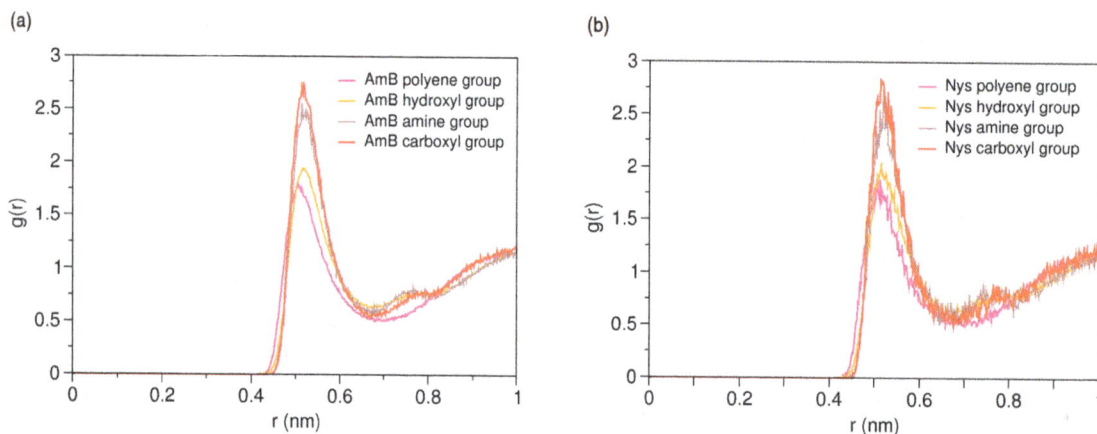

Figure 3. Average radial pair distribution functions for various moieties of (**a**) a single Amphotericin B (AmB) molecule; and (**b**) a single Nystatin (Nys) molecule against water particles.

The 3-dimensional and lateral diffusion constants of AmB were calculated to be 3.1×10^{-6} cm^2/s^{-1} and 4.6×10^{-6} cm^2/s^{-1}, respectively. The corresponding coefficients for Nys were estimated to be

$3.4 \times 10^{-6} \text{ cm}^2/\text{s}^{-1}$ and $4.9 \times 10^{-6} \text{ cm}^2/\text{s}^{-1}$. Although the lack of parallel experimental data does not allow comparison, the calculated diffusion constants can be used to compare the dynamics of PAs before and after entrapment in P80 micelle, as will be discussed below.

2.2.2. Aggregation of PAs in the Aqueous Solution

The toxic side effects of PAs are largely related to their presence in the aggregated form [17–19]. To monitor the drug self-association effect in the AmB-W and Nys-W solutions we extended the simulations of single PA molecule in water to an aqueous solution of 9 PA molecules at 1 w% (Table 1, Simulations 12 and 14, respectively). To consider the possible effect of initial system configuration on aggregation behavior, each simulation was carried out for 10 times at different random initial configurations. The selected trajectory snapshots from sample simulations of AmB and Nys in water are given in Figure 4 and Supplementary Materials 1, Figure S6, respectively. The trajectories show that self-association of PAs in water proceeds initially by formation of small oligomeric clusters, followed by association of these aggregates to form a large oligomeric assembly.

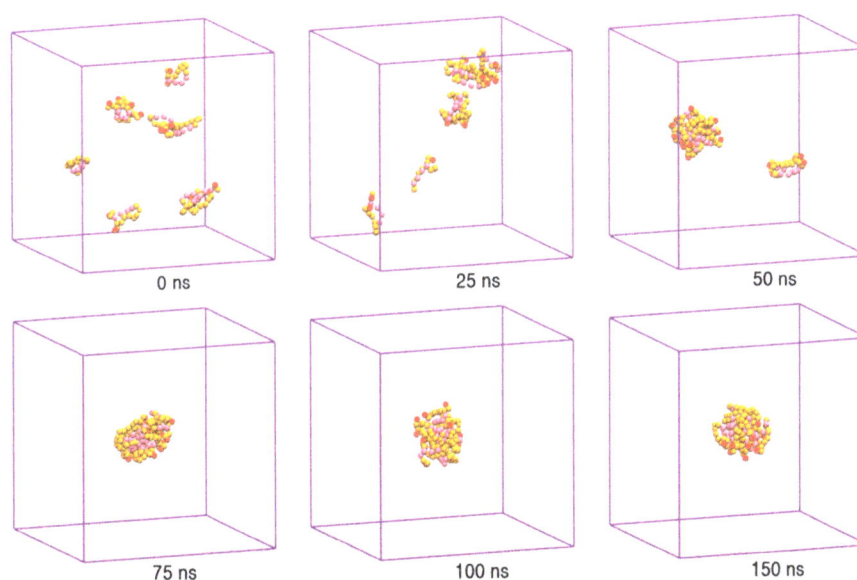

Figure 4. Trajectory snapshots of aggregation of AmB in the aqueous medium at different simulation times (water particles have been removed for clarity).

Aggregation of hydrophobic molecules in the aqueous media is known to be driven by unfavorable contacts between water and their hydrophobic domains [1,3,84]. It has been shown that as the self-association proceeds, the solvent accessible surface area (SASA) of the hydrophobic solute is constantly limited in a linear correlation with the unfavorable contacts [46,84,85]. Thereby, the SASA can be used for semi-quantitatively tracking the self-aggregation of non-soluble agents in the aqueous solvent. To quantify the progress of AmB and Nys self-association (1 w%), we calculated their SASA and the average number of water particles around each PA molecule within the first hydration shell (7 Å) over the simulation period. Figure 5a presents the SASA profile of PAs and Figure 5b shows average number of water particles around PA molecules observed from a sample simulation of 9 molecules of each drug at identical initial configurations. The SASA profiles exhibit declining trend with the simulation time, indicating that the self-aggregation proceeds by the tendency of PAs to protect their hydrophobic moieties from water.

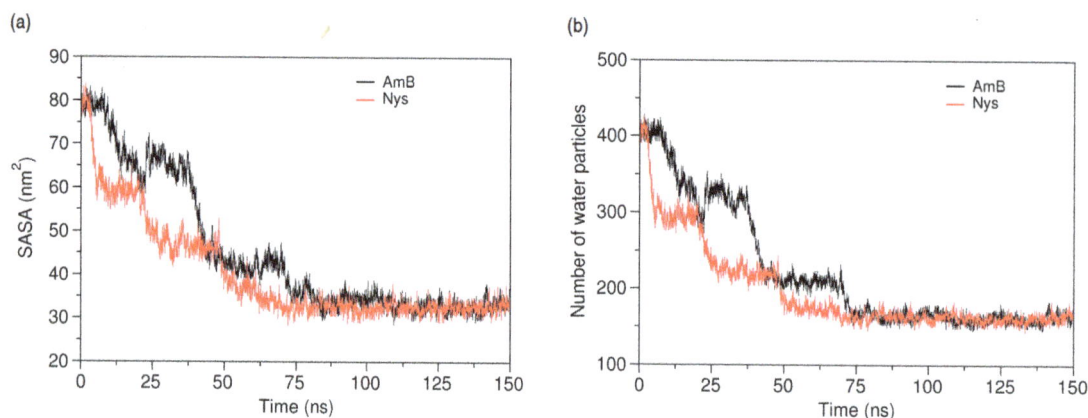

Figure 5. Comparison of the time-evolution of (**a**) solvent accessible surface area (SASA) of AmB and Nys; and (**b**) average number of water particles around AmB and Nys molecules within 7 Å, during self-association process.

This is confirmed by the progressive decrease of average number of water particles around each PA molecule (Figure 5b) as SASA becomes increasingly limited. The time for evolution of SASA of AmB (Nys) to equilibrium in 10 simulations at various initial configurations ranged from 75 ns (70 ns) to 210 ns (185 ns) and averaged at 127 ± 52 ns (113 ± 44 ns). The pattern of SASA variations indicates the stepwise formation of larger aggregates from the smaller ones, as also observed from the trajectories. The formation of the intermediate and final aggregates proceeds marginally faster for Nys compared with AmB. Although self-aggregation kinetics is sensitive to the initial system setup, considering the identical initial configuration of both AmB-W and Nys-W systems, the slightly faster aggregation of Nys may reflect its observed higher diffusivity.

Supplementary Materials 1 Figure S7 illustrates the variation of PA-PA and PA-W interaction energies during aggregation process, observed from the above-mentioned sample simulations. As seen, the interaction energy of each of PAs with itself and water falls and increases, respectively, to equilibrium level, suggesting gradual replacement of PA-W interactions by PA-PA interactions along the aggregation process. Despite the absence of hydrogen bond interactions which are cardinal to the hydrophobic effect [1,3,84], the above data confirm that CG simulations can mimic this effect sufficiently enough to capture the behavior of amphiphiles in aqueous medium.

2.3. Simulation of P80-PA Systems

2.3.1. Interaction of PAs with a Single P80 Molecule

The main purpose of this study was to characterize the interactions AmB and Nys with P80 in aqueous environment. This investigation was initiated by simulating the interaction of a single molecule of each drug in presence of a single P80 molecule in water (Table 1, Simulations 15 and 27, respectively). Radial pair distribution functions between various moieties of P80 and PAs can give a detailed account of drug-surfactant interactions. From RDFs for molecular groups of AmB and Nys against P80 (Figure 6a,b, respectively) it is seen that the polyene groups of PAs tend to have by far the highest level of interaction with the surfactant, compared with other moieties. Accordingly, the chromophores of PAs are preliminarily responsible for driving the interaction of these antibiotics with P80. The carboxyl group shows a little tendency to interact with the detergent, which is due to its preferred affinity for the aqueous surrounding (Figure 4).

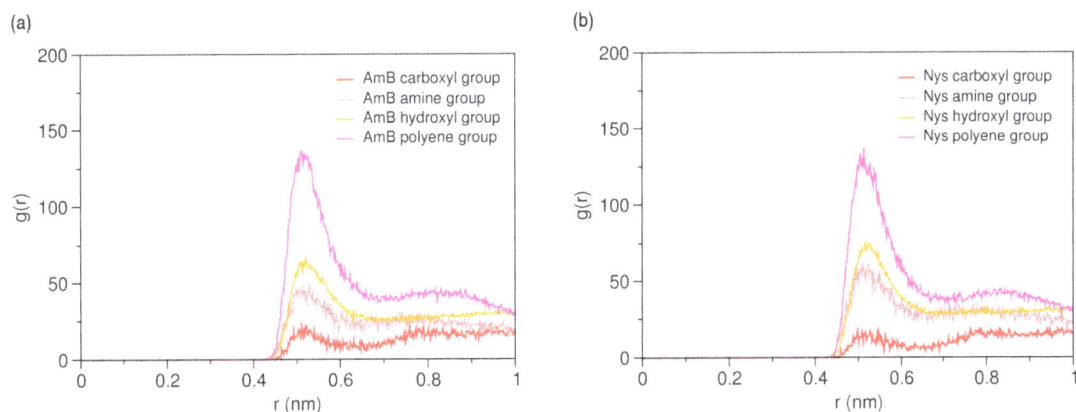

Figure 6. Average radial pair distribution functions for various moieties of (**a**) AmB; and (**b**) Nys against single P80 molecule in aqueous medium.

On the other hands, the RDFs of P80 head and tail against AmB and Nys (Figure 7a,b, respectively) reveals the comparable level of interactions of head and tail with PAs. Particularly, head and tail groups show very similar radial distribution pattern around Nys molecule. This is surprising given the larger population of head group interactions sites and the stronger affinity of drug molecules for the surfactant head group compared with the tail. This observation may in part be due to the lower density of the polar solvent nearby the hydrophobic alkyl chain, which would allow for fewer unfavorable interactions of PAs with water while interacting with the tail. Further possible explanation for this observation is the preferred integration of PAs with the branch 1 (beads 7–11) of the POE chain among all polyethylene oxide branches of P80, as described in Supplementary Materials 1 (Figure S8).

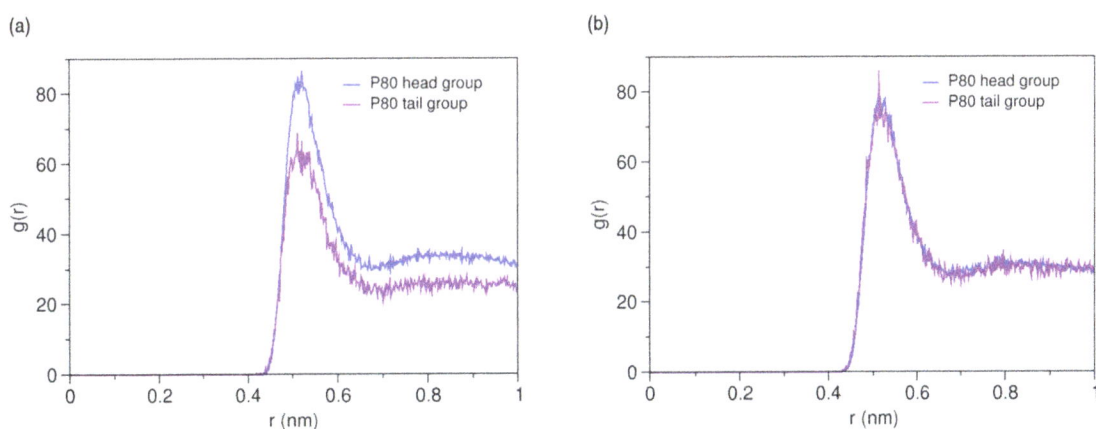

Figure 7. Average radial pair distribution functions for head and tail of P80 against (**a**) AmB; and (**b**) Nys in aqueous medium.

2.3.2. Encapsulation of PAs into P80 Micelle

Solubilizing capability of surfactants is largely dependent on their presence in micellar conformation [32,86–89]. We simulated the interaction of PAs with P80 micelle to explore the solubilization behaviors of these drugs, and properties of the mixed micelles form. Firstly, the interaction of a single PA molecule with a pre-formed equilibrium micelle of 60 P80 molecules (P80$_{60}$) was simulated (Table 1, Simulations 16 and 28, respectively) in order to characterize the hydration and dynamics of the drug within the micelle. After equilibrium was reached, the presence of PAs in micelles was confirmed by the sharp peak of P80 RDFs against AmB. The RDFs for different molecular groups of $P80_{60}$-AmB_1 and $P80_{60}$-Nys_1 mixed micelles against water are illustrated in Figure 8a,b, respectively.

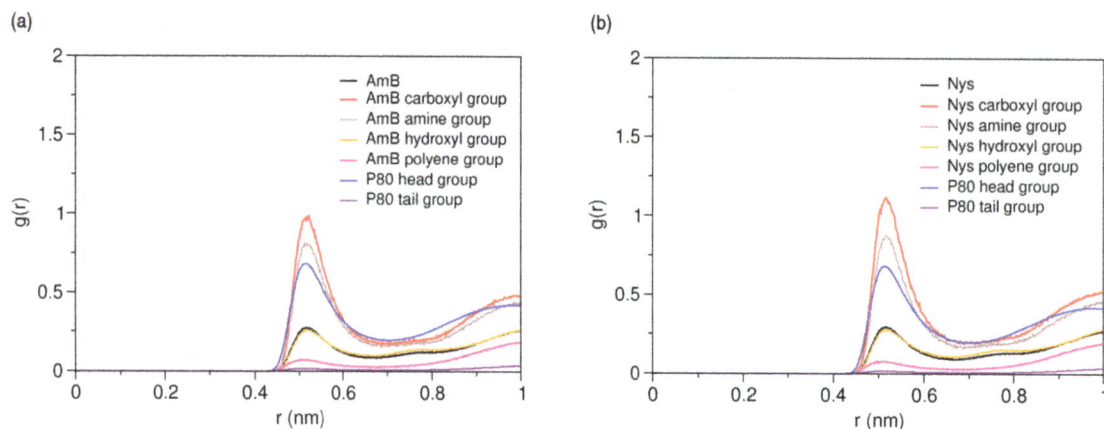

Figure 8. Average radial pair distribution functions for different molecular groups of (**a**) P80-AmB mixed micelle; and (**b**) P80-Nys mixed micelle against water particles.

The hydration level of PAs in the micelle falls in-between that of P80 head and tail groups. In comparison with the free state (Figure 5), the hydration of the polyol and polyene chains of PAs in the micelle is significantly limited. However, the charged groups of PAs are more strongly hydrated than the surfactant head. The above figure represents a competition between hydrophobic and hydrophilic moieties of PAs to modulate the interactions of the encapsulated drugs with water, leading to the dynamic residence of the PAs in micelle. To quantify the net dynamics of PAs within the micelle phase, the 3-dimentional and lateral diffusion coefficients of the entrapped drug molecules were calculated by eliminating contributions of translation of the COM and rotation of the micelle. Three-dimensional diffusion constant was estimated using Einstein equation (see Methods Section) by calculating the MSD of the remaining translational motion of PAs. For determining lateral diffusion constant, the MSD was calculated for projection of PAs displacements on the micelle spherical surface. The 3-dimensional and lateral diffusion constants for AmB were estimated 3.01×10^{-7} cm^2s^{-1} and 4.61×10^{-7}, respectively, whereas the corresponding values for Nys were obtained 3.06×10^{-7} cm^2s^{-1} and 4.38×10^{-7}. Comparison of these values with the respective values for free PA molecules in water (see Section 2.2.1) shows an order-of-magnitude decline in the freedom of motion of the PAs at encapsulated state. Such a suppression of dynamics, though not the only influencing factor, is important to the residence time of drugs in the micelle and prevention of rapid release [63,90,91]. We also calculated the average distance between the polyol and polyene chains of the encapsulated AmB (Nys) to be 4.79 ± 0.06 Å (4.78 ± 0.04 Å) and the average distance between the most distant sites on the macrolide ring (beads 8 and 15) to be 15.2 ± 0.11 Å (15.3 ± 0.11 Å). Both values are significantly larger than the corresponding values for free single AmB (Nys) molecule in water ($p < 0.01$), indicating that the structure of PAs is stretched in the P80 micelle. The attenuated effect of water repulsive forces and the augmented effect of surfactant attractive forces on the encapsulated drugs may be the reason.

The properties of pharmaceutical formulations emerge from the collective interaction of the drug molecules with the excipients [1–3,28,30,32,63]. To explore the collective interaction of AmB and Nys with P80 micelle, for each PA, a simulation setup consisted of 9 drug molecules in presence of 60 P80 molecules (1 w% of PA, 10 w% of P80, and drug-to-surfactant ratio of 0.1) was configured and run (Table 1, Simulations 17 and 27, respectively). To take into account the possible effect of initial system configuration, the simulation was carried for 10 times at different random initial configurations. Selected snapshots of the trajectories from sample simulations of $P80_{60}$-AmB_9-W and $P80_{60}$-Nys_9-W systems are presented in Figure 9 and Supplementary Materials 1, Figure S9, and the corresponding animations can be viewed from Supplementary Materials 2, Videos S2 and S3, respectively. The trajectories show a spontaneous and relatively rapid migration of PA monomers from the bulk towards the micelle phase; the time for incorporation of all AmB (Nys) molecules into the micelle ranged from 13 ns (10 ns) to 64 ns (60 ns) in 10 simulations. The average time for encapsulation of AmB (Nys) over 10 simulations was calculated to be 38.5 ± 18.4 ns (35.2 ± 16.4 ns) which is significantly ($p < 0.01$) lower than the average time for aggregation of the drug molecules (127 ± 51.8 ns (113 ± 44 ns)) at the same concentration.

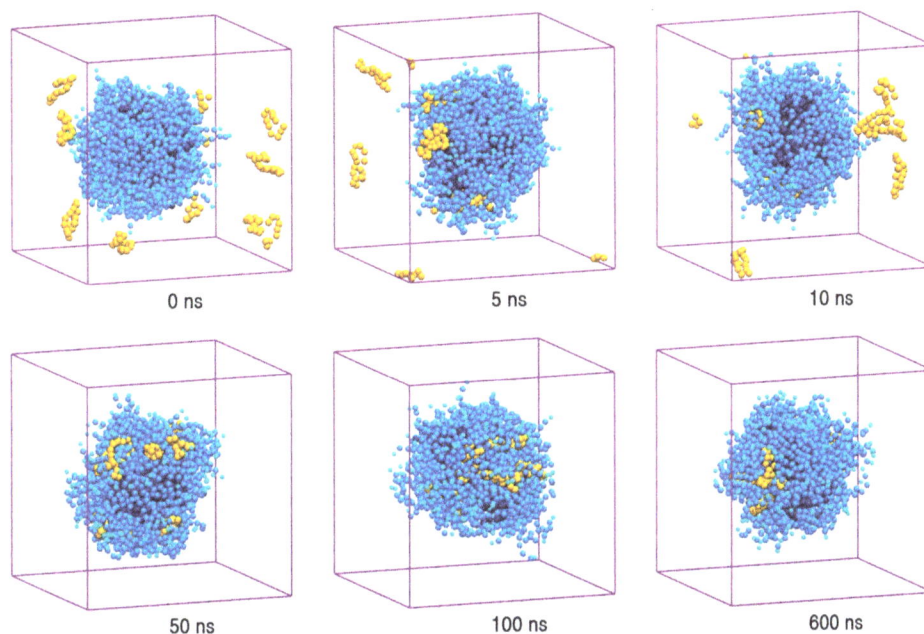

Figure 9. The trajectory snapshots of encapsulation of AmB molecules into P80 micelle at different simulation times (water particles have been removed for clarity).

The dominance of encapsulation kinetics over aggregation kinetics is crucial for inhibiting drug aggregation and achieving high drug loading efficiency [92]. Therefore, the observed faster speed of PAs encapsulation than that of their aggregation, at the same drug concentration, indicates the potential of P80 to solubilize PAs in water. Profiling SASA of AmB and Nys over simulation time (Figure 10a) reveals that SASA of AmB (Nys) significantly increases ($p < 0.05$) from the 80.85 ± 2.33 nm^2 (80.91 ± 2.42 nm^2) corresponding to free monomeric state of 9 drug molecules in water to 83.02 ± 2.27 nm^2 (83.05 ± 2.49 nm^2) corresponding to the encapsulated state of the drug. The SASA increase would be due to the aforementioned expanded conformation of PAs in the micelle. Such an increase in SASA represents the capability of P80 for monomeric solubilization of PAs.

Figure 10b compares the changes in the average number of water particles surrounding AmB and Nys molecules, within 7 Å. As can be seen, the number of water particles around PAs dramatically decreases during the encapsulation of the drugs. From comparison of Figures 5b and 10b, it is revealed that the number of water particles around the AmB (62.13) and Nys (62.12) in the micelle is lower (by approximately 62%) than that in their respective aggregates (162.8 and 162.6, respectively). Hence, the encapsulated state of PAs allows significantly more water protection for the drugs than their aggregated state. This finding together with faster kinetics of PAs encapsulation than aggregation and the monomeric solubilization of PAs in the micelle consistently reflects the experimentally documented capability of P80 micelles to remarkably increase the aqueous solubility of Nys [27], and explains the underlying molecular-level mechanism.

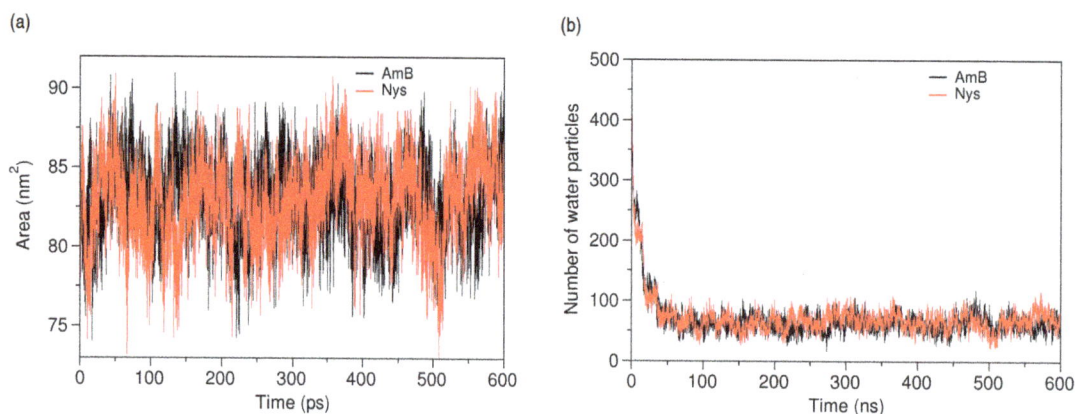

Figure 10. (**a**) Solvent accessible surface area of AmB and Nys molecules; and (**b**) average number of water particles around AmB and Nys molecules within 7 Å, during and after encapsulation into the P80 micelle.

Further insight into the forces controlling the encapsulation behaviors of PAs may be obtained from analysis of interaction energies in the P80-PA-W systems. As illustrated in Figure 11a,b, respectively, the average energy of AmB-AmB and Nys-Nys interactions remains unchanged before and after encapsulation, confirming the monomeric solubilization of these drugs, already implicated from SASA profiles. The increasing energy of AmB-W and Nys-W interactions during encapsulation depicts the solubilization process as a hydrophobic phenomenon. The equilibrium energy of PA-W interactions remains below zero (~-500 KJ/mol), suggesting that PAs tend to preserve a certain level of interaction with water. Considering the diminishing availability of water in the direction of micelle COM (Figure 3), this tendency translates into the restricted insertion of PAs into the micelle (see below for further discussion). The energy profiles also show the progressive substitution of PA-P80 interactions with PA-W ones, as the encapsulation proceeds. The absolute variation in the energy of PA-P80 interactions is approximately 1000 KJ/mol larger than that of PA-W ones, indicating that PAs establish more stable interactions with the surfactant as compared with water. Therefore, the encapsulation of PAs into the P80 micelle is not only driven by the tendency of drugs to escape from water, but by their stronger interactions with P80 compare with water as well. While contribution of drug-surfactant interactions to micellar solubilization and their stabilizing effect on the incorporated solutes has been mostly reported in polymeric surfactant solutions [32,93], our observations provide evidence for this to be the case in detergent micellar systems, as well.

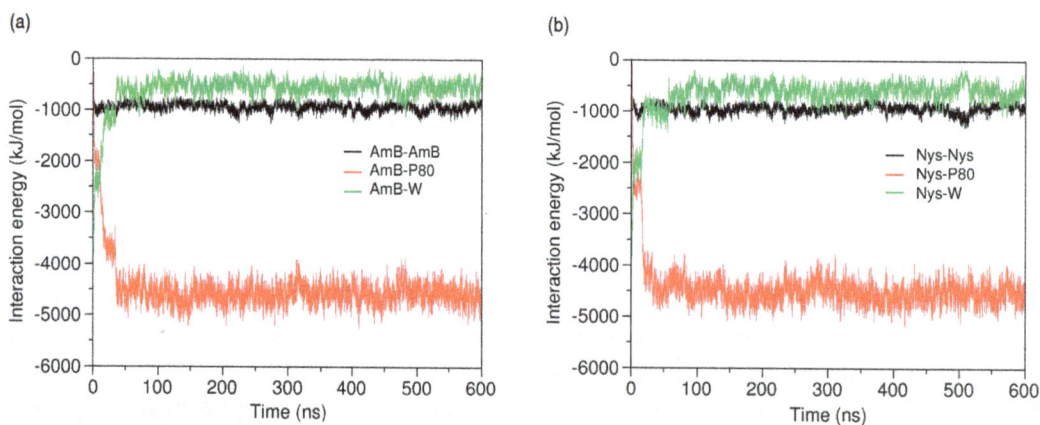

Figure 11. Time-evolution of average energy of interactions (**a**) between AmB and AmB, P80, and water; and (**b**) between Nys and Nys, P80, and water.

2.3.3. Structure of P80-PA nanomicelles

Structural properties of $P80_{60}$-AmB_9 and $P80_{60}$-Nys_9 nanomicelles were investigated by analysis of the corresponding simulation trajectories. The R_g of the both micelles was calculated to be 27.8 Å, which is slightly (0.3 Å) larger than that of drug-free micelle. The cores of mixed micelles have a R_g of 16.9 Å which is identical to that of the drug-free micelle. Based on this data, the effective radius of both mixed micelles is determined to be 35.9 Å with a R_c 21.8 Å and ST of 14.1 Å.

More details on the internal structure of P80-PA nanomicelles can be obtained by analysis of the radial density of their components. The density profiles for $P80_{60}$-AmB_9 and $P80_{60}$-Nys_9 micelles are presented in Figure 12a,b, respectively. As can be seen, the entrapped PAs are partitioned between the shell and the core-shell interface of the micelle, with most of the drugs preferentially located in the interface. This observation agrees well with the experimentally reported solubilization of Nys in the core-corona interface of P80 micelles [27]. Additionally, the close density peaks of shell and PAs together with the complete surround of PAs by the shell and the broad tail of shell in the core suggests that the presence of drugs in the core-dominant region of the interface is largely mediated by the dynamics of core-shell interactions.

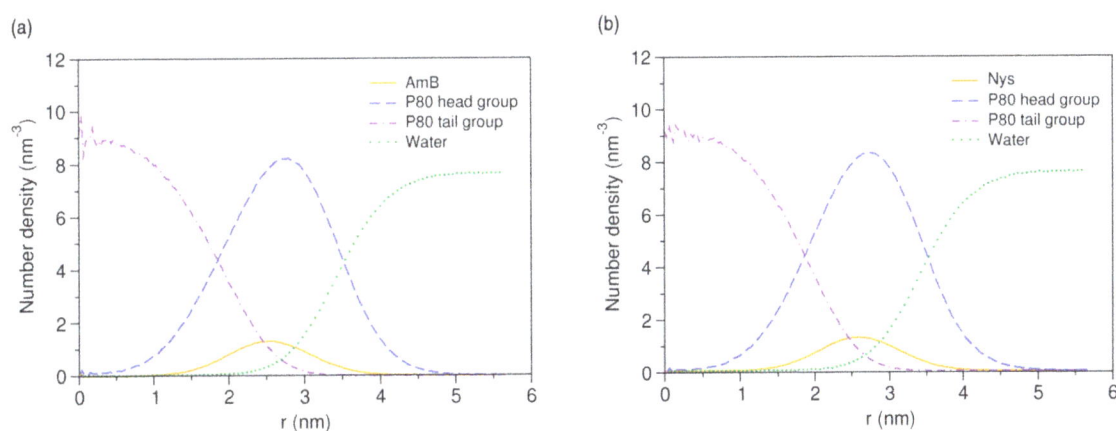

Figure 12. Radial density distributions for various molecular groups of (**a**) P80-AmB mixed micelle; and (**b**) P80-Nys mixed micelle.

The average location of both PAs is determined to be 27 Å from the micelle COM. Given the abovementioned geometrical characteristics of the mixed micelles, the drugs are located on average 5.2 Å away from the core surface and 8.9 Å before the shell surface. The ratio of drug distance from the COM to the micelle radius is 0.75 which shows the relatively shallow insertion of PAs into the micelle. In addition, the density profile displays the wide radial distribution of PAs ranging from 12 to 42 Å from COM, indicating that a portion of the drug molecules frequently leave the micelle up to 6.1 Å and re-renters to it (as could also be seen from the trajectories). All these observations may be interpreted as the limited capability of P80 for stable solubilization of PAs and possible burst release or precipitation of the drugs.

Location of drugs in the matrix of pharmaceutical carriers is an important parameter in developing micellar drug delivery systems, as it affects various aspects of the formulation efficiency, including the drug loading capacity, formulation stability, and drug release behavior [39,94,95]. From the energetics points of view, the limited diffusion of PAs into the P80 micelle can be attributed to three possible factors: (1) the repulsive interactions between the drugs and the core; (2) attractive interactions between drugs and the shell; and (3) attractive interactions between drugs and water. The first hypothesis is ruled out by the findings from simulation of single-P80-single-PA interactions, where the drugs showed comparable level of interaction with both head and tail groups. Regarding the second hypothesis, the broad radial distribution of head moieties and the thickness of micelle head imply that PAs have enough room to insert deeper into the micelle and still largely interact with the corona, undermining the role of drug-surfactant interaction as a limiting factor for drug diffusion. Thereby, the affinity for water remains the only energetic factor significantly constraining radial insertion of the drugs. As such, the depth at which PAs may penetrate in the micelle will inevitably depend on the permeability of water across the micelle. The implications of this conclusion for the stability of PAs micellar formulation will be discussed in the Section 2.4.

The localization behavior of PAs can also be understood in terms of their topological properties. Molecular length, branching, size, shape, and structure have been shown to affect both preferred localization site and degree of solublization of the solubilizates [86]. Recently, from a systematic DPD simulation study, Guo et al. [94] observed that the insertion of large and/or linearly shaped molecules into the micelle core is more challenging as compared with that of the small and/or branched molecules, due to their lower diffusion constant and the larger space required for penetration. Given that AmB and Nys are characterized by a relatively large and linear topology, the crowding of the chains and the lack of space in the regions closer to the core can hinder deep penetration of these molecules.

The extent to which the topological factors may contribute to the localization behavior of AmB and Nys may depend on their orientation in the micelle. To interrogate this issue, the orientation of these PAs at encapsulated state was characterized by calculating the angle between polyene chromophore chain of each drug molecule and the micelle radius. This was done by calculating the average distribution of angles between the vector extending from micelle COM to bead 7 and the vector between the beads 7 and 4 of each drug molecule. Figure 13 compares the frequency distribution of these angles between AmB and Nys. The average of angle distribution for AmB appears at around 95°, indicating that the hydrophobic face of the drug orients virtually perpendicular to the radius of the micelle. The inclination as high as 5° toward the surface is representative of higher affinity of the carboxyl group for water compared with the hydroxyl chain. The angle distribution of Nys averages at 98°, which is marginally (3°) higher than the corresponding value for AmB. This can be explained by the closer position of the P4 particle to the carboxyl group in Nys compared with AmB, making the polar head of Nys more inclined towards the polar solvent.

The perpendicular orientation of PAs polyene face with respect to micelle radius reveals the tendency of PAs to maintain their hydrophilic and hydrophobic interactions balanced with the aqueous environment, demonstrating that the encapsulation behavior of PAs is fine-tuned by water influence. Considering the rectangular shape of PAs, the observed angular position of the drug molecules corresponds to the widest lateral space they may occupy in the micelle which is already packed along the orthogonal direction. This mode of orientation together with the hydrophilic interactions of the drugs with water hamper deep and firm residence of PAs in the micelle in a synergistic manner, further elucidating the molecular-level challenges to stable micellar solubilization AmB and Nys.

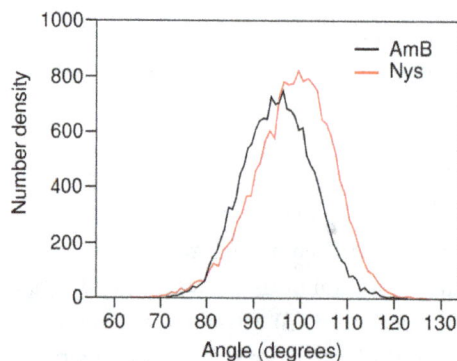

Figure 13. Bead4-Bead7-COM angle distribution for the encapsulated AmB and Nys.

2.3.4. Micellar Solvation of PAs

Characterizing the local environment around drug molecule in carrier using MD simulations may provide insight into solvation behavior of the drug in the excipient medium [96]. Figure 14a compares the RDFs for AmB against P80 head and tail groups at the solubilized state and the corresponding illustration for Nys is given in Figure 14b. The sharp peak of P80 hydrophilic head in these figures shows that the drug molecules are densely surrounded by the POE-chain-dominated corona. Conversely, lack of a clear peak on RDF plots of P80 hydrophobic tail is indicative of poor concentration of hydrocarbon groups around the drug molecules. The RDF analysis hence suggests that AmB and Nys are almost exclusively solvated in an aqueous POE medium provided within the micelle. The poor interaction of PAs with the P80 tail in the micelle is not due to the lack of miscibility, as it was already seen from simulation of single-PA-single-P80 systems that the AmB and Nys can interact with both POE and hydrocarbon domains at comparable levels. The lack of such interactions at encapsulated state would result rather from the distance between the preferred location of PAs and the hydrocarbon moieties concentrated in the core, preventing effective intermolecular interactions. The impact of distance becomes more apparent by comparing the R_g of AmB (25.5 Å) and Nys (25.7 Å) at encapsulated state with that of micelle hydrophobic chains (16.5 Å). The average distance of PAs from the core was calculated to be 5 Å which remained steady throughout the simulation period. These observations predict limited effect of drug-core compatibility on solubilization of PAs in detergents similar to P80, due to distance related lack of drug interaction with the core. This hypothesis finds support by the work of Croy and Kwon [27] in which Nys exhibited no significant micelle/water partitioning difference between P80 and a structurally similar surfactant, the Cremophor EL, with lower core polarity compared with P80.

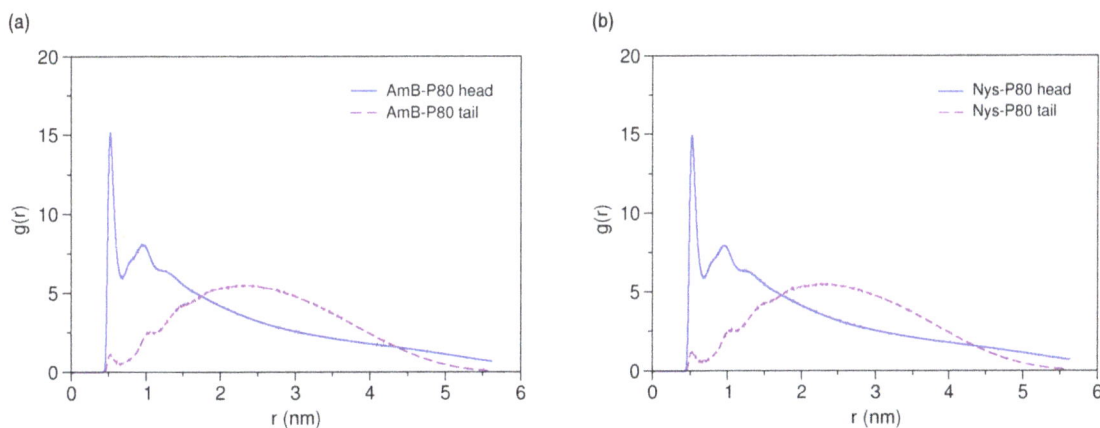

Figure 14. Radial distribution functions for head and tail of P80 against (**a**) AmB; and (**b**) Nys in $P80_{60}$-AmB_9 and $P80_{60}$-Nys_9 mixed micelles, respectively.

2.3.5. Complementary Analysis of Properties of P80-PA Micellar Systems

We further analyzed the properties of P80-PA nanomicellar solution by characterizing the effect of PAs on the conformational properties of P80 micelles (Table 1, Simulations 17–21 and 29–33), the effect of drug load and micelle size on localization of PAs (Table 1, Simulations 17–21 and Simulations 29–33), the effect of PAs on the stability of P80 micellar system (Table 1, Simulations 24–26 and Simulations 36–38), pattern of PAs distribution in micellar solution (Table 1, Simulation 24), and the drug loading capacity of P80 as a carrier of PAs (Table 1, Simulations 17–21 and Simulations 29–33). Briefly, conformational analysis of P80-PA micelle indicated the structural robustness of P80 micelle against a wide range of drug load; no significant impact of drug load and micelle size on localization of the drugs was identified; the stability of P80 multi-micellar system was found to be sensitive to drug load; PAs exhibited a heterogeneous distribution among micelles; and P80 micelle showed a relatively high capacity to encapsulate PAs at monomeric state. Detailed description of the above investigations is provided in Supplementary Material 1 (Figures S10 and S11).

2.4. Implications for Drug Toxicity and Formulation

Experimental data have shown a significant capability of P80 as a solubilizing carrier of PAs, either in single use or in combination with other excipients [27,33,66,70]. Our simulation study both confirms and rationalizes this potential by indicating the dominant kinetics of drugs encapsulation over self-association, lower hydration of the drugs at encapsulated state compared with aggregated state, monomeric solubilization of the drugs in the micelle, stabilization of the incorporated drugs by drug-surfactant interactions, significantly reduced diffusion of the entrapped drugs, high loading capacity of the micelle, and stability of micelle structure against drug loads.

Despite these advantages, some drawbacks of P80-based solubilization of PAs were also revealed. These include limited penetration of PAs towards the core and their preferred localization at the interface, the broad distribution of drug molecules over micelle radius, orientation of PAs corresponded to the largest lateral occupation space, limited interaction of PAs with the core, and sensitivity of micellar system stability to drug introduction. Most of these disadvantages are associated with the kinetic stability of the P80-PA mixed micelles, which would influence the drug release behavior. Specifically, the residence of drugs at the corona or core-corona interface has been associated with the burst release which negatively affects delivery performance [90,97,98] and may also induce local or systemic side effects [90,99–102]. Burst release has been reported in some micellar preparations of AmB [103–105] and has been associated with entrapment of this drug in shell or core-shell interface [105]. Burst release of AmB and Nys is particularly important regarding the dose-dependent nature of most side-effects of these drugs, including nephrotoxicity [106]. Therefore, an effective P80-based formulation of a PAs entails complementary strategies for suppressing the envisioned premature release.

The persistent affinity of PAs for water in the P80 micelle matrix renders localization of these drugs directly dependent on the extents of water diffusion. Because the deep interior of the conventional detergent micelles is typically void of water [107–109], the amphiphilic drugs have to remain sufficiently close to the surface to preserve their required interactions with the aqueous solvent. This may not only limit the residence time of drug in the micelle due to the short path of outward diffusion, but may also frustrate the contribution of drug-core interactions to the solubilization, as observed in this theoretical and previous experimental studied [27], due to the distance of the preferred location of the drug from the core. These notions brings up the idea that controlled promotion of micelle permeability may help deeper penetration of the amphiphilic drugs by enabling interaction of drug with water at higher depths. This hypothesis is corroborated by the evidence that block-copolymeric micelles which have proven superior over detergents in incorporating PAs into the core [29,104,106,110,111] and controlling their aggregation state drug during release [112–114], also allow penetration of water up to their cores [107,115]. In this view, water permeability of micelle may be regarded as a mechanism for passive delivery of amphiphiles to the core region, whereby drug-core interaction if adequately strong, could exert its solubilizing/stabilizing effect. Further confirmation of this hypothesis in future studies may introduce new approaches to stable solubilization of PAs in conventional non-toxic detergents.

Pharmaceutical formulations are often based on combination solubilizing strategies rather than relying only on single surfactant, to realize the potential solubilizing synergies of multiple agents. A notable exception is the Fungizone®, in which a single detergent surfactant (SDC) is used to solubilize AmB, leading thus to the administration of the formulation being associated with dose-dependent adverse effects [9]. The study of Tancrèdef et al. also reported a synergistic toxic effect of AmB when administered with P80 as the only excipient [116]. Our observation in silico that the solubilizer-drug interactions contribute to stabilization of the drugs signifies the notion that using co-surfactants or stabilizers in detergent-based formulations of PAs may yield more control over drug release behavior, which is poor in preparations such as Fungizone® [112].

Our results also revealed the significance of orientation-induced lack of lateral space for penetration of PAs. Therefore, consideration is due to the crowding of the chains in the palisade layer in formulation design for these relatively large rectagularly-shaped drug molecules. At least theoretically, both permeability enhancement and chain crowing control are procurable by tuning the surfactant hydrophobic to hydrophilic chain.

3. Experimental Section

3.1. Study Design

Although the particular objective of the present work was to investigate interaction of PAs with P80 in the aqueous medium, a detailed description of the molecular behavior of each substance in water (W) is required for interpretation of P80-PA-W systems simulation results. Therefore, a study was designed in three phases to address (i) simulation of P80; (ii) simulation of PAs; (iii) simulation of P80-PAs interactions in aqueous medium. Due to the very large computer demand otherwise, all simulations were carried out far above the critical micelle concentration (CMC) of both of the surfactant and drugs. Table 1 summarizes the simulation setups in each phase of the study.

3.2. Modeling of P80 and PAs

The MARTINI coarse-graining method introduced by Marrink *et al.* [71,72] was adopted to develop CG models of P80 and PAs. The CG mapping of P80 molecular structure is presented in Figure 15a. P80 was modeled by 30 CG sites, 24 of which representing the hydrophilic head group (N- and P-type particles), and 6 representing the hydrophobic tail (C-type particles). The CG mapping of AmB and Nys molecular structures is illustrated in Figure 15b,c respectively. PAs were represented by 17 CG sites; 4 of which representing polyene chain (C-type particles), 2 for charged groups (Qda-type particle), 8 for hydroxyl groups (P-type particles) and 3 for other groups (C and N particle types). The detailed description of the CG mapping of P80 and PAs is given in the Supplementary Materials 1. The CG models of both surfactant and drug molecules were initially parameterized according to the default MARTINI force field parameters and then optimized to reproduce the structural properties of the corresponding united-atom models, including distances, angles, and dihedrals.

3.3. Simulation and Analysis

MD simulations and the subsequent analyses were carried out using GROMACS v. 4.6.3 simulation package [117] and the simulation trajectories were visualized using VMD [118]. The full description of simulation details and methods of analysis of the simulation trajectories are provided in the Supplementary Materials 1.

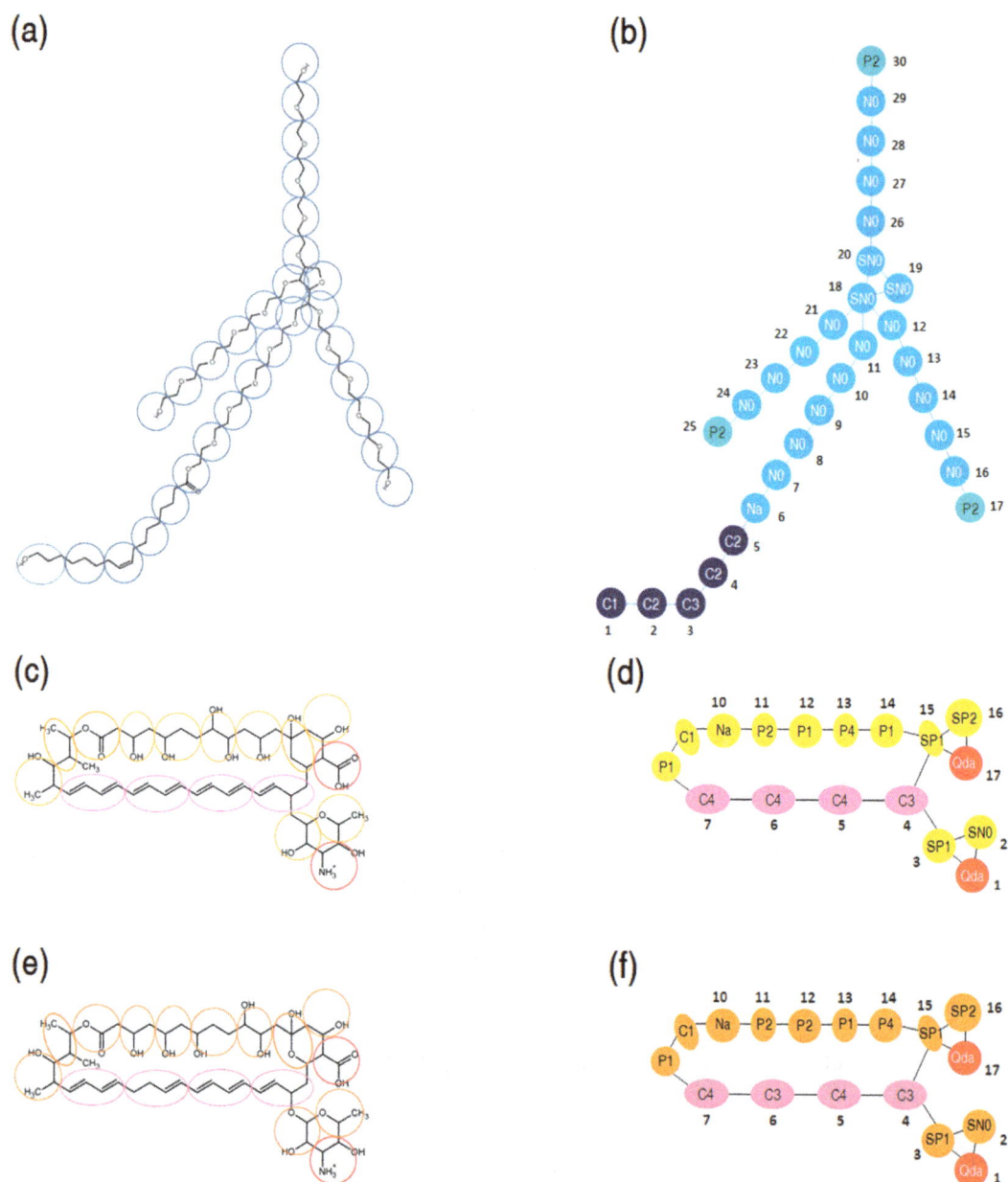

Figure 15. (a) Course-grain mapping of P80; (b) coarse-grain model of P80, comprising 30 interaction sites: 21 sites for the poly-oxyethylene (POE) chain, 5 for alkyl chain, and 3 for terminal polar groups and 1 for the ester group; (c,e) coarse-grain mapping of AmB and Nys, respectively; (d,f) coarse-grain model of AmB and Nys, respectively, with 17 interaction sites: 4 sites for the polyene chain, 1 for carboxyl group, 1 for amine group, and 11 for the hydroxyl and other groups.

4. Conclusions

Molecular knowledge of drug interaction with the excipients is crucial for overcoming the challenge of amphiphilic drug solubilization. Our CG MD simulation study provided a series of relevant clues for understanding the solubilization behaviors of PAs in micellar system and their underling mechanisms. Based on the obtained results, although the encapsulation process is predominantly driven by hydrophobic forces, various aspects of PAs solubilization including dynamics, hydration, localization, orientation, and solvation are largely influenced by the drug's hydrophilic interactions with water. Our simulations confirmed the experimentally evident capability of P80 for solubilizing PAs, and rationalized it by (i) dominant kinetics of drugs encapsulation over self-association; (ii) lower hydration of the drugs at encapsulated state compared with aggregated state; (iii) monomeric solubilization of drugs; (iv) stabilization of the incorporated drugs via drug-surfactant interactions; (v) significantly reduced diffusion of the drugs at entrapped state compared with free sate in water; (vi) high loading capacity of the micelle; and (vii) stability of micelle structure against drug overload. On the other hands, consistent with experimental data, PAs were found to preferentially localize at the core-shell interface of P80 micelle and exhibit a broad radial distribution extending beyond the micelle radius. This localization behavior was elucidated to stem from the combined effects of PAs partial affinity for water, limited micellar penetration of water, and perpendicular orientation of PAs with respect to the micelle radius. In addition, PAs were distributed heterogeneously among the micelles and the stability of multi-micellar system was found to be sensitive to the incorporation of the drugs. These data imply that in spite of high solubilizing potential of P80, stable solubilization of PAs in P80 micelle is a challenging task requiring combination approach.

The observation that encapsulation behavior of PAs is largely dependent on their interaction with water suggests that the dearth of water in interior of the P80 micelle may be a major factor restricting deep and stable encapsulation of PAs. Hence, the controlled promotion of micelle permeability may enable interaction of PAs with water at deeper sites, thereby their improved interactions with the core. Further confirmation of this hypothesis may introduce novel strategies for stable detergent-based solubilization of the amphiphilic drugs. Moreover, the particular orientation of PAs which corresponds to their largest lateral occupation space highlights the limiting effect of chain crowing on penetration of these relatively large and rectangularly-shaped drugs. Ideally, both permeability enhancement and chain crowding control are approachable by modulating the surfactant hydrophobic to hydrophilic chains.

By characterizing both advantages and drawbacks of P80 as a potential carrier of PAs, addressing a wide range of solubilization-related phenomena consistent with experimental results, and generating experimentally testable hypotheses for drug solubilization, our study demonstrates that MD simulations even at CG level can be used as an efficient explorative tool in pharmaceutical formulation engineering.

Acknowledgments: The authors would like to appreciate Tofigh Daru Research and Engineering Company (TODACO, Tehran, Iran) for financial support of this study.

Author Contributions: Mahmoud Reza Jaafari and Hossein Attar jointly conceived the study. Meysam Mobasheri made the major contribution to conduction of the study, including building the molecular models, performing the simulations, analyzing and interpreting the results, and drafting the manuscript. Mahmoud Reza Jaafari, Hossein Attar, Seyed Mehdi Rezayat Sorkhabadi, and Ali Khamesipour participated in interpretation of the data and drafting the manuscript. All authors read and approved the final manuscript.

Conflicts of Interest: The authors declare no conflict of interests.

References

1.   Schreier, S.; Malheiros, S.V.P.; de Paula, E. Surface active drugs: Self-association and interaction with membranes and surfactants. Physicochemical and biological aspects. *Biochim. Biophys. Acta Biomembr.* **2000**, *1508*, 210–234. [CrossRef]

2.   Kim, S.; Shi, Y.; Kim, J.Y.; Park, K.; Cheng, J.-X. Overcoming the barriers in micellar drug delivery: Loading efficiency, *in vivo* stability, and micelle–cell interaction. *Expert. Opin. Drug Deliv.* **2009**, *7*, 49–62. [CrossRef] [PubMed]

3.   Messina, P.; Besada-Porto, J.; Ruso, J. Self-assembly drugs: From micelles to nanomedicine. *Curr. Top. Med. Chem.* **2014**, *14*, 555–571. [CrossRef] [PubMed]

4.   Kagan, S.; Ickowicz, D.; Shmuel, M.; Altschuler, Y.; Sionov, E.; Pitusi, M.; Weiss, A.; Farber, S.; Domb, A.J.; Polacheck, I. Toxicity mechanisms of amphotericin B and its neutralization by conjugation with arabinogalactan. *Antimicrob. Agents Chemother.* **2012**, *56*, 5603–5611. [CrossRef] [PubMed]

5.   Kleinberg, M. What is the current and future status of conventional amphotericin B? *Int. J. Antimicrob. Agents* **2006**, 2712–2716. [CrossRef] [PubMed]

6.   Sundar, S. Drug resistance in Indian visceral leishmaniasis. *Trop. Med. Int. Health.* **2001**, *6*, 849–854. [CrossRef] [PubMed]

7.   Sundar, S.; Arora, R.; Singh, S.P.; Boelaert, M.; Varghese, B. Household cost-of-illness of visceral leishmaniasis in Bihar, India. *Trop. Med. Int. Health.* **2010**, *15*, 50–54. [CrossRef] [PubMed]

8.   Chattopadhyay, A.; Jafurulla, M. A novel mechanism for an old drug: Amphotericin B in the treatment of visceral leishmaniasis. *Biochem. Biophys. Res. Commun.* **2011**, *416*, 7–12. [CrossRef] [PubMed]

9.   Hamill, R. Amphotericin B Formulations: A comparative review of efficacy and toxicity. *Drugs* **2013**, *73*, 919–934. [CrossRef] [PubMed]

10. Hamilton-Miller, J.M. Chemistry and biology of the polyene macrolide antibiotics. *Bacteriol. Rev.* **1973**, *37*, 166–196. [PubMed]

11. Semis, R.; Nili, S.S.; Munitz, A.; Zaslavsky, Z.; Polacheck, I.; Segal, E. Pharmacokinetics, tissue distribution and immunomodulatory effect of intralipid formulation of nystatin in mice. *J. Antimicrob. Chemother.* **2012**, *67*, 1716–1721. [CrossRef] [PubMed]

12. Carrillo-Muñoz, A.J.; Quindós, G.; Tur, C.; Ruesga, M.T.; Miranda, Y.; Valle, O.d.; Cossum, P.A.; Wallace, T.L. *In-vitro* antifungal activity of liposomal nystatin in comparison with nystatin, amphotericin B cholesteryl sulphate, liposomal amphotericin B, amphotericin B lipid complex, amphotericin B desoxycholate, fluconazole and itraconazole. *J. Antimicrob. Chemother.* **1999**, *44*, 397–401. [CrossRef] [PubMed]

13. Arikan, S.; Ostrosky-Zeichner, L.; Lozano-Chiu, M.; Paetznick, V.; Gordon, D.; Wallace, T.; Rex, J.H. *In vitro* activity of nystatin compared with those of liposomal nystatin, amphotericin B, and fluconazole against clinical candida isolates. *J. Clin. Microbiol.* **2002**, *40*, 1406–1412. [CrossRef] [PubMed]

14. Johnson, E.M.; Ojwang, J.O.; Szekely, A.; Wallace, T.L.; Warnock, D.W. Comparison of *in vitro* antifungal activities of free and liposome-encapsulated nystatin with those of four amphotericin B formulations. *Antimicrob. Agents Chemother.* **1998**, *42*, 1412–1416. [PubMed]

15. Larson, J.L.; Wallace, T.L.; Tyl, R.W.; Marr, M.C.; Myers, C.B.; Cossum, P.A. The reproductive and developmental toxicity of the antifungal drug Nyotran® (liposomal nystatin) in rats and rabbits. *Toxicol. Sci.* **2000**, *53*, 421–429. [CrossRef] [PubMed]

16. Lemke, T.L.; Williams, D.A. *Foye's Principles of Medicinal Chemistry*, 6th ed.; Wolters Kluwer Health: Philadelphia, PA, USA, 2012; p. 1115.

17. Barwicz, J.; Christian, S.; Gruda, I. Effects of the aggregation state of amphotericin B on its toxicity to mice. *Antimicrob. Agents Chemother.* **1992**, *36*, 2310–2315. [CrossRef] [PubMed]

18. Barwicz, J.; Tancrède, P. The effect of aggregation state of amphotericin-B on its interactions with cholesterol- or ergosterol-containing phosphatidylcholine monolayers. *Chem. Phys. Lipids.* **1997**, *85*, 145–155. [CrossRef]

19. Espada, R.; Valdespina, S.; Alfonso, C.; Rivas, G.; Ballesteros, M.P.; Torrado, J.J. Effect of aggregation state on the toxicity of different amphotericin B preparations. *Int. J. Pharm.* **2008**, *361*, 64–69. [CrossRef] [PubMed]

20. Legrand, P.; Romero, E.A.; Cohen, B.E.; Bolard, J. Effects of aggregation and solvent on the toxicity of amphotericin B to human erythrocytes. *Antimicrob. Agents Chemother.* **1992**, *36*, 2518–2522. [CrossRef] [PubMed]

21. Wasko, P.; Luchowski, R.; Tutaj, K.; Grudzinski, W.; Adamkiewicz, P.; Gruszecki, W.I. Toward Understanding of Toxic Side Effects of a polyene antibiotic amphotericin B: Fluorescence spectroscopy reveals widespread formation of the specific supramolecular structures of the drug. *Mol. Pharm.* **2012**, *9*, 1511–1520. [CrossRef] [PubMed]

22. Gruszecki, W.I.; Gagoś, M.; Hereć, M. Dimers of polyene antibiotic amphotericin B detected by means of fluorescence spectroscopy: Molecular organization in solution and in lipid membranes. *J. Photochem. Photobiol. B Biol.* **2003**, *69*, 49–57. [CrossRef]

23. Barwicz, J.; Gruszecki, W.I.; Gruda, I. Spontaneous organization of amphotericin B in aqueous medium. *J. Colloid. Interface Sci.* **1993**, *158*, 71–76. [CrossRef]

24. Gruszecki, W.I.; Gagoś, M.; Hereć, M.; Kernen, P. Organization of antibiotic amphotericin B in model lipid membranes. A mini review. *Cell. Mol. Biol. Lett.* **2003**, *8*, 161–170. [PubMed]

25. Coutinho, A.; Prieto, M. Self-association of the polyene antibiotic nystatin in dipalmitoylphosphatidylcholine vesicles: A time-resolved fluorescence study. *Biophys. J.* **1995**, *69*, 2541–2557. [CrossRef]

26. Bolard, J.; Legrand, P.; Heitz, F.; Cybulska, B. One-sided action of amphotericin B on cholesterol-containing membranes is determined by its self-association in the medium. *Biochemistry* **1991**, *30*, 5707–5715. [CrossRef] [PubMed]

27. Croy, S.R.; Kwon, G.S. Polysorbate 80 and Cremophor EL micelles deaggregate and solubilize nystatin at the core-corona interface. *Int. J. Pharm.* **2005**, *94*, 2345–2354. [CrossRef] [PubMed]

28. Brajtburg, J.; Bolard, J. Carrier effects on biological activity of amphotericin B. *Clin. Microbiol. Rev.* **1996**, *9*, 512–531. [PubMed]

29. Croy, S.R.; Kwon, G.S. The effects of Pluronic block copolymers on the aggregation state of nystatin. *J. Control. Release* **2004**, *95*, 161–171. [CrossRef] [PubMed]

30. Narang, A.S.; Delmarre, D.; Gao, D. Stable drug encapsulation in micelles and microemulsions. *Int. J. Pharm.* **2007**, *345*, 9–25. [CrossRef] [PubMed]

31. Strickley, R. Solubilizing excipients in oral and injectable formulations. *Pharm. Res.* **2004**, *21*, 201–230. [CrossRef] [PubMed]

32. Williams, H.D.; Trevaskis, N.L.; Charman, S.A.; Shanker, R.M.; Charman, W.N.; Pouton, C.W.; Porter, C.J.H. Strategies to address low drug solubility in discovery and development. *Pharmacol. Rev.* **2013**, *65*, 315–499. [CrossRef] [PubMed]

33. Esposito, E.; Bortolotti, F.; Menegatti, E.; Cortesi, R. Amphiphilic association systems for Amphotericin B delivery. *Int. J. Pharm.* **2003**, *260*, 249–260. [CrossRef]

34. Pham, T.T.H.; Loiseau, P.M.; Barratt, G. Strategies for the design of orally bioavailable antileishmanial treatments. *Int. J. Pharm.* **2013**, *454*, 539–552. [CrossRef] [PubMed]

35. Lasic, D. Mixed micelles in drug delivery. *Nature* **1992**, *355*, 279–280. [CrossRef] [PubMed]

36. El-Ridy, M.S.; Abdelbary, A.; Essam, T.; Abd EL-Salam, R.M.; Aly Kassem, A.A. Niosomes as a potential drug delivery system for increasing the efficacy and safety of nystatin. *Drug Dev. Ind. Pharm.* **2011**, *37*, 1491–1508. [CrossRef] [PubMed]

37. Racles, C.; Mares, M.; Sacarescu, L. A polysiloxane surfactant dissolves a poorly soluble drug (nystatin) in water. *Colloids. Surf. A. Physicochem. Eng. Asp.* **2014**, *443*, 233–239. [CrossRef]

38. Li, Y.; Hou, T. Computational simulation of drug delivery at molecular level. *Curr. Med. Chem.* **2010**, *17*, 4482–4491. [CrossRef] [PubMed]

39. Huynh, L.; Neale, C.; Pomès, R.; Allen, C. Computational approaches to the rational design of nanoemulsions, polymeric micelles, and dendrimers for drug delivery. *Nanomedicine* **2012**, *8*, 20–36. [CrossRef] [PubMed]

40. Wang, S.; Zhou, Y.; Tan, J.; Xu, J.; Yang, J.; Liu, Y. Computational modeling of magnetic nanoparticle targeting to stent surface under high gradient field. *Comput. Mech.* **2014**, *53*, 403–412. [CrossRef] [PubMed]

41. Haddish-Berhane, N.; Rickus, J.L.; Haghighi, K. The role of multiscale computational approaches for rational design of conventional and nanoparticle oral drug delivery systems. *Int. J. Nanomed.* **2007**, *2*, 315–331.

42. Ahmad, S.; Johnston, B.F.; Mackay, S.P.; Schatzlein, A.G.; Gellert, P.; Sengupta, D.; Uchegbu, I.F. *In silico* modelling of drug–polymer interactions for pharmaceutical formulations. *Interface Focus* **2010**, *7*, S423–S433. [CrossRef] [PubMed]

43. Abedi Karjiban, R.; Basri, M.; Abdul Rahman, M.B.; Salleh, A.B. Molecular dynamics simulation of palmitate ester self-assembly with diclofenac. *Int. J. Mol. Sci.* **2012**, *13*, 9572–9583. [CrossRef] [PubMed]

44. Prakash, P.; Sayyed-Ahmad, A.; Zhou, Y.; Volk, D.E.; Gorenstein, D.G.; Dial, E.; Lichtenberger, L.M.; Gorfe, A.A. Aggregation behavior of ibuprofen, cholic acid and dodecylphosphocholine micelles. *Biochim. Biophys. Acta Biomembr.* **2012**, *1818*, 3040–3047. [CrossRef] [PubMed]

45. Luo, Z.; Jiang, J. pH-sensitive drug loading/releasing in amphiphilic copolymer PAE–PEG: Integrating molecular dynamics and dissipative particle dynamics simulations. *J. Control. Release* **2012**, *162*, 185–193. [CrossRef] [PubMed]

46. Wang, X.-Y.; Zhang, L.; Wei, X.-H.; Wang, Q. Molecular dynamics of paclitaxel encapsulated by salicylic acid-grafted chitosan oligosaccharide aggregates. *Biomaterials* **2013**, *34*, 1843–1851. [CrossRef] [PubMed]

47. Benson, S.P.; Pleiss, J. Molecular dynamics simulations of self-emulsifying drug-delivery systems (SEDDS): Influence of excipients on droplet nanostructure and drug localization. *Langmuir* **2014**, *30*, 8471–8480. [CrossRef] [PubMed]

48. Mazerski, J.; Borowski, E. Molecular dynamics of amphotericin B I. Single molecule in vacuum and water. *Biophys. Chem.* **1995**, *54*, 49–60. [CrossRef]

49. Mazerski, J.; Borowski, E. Molecular dynamics of amphotericin B. II. Dimer in water. *Biophys. Chem.* **1996**, *57*, 205–217. [CrossRef]

50. Czub, J.; Neumann, A.; Borowski, E.; Baginski, M. Influence of a lipid bilayer on the conformational behavior of amphotericin B derivatives—A molecular dynamics study. *Biophys. Chem.* **2009**, *141*, 105–116. [CrossRef] [PubMed]

51. Baginski, M.; Resat, H.; McCammon, J.A. Molecular properties of amphotericin b membrane channel: A molecular dynamics simulation. *Mol. Pharmacol.* **1997**, *52*, 560–570. [PubMed]

52. Silberstein, A. Conformational analysis of amphotericin B—cholesterol channel complex. *J. Membr. Biol.* **1998**, *162*, 117–126. [CrossRef]

53. Czub, J.; Baginski, M. Modulation of amphotericin B membrane interaction by cholesterol and ergosterola molecular dynamics study. *J. Phys. Chem. B* **2006**, *110*, 16743–16753. [CrossRef] [PubMed]

54. Sternal, K.; Czub, J.; Baginski, M. Molecular aspects of the interaction between amphotericin B and a phospholipid bilayer: Molecular dynamics studies. *J. Mol. Model.* **2004**, *10*, 223–232. [PubMed]

55. Baginski, M.; Resat, H.; Borowski, E. Comparative molecular dynamics simulations of amphotericin B-cholesterol/ergosterol membrane channels. *Biochim. Biophys. Acta Biomembr.* **2002**, *1567*, 63–78. [CrossRef]

56. Baran, M.; Mazerski, J. Molecular modelling of amphotericin B–ergosterol primary complex in water. *Biophys. Chem.* **2002**, *95*, 125–133. [CrossRef]

57. Anachi, R.B.; Bansal, M.; Easwaran, K.R.K.; Namboodri, K.; Gaber, B.P. Molecular modeling studies on amphotericin B and its complex with phospholipid. *J. Biomol. Struct. Dyn.* **1995**, *12*, 957–970. [CrossRef] [PubMed]

58. Neumann, A.; Czub, J.; Baginski, M. On the possibility of the amphotericin B-Sterol complex formation in cholesterol- and ergosterol-containing lipid bilayers: A molecular dynamics study. *J. Phys. Chem. B* **2009**, *113*, 15875–15885. [CrossRef] [PubMed]

59. Neumann, A.; Baginski, M.; Winczewski, S.; Czub, J. The effect of sterols on amphotericin B self-aggregation in a lipid bilayer as revealed by free energy simulations. *Biophys. J.* **2013**, *104*, 1485–1494. [CrossRef] [PubMed]

60. Czub, J.; Borowski, E.; Baginski, M. Interactions of amphotericin B derivatives with lipid membranes—A molecular dynamics study. *Biochim. Biophys. Acta Biomembr.* **2007**, *1768*, 2616–2626. [CrossRef] [PubMed]

61. Baran, M.; Borowski, E.; Mazerski, J. Molecular modeling of amphotericin B-ergosterol primary complex in water II. *Biophys. Chem.* **2009**, *141*, 162–168. [CrossRef] [PubMed]

62. Resat, H.; Sungur, F.; Baginski, M.; Borowski, E.; Aviyente, V. Conformational properties of amphotericin B amide derivatives—impact on selective toxicity. *J. Comput. Aided Mol.* **2000**, *14*, 689–703.

63. Loverde, S.M.; Klein, M.L.; Discher, D.E. Nanoparticle shape improves delivery: Rational coarse grain molecular dynamics (rCG-MD) of taxol in worm-like PEG-PCL micelles. *Adv. Mater.* **2012**, *24*, 3823–3830. [CrossRef] [PubMed]

64. Borowski, E.; Zieliński, J.; Ziminski, T.; Falkowski, L.; Kołodziejczyk, P.; Golik, J.; Jereczek, E.; Adlercreutz, H. Chemical studies with amphotericin B III. The complete structure of the antibiotic. *Tetrahedron Lett.* **1970**, *11*, 3909–3914. [CrossRef]

65. Iman, M.; Huang, Z.; Szoka, F.C., Jr.; Jaafari, M.R. Characterization of the colloidal properties, *in vitro* antifungal activity, antileishmanial activity and toxicity in mice of a distigmasterylhemisuccinoyl-glycero-phosphocholine liposome-intercalated amphotericin B. *Int. J. Pharm.* **2011**, *408*, 163–172. [CrossRef] [PubMed]

66. Silva, A.E.; Barratt, G.; Chéron, M.; Egito, E.S.T. Development of oil-in-water microemulsions for the oral delivery of amphotericin B. *Int. J. Pharm.* **2013**, *454*, 641–648. [CrossRef] [PubMed]

67. Hussain, A.; Samad, A.; Nazish, I.; Ahmed, F.J. Nanocarrier-based topical drug delivery for an antifungal drug. *Drug Dev. Ind. Pharm.* **2014**, *40*, 527–541. [CrossRef] [PubMed]

68. Hussain, A.; Samad, A.; Singh, S.K.; Ahsan, M.N.; Haque, M.W.; Faruk, A.; Ahmed, F.J. Nanoemulsion gel-based topical delivery of an antifungal drug: *In vitro* activity and *in vivo* evaluation. *Drug Deliv.* **2015**, 1–16.

69. Wasan, E.K.; Bartlett, K.; Gershkovich, P.; Sivak, O.; Banno, B.; Wong, Z.; Gagnon, J.; Gates, B.; Leon, C.G.; Wasan, K.M. Development and characterization of oral lipid-based Amphotericin B formulations with enhanced drug solubility, stability and antifungal activity in rats infected with Aspergillus fumigatus or Candida albicans. *Int. J. Pharm.* **2009**, *372*, 76–84. [CrossRef] [PubMed]

70. Bhattacharyya, A.; Bajpai, M. Oral bioavailability and stability study of a self-emulsifying drug delivery system (SEDDS) of amphotericin B. *Curr. Drug Deliv.* **2013**, *10*, 542–547. [CrossRef] [PubMed]

71. Marrink, S.J.; Risselada, H.J.; Yefimov, S.; Tieleman, D.P.; de Vries, A.H. The MARTINI force field: Coarse grained model for biomolecular simulations. *J. Phys. Chem. B* **2007**, *111*, 7812–7824. [CrossRef] [PubMed]

72. López, C.A.; Rzepiela, A.J.; de Vries, A.H.; Dijkhuizen, L.; Hünenberger, P.H.; Marrink, S.J. Martini coarse-grained force field: Extension to carbohydrates. *J. Chem. Theory. Comput.* **2009**, *5*, 3195–3210. [CrossRef] [PubMed]

73. Risselada, H.J.; Marc, F.; Xavier, P.; Siewert, J.M. The MARTINI force field. In *Coarse-Graining of Condensed Phase and Biomolecular Systems*; Voth, G.A., Ed.; CRC Press: Boca Raton, FL, USA, 2008; pp. 5–19.

74. Kumari, H.; Kline, S.R.; Atwood, J.L. Aqueous solubilization of hydrophobic supramolecular metal-organic nanocapsules. *Chem. Sci.* **2014**, *5*, 2554–2559. [CrossRef]

75. Amani, A.; York, P.; de Waard, H.; Anwar, J. Molecular dynamics simulation of a polysorbate 80 micelle in water. *Soft Matter* **2011**, *7*, 2900–2908. [CrossRef]

76. Chen, S.H. Small angle neutron scattering studies of the structure and interaction in micellar and microemulsion systems. *Annu. Rev. Phys. Chem.* **1986**, *37*, 351–399. [CrossRef]

77. Tang, X.; Huston, K.J.; Larson, R.G. Molecular Dynamics simulations of structure–property relationships of tween 80 surfactants in water and at interfaces. *J. Phys. Chem. B* **2014**, *118*, 12907–12918. [CrossRef] [PubMed]

78. Verbrugghe, M.; Cocquyt, E.; Saveyn, P.; Sabatino, P.; Sinnaeve, D.; Martins, J.C.; Van der Meeren, P. Quantification of hydrophilic ethoxylates in polysorbate surfactants using diffusion NMR spectroscopy. *J. Pharm. Biomed. Anal.* **2010**, *51*, 583–589. [CrossRef] [PubMed]

79. Velinova, M.; Sengupta, D.; Tadjer, A.V.; Marrink, S.-J. Sphere-to-rod transitions of nonionic surfactant micelles in aqueous solution modeled by molecular dynamics simulations. *Langmuir* **2011**, *27*, 14071–14077. [CrossRef] [PubMed]

80. Al-Anber, Z.A.; Bonet i Avalos, J.; Floriano, M.A.; Mackie, A.D. Sphere-to-rod transitions of micelles in model nonionic surfactant solutions. *J. Chem. Phys.* **2003**, *118*, 3816–3826. [CrossRef]

81. Rosen, M.J. *Surfactants and Interfacial Phenomena*, 2nd ed.; Wiley-Interscience: New York, NY, USA, 1989; pp. 108–143.

82. Christian, D.A.; Cai, S.; Garbuzenko, O.B.; Harada, T.; Zajac, A.L.; Minko, T.; Discher, D.E. Flexible filaments for *in vivo* imaging and delivery: Persistent circulation of filomicelles opens the dosage window for sustained tumor shrinkage. *Mol. Pharm.* **2009**, *6*, 1343–1352. [CrossRef] [PubMed]

83. Shinoda, W.; DeVane, R.; Klein, M.L. Coarse-grained molecular modeling of non-ionic surfactant self-assembly. *Soft Matter* **2008**, *4*, 2454–2462. [CrossRef]

84. Chandler, D. Interfaces and the driving force of hydrophobic assembly. *Nature* **2005**, *437*, 640–647. [CrossRef] [PubMed]

85. Stephenson, B.C.; Goldsipe, A.; Blankschtein, D. Molecular dynamics simulation and thermodynamic modeling of the self-assembly of the triterpenoids asiatic acid and madecassic acid in aqueous solution. *J. Phys. Chem. B* **2008**, *112*, 2357–2371. [CrossRef] [PubMed]

86. Attwood, D.; Florence, A.T. *Surfactant Systems: Their Chemistry, Pharmacy and Biology*, 1st ed.; Chapman and Hall: Orange, CA, USA, 1983; pp. 229–381.

87. Mahato, R.I.; Narang, A.S. *Pharmaceutical Dosage Forms and Drug Delivery*, 2nd ed.; CRC Press: Boca Raton, FL, USA, 2011; pp. 179–195.

88. Han, K.; Miah, J.; Shanmugam, S.; Yong, C.; Choi, H.-G.; Kim, J.; Yoo, B. Mixed micellar nanoparticle of amphotericin b and poly styrene-block-poly ethylene oxide reduces nephrotoxicity but retains antifungal activity. *Arch. Pharm. Res.* **2007**, *30*, 1344–1349. [PubMed]

89. Espuelas, M.S.; Legrand, P.; Cheron, M.; Barratt, G.; Puisieux, F.; Devissaguet, J.P.; Irache, J.M. Interaction of amphotericin B with polymeric colloids: A spectroscopic study. *Colloids. Surf. B Biointerfaces* **1998**, *11*, 141–151. [CrossRef]

90. Sun, Q.; Radosz, M.; Shen, Y. Challenges in design of translational nanocarriers. *J. Control. Release* **2012**, *164156*–164169. [CrossRef] [PubMed]

91. Amiji, M.M. *Nanotechnology for Cancer Therapy*; CRC Press: Boca Raton, FL, USA, 2006; pp. 315–356.

92. Cai, K.; He, X.; Song, Z.; Yin, Q.; Zhang, Y.; Uckun, F.M.; Jiang, C.; Cheng, J. Dimeric drug polymeric nanoparticles with exceptionally high drug loading and quantitative loading efficiency. *J. Am. Chem. Soc.* **2015**, *137*, 3458–3461. [CrossRef] [PubMed]

93. Choucair, A.; Eisenberg, A. Interfacial solubilization of model amphiphilic molecules in block copolymer micelles. *J. Am. Chem. Soc.* **2003**, *125*, 11993–12000. [CrossRef] [PubMed]

94. Guo, X.D.; Qian, Y.; Zhang, C.Y.; Nie, S.Y.; Zhang, L.J. Can drug molecules diffuse into the core of micelles? *Soft Matter* **2012**, *8*, 9989–9995. [CrossRef]

95. Lavasanifar, A.; Samuel, J.; Kwon, G.S. Poly(ethylene oxide)-block-poly(l-amino acid) micelles for drug delivery. *Adv. Drug Deliv. Rev.* **2002**, *54*, 169–190. [CrossRef]

96. Xiang, T.-X.; Anderson, B.D. Liposomal drug transport: A molecular perspective from molecular dynamics simulations in lipid bilayers. *Adv. Drug Deliv. Rev.* **2006**, *58*, 1357–1378. [CrossRef] [PubMed]

97. Teng, Y.; Morrison, M.E.; Munk, P.; Webber, S.E.; Procházka, K. Release Kinetics studies of aromatic molecules into water from block polymer micelles. *Macromolecules* **1998**, *31*, 3578–3587. [CrossRef]

98. Bromberg, L.; Magner, E. Release of hydrophobic compounds from micellar solutions of hydrophobically modified polyelectrolytes. *Langmuir* **1999**, *15*, 6792–6798. [CrossRef]

99. Wang, J.; Mongayt, D.; Torchilin, V.P. Polymeric micelles for delivery of poorly soluble drugs: Preparation and anticancer activity *in vitro* of paclitaxel incorporated into mixed micelles based on poly(ethylene glycol)-lipid conjugate and positively charged lipids. *J. Drug Target* **2005**, *13*, 73–80. [CrossRef] [PubMed]

100. Khalid, M.; Simard, P.; Hoarau, D.; Dragomir, A.; Leroux, J.-C. Long circulating poly(ethylene glycol)-decorated lipid nanocapsules deliver docetaxel to solid tumors. *Pharm. Res.* **2006**, *23*, 752–758. [CrossRef] [PubMed]

101. Mitragotri, S.; Burke, P.A.; Langer, R. Overcoming the challenges in administering biopharmaceuticals: Formulation and delivery strategies. *Nat. Rev. Drug Discov.* **2014**, *13*, 655–672. [CrossRef] [PubMed]

102. Mashayekhi, R.; Mobedi, H.; Najafi, J.; Enayati, M. *In-vitro/In-vivo* comparison of leuprolide acetate release from an in-situ forming plga system. *Daru* **2013**, *21*, 57. [CrossRef] [PubMed]

103. Vakil, R.; S. Kwon, G. Effect of cholesterol on the release of Amphotericin B from PEG-phospholipid micelles. *Mol. Pharm.* **2008**, *5*, 98–104. [CrossRef] [PubMed]

104. Yang, Z.L.; Li, X.R.; Yang, K.W.; Liu, Y. Amphotericin B-loaded poly(ethylene glycol)–poly(lactide) micelles: Preparation, freeze-drying, and *in vitro* release. *J. Biomed. Mater. Res. A* **2008**, *85A*, 539–546. [CrossRef] [PubMed]

105. Kumar, V.; Gupta, P.K.; Pawar, V.K.; Verma, A.; Khatik, R.; Tripathi, P.; Shukla, P.; Yadav, B.; Parmar, J.; Dixit, R.; *et al.* *In-Vitro* and *In-Vivo* Studies on Novel Chitosan-g-pluronic F-127 copolymer based nanocarrier of Amphotericin B for improved antifungal activity. *J. Biomater. Tissue Eng.* **2014**, *4*, 210–216. [CrossRef]

106. Jee, J.-P.; McCoy, A.; Mecozzi, S. Encapsulation and release of Amphotericin B from an ABC triblock fluorous copolymer. *Pharm. Res.* **2012**, *29*, 69–82. [CrossRef] [PubMed]

107. Gadelle, F.; Koros, W.J.; Schechter, R.S. Solubilization of aromatic solutes in block copolymers. *Macromolecules* **1995**, *28*, 4883–4892. [CrossRef]

108. Podo, F.; Ray, A.; Nemethy, G. Structure and hydration of nonionic detergent micelles. High resolution nuclear magnetic resonance study. *J. Am. Chem. Soc.* **1973**, *95*, 6164–6171. [CrossRef]

109. Dill, K.A.; Flory, P.J. Molecular organization in micelles and vesicles. *Proc. Natl. Acad. Sci. USA* **1981**, *78*, 676–680. [CrossRef] [PubMed]

110. Yu, B.G.; Okano, T.; Kataoka, K.; Kwon, G. Polymeric micelles for drug delivery: Solubilization and haemolytic activity of amphotericin B. *J. Control. Release* **1998**, *53*, 131–136. [CrossRef]

111. Vandermeulen, G.; Rouxhet, L.; Arien, A.; Brewster, M.E.; Préat, V. Encapsulation of amphotericin B in poly(ethylene glycol)-block-poly(ε-caprolactone-co-trimethylenecarbonate) polymeric micelles. *Int. J. Pharm.* **2006**, *309*, 234–240. [CrossRef] [PubMed]

112. Adams, M.; Kwon, G.S. Spectroscopic investigation of the aggregation state of amphotericin B during loading, freeze-drying, and reconstitution of polymeric micelles. *J. Pharm. Pharm. Sci.* **2004**, *7*, 1–6. [PubMed]

113. Lavasanifar, A.; Samuel, J.; Sattari, S.; Kwon, G. Block Copolymer micelles for the encapsulation and delivery of amphotericin B. *Pharm. Res.* **2002**, *19*, 418–422. [CrossRef] [PubMed]

114. Adams, M.L.; Kwon, G.S. Relative aggregation state and hemolytic activity of amphotericin B encapsulated by poly(ethylene oxide)-block–poly(N-hexyl-l-aspartamide)-acyl conjugate micelles: Effects of acyl chain length. *J. Control. Release* **2003**, *87*, 23–32. [CrossRef]

115. Hezaveh, S. Study the Interaction Mechanisms of Block Copolymers with Biological Interfaces. Ph.D. Thesis, The Jacobs University, Bremen, Germany, 2012.

116. Tancrède, P.; Barwicz, J.; Jutras, S.; Gruda, I. The effect of surfactants on the aggregation state of amphotericin B. *Biochim. Biophys. Acta* **1990**, *1030*, 289–295. [CrossRef]

117. Berendsen, H.J.C.; van der Spoel, D.; van Drunen, R. GROMACS: A message-passing parallel molecular dynamics implementation. *Comput. Phys. Commun.* **1995**, *91*, 43–56. [CrossRef]

118. Humphrey, W.; Dalke, A.; Schulten, K. VMD: Visual molecular dynamics. *J. Mol. Graph.* **1996**, *14*, 33–38. [CrossRef]

14

Synthesis and Anti-Tumor Activities of 4-Anilinoquinoline Derivatives

Dan Liu *, Tian Luan, Jian Kong, Ying Zhang and Hai-Feng Wang

Academic Editor: D. Hadjipavlou-Litina

Department of Pharmaceutical Engineering, College of Parmaceutical and Biological Engineering, Shenyang University of Chemical Technology, Shenyang 110142, China; sapphire0614@163.com (T.L.); kongjianwm2006@163.com (J.K.); zhingky621@126.com (Y.Z.); 38whf@163.com (H.-F.W.)
* Correspondence: liudan20040318@163.com

Abstract: Twenty-two 7-fluoro (or 8-methoxy)-4-anilinoquinolines compounds were designed and synthesized as potentially potent and selective antitumor inhibitors. All the prepared compounds were evaluated for their *in vitro* antiproliferative activities against the HeLa and BGC823 cell lines. Ten compounds (**1a–g**; **2c**; **2e** and **2i**) exhibited excellent antitumor activity superior to that of gefitinib. Among the ten compounds; seven (**1a–c**; **1e–1g** and **2i**) displayed excellent selectivity for BGC823 cells. In particular; **1f** and **2i** exhibited potent cytotoxic activities against HeLa cells and BGC823 cells with better IC_{50} values than gefitinib.

Keywords: 4-anilinoquinolines; EGFR; antitumor; inhibitor

1. Introduction

The epidermal growth factor receptor (EGFR) and its closely related family member HER2 play a critical role in mediating growth factor signaling, which make them interesting drug targets for oncology. EGFR and HER2 have been extensively investigated, and various classes of small molecule kinase inhibitors have emerged as promising strategies to inhibited EGFR and HER2 kinase activity.

A considerable number of 4-anilinoquinazoline-based kinase inhibitors are known, including gefitinib (**1**), erlotinib (**2**) [1,2], lapatinib (**3**) [3,4], and afatinib (**4**) [5,6] (Figure 1). Previous SAR studies suggest that the quinazoline core was the best scaffold for the development of EGFR inhibitors and the quinazoline N3 interacts with the kinase domain via a water-mediated hydrogen bond to the side chain of the gatekeeper Thr790 of EGFR [7,8]. The quinazoline-based kinase inhibitors are widely used in medicinal chemistry and chemical biology research. For instance, Some C2 position modification of quinazoline-based compounds resulted in potent cytotoxic activities [9,10]. On the basis of the SAR studies of quinazolines, a series of compounds were developed where the N3 of the quinazoline was replaced by a C-CN group, such as neratinib (**5**, Figure 1) [11,12]. However, Rauh *et al.* have found that there was no evidence with the existence of a water molecule mediating the binding of N3 of the quinazoline core to the side chain of Thr790 by calculating the corresponding electron density maps. Quinolines **6** and **7** (Figure 1) were found to be highly active kinase inhibitors in biochemical assays and were further investigated for their biological effect on EGFR-dependent Ba/F3 cells and non-small cell lung cancer (NSCLC) cell lines [13]. Furthermore, studies have shown that the 4-anilino group can interact with the hydrophobic pocket of EGFR, the introduction of electron-donating groups in benzene ring of 4-anilinoquinazoline can increase the density of electron cloud on the quinazoline N1, and follow by increase the interaction with EGFR [14].

The aforementioned findings stimulated our interest in designing and synthesizing a series of 7-fluoro or 8-methoxy 4-anilinoquinolines which were acticipated to be as potent as their quinazoline counterparts. The activity of the target compounds were evaluated by human cervical cancer cell line (HeLa) and human gastric carcinoma cell line (BGC-823), and both of the cell lines had been proved to be highly expressed cell line of EGFR [15,16].

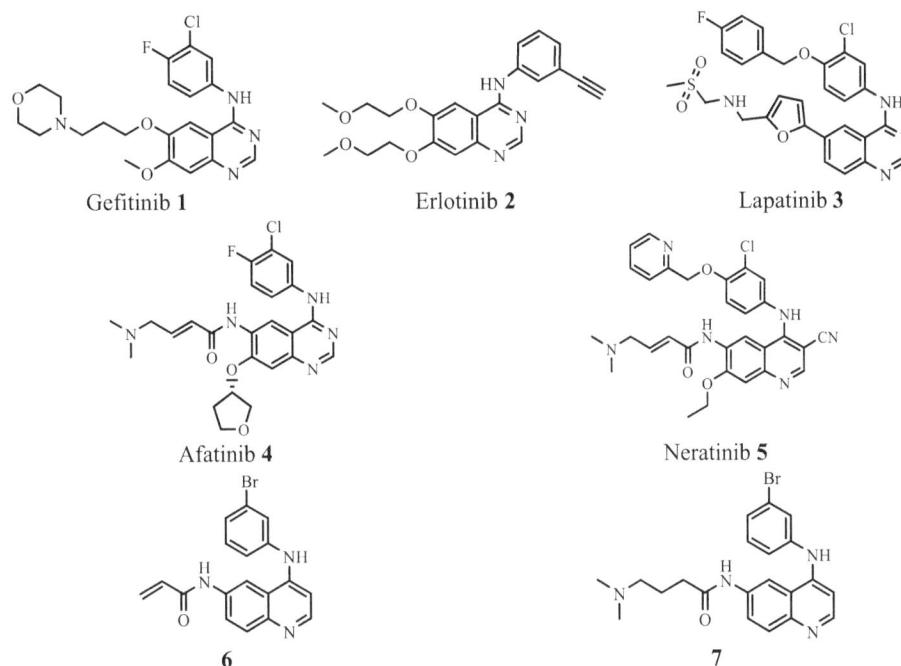

Figure 1. Several inhibitors of EGFR tyrosine kinases.

2. Results and Discussion

2.1. Chemistry

The target compounds **1a–h** and **2a–n** were synthesized via a convenient six-step reaction depicted in Scheme 1 and the respective experimental details are given in Section 3.1. Twenty-two compounds were obtained and their MS, ^{1}H-NMR spectroscopy data are provided in Section 3.1.

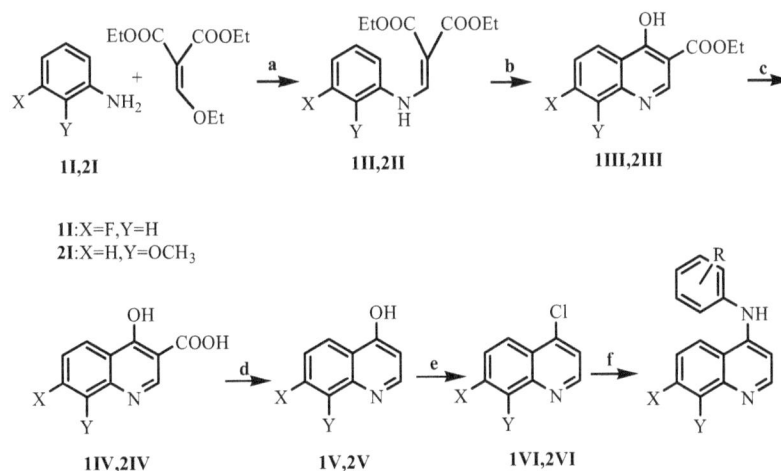

Scheme 1. General procedure for the synthesis of target compounds **1a–h** and **2a–n**. *Reagents and conditions*: (**a**) EtOH, reflux; (**b**) Dowtherm A, 250–260 °C; (**c**) 10% NaOH, EtOH, reflux; (**d**) Dowtherm A, 240 °C; (**e**) POCl$_3$; (**f**) *i*-PrOH, pyridine-HCl, substituted aniline, reflux.

2.2. Biological Evaluation

The biological activities of all the target compounds **1a–h** and **2a–n** were evaluated *in vitro* by MTT assay against HeLa and BGC-823 cell lines with gefitinib (**1**) as the positive control. Their inhibition rate and IC_{50} values are listed in Table 1.

Table 1. Antiproliferactive activity of the target compounds on HeLa and BGC-823 cell lines.

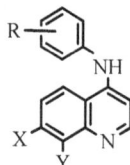

Compound	Substituents	Inhibition Rate % [a]		IC_{50} (μM) [b]	
		HeLa	BGC-823	HeLa	BGC-823
1a	X = F, Y = H, R = 3'-Cl	46.9	77.5	17.29	3.63
1b	X = F, Y = H, R = 4'-Cl	40.3	63.1	20.39	7.83
1c	X = F, Y = H, R = 3'-F	37.0	65.3	19.82	9.10
1d	X = F, Y = H, R = 4'-F	20.4	49.5	43.25	11.10
1e	X = F, Y = H, R = 3'-Cl,4'-Cl	40.8	56.9	36.69	8.29
1f	X = F, Y = H, R = 3'-Cl,4'-F	46.8	67.5	10.18	8.32
1g	X = F, Y = H, R = 4'-CH₃	41.7	59.4	18.69	7.08
1h	X = F, Y = H, R = 4'-OCH₃	36.1	35.1	55.76	21.61
2a	X = H, Y = OCH₃, R = 3'-Cl	35.3	19.5	66.69	>10
2b	X = H, Y = OCH₃, R = 4'-Cl	31.5	13.1	>10	>10
2c	X = H, Y = OCH₃, R = 3'-F	31.9	48.1	75.87	11.67
2d	X = H, Y = OCH₃, R = 4'-F	18.9	37.5	>10	46.41
2e	X = H, Y = OCH₃, R = 3'-Cl,4'-Cl	30.8	48.7	45.87	11.45
2f	X = H, Y = OCH₃, R = 3'-Cl,4'-F	37.2	20.2	45.84	>10
2g	X = H, Y = OCH₃, R = 4'-CH₃	30.5	31.1	99.54	61.52
2h	X = H, Y = OCH₃, R = 4'-OCH₃	21.9	32.0	>10	74.60
2i	X = H, Y = OCH₃, R = 4'-CH(CH₃)₂	59.1	78.4	7.15	4.65
2j	X = H, Y = OCH₃, R = 3'-Cl,4'-CH₃	32.7	24.5	71.26	>10
2k	X = H, Y = OCH₃, R = 2'F, 3'-F, 4'-F	32.5	27.6	57.88	>10
2l	X = H, Y = OCH₃, R = 4'-NO₂	35.8	36.8	13.48	>10
2m	X = H, Y = OCH₃, R = 4'-OH	32.7	25.6	39.58	>10
2n	X = H, Y = OCH₃, R = 3'-CN	28.6	16.0	30.61	>10
Gefitinib		42.6	46.0	17.12	19.27

[a] Inhibitory percentage of cells treated with each compound at a concentration of 10 μM for 96 h; [b] The agent concentration that inhibited HeLa and BGC-823 cells growth by 50%.

As shown in Table 1, 7-fluoro-4-anilinoquinolines **1a–g** displayed better cytotoxic activities against BGC-823 cells than HeLa cells and exhibited IC_{50} values within the 3.63–11.10 μmol/L range (on BGC-823 cells). It seems that substituent changes on the quinoline ring at the C7 or C8 carbons have a great influence on the antiproliferactive activity. In general, 7-fluoro-4-anilino-quinolines of these twenty-two compounds, were more active than those of the corresponding 8-methoxy-4-anilinoquinolines, as demonstrated by comparison of **1a–h** and **2a–h**. Thirdly, it was found that substituent changes on the benzene ring had little influence on antiprolifeactive activity, whether it was electron-donating or electron-withdrawing substituent at any position (**1a** *vs.* **1b** and **1d**; **2g** *vs.* **2b** and **2d**; **2g** and **2m** *vs.* **2l** and **2n**). Lastly, most of the 8-methoxy-4-anilinoquinolines exhibited moderate antiproliferactive activity against the HeLa and BGC-823 cell lines, except for compound **2i**. It was noteworthy that compound **2i** bearing **a** methoxy group on the quinoline ring and an isopropyl group on the benzene ring showed remarkable inhibitory effects on HeLa cells and BGC-823 cells (IC_{50} = 7.15 μM, IC_{50} = 4.65 μM), which represented a 2.4 and 4.1-fold increase in antitumor activity compared to gefitinib (IC_{50} = 17.12 μM against HeLa cells, IC_{50} = 19.27 μM against BGC-823 cells), respectively.

3. Experimental Section

3.1. Chemistry

All reagents and solvents were commercially available and were used without further purification. The melting points were determined on an electrically heated X-4 digital visual melting point apparatus (Tech, Beijing, China) and were uncorrected. ^{1}H-NMR spectra were recorded on an ARX-300 or AV-600 spectrometer (Bruker, Fällanden, Switzerland) at room temperature, and chemical shifts were measured in ppm downfield from TMS as internal standard. Mass spectra were recorded on Thermo-Finnigan LCQ equipment (ThermoFinnigan, San Francisco, CA, USA) with the positive Electron Spray Ionization (ESI) mode and are reported as m/z. Analytical thin layer chromatography (TLC) on silica gel plates containing UV indicator (YuHua, Gongyi, China) was routinely employed to follow the course of reactions and to check the purity of products.

3.1.1. Preparation of the Intermediates 1II–1VI

Diethyl (ethoxymethylene) malonate (6.48 g, 30.00 mmol) and m-fluroaniline (2.78 g, 25.00 mmol) was stirred in refluxing ethanol (10 mL) for 1.5 h. The mixture was concentrated under reduced pressure. The residue was crystallized in petroleum ether, filtered off, and air-dried to get 1II. Compound 1II (2.80 g, 9.96 mmol) was stirred in Dowtherm-A (8 mL) for 0.5 h at 260 °C. After the reaction was over by TLC, the mixture was cooled to room temperature and petroleum ether was added to get the crude ester, which was further washed with petroleum ether to afford 1III, as a white solid. 1III was hydrolyzed in refluxing NaOH solution (10%, 20 mL) for 1.5 h to give 1IV. Compound 1IV (1.24 g, 6 mmol) was stirred in Dowtherm-A (10 mL) for 1 h at 240 °C. After the reaction was over as monitored by TLC, the reaction mixture was cooled to room temperature and the crude was washed with petroleum ether to afford 1V as a white solid. To a solution of 1V (1.20 g, 7.36 mmol) in 1,2-dichloroethane (30 mL), $POCl_3$ (1.35 g, 8.83 mmol) was added dropwise. The mixture was refluxed for 1 h. Saturated $NaHCO_3$ solution was added to neutralize the reaction mixture, which was worked up with 1,2-dichloroethane. The organic layer was dried over anhydrous $MgSO_4$, filtered and concentrated *in vacuo*. The residue was purified by column chromatography (silica gel) using petroleum ether/ethyl acetate as an eluent (5:1) to produce 1VI as a white solid [17]. 1VI: ^{1}H-NMR (300 MHz, $CDCl_3$): 8.79–8.78 (d, 2-H, 1H), 8.27–8.24 (q, 5-H, 1H), 7.77–7.75 (q, 8-H, 1H), 7.47–7.46 (d, 3-H, 1H), 7.44–7.42 (q, 6-H, 1H), ESI-MS (m/z): 182.4[M + H]$^+$.

3.1.2. Preparation of the Intermediates 2II–2VI

Compounds 2II–2VI were prepared in the same manner as 2II–2VI. 2VI: ^{1}H-NMR (300 MHz, $CDCl_3$): 4.11 (3H, s, OCH_3), 7.11 (1H, d, J = 7.2 Hz, Ar-H-7), 7.53–7.60 (2H, m, Ar-H-3, 6), 7.81 (1H, dd, J = 8.7 Hz, J = 1.2 Hz, Ar-H-5), 8.80 (1H, d, J = 4.8 Hz, Ar-H-2); ESI-MS (m/z): 194.4 [M + H]$^+$.

3.1.3. Preparation of the Title Compounds 1 and 2

A mixture of compound 1VI (0.040 g, 0.22 mmol), m-chloroaniline (0.036 g, 0.31 mmol) and pyridine hydrochloride was heated at reflux for 45 min in isopropanol (6 mL), after the reaction is over by TLC, it was cooled to room temperature and the petroleum ether (4 mL) and $NaHCO_3$ (10 mL) were added into the reaction mixture. The product was filtered and recrystallised from ethanol to give the title compound 1. Compound 2 was prepared in the same manner as 1.

4-(3′-Chlorophenylamino)-7-fluoroquinoline (**1a**). Yield: 96.25%; white solid. m.p. 196–197 °C; ^{1}H-NMR (600 MHz, $CDCl_3$) δ: 8.62 (d, J = 4.8 Hz, 1H, ArH), 7.94 (dd, J = 9.0, 6.0 Hz, 1H, ArH), 7.71 (dd, J = 4.2, 2.4 Hz, 1H, ArH), 7.37–7.27 (m, 3H, ArH), 7.19–7.17 (m, 2H, ArH), 7.01 (d, 1H, J = 5.4 Hz, ArH). ESI-MS (m/z): 273.4 [M + H]$^+$.

4-(4′-Chlorophenylamino)-7-fluoroquinoline (**1b**). Yield: 92.93%. m.p. 194–195 °C, ^{1}H-NMR (600 MHz, CDCl$_3$) δ: 8.57 (s, 1H, ArH), 7.96 (dd, J = 8.4, 6.0 Hz, 1H, ArH), 7.70 (d, 1H, J = 10.2 Hz, ArH), 7.41–7.40 (m, 2H, ArH), 7.31 (t, J = 8.4 Hz, 1H, ArH), 7.27–7.25 (m, 2H, ArH), 6.90 (d, J = 4.8 Hz, 1H, ArH). ESI-MS (m/z): 273.4 [M + H]$^+$.

4-(3′-Fluorophenylamino)-7-fluoroquinoline (**1c**). Yield: 95.37%; m.p. 173–174 °C; ^{1}H-NMR (600 MHz, CDCl$_3$) δ: 8.62 (s, 1H, ArH), 7.94 (dd, J = 9.0, 6.0 Hz, 1H, ArH), 7.70 (dd, J = 6.6, 2.1 Hz, 1H, ArH), 7.38 (dd, J = 14.7, 8.1 Hz, 1H, ArH), 7.34–7.31 (m, 1H, 1H, ArH), 7.06 (t, J = 4.5 Hz, 1H, ArH), 7.04–7.02 (m, 2H, ArH), 6.91–6.88 (m, 1H, ArH). ESI-MS (m/z): 257.4 [M + H]$^+$.

4-(4′-Fluorophenylamino)-7-fluoroquinoline (**1d**). Yield: 93.66%; m.p. 178–179 °C; ^{1}H-NMR (600 MHz, CDCl$_3$) δ: 8.54–8.53(s, 1H, ArH), 7.94 (d, J = 6.0 Hz, 1H, ArH), 7.68 (dd, J = 10.2, 3.0 Hz, 1H, ArH), 7.32–7.28 (m, 3H, ArH), 7.17–7.13 (m, 2H, ArH), 6.74 (d, J = 5.4 Hz 1H, ArH). ESI-MS (m/z): 257.4 [M + H]$^+$.

4-(3′-Chloro-4′-chlorophenylamino)-7-fluoroquinoline (**1e**). Yield: 91.34%; m.p. 191–192 °C; ^{1}H-NMR (600 MHz, CDCl$_3$) δ: 8.61 (d, J = 4.8 Hz, 1H, ArH), 7.97 (dd, J = 9.0, 6.0 Hz, 1H, ArH), 7.71 (dd, J = 9.6, 2.4 Hz, 1H, ArH), 7.48 (d, J=9.0Hz, 1H, ArH), 7.41 (d, J = 2.4 Hz, 1H, ArH), 7.34–7.27 (m, 1H, ArH), 7.16 (dd, J = 8.7, 2.7 Hz, 1H, ArH), 6.97 (dd, J = 5.4 Hz, 1H, ArH). ESI-MS (m/z): 307.3 [M + H]$^+$.

4-(3′-Chloro-4′-fluoroamino)-7-fluoroquinoline (**1f**). Yield: 74.72%; m.p. 193–194 °C; ^{1}H-NMR (600 MHz, CDCl$_3$) δ: 8.58 (s, 1H, ArH), 7.93 (dd, J = 9.0, 5.4 Hz, 1H, ArH), 7.70 (dd, J = 10.2, 2.4 Hz, 1H, ArH), 7.38 (dd, J = 10.0, 2.4 Hz, 1H, ArH), 7.34–7.31 (m, 1H, ArH), 7.27–7.19 (m, 2H, ArH), 6.82 (d, J = 5.4 Hz, 1H, ArH). ESI-MS (m/z): 291.4 [M + H]$^+$.

4-(4′-Methylphenylamino)-7-fluoroquinoline (**1g**). Yield: 96.56%, m.p. 171–172 °C; ^{1}H-NMR (600 MHz, CDCl$_3$) δ: 8.50 (d, J = 5.4 Hz, 1H, ArH), 7.96 (dd, J = 9.3, 5.7 Hz, 1H, ArH), 7.68 (dd, J = 10.2, 2.4 Hz, 1H, ArH), 7.31–7.27 (m, 1H, ArH), 7.26–7.25 (m, 2H, ArH), 7.22–7.20 (m, 2H, ArH), 6.83 (d, J = 4.2 Hz, 1H, ArH), 2.40 (s, 3H, CH$_3$). ESI-MS (m/z): 253.4 [M + H]$^+$.

4-(4′-Methoxyphenylamino)-7-fluoroquinoline (**1h**). Yield: 76.15%, m.p. 181–182 °C; ^{1}H-NMR (600 MHz, CDCl$_3$) δ: 8.46 (d, J = 5.4 Hz, 1H, ArH), 7.98 (dd, J = 9.0, 5.4 Hz, 1H, ArH), 7.68 (dd, J = 9.9, 2.7 Hz, 1H, ArH), 7.30–7.27 (m, 1H, ArH), 7.26–7.25 (m, 2H, ArH), 7.00–6.98 (m, 2H, ArH), 6.65 (d, J = 5.4 Hz, 1H, ArH). 3.88 (s, 3H, OCH$_3$); ESI-MS (m/z): 269.3 [M + H]$^+$.

4-(3′-Chlorophenylamino)-8-methoxyquinoline (**2a**). Yield: 76.16%, m.p. 250–251 °C; ^{1}H-NMR (300 MHz, CDCl$_3$) δ: 8.58 (d, J = 5.4 Hz, 1H, ArH), 7.45–7.50 (m, 2H, ArH), 7.40–7.30 (m, 1H, ArH), 7.14–7.10 (m, 2H, ArH), 7.06–6.98 (m, 2H, ArH), 6.88 (d, J = 5.4 Hz, 1H, ArH), 4.08 (s, 3H, OCH$_3$); ESI-MS (m/z): 285.5 [M + H]$^+$.

4-(4′-Chlorophenylamino)-8-methoxyquinoline (**2b**). Yield: 77.52%; m.p. 254–258 °C. ^{1}H-NMR (300 MHz, CDCl$_3$) δ: 8.60 (d, J = 5.1 Hz, 1H, ArH), 7.49–7.52 (m, 2H, ArH), 7.40–7.30 (m, 2H, ArH), 7.14–7.10 (m, 1H, ArH), 7.06–6.98 (m, 2H, ArH), 6.84 (d, J = 5.1 Hz, 1H, ArH), 4.08 (s, 3H, OCH$_3$); ESI-MS (m/z): 285.5 [M + H]$^+$.

4-(3′-Fluorophenylamino)-8-methoxyquinoline (**2c**). Yield: 75.06%; m.p. 267–269 °C. ^{1}H-NMR (300 MHz, CDCl$_3$) δ: 8.65 (s, 1H, ArH), 7.49–7.44 (m, 2H, ArH), 7.39–7.31 (m, 1H, ArH), 7.15 (dd, J = 5.1, 2.1 Hz, 1H, ArH), 7.08–6.99 (m, 3H, ArH), 6.88–6.82 (m, 1H, ArH), 4.08 (s, 3H, OCH$_3$); ESI-MS (m/z): 269.3 [M + H]$^+$.

4-(4′-Fluorophenylamino)-8-methoxyquinoline (**2d**). Yield: 84.59%; m.p. 256–258 °C. ^{1}H-NMR (300 MHz, CDCl$_3$) δ: 8.56 (d, J = 5.1 Hz, 1H, ArH), 7.52–7.41 (m, 2H, ArH), 7.30–7.27 (m, 2H, ArH), 7.15–7.05 (m, 3H, ArH), 6.82(d, J = 5.1 Hz, 1H, ArH), 4.08 (s, 3H, OCH$_3$); ESI-MS (m/z): 269.3 [M + H]$^+$.

4-(3′-Chloro-4′-chlorophenylamino)-8-methoxyquinoline (**2e**). Yield: 79.18%; m.p. 244–246 °C. ^{1}H-NMR (300 MHz, CDCl$_3$) δ: 8.63 (d, J = 5.1 Hz, 1H, ArH), 7.44–7.49 (m, 3H, ArH), 7.41 (d, J = 2.4 Hz, 1H,

ArH), 7.16 (dd, J = 8.7, 2.4 Hz, 1H, ArH), 7.08 (d, J = 8.4 Hz, 1H, ArH), 7.03 (d, J = 5.1Hz, 1H, ArH), 4.08 (s, 3H, OCH_3); ESI-MS (m/z): 319.3 [M + H]$^+$.

4-(3'-Chloro-4'-fluorophenylamino)-8-methoxyquinoline (**2f**). Yield: 80.14%; m.p. 247–249 °C. ^{1}H-NMR (300 MHz, CDCl$_3$) δ: 8.59 (d, J = 5.1 Hz, ArH), 7.38–7.36 (m, 1H, ArH), 7.20–7.17 (m, 2H, ArH), 7.07 (dd, J = 6.6, 2.4 Hz, 1H, ArH), 6.88 (d, J = 5.1 Hz, 1H, ArH), 4.08 (s, 3H, OCH_3); ESI-MS (m/z): 303.3 [M + H]$^+$.

4-(4'-Methylphenylamino)-8-methoxyquinoline (**2g**). Yield: 78.46%; m.p. 258–260 °C. ^{1}H-NMR (300 MHz, CDCl$_3$) δ: 8.56 (d, J = 5.4 Hz, 1H, ArH), 7.43–7.47 (m, 2H, ArH), 7.22–7.17 (m, 4H, ArH), 7.06 (dd, J = 6.9, 1.5 Hz, 1H, ArH), 6.93 (d, J = 5.4 Hz, 1H, ArH), 4.08 (s, 3H, OCH_3), 2.39 (s, 3H, CH_3); ESI-MS (m/z): 265.4 [M + H]$^+$.

4-(3'-Methoxyphenylamino)-8-methoxyquinoline (**2h**). Yield: 80.96%; m.p. 231–233 °C. ^{1}H-NMR (300 MHz, CDCl$_3$) δ: 8.61 (d, J = 5.1 Hz, ArH), 7.49–7.41 (m, 2H, ArH), 7.32 (t, J = 5.4 Hz, 1H, ArH), 7.13–7.10 (m, 1H, ArH), 7.06 (dd, J = 6.9, 1.8 Hz, ArH), 6.90–6.84 (m, 2H, ArH), 6.73 (dd, J = 8.1, 2.1 Hz, 1H, ArH), 4.08 (s, 3H, OCH_3), 3.83 (s, 3H, OCH_3); ESI-MS (m/z): 281.2 [M + H]$^+$.

4-(4'-Isopropylphenylamino)-8-methoxyquinoline (**2i**). Yield: 79.47%; m.p. 236–238 °C. ^{1}H-NMR (300 MHz, CDCl$_3$) δ: 8.53 (d, J = 5.1 Hz, 1H, ArH), 7.48–7.40 (m, 2H, ArH), 7.30–7.21 (m, 4H, ArH), 7.05 (dd, J = 6.9, 2.1 Hz, 1H, ArH), 6.99–6.96 (m, 1H, ArH), 4.08 (s, 3H, OCH_3), 2.99–2.90 (m, 1H, CH), 1.29 (d, J = 6.9Hz, 6H, CH_3) ; ESI-MS (m/z): 293.6 [M + H]$^+$.

4-(3'-Chloro-4'-methylphenylamino)-8-methoxyquinoline (**2j**). Yield: 76.48%; m.p. 270–272 °C. ^{1}H-NMR (300 MHz, CDCl$_3$) δ: 8.61 (d, J = 5.4 Hz, 1H, ArH), 7.47–7.41 (m, 2H, ArH), 7.31(d, J = 2.4 Hz, 1H, ArH), 7.25 (d, J = 7.8 Hz, 1H, ArH), 7.12–7.04 (m, 2H, ArH), 6.99 (d, J = 5.1 Hz, 1H, ArH), 4.08 (s, 3H, OCH_3), 2.39 (s, 3H, CH_3); ESI-MS (m/z): 299.3[M + H]$^+$.

4-(2,3,4-Trifluorophenylamino)-8-methoxyquinoline (**2k**). Yield: 70.00%; m.p. 238–239 °C. ^{1}H-NMR (300 MHz, CDCl$_3$) δ: 8.61 (d, J = 5.1 Hz, 1H, ArH), 7.57–7.45 (m, 2H, ArH), 7.24–7.16 (m, 1H, ArH), 7.10–6.84 (m, 2H, ArH), 6.84 (d, J = 5.4 Hz, 1H, ArH), 4.10 (s, 3H, OCH_3) ; ESI-MS (m/z): 305.2 [M + H]$^+$.

4-(4'-Nitrophenylamino)-8-methoxyquinoline (**2l**). Yield: 78.18%; m.p. 265–267 °C. ^{1}H-NMR (300 MHz, CDCl$_3$) δ: 8.79 (d, J = 4.8 Hz, 1H, ArH), 8.25 (d, J = 9 Hz, 2H, ArH), 7.51–7.46 (m, 2H, ArH), 7.37 (d, J = 4.8 Hz, 1H, ArH), 7.29–7.26 (m, 2H, ArH), 7.11 (t, J = 4.2 Hz, 1H, ArH), 4.10 (s, 3H, OCH_3); ESI-MS (m/z): 296.2 [M + H]$^+$.

4-(4'-Hydroxyphenylamino)-8-methoxyquinoline (**2m**). Yield: 58.17%; m.p. 236–238 °C. ^{1}H-NMR (300 MHz, DMSO) δ: 9.44 (s, 1H, OH), 8.61 (s, 1H, ArH), 8.29 (d, J = 5.1 Hz, 1H, ArH), 7.89 (d, J = 8.7 Hz, 1H, ArH), 7.38 (t, J = 8.4 Hz, 1H, ArH), 7.12 (t, J = 8.4 Hz, 2H, ArH), 6.84 (d, J = 8.7 Hz, 2H, ArH), 6.56 (d, J = 5.1 Hz, 1H, ArH), 3.91 (s, 3H, OCH_3); ESI-MS (m/z): 267.3 [M + H]$^+$.

4-(3'-Cyanophenylamino)-8-methoxyquinoline (**2n**). Yield: 91.43%; m.p. 255–256 °C. ^{1}H-NMR (300 MHz, CDCl$_3$) δ: 8.68 (d, J = 5.1 Hz, 1H, ArH), 7.55–7.40 (m, 5H, ArH), 7.11–7.08 (m, 2H, ArH), 6.74 (s, 1H, ArH), 4.10 (s, 3H, OCH_3); ESI-MS (m/z): 276.2 [M + H]$^+$.

3.2. Cell Proliferative Assay

The antiproliferative activities of the prepared 4-anilinoquinolines against HeLa and BGC823 cell lines were evaluated by MTT assay *in vitro*, with gefitinib as the positive control. The negative control contains cells, culture medium, MTT and DMSO. All human tumor cells were cultured in RPMI 1640 medium supplemented with 10% fetal bovine serum (FBS). Cells were detached by trypsinisation, seeded at 1.0–2.0 × 10^3 cells each well in a 96-well plate and incubated in 5% CO_2 at 37 °C overnight, then treated with the test compounds at different concentration and incubated for 96 h. Fresh MTT solution was added to each well and incubated at 37 °C for 4 h. The MTT-formazan formed by metabolically viable cells was dissolved in 150 μL DMSO each well, and monitored by a microplate reader at dual-wavelength of 490 nm; IC_{50} was defined as the drug concentrations that inhibited the cell number to 50% after 96 h. Each test was performed three times.

4. Conclusions

In summary, two novel series of 4-anilinoquinolines were designed and synthesized as potentially potent and selective antitumor inhibitors. All of the final compounds were generated from aniline derivatives via six step reaction sequences including nucleophilic substitution, cyclization, hydroxylation, decarboxylation, chlorination and nucleophilic substitution. Among the 7-fluoro-4-anilinoquinolines, all the prepared compounds displayed some cytotoxic activity against the HeLa and BGC823 cell lines. Compounds **1a–g**, **2c** and **2e** displayed superior cytotoxic activities against the BGC823 cell line than gefitinib. Furthermore, compound **1f** displayed good cytotoxic activities against HeLa and BGC823 cells (IC_{50} value of 10.18 μM and 8.32 μM against HeLa and BGC823 cells, respectively). In particular, compound **2i** exhibited the most potent inhibitory activity against BGC823 cells (IC_{50} value of 7.15 μM and 4.65 μM against HeLa and BGC823 cells, respectively). The drug-like structural optimization based on the 4-anilinoquinoline skeleton will be reported in the future.

Acknowledgments: This work was supported by the grant from the Scientific Research Project of the Department of Education of Liaoning Province (Grant No. L2013171), the Natural Science Foundation of Liaoning Province of China (Grant No. 2015020695), the Natural Science Foundation of China (Grant No. 21372156). Additionally, we thank Weiling Wang, at the Experimental Therapeutics Center, Agency for Science, Technology and Research of Singapore, for her English revisions and constructive criticism.

Author Contributions: T.L., J.K., Y.Z. and H.-F.W. performed the experiments. D.L. and T.L. designed the experiments and wrote the paper. All authors took part in data analysis and discussion. All authors read and approved the final manuscript.

Conflicts of Interest: The authors declare no conflict of interest.

References

1. Pao, W.; Miller, V.; Zakowski, M.; Doherty, J.; Politi, K.; Sarkaria, I.; Singh, B.; Heelan, R.; Rusch, V.; Fulton, L.; *et al*. EGF receptor gene mutations are common in lung cancers from "never smokers" and are associated with sensitivity of tumors to gefitinib and erlotinib. *Proc. Natl. Acad. Sci. USA* **2004**, *101*, 13306–13311. [CrossRef] [PubMed]

2. Bikker, J.A.; Brooijmans, N.; Wissner, A.; Mansour, T.S. Kinase domain mutations in cancer: Implications for small molecule drug design strategies. *J. Med. Chem.* **2009**, *52*, 1493–1509. [CrossRef] [PubMed]

3. Feng, Y.F.; Zeng, C.L. Advance in research for lapatinib, a dual inhibitor of EGFR and ErbB2 tyrosine kinase activity. *Chin. New Drugs J.* **2007**, *16*, 1990–1993.

4. Tao, L.Y.; Fu, L.W. Progress of study on lapatinib. *Chin. Pharmacol. Bull.* **2008**, *24*, 1541–1544.

5. Solca, F.; Dahl, G.; Zoephel, A.; Bader, G.; Sanderson, M.; Klein, C.; Kraemer, O.; Himmelsbach, F.; Haaksma, E.; Adolf, G.R. Target binding properties and cellular activity of afatinib (BIBW 2992), an irreversible ErbB family blocker. *J. Pharmacol. Exp. Ther.* **2012**, *343*, 342–350. [CrossRef] [PubMed]

6. Chen, X.; Zhu, Q.; Zhu, L.; Pei, D.; Liu, Y.; Yin, Y.; Schuler, M.; Shu, Y. Clinical perspective of afatinib in non-small cell lung cancer. *Lung Cancer.* **2013**, *81*, 155–161. [CrossRef] [PubMed]

7. Rewcastle, G.W.; Denny, W.A.; Bridges, A.J.; Zhou, H.R.; Cody, D.R.; McMichael, A.; Fry, D.W. Tyrosine kinase inhibitors. 5. Synthesis and structure-activity relationships for 4-[(phenylmethyl)amino]- and 4-(phenylamino) quinazolines as potent adenosine 5′-triphosphate binding site inhibitors of the tyrosine kinase domain of the epidermal growth factor receptor. *J. Med. Chem.* **1995**, *38*, 3482–3487. [PubMed]

8. Wissner, A.; Berger, D.M.; Boschelli, D.H.; Floyd, M.B.; Greenberger, L.M.; Gruber, B.C.; Johnson, B.D.; Mamuya, N.; Nilakantan, R.; Reich, M.F.; *et al*. 4-Anilino-6,7-dialkoxyquinolime-3-carbonitrile inhibitors of epidermal growth factor receptor kinase and their bioisosteric relationship to the 4-anilino-6,7-dialkoxyquinazoline inhibitors. *J. Med. Chem.* **2000**, *3*, 3244–3256. [CrossRef]

9.   Barbosa, M.L.D.C.; Lima, L.M.; Tesch, R.; Sant'Anna, C.M.R.; Totzke, F.; Kubbutat, M.H.G.; Schächtele, C.; Laufer, S.A.; Barreiro, E.J. Novel 2-chloro-4-anilino-quinazoline derivatives as EGFR and VEGFR-2 dual inhibitors. *Eur. J. Med. Chem.* **2014**, *71*, 1–14. [CrossRef] [PubMed]

10. Jiang, N.; Zhai, X.; Zhao, Y.F.; Liu, Y.J.; Qi, B.H.; Tao, H.Y.; Gong, P. Synthesis and biological evaluation of novel 2-(2-arylmethylene) hydrazinyl-4-aminoquinazoline derivatives as potent antitumor agents. *Eur. J. Med. Chem.* **2012**, *54*, 534–541. [CrossRef] [PubMed]

11. Minami, Y.; Shimamura, T.; Shah, K.; LaFramboise, T.; Glatt, K.A.; Liniker, E.; Borgman, C.L.; Haringsma, H.J.; Feng, W.; Weir, B.A.; *et al.* The major lung cancer-derived mutants of ERBB2 are oncogenic and are associated with sensitivity to the irreversible EGFR/ERBB2 inhibitor HKI-272. *Oncogene* **2007**, *26*, 5023–5027. [CrossRef] [PubMed]

12. Wissner, A.; Mansour, T.S. The development of HKI-272 andrelated compounds for the treatment of cancer. *Arch. Pharm.* **2008**, *341*, 465–477. [CrossRef] [PubMed]

13. Pawar, V.G.; Sos, M.L.; Rode, H.B.; Rabiller, M.; Heynck, S.; Otterlo, W.A.L.V.; Thomas, R.K.; Rauh, D. Synthesis and biological evaluation of 4-anilinoquinolines as potent inhibitors of epidermal growth factor receptor. *J. Med. Chem.* **2010**, *53*, 2892–2901. [CrossRef] [PubMed]

14. Roskoski, R., Jr. The ErbB/HER family of protein-tyrosine kinases and cancer. *Pharmacol. Res.* **2014**, *79*, 34–74. [CrossRef] [PubMed]

15. Zhou, S.Z.; Li, X.H.; Zhang, Y.X. Methylation patterns of the human EGFR gene promoter region. *Chin. J. Biochem. Mol. Biol.* **2010**, *26*, 568–574.

16. Long, H.; Li, H.; Wu, Q.M. RNA interference targeting EGFR on proliferation and apoptosis of human gastric carcinama cell line BCG823. *China J. Mod. Med.* **2012**, *22*, 59–62.

17. Ray, A.W.; Joseph, C.M.; Victor, K.E.; Phil, J.G.; George, S.R. Quinoline, Quinazoline, and Cinnoline Fungicides. E.P. Patent 0,326,330(A2), 2 August 1989.

15

Anti-Diabetic, Anti-Oxidant and Anti-Hyperlipidemic Activities of Flavonoids from Corn Silk on STZ-Induced Diabetic Mice

Yan Zhang, Liying Wu, Zhongsu Ma, Jia Cheng and Jingbo Liu *

Academic Editor: Derek J. McPhee

College of Food Science and Engineering, Jilin University, Changchun 130062, China; zy01@jlu.edu.cn (Y.Z.); wuliying-lisa@hotmail.com (L.W.); zsma@jlu.edu.cn (Z.M.); chengj_369@sohu.com (J.C.)
* Correspondence: ljb168@sohu.com

Abstract: Corn silk is a well-known ingredient frequently used in traditional Chinese herbal medicines. This study was designed to evaluate the anti-diabetic, anti-oxidant and anti-hyperlipidemic activities of crude flavonoids extracted from corn silk (CSFs) on streptozotocin (STZ)-induced diabetic mice. The results revealed that treatment with 300 mg/kg or 500 mg/kg of CSFs significantly reduced the body weight loss, water consumption, and especially the blood glucose (BG) concentration of diabetic mice, which indicated their potential anti-diabetic activities. Serum total superoxide dismutase (SOD) and malondialdehyde (MDA) assays were also performed to evaluate the anti-oxidant effects. Besides, several serum lipid values including total cholesterol (TC), triacylglycerol (TG), low density lipoprotein cholesterol (LDL-C) were reduced and the high density lipoprotein cholesterol level (HDL-C) was increased. The anti-diabetic, anti-oxidant and anti-hyperlipidemic effect of the CSFs suggest a potential therapeutic treatment for diabetic conditions.

Keywords: anti-diabetic; anti-oxidant; anti-hyperlipidemic; corn silk; flavonoids

1. Introduction

Diabetes mellitus [DM] is a serious chronic metabolic complication that results from abnormal insulin production or metabolism and chronic hyperglycemia [1]. DM is characterized by carbohydrate, lipid and protein metabolism disturbances and hyperglycemia, as well as oxidative stress accompanied with the main clinical symptoms polydipsia, polyuria, polyphagia, high urine glucose level and weight loss [2,3]. It reported that 7% of the adults around the world suffer from DM [4]. Recently, there has been a sharp increase in DM levels which parallels that of obesity and overweight. It projected by the International Diabetes Federation that by the year 2030 the number of diabetic patients will be approximately 552 million [5]. There are two major classes of DM, which are Type 1 DM (T1DM) and Type 2 DM (T2DM) [6]. Of the two types, T2DM cases are prevalent, with only 5%–10% corresponding to T1DM [7].

At present, insulin and oral anti-diabetic chemical agents (*i.e.*, glucosidase inhibitor, biguanides, insulin sensitizer and sulfonylureas, *etc.*) are used in clinical practice as therapies for DM [8]. Many many of them have some limitations and side effects, such as liver and kidney failure, hypoglycemia, diarrhea and lactic acidosis which are difficult to tolerate [9,10]. Therefore, in order to protect patients from these negative effects of synthetic agents, the search for new compounds with better effectiveness and lower toxicity has received more and more attention as a potential source of new therapeutic anti-diabetic drugs for DM patients. This has encouraged investigation searching for alternative remedies derived from traditional herbal medicines which are accepted as valuable resources for primary healthcare by the World Health Organization (WHO) [11,12].

Corn is one of the top three most widely cultivated cereal crops in the world. Corn silk (*Zea mays* L.) is the style and stigma of corn fruit and is a waste material from corn cultivation and thus available in abundance throughout the world [13]. Corn silk is known as a traditional Chinese herbal medicine which has been widely used to treat edema, cystitis, gout, nephritis, kidney stones, obesity, as well as prostatitis and similar ailments [14]. It also reported that corn silk possesses hypoglycemic, anti-tumor, antioxidant, anti-fatigue and anti-fungal properties [15]. Meanwhile, corn silk contains various chemical components including polysaccharides, proteins, flavonoids, vitamins, minerals, alkaloids and tannins, as well as steroids, *etc.* [16]. Previous studies have showed that among all the components, flavonoids can be regarded as the main contributors to most of the therapeutic effects, including anti-oxidant, anti-aging, diuretic, and anti-proliferative activity on human cancer cell lines, *etc.* [15]. As mentioned, oxidative stress, as well as lipid metabolism disturbances play an important role in diabetes besides hyperglycemia, hence, drugs with several properties would be much more effective in the treatment of diabetes [17]. However, data regarding the corn silk flavonoids' *in vivo* anti-diabetic, anti-oxidant and anti-hyperlipidemic activities are very limited, with only a few studies performed that demonstrated their anti-oxidant capacity. Flavonoids from some rare or regionally limited natural plant materials, such as *Sanguis draxonis* [18], *Malus toringoides* (Rehd.) Hughes leaves [19], and *Pilea microphylla* (L.) [20] were proved to possess anti-diabetic activity, which implies that there is a good chance that flavonoids from corn silk also have anti-diabetes capability.

In this regard, it made great sense to evaluate the anti-diabetic, anti-oxidant and anti-hyperlipidemic activities of flavonoids from corn silk in a STZ-induced diabetic mice model to identify a more abundant natural source for discovering new DM therapies which might be more effective with less side effects and readily accessible to all the diabetic population.

2. Results and Discussion

2.1. Total Flavonoids and Total Phenolic Content

Previous studies reported that the total phenolic and the total flavonoids content of its extracts were associated with the pharmacological effects of corn silk, such as the antioxidant, anti-inflammatory, and antioxidant activities or diuretic activity [21,22]. Hence, in this study, ethanol was used as our extraction solvent to obtain the crude corn silk extract. The total phenolic content of the corn silk extracts was determined through a linear gallic acid standard curve (y = 0.1896x + 0.4326; R^2 = 0.9984) and the total phenolic content was 34.6 ± 0.2 milligram of gallic acid equivalents per gram of corn silk extract. The total flavonoids content of the corn silk extracts was evaluated by an aluminium chloride colorimetric assay, using rutin as a standard (y = 28.42x − 0.008, R^2 = 0.9991) and the total flavonoids content was 16.8 ± 0.4 milligram of rutin equivalents per gram of corn silk extract.

2.2. Effect of CSFs on Body Weight of Normal and STZ-Induced Diabetic Mice

An earlier investigation has demonstrated the anti-oxidant and free radicals scavenging activity [23] of the components of corn silk flavonoids. In the present study, STZ rats showed slight anti-hyperglycemic and anti-hyperlipidemic activity. The acute effect of STZ-Induced T2DM model (MI) on body weight was determined by measuring the body weights of mice in the week following the MI process. Body weights of mice in the diabetic control (DC) group were significantly decreased by 7 days after the STZ treatment, while that of the non-diabetic control (NC) animals were significantly elevated with a statistical significance level at $p < 0.01$. Figure 1a shows that during the first two days, there was no obvious difference on body weights amongst all seven groups and on day 3, the situation had changed, with no significant difference noted between the two NC groups and among the DC groups, while body weights of the NCs significantly differed from those of the DC groups.

Weight loss is one of the most important symptoms of DM, so we observed the body weights of mice by measuring them weekly. As shown in Figure 1b, when compared to NCs, no significant difference was found between the NC and CS groups, which implied that CSFs did not have any obvious effect on the body weight of normal mice, while some groups, including the DC and LD groups, showed an opposite result. A constant weight loss was noted in the DC group during the whole experimental period, while mice in both the PC and the groups administrated with CSFs gained weight during the four weeks. This suggested that CSFs and dimethylbiguanide can minimize the body weight loss of DM mice to a different extent. The effect of dimethylbiguanide on weight loss of DM mice was better than that of CSFs and the effect of CSFs was positively dose correlated through no significant diffidence was observed. The weight loss of STZ-induced diabetic mice is a symptom of diabetes, which is in agreement with the reported anti-diabetic activity of embelin in STZ-treated rats [17].

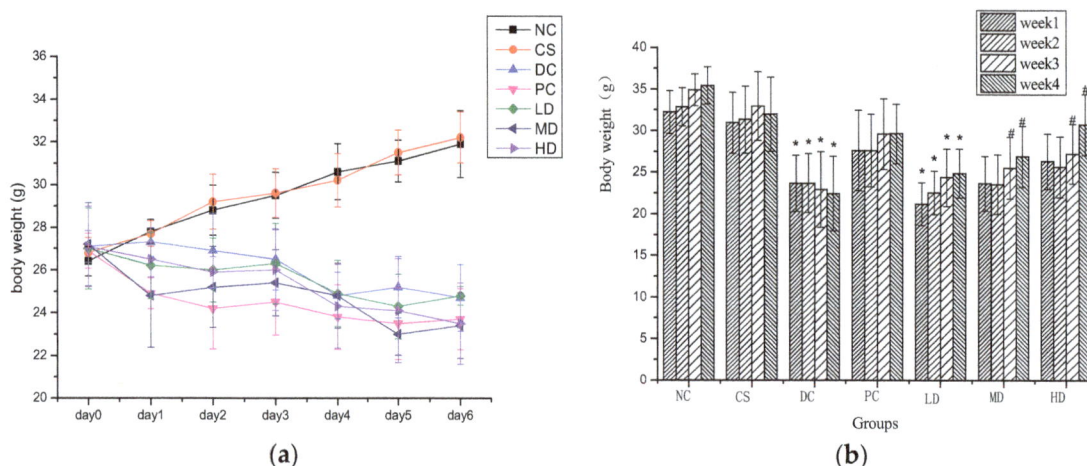

Figure 1. (a) Body weight of normal and STZ-induced diabetic mice in first 6 days after the STZ treatment; (b) Body weight of normal and STZ-induced diabetic mice in four weeks. Results were presented as means ± SD (n = 10). The columns of each index in have * $p < 0.05$, *vs.* NC; # $p < 0.05$, *vs.* week 1. NC, CS, DC, PC, LD, MD and HD are abbreviations for non-diabetic control group, non-diabetic CSFs high dose group, diabetic control group, diabetic CSFs low dose group, diabetic CSFs medium dose group, and diabetic CSFs high dose group, respectively.

2.3. Effect of MI and CSFs on Fasting BG of Normal and STZ-Induced Diabetic Mice

Figure 2a displays the fasting BG level of mice on the sixth day after the MI. According to Figure 2a the fasting BG level of DC mice were 17.67 mmol/L which was significantly ($p < 0.01$) higher than that of NCs and above the afore mentioned standardized DM value, 11.1 mmol/L, so that we considered the diabetes induction succeeded.

Figure 2b reveals the alteration of BG concentration in different groups correlated with the duration of the experiments. BG values of mice in NCs remained almost unchanged during the four weeks, which differed significantly from that of DCs with a statistical significance level at $p < 0.01$. BG values of DC and LD were continuously rising, which indicated that low dose (100 mg/kg) CSFs had no observable effect on lowering the BG level of DM mice, while BG of PC decreased after being treated with dimethylbiguanide and BG level of MD and HD decreased in week 3 and week 4, but the BG value of these three groups did not recover to the normal level at the end of this experiment ($p < 0.05$). This suggested that dimethylbiguanide and the administration of 300 mg/kg and 500 mg/kg CSFs can lower the BG concentration of diabetic mice. This observation was consistent with previous reports about otherphy to chemicals [24,25].

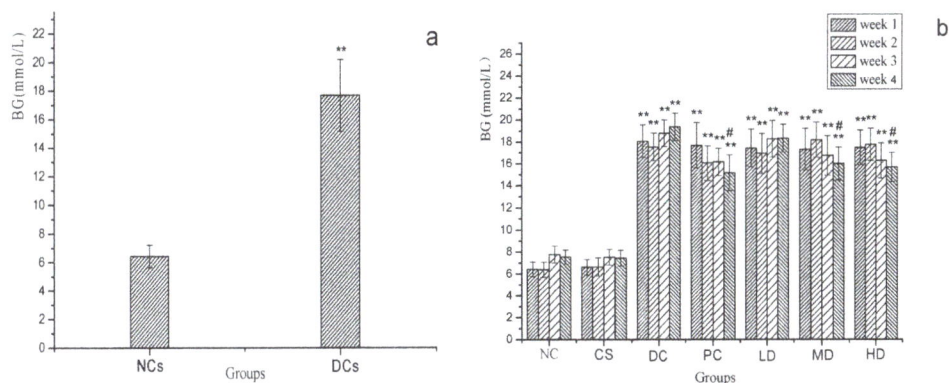

Figure 2. (**a**) Fasted BG levels of NCs [normal mice] and DCs [streptozotocin-induced diabetic mice] on the sixth day; (**b**) Fasted BG levels of NC, CS, DC, PC, LD, MD and HD in four weeks. Results were presented as means $\pm$ SD (n = 10). The columns of each index in Figure 2a have ** $p < 0.01$ *vs.* NCs. The columns of each index in Figure 2b have ** $p < 0.01$ *vs.* NC; # $p < 0.05$ *vs.* week1. Group abreviations are as given in the caption of Figure 1.

2.4. Effect of CSFs on Water Consumption and Food Intake of Normal and STZ-Induced Diabetic Mice

The effect of CSFs on daily water consumption and food intake of mice is shown in Figure 3. Water consumption of NC mice increased continuously and slightly during the 4 weeks, but the rise was not significant with water consumption remaining between 16.6 to 20.4 mL/day each. Both water intake and urinary output of DCs on the other hand were significantly elevated. The maximum water consumption increase was noted in the DC group which went from 28.2 to 47.7 mL/day each, while, the cage of DC animals got very wet, which was in accordance with the typical symptoms of diabetic polydipsia and polyuria. According to Figure 3a, the water consumptions of both the PC and the groups which received CSFs were all decreased in the fourth week when compared with that in the first three weeks, which indicated that the polydipsia and polyuria symptoms were improved to a certain degree. The water intake values of the PC and groups treated with CSFs were significantly lower than that of DC ($p < 0.05$) which suggested that both dimethylbiguanide and CSFs can effectively inhibit the increase of water consumption. No significant differences were found between the NC and CS groups. As shown in Figure 3b, food consumptions in all groups were elevated as time elapsed with that of NCs's and PC's increasing more slowly, though there was no significant variation observed among the food intakes of the seven groups during the four-week experimental period.

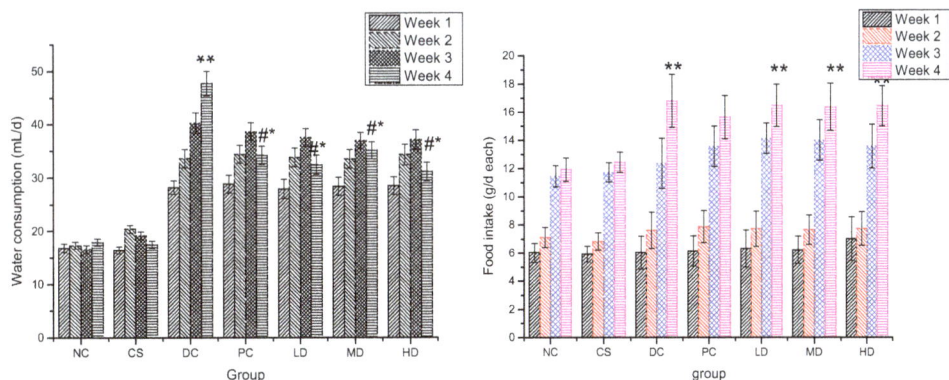

Figure 3. Water consumption (**a**) and food intake (**b**) of normal and STZ-induced diabetic mice. Results are presented as means $\pm$ SD (n = 10). The columns of each index have * $p < 0.05$, ** $p < 0.01$ *vs.* NC; # $p < 0.05$ *vs.* week 1. Group abbreviations are as given in the caption of Figure 1.

2.5. Effect of CSFs on Related Organ Weight and Liver Glycogen of Normal and STZ-Induced Diabetic Mice

As shown in Figure 4a, STZ led to the damage of the liver, kidney, and pancreas. Livers of mice in the DC group were much more damaged than in any other group ($p < 0.05$) and the pancreases of DC mice were also seriously damaged. When compared to DC, no significant damage was observed in kidneys in both the NC and PC groups and in pancreases in the CS and HD groups ($p < 0.05$). The related liver and kidney weights were elevated in the PC and HD groups, which indicated that both dimethylbiguanide and CSFs at high concentration can protect diabetic mice from liver and kidney damage.

The effect of CSFs on liver glycogen of mice is shown in Figure 4b. Liver glycogen levels in the NC, CS, PC, as well as HD groups significantly differed from those of the DC, LD, and MD ($p < 0.05$) groups, with NC showing the highest and DC the lowest and values between NC and CS were close to each other and no significant difference was observed, which suggested that the ingestion of 500 mg/kg CSFs not only had no adverse effect on liver glycogen metabolism but could also prevent the liver glycogen of diabetic mice from decreasing.

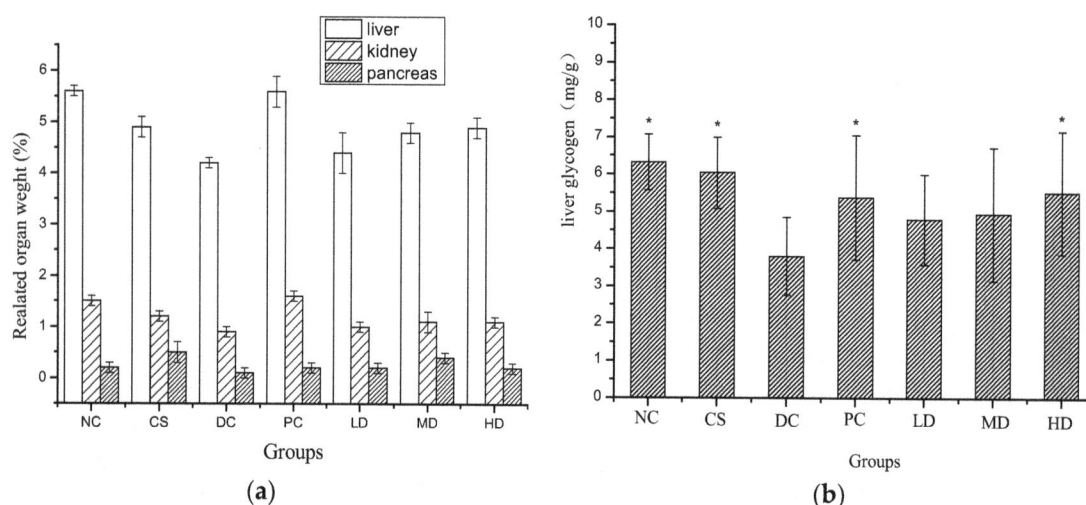

Figure 4. Related organ weight (**a**) and liver glycogen (**b**) of normal and STZ induced diabetic mice. Results were presented as means ± SD ($n = 10$). The columns of each index have * $p < 0.05$ *vs.* NC; Relative liver weight (%) = absolute kidney weight (g)/final body weight (g); relative kidney weight (%) = absolute kidney weight (g)/final body weight (g); relative pancreas weight (%) = absolute pancreas weight (g)/final body weight (g). Group abbreviations are as given in the caption of Figure 1.

2.6. Effect of CSFs on Serum SOD and MDA of Normal and STZ-Induced Diabetic Mice

Shown in Figure 5 are the SOD and MDA data, with Figure 5a showing the SOD value and Figure 5b the MDA value, respectively. When compared to NC, the SOD values of all the DC groups were significantly lower at $p < 0.05$ and the MDA values of the DC groups were observed to display the opposite trend, while both those of CS had almost the same value as the NC group, which revealed that the anti-oxidant capacity of diabetic mice was damaged.

No significant difference was found between the SOD value of the DC group ($p < 0.05$) and the serum SOD level was rising slightly with the higher concentration of CSFs. The level of SOD increased due to the production of superoxide, which has been implicated in cell dysfunction [24]. When compared to DC, the MDA values of PC, MD, and HD were significantly reduced ($p < 0.05$). It suggested that dimethylbiguanide and CSFs were able to repair the anti-oxidant capacity of diabetic mice, but the effects were not significant.

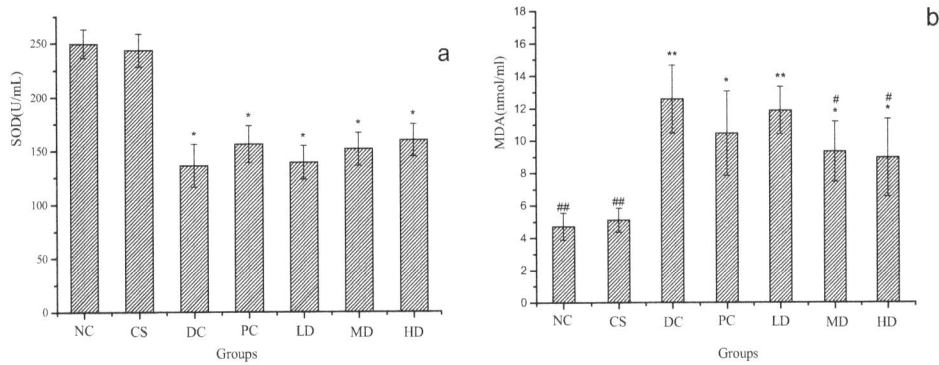

Figure 5. SOD (**a**) and MDA (**b**) of normal and STZ induced diabetic mice. Results were presented as means ± SD ($n = 10$). The columns of each index have * $p < 0.05$, ** $p < 0.01$ *vs.* NC; # $p < 0.05$, ## $p < 0.01$ *vs.* DC. Group abbreviations are as given in the caption of Figure 1.

2.7. Effect of CSFs on Serum TC, TG HDL-C and LDL-C of Normal and STZ-Induced Diabetic Mice

Figure 6 displays the effect of CSFs on serum TC, TG, HDL-C and LDL-C levels of normal and streptozotocin-induced diabetic mice. The serum TC value obtained from the experiment is shown in Figure 6a. Compared to NC, significant variances were found in DC, LD, and MD groups with NC showing a lowest value ($p < 0.05$), and TC concentrations of diabetic mice treated with dimethylbiguanide and 500 mg/kg CSFs were also significantly lower than that of DC and close to the normal level at $p < 0.05$. Figure 6b displays the effect of CSFs on serum TG of mice. As shown, the TG value of DC animals were significantly increased when compared to NC at the significant level $p < 0.01$. When compared to DC, TG concentrations of mice in the PC group and mice administrated with CSFs obviously decreased, but it did not reach the normal value ($p < 0.05$).

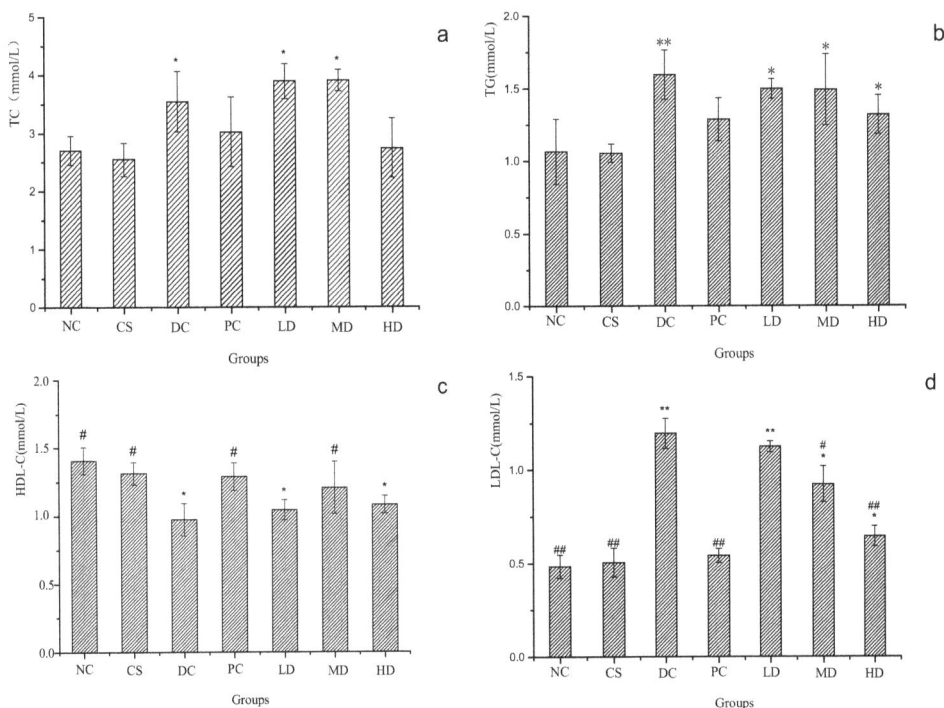

Figure 6. TC (**a**); TG (**b**); HDL-C (**c**) and LDL-C (**d**) of normal and STZ induced diabetic mice. Results were presented as means ± SD ($n = 10$). The columns of each index have * $p < 0.05$, ** $p < 0.01$ *vs.* NC; # $p < 0.05$, ## $p < 0.01$ *vs.* DC. Group abbreviations are as given in the caption of Figure 1.

The TG value in the HD group was similar to that of the PC one, all of which decreased and were the lowest values among the DC groups. This suggested that the ingestion of both 140 mg/kg dimethylbiguanide and 500 mg/kg CSFs was able to lower the serum TC and TG levels of diabetic mice to a similar extent. Serum HDL-C and LDL-C of mice are shown in Figure 6c,d, respectively. Compared to NC, HDL-C values of DC, LD, and HD were significantly lowered ($p < 0.05$), and that of NC, CS, PC, and MD were obviously higher than that of DC, $p < 0.05$. Besides, the HDL-C concentration of MD has no significance difference with that of PC, which indicated that 300 mg/kg CSFs increased the serum HDL-C as effectively as dimethylbiguanide did. As shown in Figure 6d, the LDL-C levels of DC and LD were significantly elevated when compared with that of NC with a significance level at 0.01. The LDL-C concentrations of MD and HD were lower than that in DC and LD, but higher than the value of NC and significant differences were observed between every two groups among these five groups ($p < 0.05$). Compared to DC, significant decreases of LDL-C concentration were noted in NC, CS, PC, and HD at $p < 0.01$ and in MD at $p < 0.05$ and the decreases were positively correlated with the CSFs concentration. As Figure 6 reveals, no apparent difference was found between PC and HD, which implied that CSFs were capable of reducing serum LDL-C and regulating the lipid metabolism of diabetic mice.

3. Experimental Section

3.1. Materials and Chemicals

Corn silk (958 species of Zheng Dan, collected in early October 2013) was obtained from Tianjing food company of Jilin Province (China), dried at 40 °C, ground and stored in a dry environment. Streptozotocin, citrate sodium and citric acid were obtained from Sigma Chemical Co. (St. Louis, MO, USA). Dimethyl biguanide was obtained from Feihong Drug Co. (Nanchang, China). Total cholesterol (TC), triacylglycerol (TG), high density lipoprotein cholesterol (HDL-C), and low density lipoprotein cholesterol (LDL-C), superoxide dismutase (SOD), malondialdehyde (MDA), as well as liver glycogen analytical kits were purchased from Shanghai Yuanmu Bioengineering Co. Ltd. (Shanghai, China). All the other reagents used in the investigation were of analytical grade.

3.2. Preparation of Corn Silk Flavonoids (CSFs)

Five kilograms of corn silk was cut into small pieces and ground, then extracted with 20 L of ethanol (80% (v/v)) in a rotary shaker at 60 °C for 3.5 h and thereafter filtered immediately. The corn silk residue was subjected to the aforementioned process in triplicate to extract more flavonoid components. The filtrates were concentrated to 2.5 L under low pressure and filtered again, then the enriched filtrates were freeze-dried and kept at 0–4 °C for the further study.

3.3. Determination of Total Phenolic Content

The contents of total phenolics in samples were analyzed by the Folin-Ciocalteu colorimetric method described previously [26,27], using gallic acid as a standard. Briefly, the appropriate dilutions of extracts were oxidized with Folin-Ciocalteu reagent and the reaction was neutralized with sodium carbonate. The absorbance of the resulting blue color was measured at 760 nm after 90 min by an ultraviolet-visible spectrophotometer (UV-2550, Shimadzu Corporation, Kyoto, Japan). The total phenolic content was determined using the standard gallic acid calibration curve and results were expressed as milligram gallic acid equivalents per gram dry mass of corn silk.

3.4. Determination of Total flavonoid Content

The contents of total flavonoid in samples were analyzed by the modified colourimetric aluminium chloride method [28]. In brief, a dilute solution of the extracts in methanol was mixed with 0.01 mol/L aluminium chloride in methanol. Then the mixture was allowed to stand for 10 min at room temperature. The absorbance of the reaction mixture was measured at 400 nm with an ultraviolet visible spectro photometer (UV-2550, Shimadzu Corporation). Again the blank consisted of all reagents and solvents, but without the sample. The total flavonoids content was determined using the rutin calibration curve, at concentrations from 0.005 to 0.125 mg/mL in methanol, and expressed as milligram of rutin equivalents per gram dry mass of corn silk.

3.5. Animals and Diets

Seventy male mice weighted from 18 g to 20 g were obtained from Jilin University Animal Center, Jilin province, China. All the experimental mice were kept at the animal care room with the temperature 16–20 °C, the humidity 50%–65% and 12 h light and 12 h dark cycle. The animals were acclimatized to environment for a week with free access to normal commercial diets and water. The diets were purchased from Jilin University Animal Center consisted of crude protein $\geqslant 18.0\%$, crude fat $\geqslant 4.0\%$ and moisture content $\leqslant 10.0\%$, crude ash $\leqslant 8.0\%$, crude fiber $\leqslant 5.0\%$, calcium 1%–1.8%, as well as phosphor 0.6%–1.2%. All animal experiments were conducted in compliance with "Guide of the care and use of laboratory animals" [29].

3.6. Induction of T2DM model [MI]

Streptozotocin (STZ) was used to induce diabetes in this study by being injected to the abdominal cavity of overnight fasted mice at the dose of 160 mg/kg body weight which was freshly dissolved in 0.1 mol/L cold citrate buffer (pH 4.2) [25]. The blood glucose value was evaluated in the next 6th day after the induction. Only mice with a fasting blood glucose concentration above 11.1 mmol/L were considered diabetic and were used in the corresponding groups in the experiment.

3.7. Experimental Design

Firstly, the mice were divided into two groups after the acclimation. One was the non-diabetic control group (NC) and another was the diabetic control group (DC). When performing the diabetic induction procedure, mice in the NC group were subjected to citrate buffer, while the DC mice received STZ instead. After the completion of MI, mice both in the NC and DC groups were randomly grouped again with NCs divided into two groups that were non-diabetic control group (NC) and non-diabetic CSFs high dose group (CS) namely, DCs into five groups, including diabetic control group (DC), diabetic dimethylbiguanide group (PC), diabetic CSFs low dose group (LD), diabetic CSFs medium dose group (MD), as well as diabetic CSFs high dose group (HD) and with 10 in each group. During the experimental process, they were assigned to the following treatment, respectively. Group 1 (NC): normal control, non-diabetic control mice administrated tap water only; Group 2 (CS): non-diabetic control mice administrated 500 mg/kg body weight of CSFs; Group 3 (DC): diabetic control mice administrated tap water only; Group 4 (PC): positive control, diabetic control mice administrated 140 mg/kg body weight of dimethylbiguanide; Group 5 (LD): low dose group, diabetic control mice administrated 100 mg/kg body weight of CSFs; Group 6 (MD): medium dose group, diabetic control mice administrated 300 mg/kg body weight of CSFs; Group 7 (HD): high dose group, diabetic control mice administrated 500 mg/kg body weight of CSFs.

All animals were intragastrically administered with the corresponding materials once a day for 28 days. Body weight, water consumption, as well as food intake were determined weekly. Blood samples were obtained by withdrawing from the tails at day 7, day 14, and day 21. At the end of the experiment procedure, mice were deprived from diets overnight and weighed before their sacrifice. Blood samples were collected from the eyes, and centrifuged to obtain the serum which was stored at −20 °C before the further analysis. Livers, kidneys, and pancreas were removed, weighted and stored at −80 °C, respectively. The related organ weight was calculated by the following formula: Related organ weight = (absolute organ weight (g)/final body weight (g)).

3.8. Biological Analysis

Serum TC, TG, HDL-C, LDL-C, MDA, and SOD values, in addition to the liver glycogen (LG) level were determined by the prescribed corresponding analytical kits with the detector, ELISA reader (BioTek Instruments, Inc., Winooski, VT, USA). Blood glucose (BG) was measured by the glucose oxidize method. They were expressed as mmol/L with several exceptions that MDA, SOD, and LG value was expressed as umol/mL, NU/mL, and mg/g, respectively.

3.9. Statistical Analysis

All the data were expressed as mean $\pm$ SD. The significant difference of data in one group was analyzed by unpaired student *t*-test, and statistical differences between different groups were compared by ANOVA followed by Dunnett's test (SPSS 10.0). *p* value under 0.05 was considered statistically significant. The figures were drawn by Origin 8.5.

4. Conclusions

The discovery of the anti-diabetic properties of corn silk, which is abundant and readily accessible to diabetic patients all over the world is attractive with the seriously growing large diabetes population. The data of this investigation showed that the ingestion of CSFs with a dose under 500 mg/kg had no observed adverse effect on normal mice and it had significant anti-diabetic potential, accompanied with anti-oxidant and anti-hyperlipidemic activities. In our present study, we found that after the administration of CSFs for four weeks, the polydipsia, polyphagia, and weight loss symptoms of diabetic mice had been relieved and 500 mg/kg obtained the best effect on preventing diabetic mice from weight loss. The serum BG concentration was ameliorated, and serum TC, TG, LDL-C, MDA and liver glycogen level in the diabetic CSFs high dose group (HD) group were lower than that of diabetic control (DC) group, and furthermore, the HDL-C and SOD value were slightly rising, which implied that CSFs had beneficial anti-diabetic effects by regulating the lipid metabolism and eliminating the oxygen radicals, which protected the organism's metabolism and repaired the anti-oxidant capacity.

Acknowledgments: This work is supported by Project of National Key Technology Research and Development Program for The 12th Five-year Plan (2012BAD33B03), Project of National Key Technology Research and Development Program (2012BAD34B07), China—Jilin Science and Technology Research Project (3R114A976604) and Scientific Research Foundation for Youth scholars of Jilin University (2014).

Author Contributions: All authors contributed to this work. Yan Zhang and Jingbo Liu initiated and designed the study; Yan Zhang and Zhongsu Ma conducted the study; Jia Cheng and Liying Wu performed the statistical analysis; Yan Zhang prepared the manuscript and all other authors approved this version of the article.

Conflicts of Interest: The authors declare no conflict of interest.

References

1.   Van Roozendaal, B.W.; Krass, I. Development of an evidence-based checklist for the detection of drug related problems in type 2 diabetes. *Pharm. World Sci.* **2009**, *31*, 580–595. [CrossRef] [PubMed]

2.   Aladag, I.; Eyibilen, A.; Guven, M.; Atis, O.; Erkorkmaz, U. Role of oxidative stress in hearing impairment in patients with type two diabetes mellitus. *J. Laryngol. Otol.* **2009**, *123*, 957–963. [CrossRef] [PubMed]

3.   Dierckx, N.; Horvath, G.; van Gils, C.; Vertommen, J.; van de Vliet, J.; de Leeuw, I.; Manuel-y-Keenoy, B. Oxidative stress status in patients with diabetes mellitus: Relationship to diet. *Clin. Nutr.* **2003**, *57*, 999–1008. [CrossRef] [PubMed]

4.   Babu, P.V.; Liu, D.; Gilbert, E.R. Recent advances in understanding the anti-diabetic actions of dietary flavonoids. *J. Nutr. Biochem.* **2013**, *24*, 1777–1789. [CrossRef] [PubMed]

5.   Ibrahim, M.A.; Islam, M.S. Anti-diabetic effects of the acetone fraction of Senna singueana stem bark in a type 2 diabetes rat model. *J. Ethnopharmacol.* **2014**, *153*. [CrossRef] [PubMed]

6.   Canivell, S.; Gomis, R. Diagnosis and classification of autoimmune diabetes mellitus. *Autoimmun. Rev.* **2014**, *13*, 403–407. [CrossRef] [PubMed]

7. Simpson, R.; Morris, G.A. The anti-diabetic potential of polysaccharides extracted from members of the cucurbit family: A review. *Bioact. Carbohydr. Diet. Fibre* **2014**, *3*, 106–114. [CrossRef]

8. Erkens, J.A.; Klungel, O.H.; Stolk, R.P.; Spoelstra, J.A. Cardiovascular drug use and hospitalizations attributable to type 2 diabetes. *Diabetes Care* **2001**, *24*, 1428. [CrossRef] [PubMed]

9. Wang, T.; Shankar, K.; Ronis, M.J.; Mehendale, H.M. Mechanisms and Outcomes of Drug- and Toxicant-Induced Liver Toxicity in Diabetes. *Crit. Rev. Toxicol.* **2007**, *37*, 413–459. [CrossRef] [PubMed]

10. Jiang, S.; Du, P.; An, L.; Yuan, G.; Sun, Z. Anti-diabetic effect of Coptis Chinensis polysaccharide in high-fat diet with STZ-induced diabetic mice. *Int. J. Biol. Macromol.* **2013**, *55*, 118–122. [CrossRef] [PubMed]

11. Girija, K.; Lakshman, K.; Udaya, C.; Sabhya Sachi, G.; Divya, T. Anti–diabetic and anti–cholesterolemic activity of methanol extracts of three species of Amaranthus. *Asian Pac. J. Trop. Biomed.* **2011**, *1*, 133–138. [CrossRef]

12. Surya, S.; Salam, A.D.; Tomy, D.V.; Carla, B.; Kumar, R.A.; Sunil, C. Diabetes mellitus and medicinal plants—A review. *Asian Pac. J. Trop. Dis.* **2014**, *4*, 337–347. [CrossRef]

13. Yang, J.; Li, X.; Xue, Y.; Wang, N.; Liu, W. Anti-hepatoma activity and mechanism of corn silk polysaccharides in H22 tumor-bearing mice. *Int. J. Biol. Macromol.* **2014**, *64*, 276–280. [CrossRef] [PubMed]

14. Wang, C.; Zhang, T.; Liu, J.; Lu, S.; Zhang, C.; Wang, E.; Wang, Z.; Zhang, Y.; Liu, J. Subchronic toxicity study of corn silk with rats. *J. Ethnopharmacol.* **2011**, *137*, 36–43. [CrossRef] [PubMed]

15. Liu, J.; Lin, S.; Wang, Z.; Wang, C.; Wang, E.; Zhang, Y.; Liu, J. Supercritical fluid extraction of flavonoids from Maydis stigma and its nitrite-scavenging ability. *Food Bioprod. Process.* **2011**, *89*, 333–339. [CrossRef]

16. Rahman, N.A.; Wan Rosli, W.I. Nutritional compositions and antioxidative capacity of the silk obtained from immature and mature corn. *J. King Saud Uni. Sci.* **2014**, *26*, 119–127. [CrossRef]

17. Naik, S.R.; Niture, N.T.; Ansari, A.A.; Shah, P.D. Anti-diabetic activity of embelin: Involvement of cellular inflammatory mediators, oxidative stress and other biomarkers. *Phytomedicine* **2013**, *20*, 797–804. [CrossRef] [PubMed]

18. Chen, F.; Xiong, H.; Wang, J.; Ding, X.; Shu, G.; Mei, Z. Antidiabetic effect of total flavonoids from Sanguis draxonis in type 2 diabetic rats. *J. Ethnopharmacol.* **2013**, *149*, 729–736. [CrossRef] [PubMed]

19. Wan, L.S.; Chen, C.P.; Xiao, Z.Q.; Wang, Y.L.; Min, Q.X.; Yue, Y.; Chen, J. *In vitro* and *in vivo* anti-diabetic activity of *Swertia kouitchensis* extract. *J. Ethnopharmacol.* **2013**, *147*, 622–630. [CrossRef] [PubMed]

20. Bansal, P.; Paul, P.; Mudgal, J.; Nayak, P.G.; Pannakal, S.T.; Priyadarsini, K.I.; Unnikrishnan, M.K. Antidiabetic, antihyperlipidemic and antioxidant effects of the flavonoid rich fraction of *Pilea microphylla* (L.) in high fat diet/streptozotocin-induced diabetes in mice. *Exp. Toxicol Pathol.* **2012**, *64*, 651–658. [CrossRef] [PubMed]

21. Maksimovic, Z.A.; Kovacevic, N. Preliminary assay on the antioxidative activity of *Maydis stigma* extracts. *Fitoterapia* **2003**, *74*, 144–147. [CrossRef]

22. Zhang, H.-E.; Xu, D.-P. Study on the chemical constituents of flavones from corn silk. *Zhong Yao Cai* **2007**, *30*, 164–166. [PubMed]

23. Liu, J.; Wang, C.; Wang, Z.; Zhang, C.; Lu, S.; Liu, J. The antioxidant and free-radical scavenging activities of extract and fractions from corn silk (*Zea mays* L.) and related flavone glycosides. *Food Chem.* **2011**, *126*, 261–269. [CrossRef]

24. Kumar, V.; Ahmed, D.; Gupta, P.S.; Anwar, F.; Mujeeb, M. Anti-diabetic, anti-oxidant and anti-hyperlipidemic activities of *Melastoma malabathricum* Linn. leaves in streptozotocin induced diabetic rats. *BMC Complement. Altern. Med.* **2013**, *13*. [CrossRef] [PubMed]

25. Zhao, W.; Yin, Y.; Yu, Z.; Liu, J.; Chen, F. Comparison of anti-diabetic effects of polysaccharides from corn silk on normal and hyperglycemia rats. *Int. J. Biol. Macromol.* **2012**, *50*, 1133–1137. [CrossRef] [PubMed]

26. Singleton, V.L.; Orthofer, R.; Lamuela-Raventos, R.M. Analysis of total phenols and other oxidation substrates and antioxidants by means of Folin-Ciocalteu reagent. In *Oxidants and Antioxidants, Pt A*; Packer, L., Ed.; Elsevier: San Diego, CA, USA, 1999; Volume 299, pp. 152–178.

27. Dewanto, V.; Wu, X.Z.; Adom, K.K.; Liu, R.H. Thermal processing enhances the nutritional value of tomatoes by increasing total antioxidant activity. *J. Agric. Food Chem.* **2002**, *50*, 3010–3014. [CrossRef] [PubMed]

28. Chang, C.C.; Yang, M.H.; Wen, H.M.; Chern, J.C. Estimation of total flavonoid content in propolis by two complementary colorimetric methods. *J. Food Drug Anal.* **2002**, *10*, 178–182.

29.   National Research Council. *Guide of the Care and Use of Laboratory Animals*; The National Academy Press: Washington, DC, USA, 1996; p. 125.

16

Synthesis and Crystal Structures of Benzimidazole-2-thione Derivatives by Alkylation Reactions

El Sayed H. El Ashry [1,*,†], Yeldez El Kilany [1,2,†], Nariman M. Nahas [2,†], Assem Barakat [1,3,†], Nadia Al-Qurashi [2,†], Hazem A. Ghabbour [4,†] and Hoong-Kun Fun [4,†]

Academic Editor: Derek J. McPhee

[1] Chemistry Department, Faculty of Science, Alexandria University, P. O. Box 426, Alexandria 21321, Egypt; yeldez244@hotmail.com (Y.E.K.); ambarakat@ksu.edu.sa (A.B.)

[2] Chemistry Department, Faculty of Applied Science, Umm Al-Qura University, Makkah 21955, Saudi Arabia; narinahas@hotmail.com (N.M.N.); maam10@hotmail.com (N.A.-Q.)

[3] Department of Chemistry, College of Science, King Saud University, P. O. Box 2455, Riyadh 11451, Saudi Arabia

[4] Department of Pharmaceutical Chemistry, College of Pharmacy, King Saud University, P. O. Box 2457, Riyadh 11451, Saudi Arabia; ghabbourh@yahoo.com (H.A.G.); hfun.c@ksu.edu.sa (H.-K.F.)

* Correspondence: eelashry60@hotmail.com

† These authors contributed equally to this work.

Abstract: Alkylated, benzylated and bromoalkylated benzimidazole-thione that intramolecularly heterocyclized to 3,4-dihydro-2*H*-[1,3]thiazino[3,2-*a*]benzimidazole were synthesized. The chemical structure of the synthesized product was characterized by Infra Red, ^{1}H-NMR, ^{13}C-NMR, and Mass spectroscopy. Furthermore, the molecular structures of **8** and **9** were confirmed by X-ray single crystallography in different space groups, *Pbca* and *P2$_1$/c*, respectively.

Keywords: benzimidazole-thione; thiazino[3,2-*a*]benzimidazole; X-ray

1. Introduction

Benzimidazole is a biologically important scaffold and it is a useful structural motif for the development of molecules of pharmaceutical or biological interest. Appropriately substituted benzimidazole derivatives have found diverse therapeutic applications. It has earned an important place as a pharmacophore in chemotherapeutic agents of pharmacological activities. The biological significance of benzimidazoles can be correlated with its close relationship with the structure of purines that have vital role in the biological system. Moreover, 5,6-dimethyl-1-(α-D-ribofuranosyl)benzimidazole is an integral part of the structure of Vitamin B12. Different pharmacological effects, including antifungal [1], anthelmintic [2], anti-HIV [3], antihistaminic [4–6], antiulcer [7,8], cardiotonic [9], antihypertensive [10,11], and neuroleptic [12], have been reported. The optimization of benzimidazole-based structures has resulted in various drugs that are currently on the market, such as omeprazole (proton pump inhibitor), pimobendan (ionodilator), and mebendazole (anthelmintic).

Having the above aspects in mind, our attention has been attracted to synthesizing benzimidazolethione ring and its *N*-acetyl derivative to study their alkylation, aralkylation and bromoalkylation, which subsequently underwent intramolecular cyclisation to give the fused tricyclic ring 3,4-dihydro-2*H*-[1,3]thiazino[3,2-*a*]benzimidazole. The X-ray crystallographic analysis of the alkylated derivatives was also investigated.

2. Results

2.1. Chemistry

Diversified methods for the preparation of 1H-benzo[d]imidazole-2(3H)-thione 1 have been reported [13–15], which can exhibit tautomerism of the type thione-thiol. Its ^{1}H-NMR spectrum showed the presence of 2NH as a singlet at δ 12.20 ppm, which disappeared upon addition of D_2O. The protons of the phenyl ring showed two sets of signals at δ 7.19 and 7.10 ppm. Thus, it existed in the thione form. Its reported [16] X-ray also agreed with the thione tautomer. The minimized energy structure agreed with that of the X-ray, which has been repeated and found to be similar to that reported earlier.

The reaction of 1H-benzo[d]imidazole-2(3H)-thione 1 with boiling acetic anhydride gave 1-(2-thioxo-2,3-dihydro-1H-benzo[d]imidazol-1-yl)ethanone 2 (Scheme 1). The product was similar to that prepared in literature [17]. The structure was confirmed from the spectral characteristics. Its IR spectrum showed the presence of broad signal at 1716 (C=O) and 3450 cm^{-1} (NH). Its ^{1}H-NMR spectrum showed the presence of CH_3 as a singlet at δ 1.92 ppm, the signal of NH disappeared because the sample was measured in a solvent mixture including D_2O. Its ^{13}C-NMR spectrum showed the presence of CH_3 as a singlet at δ 27.1 ppm, the C=O as a singlet at δ 171.1 ppm and C=S as a singlet at δ 168.9 ppm.

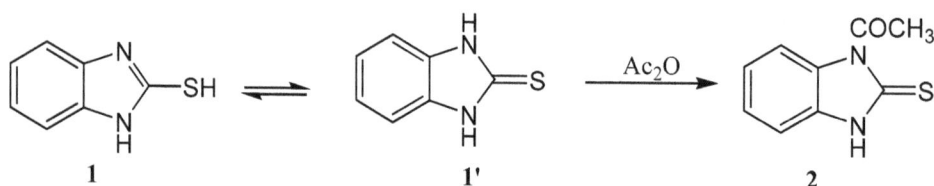

Scheme 1. Synthesis of 1-(2-thioxo-2,3-dihydro-1H-benzo[d]imidazol-1-yl)ethanone 2.

Attempted alkylation of 1 and 2 in presence of base gave the same alkylated products, indicating the loss of the acetyl group during the alkylation. Thus, reaction of 1-(2-thioxo-2,3-dihydro-1H-benzo[d]imidazol-1-yl)ethanone 2 with different bases in acetone as a solvent gave 1H-benzo[d]imidazole-2(3H)-thione 1 whose rate of formation depend on the base and time. Piperidine was found to be the most efficient base for the deacetylation. This was followed by potassium hydroxide, triethylamine and then potassium carbonate. Hydrazine hydrate led to hydrolysis of the acetyl group without further reaction. These results confirmed that our conclusion of losing the acetyl group during the alkylation was due to the base present in the reaction medium.

Similarly, reaction of 5,6-Dimethyl-1H-benzo[d]imidazole-2(3H)-thione 3 with acetic anhydride gave a product that was identified as 1-(5,6-dimethyl-2-thioxo-2,3-dihydro-1H-benzo[d]imidazol-1-yl)ethanone 4 (Scheme 2). The chemical structure was deduced from the spectral analysis. IR spectrum showed the presence carbonyl amide at 1685 cm^{-1}. Its ^{1}H-NMR spectrum showed the presence of two CH_3 as a singlet at δ 2.22 ppm and NH as a singlet at δ 13.14 ppm. Its ^{13}C-NMR spectrum showed the presence of two CH_3 as two signals at δ 18.6 and 18.9 ppm. The C=S appeared as a signal at 168.4 ppm, and the C=O as a signal at 171.1 ppm.

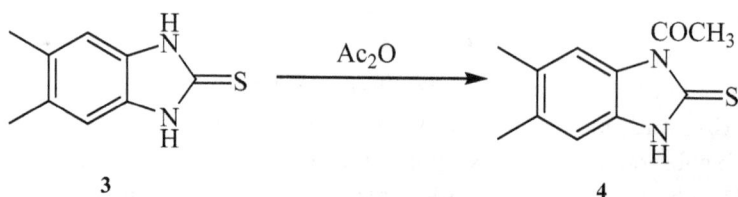

Scheme 2. 1-(5,6-Dimethyl-2-thioxo-2,3-dihydro-1H-benzo[d]imidazol-1-yl)ethanone 4.

Reaction of 1-(5,6-dimethyl-2-thioxo-2,3-dihydro-1H-benzo[d]imidazol-1-yl)ethanone **4** in ethanol with hydrazine hydrate gave a crystalline product that was identified as 5,6-dimethyl-1H-benzo[d]imidazole-2(3H)-thione **3** but 1-(2-hydrazono-5,6-dimethyl-2,3- dihydro-1H-benzo[d]imidazol-1-yl)ethanone did not form.

Reaction of 1H-benzo[d]imidazole-2(3H)-thione **1** with ethyl bromoacetate, in presence of different bases in dry acetone gave ethyl 2-(1H-benzo[d]imidazol-2-ylthio)acetate **5** (Scheme 3). When using triethylamine as base, the product was obtained in high yield, but low yield was obtained when potassium carbonate was used [18]. The structure of the compound is consistent with the expected product. The IR data showed bands at v 3457 (NH), 1739 (C=O), 1269, and 1167 cm^{-1} (C-O). Its ^{1}H-NMR spectrum showed the presence of one NH as a singlet at δ 12.60 ppm, the methylene group show one singlet at δ 4.18 ppm connected with sulfur but not nitrogen. Its ^{13}C-NMR spectrum showed the presence of one CH$_3$ as a signal at δ 13.2, two CH$_2$ as a signal at δ 32.0 and 60.4 ppm, and the presence of C=O at 167.8 ppm.

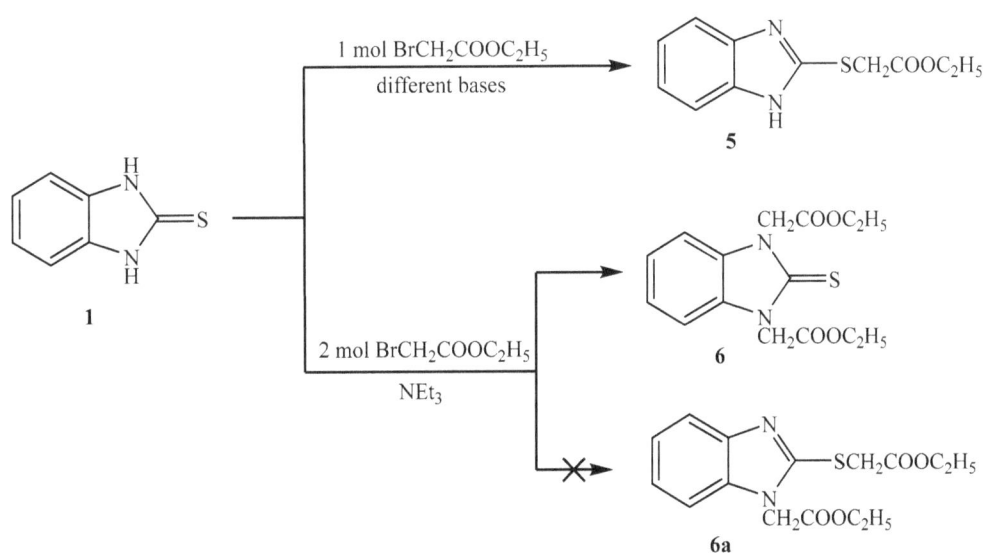

Scheme 3. Synthesis of **5** and **6**.

It was found that reaction of **1** with two equivalents of ethylbromoacetate, in the presence of triethylamine in dry acetone gave diethyl 2,2'-(2-thioxo-1H-benzo[d]imidazole-1,3(2H)-diyl)diacetate **6**. From spectral analysis, IR data showed bands at v: 1739 (C=O) and 1055 cm^{-1} (C=S). Its ^{1}H-NMR spectrum showed the presence of two methylene groups as one singlet at δ 5.16 ppm connected with two nitrogens and indicated the absence of NH signals. This confirmed that the two methylene groups were attached to the two nitrogen atoms but not attached to sulfur and nitrogen. Its ^{13}C-NMR showed the presence of two CH$_3$ as signal at δ 13.10, and two CH$_2$ as signals at δ 44.00 and 60.40 ppm. The ^{13}C-NMR showed the presence of C=S as signal at δ 166.20 and C=O as signal at δ 169.20 ppm

Reaction of 1-(2-mercapto-1H-benzo[d]imidazol-1-yl)ethanone **2** with 1-bromobutane in acetone as a solvent in the presence of triethylamine gave a product that was identified as 2-(butylthio)-1H-benzo[d]imidazole **8** (Scheme 4) [15]. Its IR spectrum showed the presence of a band at δ 3450 cm^{-1} (NH). Its ^{1}H-NMR spectrum showed the presence of NH as a singlet at δ 12.47 ppm and did not show the presence of a methyl group. Its ^{13}C-NMR spectrum showed the presence of CH$_3$ at δ 12.5, and three sets of signals belong to CH$_2$ at δ 20.3, 29.9 and 30.4 ppm. The compound 1-(2-(butylthio)-1H-benzo[d]imidazol-1-yl)ethanone **7** cannot be formed because the acetyl group was removed by base. The structure of **8** was confirmed by X-ray crystallography. The butyl group was found to exist in a zigzag conformation, as shown in Figure 1 which shows the Oak Ridge Thermal Ellipsoid Plot Program (ORTEP) for Crystal Structure.

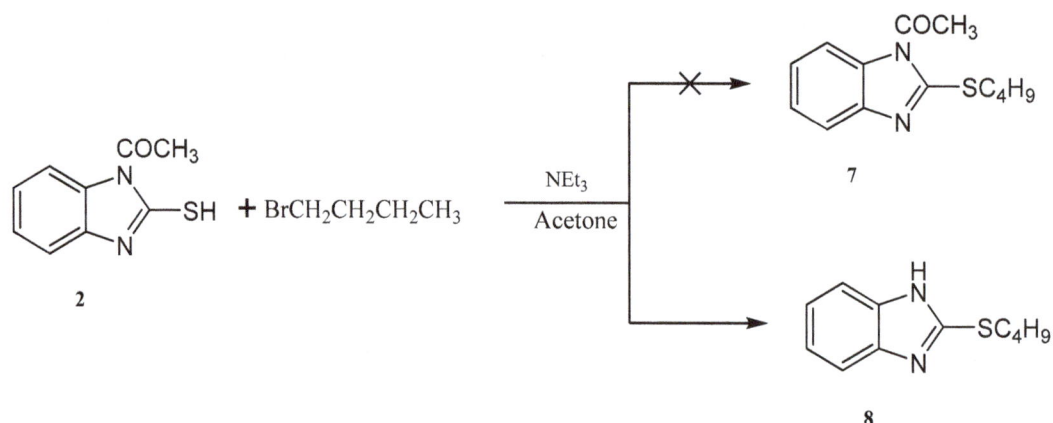

Scheme 4. Synthesis of the 2-(butylthio)-1*H*-benzo[*d*]imidazole **8**.

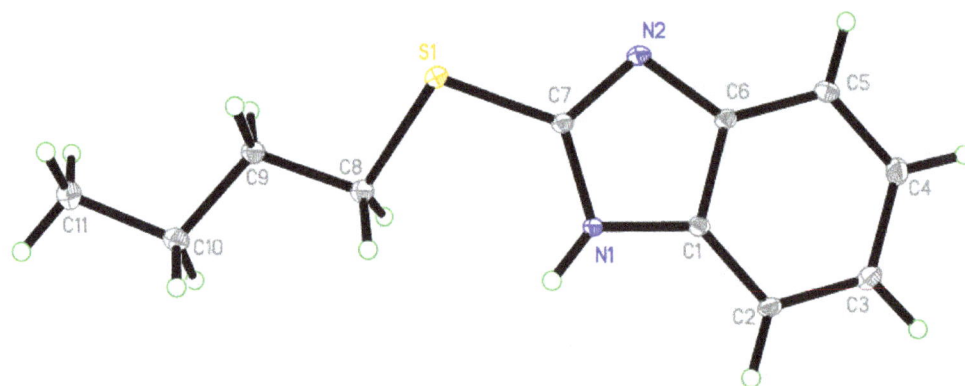

Figure 1. ORTEP diagram of the titled compound **8** drawn at 50% ellipsoids for non-hydrogen atom.

Reaction of 1*H*-benzo[*d*]imidazole-2(3*H*)-thione **1** with dibromopropane, in ethanol as solvent and in the presence of triethylamine gave 3,4-dihydro-2*H*-[1,3]thiazino[3,2-*a*]benzimidazole **9** and not the 1,3-*bis*(1*H*-benzo[*d*]imidazol-2-ylthio)propane **10** [17] (Scheme 5). The ^{1}H-NMR spectrum of the synthesized compound showed the presence of a methylene group at δ 3.25 ppm connected with sulfur, another methylene group at δ 4.18 ppm connected with nitrogen and the disappearance of signal belongs to NH. This confirmed that the alkylation occurred on the ring-sulfur that followed intramolecular cyclisation with the nitrogen atom of the same ring to give **9** and not with another ring to give **10**. The structure of **9** was confirmed by the X-ray crystallography (Figure 2). It showed that one of the CH$_2$ is out of the plane of the ring system. Thus, the envelope conformation is the one having the minimized energy structure.

Reaction of 1-(2-mercapto-1*H*-benzo[*d*]imidazol-1-yl)ethanone **2** with benzyl chloride in acetone as a solvent in the presence of triethylamine gave afford 2-(benzylthio)-1*H*-benzo[*d*]imidazole **13** (Scheme 6). The structure was confirmed from the spectral analysis [19]. Its IR spectrum showed the presence at 3400 cm^{-1} (NH), and did not show C=O band. The ^{1}H-NMR spectrum showed the presence of NH as a singlet at δ 12.56 ppm, and the CH$_2$ as a singlet at δ 4.50 ppm. Its ^{13}C-NMR spectrum showed the presence of CH$_2$ as a signal at δ 30.5, and the C=N at δ 149.1 ppm. The 1-(2-(benzylthio)-1*H*-benzo[*d*]imidazol-1-yl)ethanone **12** was not obtained because the acetyl group was removed by the base.

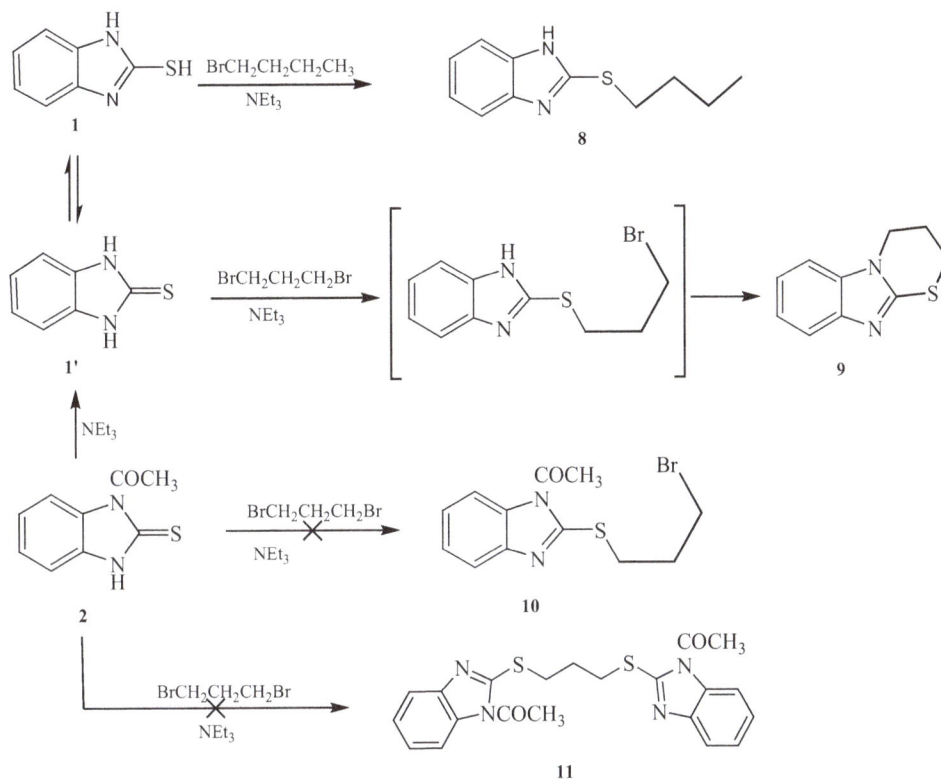

Scheme 5. Synthesis of 3,4-dihydro-2*H*-[1,3]thiazino[3,2-*a*]benzimidazole **9**.

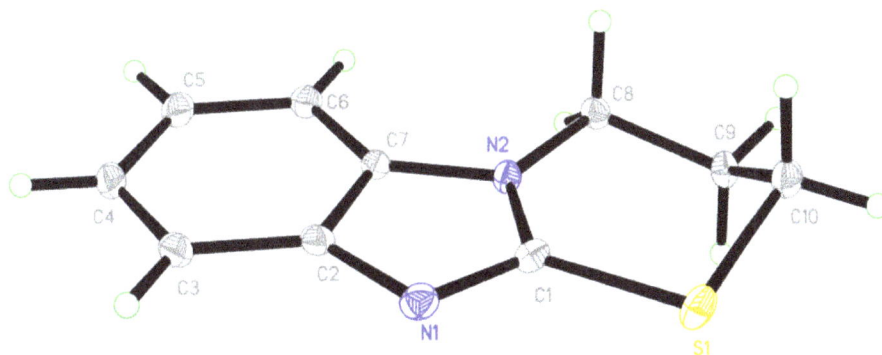

Figure 2. ORTEP diagram of the titled compound **9** drawn at 50% ellipsoids for non-hydrogen atoms.

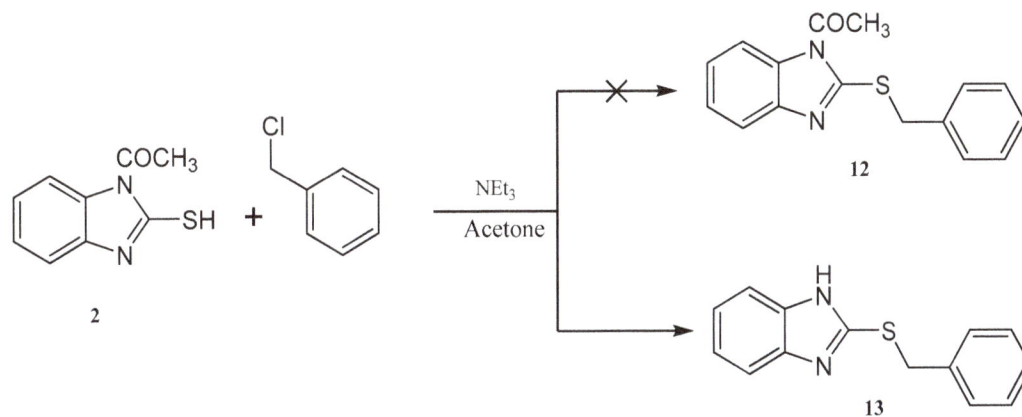

Scheme 6. Synthesis of 2-(benzylthio)-1*H*-benzo[*d*]imidazole **13**.

2.2. X-ray Single Crystal of 8 and 9

The structures of compounds **8** and **9** were unambiguously deduced by single-crystal X-ray diffraction (Figures 1–4) technique. CCDC-1433060 and CCDC-1433059 contain the supplementary crystallographic data for this paper. These data can be obtained free of charge via http://www.ccdc. cam.ac.uk/conts/retrieving.html (or from the CCDC, 12 Union Road, Cambridge CB2 1EZ, UK; Fax: +44-1223-336033; E-mail: deposit@ccdc.cam.ac.uk. The structures were resolved by direct methods using the SHELXS97 program in the SHELXTL-plus package, and refined by a full-matrix least-squares procedure on F^2 using SHELXS97 [20]. Diffraction data were collected on a Bruker SMART APEXII CCD diffractometer (Bruker AXS Advanced X-ray Solutions GmbH, Karlsruhe, Germany). The crystal structure and refinement data of compounds **8** and **9** are listed in Table 1. The selected bond distances and angles are presented in Tables 2–4. ORTEP drawings of final X-ray model of compounds **8** and **9** with the atomic numbering scheme are presented in Figures 1 and 2 while crystal packing presentation of compounds **8** and **9** are shown in Figures 3 and 4 respectively.

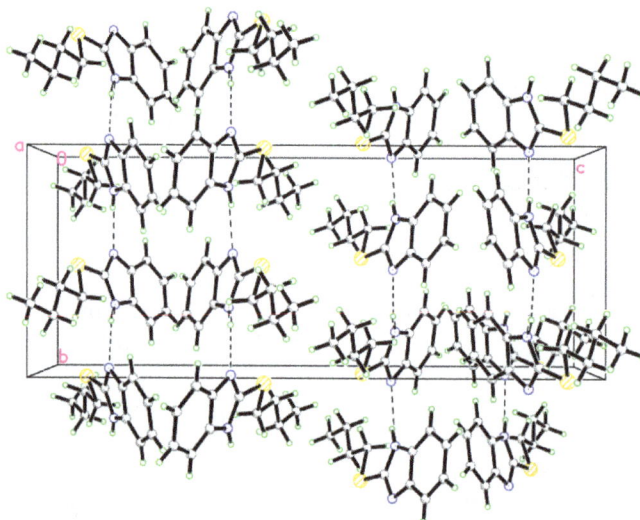

Figure 3. Crystal packing of compound **8** showing intermolecular hydrogen bonds as dashed lines along the c-axis.

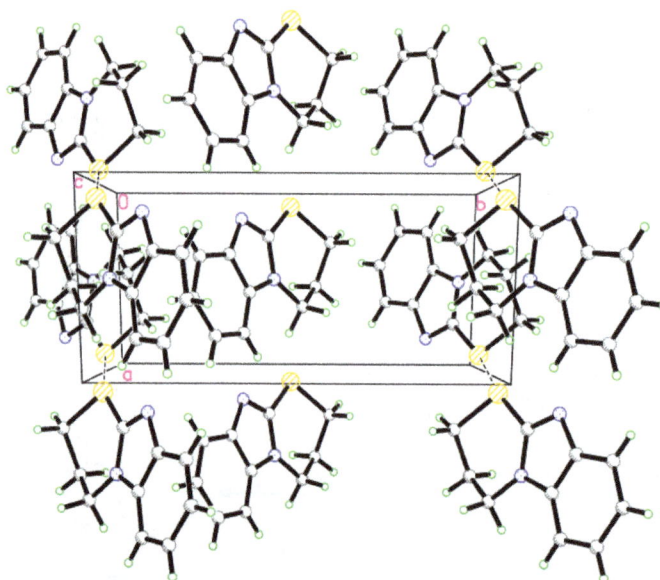

Figure 4. Crystal packing of compound **9**, dotted lines are short S···S interactions.

Table 1. The crystal and experimental data of compounds **8** and **9**.

Crystal Data	Compound 8	Compound 9
Empirical formula	$C_{11}H_{14}N_2S$	$C_{10}H_{10}N_2S$
Formula weight	206.30	190.27
Temperature	293 K	293 K
Wavelength	0.71073 Å	0.71073 Å
Crystal system	Orthorhombic	Monoclinic
Space group	*Pbca*	*P2₁/c*
a	8.9060 (4)	6.0874 (3)
b	9.6531 (4)	12.3309 (6)
c	24.8720 (13)	12.4511 (6)
β	90.00	110.087 (3)
Volume	2138.26 (17)	877.77 (8)
Z	8	4
Calculated density	$1.282 \, Mg \cdot m^{-3}$	$1.440 \, Mg \cdot m^{-3}$
Absorption coefficient	0.26	0.32
F(000)	880	400
Crystal size	$0.38 \times 0.25 \times 0.22$ mm	$0.68 \times 0.53 \times 0.40$ mm
θ range	$2.8°$ to $30.5°$	$2.4°$ to $30.6°$
Reflections Collected	3278	2691
(R_{int})	0.087	0.066
R_1 with I > 2σ (I)	0.056	0.035
R_2 with I > 2σ (I)	0.142	0.091
Goodness of fit	1.20	1.06
max/min $\rho e Å^{-3}$	0.77 and −0.29	0.42 and −0.33
CCDC number	1433060	1433059

Table 2. Selected geometric parameters (Å, °) of **8**.

Bond Length or Angle	(Å, °)	Bond Length or Angle	(Å, °)
S1—C7	1.7350 (18)	N1—C7	1.365 (2)
S1—C8	1.8070 (19)	N2—C6	1.393 (2)
N1—C1	1.385 (2)	N2—C7	1.325 (2)
C7—S1—C8	104.03 (9)	N2—C6—C1	109.87 (15)
C1—N1—C7	106.36 (15)	S1—C7—N2	120.67 (13)
C6—N2—C7	104.41 (15)	N1—C7—N2	113.81 (16)
N1—C1—C2	132.08 (16)	S1—C7—N1	125.46 (13)
N1—C1—C6	105.54 (15)	S1—C8—C9	105.48 (13)
N2—C6—C5	130.10 (16)		

Table 3. Hydrogen-bond geometry (Å, °) of **8**.

D—H⋯A	D—H	H⋯A	D⋯A	D—H⋯A
N1—H1N1⋯N2 ^{i}	0.86 (3)	2.07 (3)	2.886 (2)	159 (3)
Symmetry code: (i) − x + 1/2, y + 1/2, z.				

Table 4. Selected geometric parameters (Å, °) of **9**.

Bond Length or Angle	(Å, °)	Bond Length or Angle	(Å, °)
S1—C1	1.7402 (11)	N2—C1	1.3700 (16)
S1—C10	1.8174 (12)	N2—C7	1.3850 (14)
N1—C1	1.3226 (15)	N2—C8	1.4654 (15)
N1—C2	1.3925 (15)		
C1—N1—C2	104.02 (10)	N1—C2—C3	130.19 (11)
C1—N2—C7	105.90 (9)	N1—C2—C7	110.22 (10)
C1—N2—C8	128.83 (9)	N2—C7—C2	105.67 (10)
C7—N2—C8	125.04 (10)	N2—C7—C6	131.10 (11)
S1—C1—N1	122.49 (9)	N2—C8—C9	111.91 (10)
S1—C1—N2	123.29 (8)	S1—C10—C9	111.49 (8)

3. Materials and Methods

General Methods

The stated melting points are uncorrected and were performed on Gallenkamp melting point apparatus (Toledo, OH, USA). The purities of the compounds were checked by TLC using Merk Kieselgel 60-F254 plates (Darmstadt, Germany), and visually detected in an iodine chamber. The structures of the synthesized compounds were elucidated by using IR spectra on FT-IR (Shizmadu-series, Kyoto, Japan) using KBr disc technique at spectral laboratories in King Saud University. ^{1}H-NMR spectra were determined with Jeol spectrometer (Tokyo, Japan) at 500 MHz and expressed in δ units (ppm) relative to an internal standard of tetramethylsilane in solvent DMSO-d_6 at Alexandria University. ^{13}C-NMR spectra were recorded with Jeol spectrometer at 125.7 MHz. The chemical shifts are expressed in δ (ppm) relative to the reference tetramethylsilane. The DMSO-d_6 was used as a solvent. Elemental analyses were performed at the micro-analytical laboratory at Cairo University, Egypt.

1-(2-Thioxo-2,3-dihydro-1*H*-benzo[*d*]imidazol-1-yl)ethanone 2

Acetic anhydride (30 mL, 0.033 mol) was added to 1*H*-benzo[*d*]imidazole-2(3*H*)-thione 1 (5 g, 0.033 mol) and the stirred mixture was heated to 110–115 °C for 30 min. The solution was cooled then water (150 mL) was added, and finally kept for 30 min at room temperature. Colorless crystals were filtered off (6.21 g, 97% Yield); TLC, R_f = 0.773 (1:1, *n*-hexane:ethyl acetate). The product was recrystallized from benzene to give white crystals. m.p. 201–202 °C, (lit. [21] m.p., 195 °C); IR (KBr) ν 1716 (C=O), 1368 (CH$_3$), 2996 (CH aliphatic), and 3146 (CH aromatic), 1591 cm^{-1} (C=C). ^{1}H-NMR (CDCl$_3$/D$_2$O) δ: 1.92 (s, 3H, CH$_3$), 7.19 (m, 2H, Ar-H), and 7.26 ppm (m, 2H, Ar-H). ^{13}C-NMR (DMSO-d_6) δ: 27.1 (CH$_3$), 108.5 (CH), 114.3 (CH), 122.3 (CH), 124.3 (CH), 129.8 (C), 130.1 (C), 168.9 (C=S), and 171.1 ppm (C=O).

1-(5,6-Dimethyl-2-thioxo-2,3-dihydro-1*H*-benzo[*d*]imidazol-1-yl)ethanone 4

Acetic anhydride (0.002 mol, 4 mL) was added to (0.002 mol, 0356 g) of 5,6-dimethyl-1*H*-benzo[*d*]imidazole-2(3*H*)-thione 3 and the mixture was heated for 0.5 h. The solution was cooled, stirred and poured onto water, and finally kept for 30 min at room temperature. Colorless sparkling crystals were filtered off (0.39 g, 89% yield), TLC, R_f = 0.966 (1:1, *n*-hexane:ethyl acetate). The product was recrystallized from ethanol, m.p. 258–259 °C; IR (KBr) ν 1685 (N-C=O), 1627 (C=N), 1506 (C=C), 3192 cm^{-1} (C-H aromatic). ^{1}H-NMR (DMSO-d_6) δ: 2.22 (s, 6H, 2CH$_3$), 2.94 (s, 3H, CH$_3$), 6.88 (s, 1H, Ar-H), 7.76 (s, 1H, Ar-H), and 13.14 ppm (s, 1H, NH). ^{13}C-NMR (DMSO-d_6) δ: 18.6 (CH$_3$), 18.9 (CH$_3$), 27.1 (CH$_3$), 109.0 (CH), 115.0 (CH), 128.2 (C), 130.7 (C), 132.9 (C), 168.4 (C=S), and 171.1 ppm (C=O). Calc. for C$_{11}$H$_{12}$N$_2$OS (220.29); C, 59.97; H, 5.49; N, 12.72%, Found C, 59.66; H, 5.23; N, 12.43%

Ethyl 2-(1*H*-benzo[*d*]imidazol-2-ylthio)acetate 5

A mixture of 1*H*-benzo[*d*]imidazole-2(3*H*)-thione 1 (0.01 mol, 1.5 g), in dry acetone 25 mL and potassium carbonate (0.01 mol, 1.62 g) was stirred and heated under reflux for 1 h. Ethyl bromoacetate (0.01 mol, 1.67 g, 1.1 mL) was added to the reaction mixture and continuing stirring and heating for another 15 h until completion of the reaction. The reaction mixture was cooled then filtered off. Water was added to the filtrate and left at room temperature for 24 h. The precipitate was filtered off and washed with water to give product, yield 77%. It was recrystallized from ethanol to give white crystal, m.p. 97–98 °C (Lit. [18] m.p. 60–62 °C, Lit. [15] m.p. 117 °C), TLC, R_f = 0.554 (1:1, *n*-hexane:ethyl acetate) IR (KBr) ν 3457 (NH), 1507 (NH) IP, 1739 (C=O), 1269, 1167 (C-O), 3150 (CH aromatic), and 1591 cm^{-1} (C=C). ^{1}H-NMR (DMSO-d_6) δ: 1.13 (t, 3H, CH$_3$), 4.09 (q, 2H, CH$_2$), 4.18 (s, 2H, CH$_2$), 7.1 (s, 2H, Ar-H), and 7.40 (d, 2H, Ar-H), 12.60 ppm (s, 1H, NH). ^{13}C-NMR (DMSO-d_6) δ: 13.2 (CH$_3$), 32.0 (CH$_2$), 60.4 (CH$_2$), 120.7 (Ar-C), 148.3 (C=N), and 167.8 ppm (C=O).

Diethyl 1,3-(2-thioxo-1*H*-benzo[*d*]imidazole-1,3(2*H*)-diyl)diacetate **6**

A mixture of 1*H*-benzo[*d*]imidazole-2(3*H*)-thione **1** (0.01 mol, 1.5 g), in dry acetone (25 mL) and triethylamine (0.01 mol, 1.7 mL) was stirred and heated under reflux for 1 h. Ethylbromoacetate (0.02 mol, 2.2 mL) was added to the reaction mixture and continuing stirring and heating for another 19 h until completion of the reaction. The reaction mixture was cooled then filtered off. Water was added to the filtrate and left at room temperature for 24 h till precipitate was filtered off with water. The product was white powder, yield 76%. It was recrystallized from ethanol. m.p. 199–201 °C, TLC, R_f = 0.722 (1:1, *n*-hexane:ethyl acetate), IR (KBr) ν 3457 (NH), 1507 (NH) IP, 1739 (C=O), 1220, 1093 (C-O), 1055 (C=S), 3058 (CH aromatic), and 1617 cm^{-1} (C=C); ^{1}H-NMR (DMSO-d_6) δ: 1.18 (t, 6H, 2CH$_3$), 4.14 (q, 4H, 2CH$_2$), 5.16 (s, 4H, 2CH$_2$), 7.25 (m, 2H, Ar-H), and 7.47 ppm (m, 2H, Ar-H). ^{13}C-NMR (DMSO-d_6) δ: 13.1 (2CH$_3$), 44 (2CH$_2$), 60.4 (2CH$_2$), 108.9 (2CH), 122.3 (2CH), 130.7 (2C), 166.2 (2C=O), and 169.2 ppm (C=S). Calc. for C$_{15}$H$_{18}$N$_2$O$_4$S (322.38): C, 55.88; H, 5.63; N, 8.69%. Found; C, 55.88; H, 5.63; N, 8.69%.

2-(Butylthio)-1*H*-benzo[*d*]imidazole **8**

To a stirred solution of 1-bromobutane (0.005 mol, 0.54 mL) in 10 mL of acetone was added a solution of 1-(2-mercapto-1*H*-benzo[*d*]imidazol-1-yl)ethanone **2** (0.005 mol, 0.96 g) in 40 mL acetone containing triethylamine (0.01 mol, 1.4 mL). The reaction mixture was stirred for 28 h at room temperature, then evaporated under vacuum, water was added to the precipitate, filtered off and the product (0.64 g, 52% yield); TLC, R_f = 0.601 (1;1, *n*-hexane:ethyl acetate) and was crystallized from ethanol to give 2-(butylthio)-1*H*-benzo[*d*]imidazole **8**, m.p. 135 °C (135 °C lit. [19], 134–135 °C lit. [22]); IR (KBr) ν 3400 (NH), 1671 (C=N), 1619 (NH-IP), 1588 (C=C), 3047 (CH-aromatic), 2955 (CH aliphatic), and 1498, 1465, 1432.7 cm^{-1} (CH$_2$). ^{1}H-NMR (DMSO-d_6) δ: 0.85 (t, 3H, CH$_3$), 1.37 (sextet, 2H, CH$_2$), 1.63 (quint, 2H, CH$_2$), 3.23 (t, 2H, CH$_2$), 7.06 (m, 2H, Ar-H), 7.50 (s, 2H, Ar-H), and 12.50 ppm (s, IH, NH). ^{13}C-NMR (DMSO-d_6) δ: 12.52 (CH$_3$), 20.3 (CH$_2$), 29.9 (CH$_2$), 30.4 (CH$_2$), 120.4 (Ar), and 149.3 ppm (C=N). Calc. for: C$_{11}$H$_{14}$N$_2$S (206.31); C, 64.04; H, 6.84; N, 13.58%, Found C, 64.32; H, 6.34; N, 13.26%.

3,4-Dihydro-2*H*-[1,3]thiazino[3,2-*a*]benzimidazole **9**

A mixture of 1*H*-benzo[*d*]imidazole-2(3*H*)-thione **1** (0.026 mole, 5 gm) in 60 mL ethanol and in presence of triethylamine (0.01 mol, 1.39 mL) was refluxed for 1 h, then 1,3-dibromopropane (0.013 mol, 2.6 g) was added. The reaction mixture was further heated under reflux for 5 h. Then, ethanol was removed under vacuum, and water (20 mL) was added to the product, and kept for 24 h at room temperature to give white crystals. The product (4.68 g, 83% yield), was recrystallized from ethanol, TLC, R_f = 0.382 (1:1, *n*-hexane:ethyl acetate), m.p. 201–202 °C. ^{1}H-NMR (DMSO-d_6) δ: 2.29 (m, 2H, CH$_2$), 3.26 (t, 2H, CH$_2$), 4.17 (t, 2H, CH$_2$), 7.11 (m, 2H, Ar-H), and 7.39 ppm (m, 2H, Ar-H). ^{13}C-NMR (DMSO-d_6) δ: 21.9 (CH$_2$), 24.3 (CH$_2$), 41.7 (CH$_2$), 107.9 (CH), 116.1 (CH), 120.02 (CH), 120.94 (CH), 134.69 (C), 134.81 (C), 141.4; 145.75 ppm (C=N). Calc. for C$_{10}$H$_{10}$N$_2$S (190.26): C, 63.13; H, 5.30; N, 14.72%. Found: C, 63.42; H, 5.11; N, 14.41%.

2-(Benzylthio)-1*H*-benzo[*d*]imidazole **13**

To a stirred solution of benzyl chloride (0.005 mol, 0.57 mL) in 10 mL of acetone was added a solution of 1-(2-mercapto-1*H*-benzo[*d*]imidazol-1-yl)ethanone (0.005 mol, 0.96 g) in 40 mL acetone containing triethylamine (0.01 mol, 1.39 mL). The reaction mixture was stirred for 30 h at room temperature, then evaporated under vacuum, water was added to the precipitate, filtered off and the product (0.4 g, 74% yield), TLC, R_f = 0.644 (1:1, *n*-hexane: ethyl acetate) was recrystallized from ethanol to give 2-(benzylthio)-1*H*-benzo[*d*]imidazole, m.p. 185–186 °C (lit. [19] m.p. 184 °C, lit. [23]. m.p. 184–185 °C); IR (KBr) ν 3400 (NH), 1585 (NH-IP), 1453 (CH$_2$), 2963 (CH aliphatic), 3070 (CH aromatic), and 1610, 1515 cm^{-1} (C=C). ^{1}H-NMR (DMSO-d_6); δ: 4.50 (s, 2H, CH$_2$), 7.08 (m, 2H. Ar-H),

232 Medicinal and Natural Product Chemistry

7.21 (t, 1H, Ar-H), 7.30 (t, 2H, Ar-H), and 7.41–7.51 ppm (bd, 4H, Ar-H). ^{13}C-NMR (DMSO-d_6); δ: 30.5 (CH$_2$), 120.8 (CH), 126.7 (CH), 127.9 (CH), 128.2 (CH), 137.0 (C), and 149.1 ppm (C=N).

4. Conclusions

The benzimidazole ring has been considered a pharmacophore ring. It has active centers that can be modified by some chemical reactions such as alkylation and acetylation to provide compounds of potential biological activity. The alkylation of 1H-benzo[d]imidazole-2(3H)-thione with dibromopropane did not give the respective alkylated derivatives 2-(3-bromopropylthio)-1H-benzo[d]imidazole or 1,3-bis(1H-benzo[d]imidazol-2-ylthio)propane, but gave 3,4-dihydro-2H-[1,3]thiazino[3,2-a]benzimidazole. The formation of the last tricyclic ring indicated that the alkylation that occurred on the ring-sulfur atom was followed by intramolecular cyclisation with the nitrogen atom of the benzimidazole ring. This was confirmed by the X-ray crystallography. Alkylation of benzimidazole thione and its acetyl derivative gave the corresponding S-alkylated products, and in some cases a deacetylation process has been occurred. In the case of using ethyl bromacetate, it gave S-mono- or S-, N-dialkylated derivatives depending on the ratio of the molar equivalents of the reactants. Deacetylation of acetyl benzimidazolethione derivative by different bases gave the imidazole thione, together with the starting acetyl-benzimidazolethione. However, increasing time caused a gradual change in their ratio till complete conversion to the deacetylated derivative. Both 1H-benzo[d]imidazole-2(3H)-thione and 5,6-dimethyl-1H-benzo[d]imidazole-2(3H)-thione can exhibit tautomerism, and it has also been proven that the compounds in the solid state and in solution exist in the thione form, as confirmed by ^{1}H-NMR spectra.

Acknowledgments: The authors would like to extend their sincere appreciation to the Deanship of Scientific Research at King Saud University for funding this Research group NO (RGP-1436-038).

Author Contributions: E.S.H.E.A. and Y.E.K. conceived and designed the experiments; N.A.-Q. performed the experiments; E-S.H.E.-A., Y.E.K. and N.M.N. analyzed the data; A.B., H.A.G. and H.-K.F. contributed reagents/materials/analysis tools; and E.-S.H.E.-A. wrote the paper.

Conflicts of Interest: The authors declare no conflict of interest.

References

1. Berg, D.; Büchel, K.H.; Plempel, M.; Zywietz, A. Action mechanisms of cell-division-arresting benzimidazoles and of sterol biosynthesis-inhibiting imidazoles, 1,2,4-triazoles, and pyrimidines. *Mycoses* **1986**, *29*, 221–229. [CrossRef]
2. Saimot, A.; Cremieux, A.; Hay, J.; Meulemans, A.; Giovanangeli, M.; Delaitre, B.; Coulaud, J. Albendazole as a potential treatment for human hydatidosis. *Lancet* **1983**, *322*, 652–656. [CrossRef]
3. Chimirri, A.; Grasso, S.; Monforte, A.; Monforte, P.; Zappala, M. Anti-HIV agents. I: Synthesis and *in vitro* anti-HIV evaluation of novel 1H,3H-thiazolo [3,4-a] benzimidazoles. *Farmaco* **1991**, *46*, 817–823. [PubMed]
4. Niemegeers, C.; Awouters, F.; Janssen, P. The pharmacological profile of a specific, safe, effective and non-sedative anti-allergic, astemizole. *Agents Actions* **1986**, *18*, 141–144. [CrossRef] [PubMed]
5. Iemura, R.; Hori, M.; Ohtaka, H. Syntheses of the metabolites of 1-(2-ethoxyethyl)-2-(hexahydro-4-methyl-1H-1,4-diazepin-1-yl)-1H-benzimidazole difumarate (KG-2413) and related compounds. *Chem. Pharm. Bull.* **1989**, *37*, 962–966. [CrossRef] [PubMed]
6. Benavides, J.; Schoemaker, H.; Dana, C.; Claustre, Y.; Delahaye, M.; Prouteau, M.; Manoury, P.; Allen, J.; Scatton, B.; Langer, S. *In vivo* and *in vitro* interaction of the novel selective histamine H1 receptor antagonist mizolastine with H1 receptors in the rodent. *Arzneim. Forsch.* **1995**, *45*, 551–558.
7. Ishihara, K.; Ichikawa, T.; Komuro, Y.; Ohara, S.; Hotta, K. Effect on gastric mucus of the proton pump inhibitor leminoprazole and its cytoprotective action against ethanol-induced gastric injury in rats. *Arzneim. Forsch.* **1994**, *44*, 827–830.
8. Graham, D.Y.; McCullough, A.; Sklar, M.; Sontag, S.J.; Roufail, W.M.; Stone, R.C.; Bishop, R.H.; Gitlin, N.; Cagliola, A.J.; Berman, R.S. Omeprazole *versus* placebo in duodenal ulcer healing. *Dig. Dis. Sci.* **1990**, *35*, 66–72. [CrossRef] [PubMed]

9.   Piazzesi, G.; Morano, I.; Rüegg, J. Effect of sulmazole and pimobendan on contractility of skinned fibres from frog skeletal muscle. *Arzneim. Forsch.* **1987**, *37*, 1141–1143.

10. Wiedemann, I.; Peil, H.; Justus, H.; Adamus, S.; Brantl, V.; Lohmann, H. Pharmacokinetics of adimolol after single and multiple dose administration in healthy volunteers. *Arzneim. Forsch.* **1984**, *35*, 964–969.

11. Kubo, K.; Kohara, Y.; Imamiya, E.; Sugiura, Y.; Inada, Y.; Furukawa, Y.; Nishikawa, K.; Naka, T. Nonpeptide angiotensin II receptor antagonists. Synthesis and biological activity of benzimidazolecarboxylic acids. *J. Med. Chem.* **1993**, *36*, 2182–2195. [CrossRef] [PubMed]

12. Janssen, P.; Niemegeers, C.; Schellekens, K.; Dresse, A.; Lenaerts, F.; Pinchard, A.; Schaper, W.; van Nueten, J.; Verbruggen, F. Pimozide, a chemically novel, highly potent and orally long-acting neuroleptic drug. I. The comparative pharmacology of pimozide, haloperidol, and chlorpromazine. *Arzneim. Forsch.* **1968**, *18*, 261–279.

13. El Ashry, E.S.; Aly, A.A.; Aouad, M.R.; Amer, M.R. Revisit to the reaction of *o*-phenylene diamine with thiosemicarbazide to give benzimidazole-2-thione rather than benzotriazine-2-thione and its glycosylation. *Nucleosides Nucleotides Nucleic Acids* **2010**, *29*, 698–706. [CrossRef] [PubMed]

14. Bakavoli, M.; Seresht, E.R.; Rahimizadeh, M. Reinvestigation of *o*-phenylenediamine thermal cyclocondensation with thiosemicarbazide. *Heterocycl. Commun.* **2006**, *12*, 273–274. [CrossRef]

15. Elrayess, R.A.; Ghareb, N.; Azab, M.M.; Said, M.M. Synthesis and antimicrobial activities of some novel benzimidazole and benzotriazole derivatives containing β-lactam moiety. *Life Sci. J.* **2013**, *10*, 1784–1793.

16. Form, G.; Raper, E.; Downie, T. The crystal and molecular structure of 2-mercaptobenzimidazole. *Acta Cryst.* **1976**, *B32*, 345–348. [CrossRef]

17. Lubenets, V.; Stadnitskaya, N.; Novikov, V. Synthesis of thiosulfonates belonging to quinoline derivatives. *Russ. J. Electrochem.* **2000**, *36*, 851–853. [CrossRef]

18. Shingalapur, R.V.; Hosamani, K.M.; Keri, R.S.; Hugar, M.H. Derivatives of benzimidazole pharmacophore: Synthesis, anticonvulsant, antidiabetic and DNA cleavage studies. *Eur. J. Med. Chem.* **2010**, *45*, 1753–1759. [CrossRef] [PubMed]

19. Narkhede, H.; More, U.; Dalal, D.; Mahulikar, P. Solid supported synthesis of 2-mercaptobenzimidazole derivatives using microwaves. *J. Sci. Ind. Res.* **2008**, *67*, 374–376.

20. Sheldrick, G.M. *Shelxs 97, Program for the Solution of Crystal Structure*; University of Göttingen: Göttingen, Germany, 1997.

21. Parras, F.; Guerrero, M.D.C.; Bouza, E.; Blázquez, M.J.; Moreno, S.; Menarguez, M.C.; Cercenado, E. Comparative study of mupirocin and oral co-trimoxazole plus topical fusidic acid in eradication of nasal carriage of methicillin-resistant Staphylococcus aureus. *Antimicrob. Agents Chemother.* **1995**, *39*, 175–179. [CrossRef] [PubMed]

22. Saxena, D.; Khajuria, R.; Suri, O. Synthesis and spectral studies of 2-mercaptobenzimidazole derivatives **1**. *J. Heterocycl. Chem.* **1982**, *19*, 681–683. [CrossRef]

23. Suri, O.; Khajuria, R.; Saxena, D.; Rawat, N.; Atal, C. Synthesis and spectral studies of 2-mercaptobenzimidazole derivatives **2**. *J. Heterocycl. Chem.* **1983**, *20*, 813–814. [CrossRef]

17

Detecting and Quantifying Biomolecular Interactions of a Dendritic Polyglycerol Sulfate Nanoparticle Using Fluorescence Lifetime Measurements

Alexander Boreham [1], Jens Pikkemaat [1], Pierre Volz [1], Robert Brodwolf [1,2], Christian Kuehne [3], Kai Licha [4], Rainer Haag [2,5], Jens Dernedde [2,3] and Ulrike Alexiev [1,2,*]

Academic Editor: Didier Astruc

[1] Institut für Experimentalphysik, Freie Universität Berlin, Arnimallee 14, 14195 Berlin, Germany; alexander.boreham@fu-berlin.de (A.B.); jpikkemaat@zedat.fu-berlin.de (J.P.); pierre.volz@fu-berlin.de (P.V.); brodwolf@zedat.fu-berlin.de (R.B.)

[2] Helmholtz Virtual Institute—Multifunctional Biomaterials for Medicine, Helmholtz-Zentrum Geesthacht, Kantstr. 55, 14513 Teltow, Germany; haag@chemie.fu-berlin.de (R.H.); jens.dernedde@charite.de (J.D.)

[3] Institut für Laboratoriumsmedizin, Klinische Chemie und Pathobiochemie, Charité—Universitätsmedizin Berlin, Augustenburger Platz 1, 13353 Berlin, Germany; christian.Kuehne@charite.de

[4] Mivenion GmbH, Robert-Koch-Platz 4, 10115 Berlin, Germany; Licha@mivenion.com

[5] Institut für Chemie und Biochemie, Freie Universität Berlin, Takustrasse 3, 14195 Berlin, Germany

*   Correspondence: ulrike.alexiev@fu-berlin.de

Abstract: Interactions of nanoparticles with biomaterials determine the biological activity that is key for the physiological response. Dendritic polyglycerol sulfates (dPGS) were found recently to act as an inhibitor of inflammation by blocking selectins. Systemic application of dPGS would present this nanoparticle to various biological molecules that rapidly adsorb to the nanoparticle surface or lead to adsorption of the nanoparticle to cellular structures such as lipid membranes. In the past, fluorescence lifetime measurements of fluorescently tagged nanoparticles at a molecular and cellular/tissue level have been proven to reveal valuable information on the local nanoparticle environment via characteristic fluorescent lifetime signatures of the nanoparticle bound dye. Here, we established fluorescence lifetime measurements as a tool to determine the binding affinity to fluorescently tagged dPGS (dPGS-ICC; ICC: indocarbocyanine). The binding to a cell adhesion molecule (L-selectin) and a human complement protein (C1q) to dPGS-ICC was evaluated by the concentration dependent change in the unique fluorescence lifetime signature of dPGS-ICC. The apparent binding affinity was found to be in the nanomolar range for both proteins (L-selectin: 87 ± 4 nM and C1q: 42 ± 12 nM). Furthermore, the effect of human serum on the unique fluorescence lifetime signature of dPGS-ICC was measured and found to be different from the interactions with the two proteins and lipid membranes. A comparison between the unique lifetime signatures of dPGS-ICC in different biological environments shows that fluorescence lifetime measurements of unique dPGS-ICC fluorescence lifetime signatures are a versatile tool to probe the microenvironment of dPGS in cells and tissue.

Keywords: nanomedicine; dendritic polymers; protein corona; fluorescence lifetime

1. Introduction

The molecular basis of nanoparticle interactions with cells and tissue can only be understood with precise knowledge of molecular properties of the nanoparticle. This characterization is of paramount importance, especially with regard to the targeted delivery of drugs or the physiological action of the nanoparticle itself. New and different analytical methods have to be established to meet the needs of the rapidly growing field of nanomedicine [1–4].

Important properties of dendrimeric nanoparticles like dendritic polyglycerol sulfates (dPGS) [5–10] include the size, shape and flexibility of the nanoparticle [11]. In a recent report we showed that both size and conformational flexibility of dPGS depend on temperature [1]. The fluorescent indocarbocyanine (ICC) (Figure 1) tag proved to be an efficient sensor for environmental properties as shown by the different fluorescence lifetimes in a systematic study of dPGS-ICC in different aqueous and organic solvents [1]. In addition to being dependent on the polarity [3], the fluorescence lifetime of the ICC dye is also sensitive to steric restrictions (Figure 1). In aqueous environments the ICC methine linker can rotate freely, leading to the short fluorescence lifetime of about 0.15 ns of the fluorophore. Steric hindrance of the ICC methine linker results in longer lifetimes [1–3,12]. A longer lifetime component of about 1 ns was observed for ICC bound to dPGS [2] (Figure 1A, Table 1). Upon binding of dPGS-ICC to lipid membranes an additional 4 ns component becomes apparent [1] (Figure 1A).

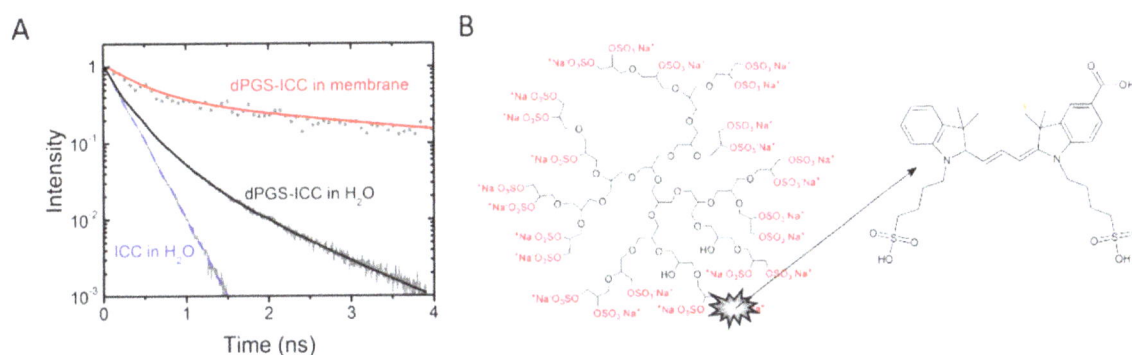

Figure 1. Structure and fluorescence lifetime characteristics of dPGS-ICC. (**A**) Fluorescence lifetime curves of ICC and dPGS-ICC in water and bound to DMPC lipid membranes; and (**B**) structural scheme of dPGS and the bound fluorescent tag ICC. A detailed description of the conjugation procedure is given by Licha *et al.* [13].

Upon administration of dPGS, the nanoparticle would interact with proteins of the blood serum or constituents of biological membranes (e.g., proteins and/or lipids). Interaction of nanoparticles with proteins is known as the "protein corona" [4,14] that determines the biological activity of the respective nanoparticle, because proteins compete with the target structures for the nanoparticle surface [15,16]. In the past several different methods were implemented to study the equilibrium and kinetic parameters of protein-nanoparticle interactions such as isothermal titration calorimetry (ITC), gel filtration, size-exclusion chromatography, surface plasmon resonance (SPR), or centrifugation based pull-down assays [4]. We recently established time-resolved fluorescence spectroscopy and fluorescence lifetime imaging microscopy (FLIM) as a versatile tool to analyze nanoparticle interactions at the molecular and cell/tissue level [1–3].

Here, we extended our previous fluorescence lifetime studies on dPGS-ICC to nanoparticle-protein interactions using L-selectin and the complement protein C1q as well as human serum. We show that fluorescence lifetime methods are well suited to detect the "protein corona" and to quantify binding affinities of individual proteins. Moreover, we show that different biomolecules display different fluorescence decay parameters underscoring our concept of using unique fluorescence lifetimes as target signatures in FLIM-based analyses of nanoparticle interactions in cells and tissue.

2. Results and Discussion

2.1. Determination of Apparent Protein Binding Constants for dPGS-ICC

2.1.1. Binding of L-Selectin

We investigated fluorescence lifetime spectroscopy as a tool to determine affinity of protein–dPGS-ICC interaction. Multivalent dPGS is known to efficiently bind L- and P-selectin [6,7,9,10]. L-selectin is one of the natural cell adhesion molecules on leukocytes that in chronic inflammation processes extravasate into inflamed tissue by interactions of L-, E-, and P- selectins and their corresponding ligands consisting of fucosylated and sialylated glycoproteins on the endothelium.

First, soluble L-selectin (L-selectin-IgG chimera) was titrated in three different concentrations into a solution of 0.1 µM dPGS to evaluate whether a change in the fluorescence lifetime curve occurs. As this was the case we titrated 11 different concentrations into the dPGS-ICC solution until no change in the fluorescence lifetime curve was observable (Figure 2).

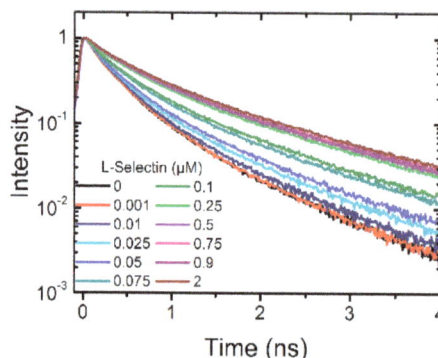

Figure 2. Fluorescence lifetime curves of 0.1 µM dPGS-ICC alone and with 11 different concentrations of soluble L-selectin-IgG chimera (0.001 µM to 2 µM) in DPBS at 20 °C.

The fluorescence decay curves were fitted with a multi-exponential decay function and the mean fluorescence lifetime was determined. The time-resolved fluorescence method belongs to the indirect methods to determine a binding constant. The assumption is made that the measured fluorescence lifetime (or better the change upon binding) is directly proportional to the concentration of the nanoparticle-bound protein, assuming that the protein exists only in two states, in the nanoparticle bound and in the free state, each state having its own unique fluorescence lifetime characteristics.

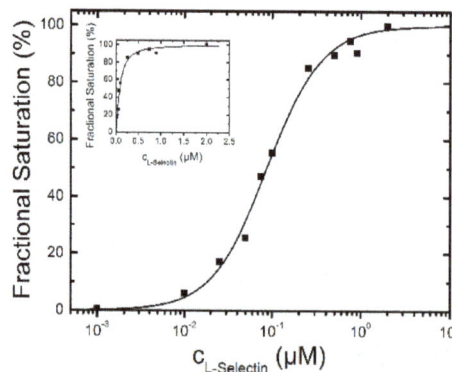

Figure 3. L-selectin-binding curve to 0.1 µM dPGS-ICC in DPBS at 20 °C. The concentrations are plotted on a logarithmic scale. The inset shows the data on a linear scale. The half maximum binding concentration was determined to be 87 ± 4 nM with a Hill-coefficient of $n = 1.4$.

To analyze the binding affinity, the difference between the mean fluorescence lifetime at 0 µM L-selectin and the mean fluorescence lifetimes at different L-selectin concentrations was calculated. The lifetime difference at 2 µM L-selectin constitutes the value at saturation of the binding reaction and was set as 100%. To determine the binding affinity the fractional saturation values (lifetime differences in %) were plotted *vs.* the concentration of L-selectin and fitted by a sigmoidal function (Figure 3). The half-maximum concentration of L-selectin binding was determined to 87 ± 4 nM with a Hill coefficient of $n = 1.4$ indicating a binding stoichiometry of about 1:1. The fluorescence decay time constants of the saturated lifetime signal are summarized in Table 1.

2.1.2. Binding of C1q

Second, the complement protein C1q is the first protein that binds to immobilized antibodies and activates the classical pathway. In addition to its beneficial role in foreign antigen targeting, C1q plays an important role in autoimmune diseases and triggers inflammation. To study the interaction with dendritic polyglycerol sulfate C1q was titrated to a solution of 0.1 µM dPGS to evaluate whether a change in the fluorescence lifetime curve occurs. This was the case and we titrated eight different concentrations into the dPGS-ICC solution (Figure 4).

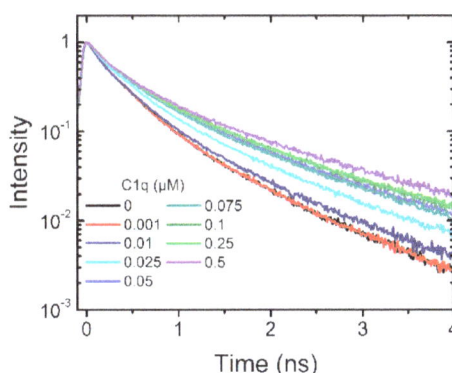

Figure 4. Fluorescence lifetime curves of 0.1 µM dPGS-ICC alone and with eight different concentrations of the complement protein C1q (0.001 µM to 0.5 µM) in DPBS at 20 °C.

Using the fractional saturation method the apparent binding affinity was determined to be 42 ± 12 nM for C1q (Figure 5). The Hill coefficient was $n = 1.1$ indicating also for C1q a 1:1 binding stoichiometry. The fluorescence decay time constants of the saturated lifetime signal are summarized in Table 1.

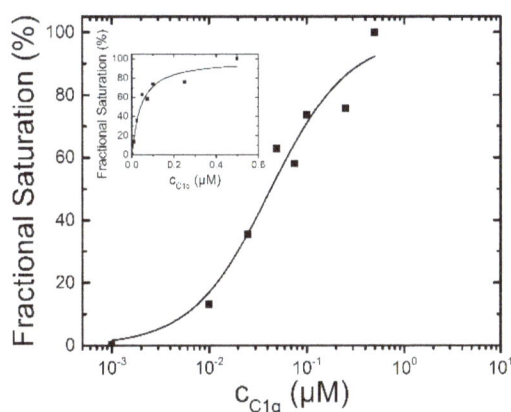

Figure 5. C1q- binding curve to 0.1 µM dPGS-ICC in DPBS at 20 °C. The half maximum binding concentration was determined to be 42 ± 12 nM and $n = 1.1$.

2.1.3. Binding of Human Serum

Next we tested at which concentration (in %) human serum with a protein content of 53 mg/mL results in a saturation of the fluorescence lifetime signal. Figure 6A shows the fluorescence lifetime curves of dPGS-ICC with 0%, 10%, 30%, 50%, and 70% serum concentration. These data clearly show that already at about 10% serum concentration (equivalent to a 5.3 mg/mL solution) the saturation fluorescence signal is reached. The fluorescence decay time constants of the saturated lifetime signal are summarized in Table 1.

To determine the binding affinity of human serum constituents to dPGS we titrated human serum at different concentrations between 0.001% and 10% serum (Figure 6B). Using the fractional saturation method, the mean fluorescence lifetime derived data were plotted in percent as a function of human serum concentration in Figure 6C,D. Analysis of the binding curve gives a mean apparent binding affinity corresponding to 0.3% serum (with a Hill coefficient $n = 1$). In contrast to the plots shown in Figures 3 and 5 the sigmoidal fit (Figure 6D) does not completely fit the data. This is understandable as the serum consists of a mixture of different proteins which could bind with different affinities. We thus added a second component to the fit and the resulting apparent binding affinities (half maximum binding concentration) were 0.2% serum (78% amplitude) and 5% serum (22% amplitude). Even though there are more than two different proteins in the human serum, the fit considerably improves. We conclude that dPGS has the potential to bind different protein species with varying binding affinity. This is an important result as it provides a rationale that, despite its protein corona, dPGS recognizes L-selectin in living tissue [6].

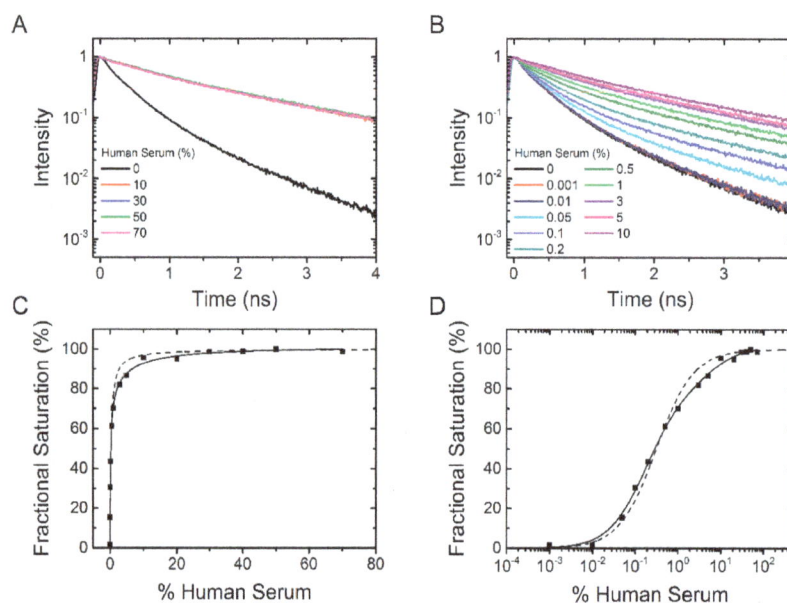

Figure 6. dPGS-ICC binding to human serum. (**A**) Fluorescence lifetime curves of 0.1 μM dPGS-ICC alone and with 10%, 30%, 50%, and 70% human serum in DPBS at 20 °C; (**B**) Fluorescence decay curves for titration of dPGS-ICC with human serum at different concentrations between 0.001% and 10% serum; (**C,D**) Fractional saturation as a function of human serum concentration both linear (**C**) and with a logarithmic x-axis (**D**). The half maximum binding concentration was determined to be 0.3% human serum (using a Hill-coefficient of $n = 1$) for a single component fit and 0.2% and 5% for a double component fit.

2.1.4. SPR Measurements of Protein Association

L-selectin binding to dPGS was also investigated by surface plasmon resonance (SPR). The dPGS was immobilized on a chip and the defined concentrations of the analytes L-selectin and C1q were passed over the functionalized surface. Binding affinity was determined from resulting binding isotherms (Figure 7) and gave values of 45 ± 17 nM for L-selectin and 62 ± 10 nM for C1q. These complementary experiments prove that binding of L-selectin and C1q to dPGS determined by either analysis of fluorescence decay curves or SPR revealed comparable affinities in the lower nanomolar range.

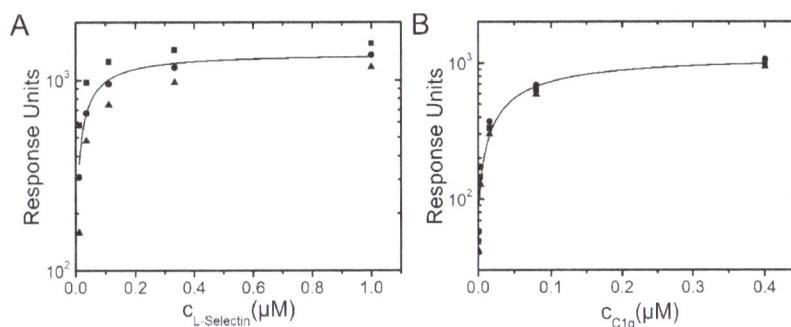

Figure 7. Langmuir binding isotherms of L-selectin (**A**) and C1q (**B**) to dPGS.

2.2. Determination of Protein Binding Kinetics to dPGS-ICC

2.2.1. Binding Kinetics of Human Serum

Binding kinetics measurements offer an additional approach to evaluate protein association. We first tested the binding kinetics of different concentrations of human serum. We choose two concentrations, near the half maximum binding concentration (0.5% serum, Figure 8A) and a concentration that leads to saturation of the fluorescence lifetime signal (10% serum, Figure 8B). Figure 8C shows the binding curves. As expected from the fractional binding signal (Figure 6), the kinetics is faster for 10% serum (34 ± 3 s) than for 0.5% serum (77 ± 4 s).

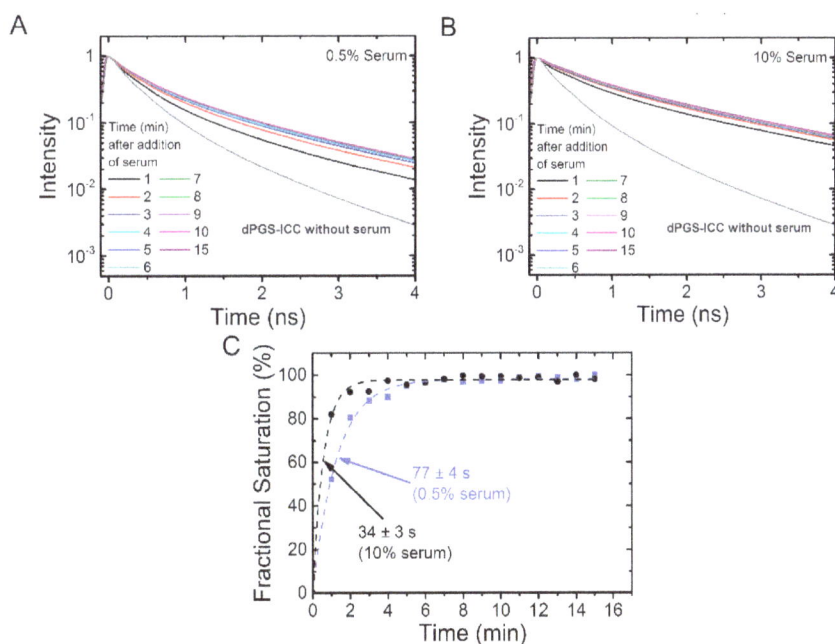

Figure 8. Kinetics of dPGS-ICC binding in human serum. Lifetime decay curves in (**A**) 0.5% serum and (**B**) in 10% human serum at different time points are shown; (**C**) Fractional saturation, based on the mean fluorescence lifetimes, as a function of time. The dashed lines indicate an exponential fit to the data and the time constant τ was 34 s for 10% serum and 77 s for 0.5% serum.

2.2.2. Binding Kinetics of L-Selectin

Next, we determined the binding kinetics of L-selectin (Figure 9A,B). At the saturation concentration of 1 μM, L-selectin binds very fast with a time constant of 26 ± 1 s. This is in the same range as the binding time constant of 10% serum (Figure 8). Subsequent addition of 10% serum to L-selectin saturated dPGS-ICC leads to further increase of the fluorescence lifetime (Figure 9C). Evaluation of the kinetics of serum binding to L-selectin saturated dPGS-ICC reveals a bi-exponential binding kinetics with time constants of 24 ± 3 s (30% amplitude) and 242 ± 36 s (70% amplitude). This is in contrast to the monoexponential binding kinetics with a time constant of 34 s observed for 10% serum alone. Clearly, L-selectin competes with the binding of other serum proteins. A comparison of the unique lifetime signatures of dPGS-ICC with 10% serum and dPGS with L-selectin/10% serum shows subtle but clear differences that indicate the presence of dPGS bound L-selectin under saturating binding concentrations of human blood serum. This result supports our conclusion from the previous section that dPGS recognizes L-selectin despite the presence of other competing proteins in blood plasma.

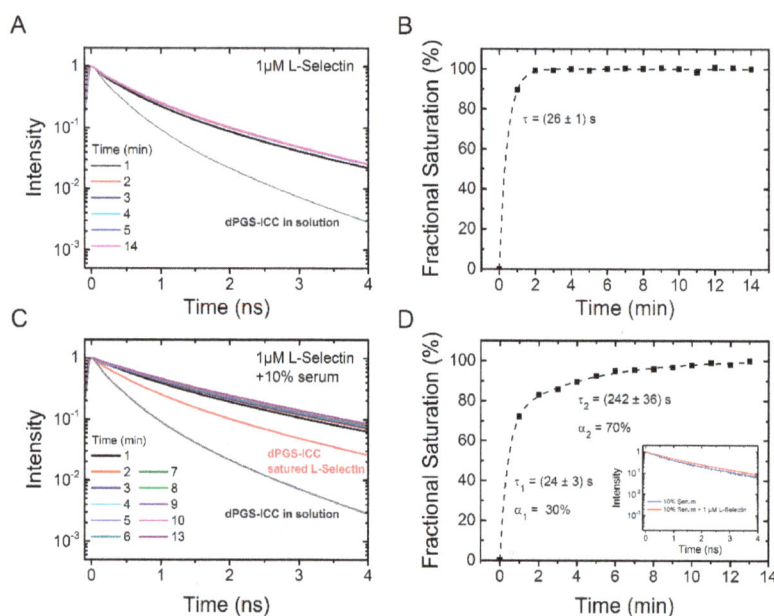

Figure 9. Time dependence of dPGS-ICC binding to L-selectin and effect of serum. (**A**) Lifetime decay curves of 0.1 μM dPGS-ICC binding to 1 μM L-selectin and (**B**) the fractional saturation, based on the mean fluorescence lifetimes, as a function of time. The value of $\tau = 26$ s was obtained from an exponential fit to the data (dashed line); (**C**) Lifetime decay curves of 0.1 μM dPGS-ICC binding to 1 μM L-selectin upon addition of human serum (final concentration: 10% serum) and (**D**) the fractional saturation, based on the mean fluorescence lifetimes, as a function of time. A bi-exponential fit to the data (dashed line) yielded two time constants of $\tau_1 = 24$ s and $\tau_2 = 242$ s with amplitudes of 30% and 70% respectively. Inset: fluorescence lifetime curves of dPGS-ICC in 10% serum and in 10% serum with 1 μM L-selectin.

Table 1. Fluorescence lifetime parameters for different dPGS-ICC interaction partners according to Equation (1). The mean fluorescence lifetime (Equation (2)) and the reduced X^2 is given.

Sample	α_1 (%)	τ_1 (ns)	α_2 (%)	τ_2 (ns)	α_3 (%)	τ_3 (ns)	τ_{mean} (ns)	X^2_R
0.1 μM dPGS-ICC	5.7	1.16	30.7	0.47	63.6	0.17	0.51	1.05
+ 0.5 μM C1q	11.2	1.85	31.3	0.67	57.5	0.18	1.04	0.95
+ 1 μM L-selectin	17.4	1.75	40.5	0.65	42.1	0.19	1.11	1.01
+ 1 μM L-selectin + 10% Serum	37.1	2.24	39.5	0.90	23.4	0.19	1.78	0.91
+ 70% Serum	39.2	2.27	35.1	0.85	25.7	0.14	1.87	0.98

3. Materials and Methods

3.1. Time-Resolved Fluorescence Setup

All measurements were performed on a time-resolved fluorescence setup [17–19] with a tunable white light laser source (SuperK Extreme EUV3, NKT, Birkerød, Denmark), a microchannel plate detector and time-correlated single photon counting (TCSPC) electronics (SPC-830, Becker & Hickl GmbH, Berlin, Germany) with picosecond time resolution. The excitation of ICC at 530 nm was selected via an acousto-optical tunable filter (SELECT UV-VIS, NKT, Birkerød, Denmark). Fluorescence emission was collected above 545 nm using a long-pass filter (HQ545 LP, Chroma, Bellows Falls, VT, USA). The excitation power was 220 µW and the repetition rate was 19.5 MHz. The time range was set to 10 ns divided into 1024 channels resulting in a resolution of 9.8 ps/channel. The instrument response function of the system was ~54 ps full width at half maximum.

3.2. Experimental Procedures

3.2.1. Dendritic Polyglycerol Sulfate Nanocarrier Labeled with a Fluorescent ICC Dye (dPGS-ICC)

Dendritic polyglycerolsulfate (dPGS) with a molecular weight of approx. 12,000 g/mol was synthesized according to literature via an anionic multi-branching ring-opening polymerization of glycidol and sulfation using sulfurtrioxide-pyridinium complex [20]. A detailed description of the dye conjugation procedure is given by Licha et al. [13]. Briefly, to attach ICC to dPGS, an azido-linker (linker 11-azido-1-undecanyl-tosylate) was conjugated to the polyglycerol scaffold at a molar ratio of approximately one linker per polymer before sulfation. After sulfation, the azido-containing polymer was conjugated with a propargyl derivative of ICC by copper-catalyzed 1,3-dipolar cycloaddition (click conjugation) in water/ethanol. After synthesis, the dPGS-ICC was lyophilized. All experiments were conducted with a freshly prepared dPGS-ICC sample from the lyophilisate of the same batch.

3.2.2. Determination of Apparent Protein Binding Constants for dPGS-ICC

To determine the apparent protein binding constants for dPGS-ICC, samples containing 0.1 µM dPGS-ICC and varying concentrations of either L-selectin (L-selectin-IgG chimera, R & D Systems, Wiesbaden, Germany) or C1q (Fitzgerald Industries International, Acton, MA, USA) were prepared in DPBS buffer (PAA; 10 mM phosphate, 1 mM calcium, 1 mM magnesium, 3 mM KCl, 137 mM NaCl, pH 7.4). The measurements were conducted after incubating 0.1 µM dPGS-ICC and the different protein concentrations for 2 min at room temperature. In addition the binding behavior of dPGS-ICC in human serum was analyzed. Here, fluorescence lifetime measurements were conducted after incubating 0.1 µM dPGS-ICC and different concentrations of serum for 5 min at room temperature. The fluorescence decay was recorded for 180 s.

3.2.3. Determination of Protein Binding Kinetics to dPGS-ICC

To determine the kinetics of dPGS-ICC binding in human serum, a buffer solution containing dPGS-ICC was prepared. Directly before the start of the measurement human serum was added to either 10% or 0.5% final concentration. The concentration of dPGS-ICC in the final volume was 0.1 µM in both cases. The fluorescence decay was recorded over a time period of 15 min, with a duration of 60 s for each individual recording.

3.2.4. Determination of Protein Exchange

To determine the influence of human serum on preincubated L-selectin, first the kinetics of dPGS-ICC binding to L-selectin were determined and then the effect of human serum addition was followed. A buffer solution containing dPGS-ICC was prepared. Directly before the start of the measurement L-selectin was added at a concentration of 1 µM. The concentration of dPGS-ICC in the final volume was 0.1 µM. The fluorescence decay was recorded over a time period of 14 min, with a duration of 60 s for each individual recording. Then, human serum was added, final concentration 10%, and further measurements were conducted for the next 13 min, again with a duration of 60 s for each individual recording.

3.2.5. SPR Measurements of Protein Association

Experiments were carried out on a Biacore X100 device (GE Healthcare, Freiburg, Germany). A streptavidin coated chip (SA-Chip, GE Healthcare) was coupled with dPGS-biotin to a level of ~400 resonance units (RU) in HBS-Ca (10mM HEPES pH 7.4 + 150 mM NaCl + 1 mM $CaCl_2$) as a running buffer using standard procedures. Briefly, the chip was conditioned using three consecutive injections of 60 s 1 M NaCl + 50 mM NaOH before injection of dPGS-biotin and washing with three consecutive injections of running buffer. Affinities were measured using a kinetic titration series (single cycle kinetics) at 25 °C in which five ascending concentrations of the analyte (C1q or L-Sel-IgG chimera) were injected consecutively for 120 s at 30 μL/min followed by a dissociation time of 600 s and one regeneration step with 4 M $MgCl_2$ for 120 s. L-selectin was diluted in HBS-Ca to final concentrations of 1000, 333, 111, 37, and 12.34 nM; C1q in surfactant P20 supplemented (+0.005%) running buffer (HBSP) at concentrations of 400, 80, 16, 3.2, and 0.64 nM. The signal of the untreated flow cell was subtracted from the binding signal. Additionally, blank injects of running buffer only were also subtracted (double referencing) for each run. Sensorgrams were analyzed by plotting the analyte concentration against the binding signal at the end of inject. The resulting isotherm was fitted using the steady state model.

3.3. Data Analysis

3.3.1. Determination of Fluorescence Decay Parameters

The fluorescence decay profiles were analyzed using the software package Globals Unlimited V2.2 (Laboratory for Fluorescence Dynamics). An algorithm based on a Marquardt–Levenberg type of nonlinear least-squares analysis was used. The time course of the fluorescence was fitted with a sum of exponentials:

$$I(t) = \sum_i a_i * \exp\left(-\frac{t}{\tau_i}\right) \tag{1}$$

where a_i are the amplitudes and τ_i are the lifetimes of the ith decay component. α_i are the corresponding relative amplitudes, with $\alpha_i = a_i / \sum a_i$ [17,21].

3.3.2. Determination of Apparent Binding Constants

To determine the apparent binding constants of dPGS-ICC, the mean lifetime of the respective fluorescence decay curves was calculated as follows:

$$\tau_{mean} = \sum_i \tau_i * \left(\frac{\alpha_i * \tau_i}{\sum_i \alpha_i * \tau_i}\right) \tag{2}$$

where α_i are the relative amplitudes and τ_i are the lifetimes of the ith decay component.

The fractional saturation (in %) was determined as follows:

$$fractional\ saturation\ (\%) = \left(\frac{\tau_{mean} - \tau_0}{\tau_{max}}\right) * 100 \tag{3}$$

where τ_{mean} is the mean lifetime of the respective lifetime decay curve, τ_0 is the mean lifetime of the lifetime decay curve of dPGS-ICC in solution, and τ_{max} is the highest mean lifetime.

The resulting data points were then fit according to the model function

$$y(x) = \frac{S * x^n}{(K_{50}{}^n + x^n)} \tag{4}$$

where S is the saturation of binding, K_{50} is the half maximum binding concentration (apparent binding affinity), and the Hill-coefficient n, a cooperativity factor.

3.3.3. Determination of Protein Binding Kinetics to dPGS-ICC

To determine the binding kinetics of dPGS-ICC, the mean lifetime of the respective fluorescence decay curves was calculated as described in Equation (2) and the fractional saturation (in %) was calculated according to Equation (3). The kinetic data was then fit with an exponential model function.

$$y(t) = A * e^{\frac{t}{\tau}} \tag{5}$$

3.3.4. Determination of Protein Exchange

The binding kinetics of dPGS-ICC to L-selectin were determined as described in 3.3.2. To evaluate the kinetics of serum addition the fractional saturation (in %) was determined using Equation (3), however with τ_0 the mean lifetime of the lifetime decay curve of dPGS-ICC bound to L-selectin. The kinetic data required a fit with a biexponential model function according to:

$$y(t) = A_1 * e^{\frac{t}{\tau_1}} + A_2 * e^{\frac{t}{\tau_2}} \tag{6}$$

4. Conclusions

dPGS is a multivalent dendritic negatively-charged nanoparticle whose binding affinities are largely determined by electrostatic interactions. For example binding to L-selectin occurs via a patch of basic amino acid residues in the lectin binding domain [22]. However, other proteins including membrane proteins very often feature charged patches that may also undergo transient changes upon protein function [23–25]. Thus it is very likely that in a physiological environment a plethora of dPGS binding partners exists. Here, a comparison of the lifetime signatures of dPGS-ICC with different proteins present in human serum and cellular membranes, like C1q and L-selectin, respectively, shows clear differences (Figure 10).

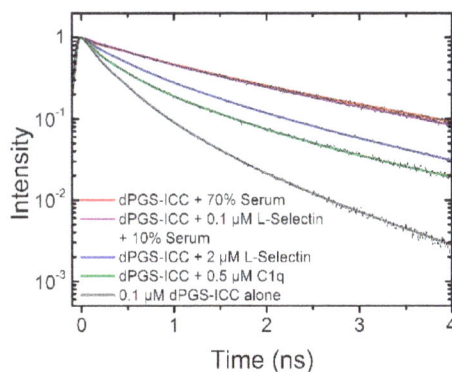

Figure 10. Lifetime signatures of dPGS-ICC with different proteins and human serum.

Previously-published data on dPGS-ICC showed that the two fastest decay components reflect on the polarity of the ICC environment while the slowest component is due to steric hindrance of the methine linker rotation caused by the PG branches [1]. Here, we observed that both the slowest component and the intermediate component show slower lifetime values upon dPGS-ICC binding to the proteins L-selectin and C1q but also in the presence of human serum, with the exact values depending on the specific protein. This clearly indicates that the binding of dPGS-ICC to the proteins L-selectin and C1q, but also to binding partners in human serum, changes both the polarity of the immediate dPGS-ICC environment and also the steric hindrance for the rotation of the ICC methine linker in a protein dependent fashion.

The measurements presented in this paper were performed in solution. However, fluorescence lifetime can also be recorded in a spatial resolved fashion on cells and tissues under a microscope, *i.e.*, with fluorescence lifetime imaging microscopy (FLIM). The use of FLIM is known to include environmental sensing of, amongst others, polarity, local pH, and calcium concentrations, as well as the study of protein interactions in living cells [26]. The results provided in this study show that the fluorescence lifetime also allows for the environmental sensing of biomolecular interactions with dPGS, as unique dPGS-ICC lifetimes exist depending on the dPGS binding partner (Figure 10). Thus, we now extend our concept of using unique fluorescence lifetime signatures for fast and reliable localization of fluorescently labeled nanoparticles in cellular systems and tissue samples [2,27–30] to the potential detection of biomolecular interactions of dPGS in physiological environments. The results presented allow for the possibility of specifically determining the dPGS interaction partners based on the specific fluorescence signature. Further, this methodology is not necessarily limited to dPGS, as in theory any nanoparticle with a fluorescent reporter group can be used, as long as the environmental sensitivity is high enough. However, the use of ICC, as in this study, offers the additional benefit of being sensitive to the steric hindrance of the dye upon nanoparticle-biomolecule interactions. This concept will be developed further in future experiments.

Acknowledgments: The authors would like to acknowledge the Helmholtz Association through the Helmholtz Virtual Institute "Multifunctional Biomaterials for Medicine" (Kantstr. 55, 14513 Teltow, Germany) and the German Research Foundation (DFG, SFB 1112 TP B03 U.A.). A.B. gratefully acknowledges an HONORS Fellowship from the Freie Universität Berlin.

Author Contributions: A.B. performed fluorescence experiments, analyzed and interpreted the data, and contributed to the writing of the manuscript, J.P. contributed to the fluorescence experiments and data analysis, P.V. contributed to the fluorescence experiments and data analysis, R.B. contributed to the data analysis, C.K. contributed to the SPR experiments and data analysis, K.L. synthesized the fluorescently labeled dPGS-ICC nanoparticle used in the study, R.H. contributed to experiment design, J.D. contributed to experiment design, data interpretation and writing of the manuscript, U.A. designed the experiments, analyzed and interpreted the data and wrote the manuscript.

Conflicts of Interest: The authors declare no conflict of interest.

Abbreviations

The following abbreviations are used in this manuscript:

dPGS	Dendritic polyglycerol sulfate
dPGS-ICC	Indocarbocyanine bound to dendritic polyglycerol sulfate
ICC	Indocarbocyanine
DMPC	Dimyrostoylphosphatidylcholine
ITC	Isothermal titration calorimetry
FLIM	Fluorescence lifetime imaging microscopy
TCSPC	Time-correlated single photon counting
IgG	Immunglobulin G
RU	Resonance units
SPR	Surface plasmon resonance

References

1. Boreham, A.; Brodwolf, R.; Pfaff, M.; Kim, T.-Y.; Schlieter, T.; Mundhenk, L.; Gruber, A.D.; Gröger, D.; Licha, K.; Haag, R.; Alexiev, U. Temperature and environment dependent dynamic properties of a dendritic polyglycerol sulfate. *Polym. Adv. Technol.* **2014**, *25*, 1329–1336. [CrossRef]

2. Boreham, A.; Kim, T.Y.; Spahn, V.; Stein, C.; Mundhenk, L.; Gruber, A.D.; Haag, R.; Welker, P.; Licha, K.; Alexiev, U. Exploiting fluorescence lifetime plasticity in flim: Target molecule localization in cells and tissues. *ACS Med. Chem. Lett.* **2011**, *2*, 724–728. [CrossRef] [PubMed]

3. Boreham, A.; Pfaff, M.; Fleige, E.; Haag, R.; Alexiev, U. Nanodynamics of dendritic core-multishell nanocarriers. *Langmuir* **2014**, *30*, 1686–1695. [CrossRef] [PubMed]

4. Cedervall, T.; Lynch, I.; Lindman, S.; Berggard, T.; Thulin, E.; Nilsson, H.; Dawson, K.A.; Linse, S. Understanding the nanoparticle-protein corona using methods to quantify exchange rates and affinities of proteins for nanoparticles. *Proc. Natl. Acad. Sci. USA* **2007**, *104*, 2050–2055. [CrossRef] [PubMed]

5. Calderon, M.; Quadir, M.A.; Sharma, S.K.; Haag, R. Dendritic polyglycerols for biomedical applications. *Adv. Mater.* **2010**, *22*, 190–218. [CrossRef] [PubMed]

6. Dernedde, J.; Rausch, A.; Weinhart, M.; Enders, S.; Tauber, R.; Licha, K.; Schirner, M.; Zügel, U.; von Bonin, A.; Haag, R. Dendritic polyglycerol sulfates as multivalent inhibitors of inflammation. *Proc. Natl. Acad. Sci. USA* **2010**, *107*, 19679–19684. [CrossRef] [PubMed]

7. Paulus, F.; Schulze, R.; Steinhilber, D.; Zieringer, M.; Steinke, I.; Welker, P.; Licha, K.; Wedepohl, S.; Dernedde, J.; Haag, R. The effect of polyglycerol sulfate branching on inflammatory processes. *Macromol. Biosci.* **2014**, *14*, 643–654. [CrossRef] [PubMed]

8. Quadir, M.A.; Haag, R. Biofunctional nanosystems based on dendritic polymers. *J. Control. Release* **2012**, *161*, 484–495. [PubMed]

9. Weinhart, M.; Gröger, D.; Enders, S.; Dernedde, J.; Haag, R. Synthesis of dendritic polyglycerol anions and their efficiency toward L-selectin inhibition. *Biomacromolecules* **2011**, *12*, 2502–2511. [CrossRef] [PubMed]

10. Weinhart, M.; Gröger, D.; Enders, S.; Riese, S.B.; Dernedde, J.; Kainthan, R.K.; Brooks, D.E.; Haag, R. The role of dimension in multivalent binding events: Structure-activity relationship of dendritic polyglycerol sulfate binding to L-selectin in correlation with size and surface charge density. *Macromol. Biosci.* **2011**, *11*, 1088–1098. [CrossRef] [PubMed]

11. Kannan, R.M.; Nance, E.; Kannan, S.; Tomalia, D.A. Emerging concepts in dendrimer-based nanomedicine: From design principles to clinical applications. *J. Intern. Med.* **2014**, *276*, 579–617. [CrossRef] [PubMed]

12. Chibisov, A.K.; Zakharova, G.V.; Gorner, H.; Sogulyaev, Y.A.; Mushkalo, I.L.; Tolmachev, A.I. Photorelaxation processes in covalently-linked indocarbocyanine and thiacarbocyanine dyes. *J. Phys. Chem.* **1995**, *99*, 886–893. [CrossRef]

13. Licha, K.; Welker, P.; Weinhart, M.; Wegner, N.; Kern, S.; Reichert, S.; Gemeinhardt, I.; Weissbach, C.; Ebert, B.; Haag, R.; Schirner, M. Fluorescence imaging with multifunctional polyglycerol sulfates: Novel polymeric near-IR probes targeting inflammation. *Bioconjugate Chem.* **2011**, *22*, 2453–2460. [CrossRef] [PubMed]

14. Monopoli, M.P.; Aberg, C.; Salvati, A.; Dawson, K.A. Biomolecular coronas provide the biological identity of nanosized materials. *Nat. Nanotechnol.* **2012**, *7*, 779–786. [CrossRef] [PubMed]

15. Walkey, C.D.; Chan, W.C. Understanding and controlling the interaction of nanomaterials with proteins in a physiological environment. *Chem. Soc. Rev.* **2012**, *41*, 2780–2799. [CrossRef] [PubMed]

16. Mu, Q.; Jiang, G.; Chen, L.; Zhou, H.; Fourches, D.; Tropsha, A.; Yan, B. Chemical basis of interactions between engineered nanoparticles and biological systems. *Chem. Rev.* **2014**, *114*, 7740–7781. [CrossRef] [PubMed]

17. Alexiev, U.; Rimke, I.; Pöhlmann, T. Elucidation of the nature of the conformational changes of the EF-interhelical loop in bacteriorhodopsin and of the helix VIII on the cytoplasmic surface of bovine rhodopsin: A time-resolved fluorescence depolarization study. *J. Mol. Biol.* **2003**, *328*, 705–719. [CrossRef]

18. Kim, T.Y.; Winkler, K.; Alexiev, U. Picosecond multidimensional fluorescence spectroscopy: A tool to measure real-time protein dynamics during function. *Photochem. Photobiol.* **2007**, *83*, 378–384. [CrossRef] [PubMed]

19. Richter, C.; Schneider, C.; Quick, M.T.; Volz, P.; Mahrwald, R.; Hughes, J.; Dick, B.; Alexiev, U.; Ernsting, N.P. Dual-fluorescence pH probe for bio-labelling. *Phys. Chem. Chem. Phys.* **2015**, *17*, 30590–30597. [CrossRef] [PubMed]

20. Türk, H.; Haag, R.; Alban, S. Dendritic polyglycerol sulfates as new heparin analogues and potent inhibitors of the complement system. *Bioconjugate Chem.* **2004**, *15*, 162–167. [CrossRef] [PubMed]

21. Alexiev, U.; Farrens, D.L. Fluorescence spectroscopy of rhodopsins: Insights and approaches. *Biochim. Biophys. Acta* **2014**, *1837*, 694–709. [CrossRef] [PubMed]

22. Woelke, A.L.; Kuehne, C.; Meyer, T.; Galstyan, G.; Dernedde, J.; Knapp, E.W. Understanding selectin counter-receptor binding from electrostatic energy computations and experimental binding studies. *J. Phys. Chem. B* **2013**, *117*, 16443–16454. [CrossRef] [PubMed]

23. Alexiev, U.; Scherrer, P.; Marti, T.; Khorana, H.G.; Heyn, M.P. Time-resolved surface-charge change on the cytoplasmic side of bacteriorhodopsin. *Febs Lett.* **1995**, *373*, 81–84. [CrossRef]

24. Möller, M.; Alexiev, U. Surface charge changes upon formation of the signaling state in visual rhodopsin. *Photochem. Photobiol.* **2009**, *85*, 501–508. [CrossRef] [PubMed]

25. Narzi, D.; Winkler, K.; Saidowsky, J.; Misselwitz, R.; Ziegler, A.; Böckmann, R.A.; Alexiev, U. Molecular determinants of major histocompatibility complex class I complex stability: Shaping antigenic features through short and long range electrostatic interactions. *J. Biol. Chem.* **2008**, *283*, 23093–23103. [CrossRef] [PubMed]

26. Festy, F.; Ameer-Beg, S.M.; Ng, T.; Suhling, K. Imaging proteins *in vivo* using fluorescence lifetime microscopy. *Mol. Biosyst.* **2007**, *3*, 381–391. [CrossRef] [PubMed]

27. Alnasif, N.; Zoschke, C.; Fleige, E.; Brodwolf, R.; Boreham, A.; Rühl, E.; Eckl, K.M.; Merk, H.F.; Hennies, H.C.; Alexiev, U.; Haag, R.; Küchler, S.; Schäfer-Korting, M. Penetration of normal, damaged and diseased skin—An *in vitro* study on dendritic core-multishell nanotransporters. *J. Control. Release* **2014**, *185*, 45–50. [CrossRef] [PubMed]

28. Ostrowski, A.; Nordmeyer, D.; Boreham, A.; Brodwolf, R.; Mundhenk, L.; Fluhr, J.W.; Lademann, J.; Graf, C.; Rühl, E.; Alexiev, U.; Gruber, A.D. Skin barrier disruptions in tape stripped and allergic dermatitis models have no effect on dermal penetration and systemic distribution of ahaps-functionalized silica nanoparticles. *Nanomedicine* **2014**, *10*, 1571–1581. [CrossRef] [PubMed]

29. Ostrowski, A.; Nordmeyer, D.; Boreham, A.; Holzhausen, C.; Mundhenk, L.; Graf, C.; Meinke, M.C.; Vogt, A.; Hadam, S.; Lademann, J.; Rühl, E.; Alexiev, U.; Gruber, A.D. Overview about the localization of nanoparticles in tissue and cellular context by different imaging techniques. *Beilstein J. Nanotechnol.* **2015**, *6*, 263–280. [CrossRef] [PubMed]

30. Witting, M.; Boreham, A.; Brodwolf, R.; Vávrová, K.; Alexiev, U.; Friess, W.; Hedtrich, S. Interactions of hyaluronic acid with the skin and implications for the dermal delivery of biomacromolecules. *Mol. Pharm.* **2015**, *12*, 1391–1401. [CrossRef] [PubMed]

Permissions

All chapters in this book were first published in Molecules, by MDPI; hereby published with permission under the Creative Commons Attribution License or equivalent. Every chapter published in this book has been scrutinized by our experts. Their significance has been extensively debated. The topics covered herein carry significant findings which will fuel the growth of the discipline. They may even be implemented as practical applications or may be referred to as a beginning point for another development.

The contributors of this book come from diverse backgrounds, making this book a truly international effort. This book will bring forth new frontiers with its revolutionizing research information and detailed analysis of the nascent developments around the world.

We would like to thank all the contributing authors for lending their expertise to make the book truly unique. They have played a crucial role in the development of this book. Without their invaluable contributions this book wouldn't have been possible. They have made vital efforts to compile up to date information on the varied aspects of this subject to make this book a valuable addition to the collection of many professionals and students.

This book was conceptualized with the vision of imparting up-to-date information and advanced data in this field. To ensure the same, a matchless editorial board was set up. Every individual on the board went through rigorous rounds of assessment to prove their worth. After which they invested a large part of their time researching and compiling the most relevant data for our readers.

The editorial board has been involved in producing this book since its inception. They have spent rigorous hours researching and exploring the diverse topics which have resulted in the successful publishing of this book. They have passed on their knowledge of decades through this book. To expedite this challenging task, the publisher supported the team at every step. A small team of assistant editors was also appointed to further simplify the editing procedure and attain best results for the readers.

Apart from the editorial board, the designing team has also invested a significant amount of their time in understanding the subject and creating the most relevant covers. They scrutinized every image to scout for the most suitable representation of the subject and create an appropriate cover for the book.

The publishing team has been an ardent support to the editorial, designing and production team. Their endless efforts to recruit the best for this project, has resulted in the accomplishment of this book. They are a veteran in the field of academics and their pool of knowledge is as vast as their experience in printing. Their expertise and guidance has proved useful at every step. Their uncompromising quality standards have made this book an exceptional effort. Their encouragement from time to time has been an inspiration for everyone.

The publisher and the editorial board hope that this book will prove to be a valuable piece of knowledge for researchers, students, practitioners and scholars across the globe.

List of Contributors

Antonia Eliene Duarte, Luiz Marivando Barros and Francisco Assis Bezerra da Cunha
Centro de Ciências Biológicas e da Saúde-CCBS, Departamento de Ciências Biológicas, Universidade Regional do Cariri (URCA), Pimenta, Crato CEP 63.100-000, CE, Brazil3
Programa de Pós-Graduação em Bioquímica Toxicológica, Departamento de Bioquímica e Biologia Molecular, Universidade Federal de Santa Maria, Santa Maria 97105-900, RS, Brazil

Emily Pansera Waczuk and João Batista Teixeira Rocha
Programa de Pós-Graduação em Bioquímica Toxicológica, Departamento de Bioquímica e Biologia Molecular, Universidade Federal de Santa Maria, Santa Maria 97105-900, RS, Brazil

Katiane Roversi and Marilise Escobar Burger
Departamento de Fisiologia e Farmacologia, Universidade Federal de Santa Maria, Santa Maria 97105-900, RS, Brazil

Maria Arlene Pessoa da Silva
Laboratório de Botânica Aplicada, Departamento de Ciências Biológicas, Universidade Regional do Cariri (URCA), Pimenta, Crato CEP 63.100-000, CE, Brazil

Irwin Rose Alencar de Menezes
Laboratório de Farmacologia e Química Molecular, Departamento de Química Biológica, Universidade Regional do Cariri, Pimenta, Crato CEP 63.100-000, CE, Brazil

José Galberto Martins da Costa
Laboratório de Pesquisas de Produtos Naturais, Departamento de Química Biológica, Universidade Regional do Cariri, Crato CEP 63.105.000, CE, Brazil

Aline Augusti Boligon
Laboratório de Fitoquímica, Departamento de Farmácia Industrial, Universidade Federal de Santa Maria, Santa Maria 97105-900, RS, Brazil

Adedayo Oluwaseun Ademiluyi
Programa de Pós-Graduação em Bioquímica Toxicológica, Departamento de Bioquímica e Biologia Molecular, Universidade Federal de Santa Maria, Santa Maria 97105-900, RS, Brazil
Functional Foods and Nutraceutical Unit, Department of Biochemistry, Federal University of Technology, P.M.B. 704, Akure 340001, Nigeria

Jean Paul Kamdem
Departamento de Bioquímica, Instituto de Ciências Básica da Saúde, Universidade Federal do Rio Grande do Sul, Porto Alegre CEP 90035-003, RS, Brazil
Programa de Pós-Graduação em Bioquímica Toxicológica, Departamento de Bioquímica e Biologia Molecular, Universidade Federal de Santa Maria, Santa Maria 97105-900, RS, Brazil

Eduardo Soriano, Cory Holder and Andrew Levitz
Department of Chemistry, Georgia State University, 50 Decatur St., Atlanta, GA 30303, USA

Maged Henary
Center for Diagnostics and Therapeutics, Georgia State University, Petit Science Center, 100 Piedmont Ave SE, Atlanta, GA 30303, USA
Department of Chemistry, Georgia State University, 50 Decatur St., Atlanta, GA 30303, USA

Shukranul Mawa, Ibrahim Jantan and Khairana Husain
Drug and Herbal Research Centre, Faculty of Pharmacy, Universiti Kebangsaan Malaysia (UKM), Jalan Raja Muda Abdul Aziz, Kuala Lumpur 50300, Malaysia

Ling-Shang Wu, Bo Zhu, Hong-Xiu Dong and Jin-Ping Si
Nurturing Station for the State Key Laboratory of Subtropical Silviculture, Zhejiang A & F University, Lin'an 311300, China

Min Jia and Ling Chen
Department of Pharmacognosy, School of Pharmacy, Second Military Medical University, Shanghai 200433, China

Wei Peng
College of Pharmacy, Chengdu University of Traditional Chinese Medicine, Chengdu 610075, China

Ting Han
Nurturing Station for the State Key Laboratory of Subtropical Silviculture, Zhejiang A & F University, Lin'an 311300, China
Department of Pharmacognosy, School of Pharmacy, Second Military Medical University, Shanghai 200433, China

Magnus Blom, Sara Norrehed, Claes-Henrik Andersson, Hao Huang, Jonas Bergquist, Helena Grennberg and Adolf Gogoll
Department of Chemistry-BMC, Uppsala University, Uppsala S-75123, Sweden

Mark E. Light
Department of Chemistry, University of Southampton, Highfield, Southampton SO17 1BJ, UK

Tobias Illg
Fraunhofer ICT-IMM, Carl-Zeiss-Straße 18-20, 55129 Mainz, Germany

Annett Knorr and Lutz Fritzsche
Federal Institute for Materials Research and Testing (BAM), Unter den Eichen 87, 12205 Berlin, Germany

Fiaz S. Mohammed and Christopher L. Kitchens
Department of Chemical and Biomolecular Engineering, Clemson University, Clemson, SC 29634, USA

Po-Jung Tsai, Wen-Cheng Huang, Ming-Chi Hsieh and Wen-Huey Wu
Department of Human Development and Family Studies, National Taiwan Normal University, Taipei 106

Ping-Jyun Sung
National Museum of Marine Biology and Aquarium, Pingtung 944, Taiwan
Graduate Institute of Marine Biology, National Dong Hwa University, Pingtung 944, Taiwan

Yueh-Hsiung Kuo
Department of Chinese Pharmaceutical Sciences and Chinese Medicine Resources, China Medical University, Taichung 404, Taiwan
Department of Biotechnology, Asia University, Taichung 413, Taiwan

Denisa Leonte and Valentin Zaharia
Department of Organic Chemistry, "Iuliu Ha țieganu" University of Medicine and Pharmacy, RO-400012 Cluj-Napoca, Victor Babeş41, Romania

László Csaba Bencze, Csaba Paizs, Monica Ioana Toşa and Florin Dan Irimie
Biocatalysis and Biotransformation Research Group, Babeş-Bolyai University, RO-400028 Cluj-Napoca, Arany János 11, Romania

Yan-Jun Sun, Li-Xin Pei, Jun-Min Wang, Yan-Li Zhang, Mei-Ling Gao and Bao-Yu Ji
Collaborative Innovation Center for Respiratory Disease Diagnosis and Treatment & Chinese Medicine Development of Henan Province, Henan University of Traditional Chinese Medicine, Zhengzhou 450046, Henan, China
School of Pharmacy, Henan University of Traditional Chinese Medicine, Zhengzhou 450046, Henan, China

Kai-Bo Wang
Key Laboratory of Structure-Based Drug Design & Discovery, Ministry of Education, Shenyang Pharmaceutical University, Shenyang 110016, Liaoning, China
School of Traditional Chinese Materia Medica, Shenyang Pharmaceutical University, Shenyang 110016, Liaoning, China

Yin-Shi Sun
Institute of Special Animal and Plant Sciences, Chinese Academy of Agricultural Sciences, Changchun 130112, Jilin, China

Sobhi M. Gomha and Huwaida M. E. Hassaneen
Department of Chemistry, Faculty of Science, Cairo University, Giza 12613, Egypt

Taher A. Salaheldin
Nanotechnology and Advanced Materials Central Lab, Agricultural Research Center, Giza 12613, Egypt

Hassan M. Abdel-Aziz
Chemistry Department, Faculty of Science, Cairo University, Bani Suef Branch, Bani Suef 62514, Egypt

Mohammed A. Khedr
Department of Pharmaceutical Chemistry, Faculty of Pharmacy, Helwan University, Ein Helwan, Cairo 11795, Egypt
Department of Pharmaceutical Sciences, College of Clinical Pharmacy, King Faisal University, P. O. 380, Al-Hasaa 31982, Saudi Arabia

Eman M. Flefel
Department of Chemistry, College of Science, Taibah University, Al-Madinah Al-Monawarah 1343, Saudi Arabia
Department of Photochemistry, National Research Centre, Dokki, Cairo 12622, Egypt

Hebat-Allah S. Abbas
Department of Photochemistry, National Research Centre, Dokki, Cairo 12622, Egypt
Department of Chemistry, College of Science, King Khalid University, Abha 9004, Saudi Arabia

Randa E. Abdel Mageid
Department of Photochemistry, National Research Centre, Dokki, Cairo 12622, Egypt

Wafaa A. Zaghary
Department of Pharmaceutical Chemistry, College of Pharmacy, Helwan University, Ain Helwan, Cairo 11795, Egypt

Jéssica K. S. Maciel and Yanna C. F. Teles
Post-Graduation Program in Development and Technological Innovation in Medicines, Health Science Center, Federal University of Paraiba, Campus I, João Pessoa, PB, 58051-900, Brazil

Otemberg S. Chaves, Severino G. Brito Filho and Marianne G. Fernandes
Post-Graduation Program in Bioactive Natural and Synthetic Products, Health Science Center, Federal University of Paraiba, Campus I, João Pessoa, PB, 58051-900 Brazil

Temilce S. Assis
Post-Graduation Program in Development and Technological Innovation in Medicines, Health Science Center, Federal University of Paraiba, Campus I, João Pessoa, PB, 58051-900, Brazil
Health Science Centre, Physiology and Pathology Department, Federal University of Paraiba, Campus I, Cidade Universitária—João Pessoa, PB, 58059-900, Brazil

Pedro Dantas Fernandes, Albericio Pereira de Andrade and Leonardo P. Felix
Department of Agroecology and Agriculture, Center of Agricultural and Environmental Sciences, University of Paraiba State, 351 Baraúnas Street, Campina Grande, PB, 58429-500, Brazil

Tania M. S. Silva, Nathalia S. M. Ramos and Girliane R. Silva
Postgraduate Program in Development and Technological Innovation in Medicines, Department of Molecular Sciences, Rural Federal University of Pernambuco, Campus Dois Irmãos, Recife, PE, 52171-900, Brazil

Maria de Fátima Vanderlei de Souza
Post-Graduation Program in Development and Technological Innovation in Medicines, Health Science Center, Federal University of Paraiba, Campus I, João Pessoa, PB, 58051-900, Brazil

Meysam Mobasheri
Department of Chemical Engineering, Science and Research Branch, Islamic Azad University, Tehran 1477893855, Iran

Hossein Attar
Department of Chemical Engineering, Science and Research Branch, Islamic Azad University, Tehran 1477893855, Iran
Tofigh Daru Research and Engineering Company (TODACO), Tehran 1397116359, Iran

Seyed Mehdi Rezayat Sorkhabadi
Department of Medical Nanotechnology, School of Advanced Technologies in Medicine, Tehran University of Medical Sciences, Tehran 1417755469, Iran
Department of Toxicology and Pharmacology, Pharmaceutical Sciences Branch, Islamic Azad University, Tehran 193956466, Iran

Ali Khamesipour
Center for Research and Training in Skin Diseases and Leprosy, Tehran University of Medical Sciences, Tehran 1416613675, Iran

Mahmoud Reza Jaafari
Biotechnology Research Center, Nanotechnology Research Center, School of Pharmacy, Mashhad University of Medical Sciences, P. O. Box: 91775-1365, Mashhad 917751365, Iran

Dan Liu, Tian Luan, Jian Kong, Ying Zhang and Hai-Feng Wang
Department of Pharmaceutical Engineering, College of Parmaceutical and Biological Engineering, Shenyang University of Chemical Technology, Shenyang 110142, China

Wen-Xuan Zhang, Hong-Na Wu, Bo Li, Hong-Lin Wu, Dong-Mei Wang and Song Wu
State Key Laboratory of Bioactive Substance and Function of Natural Medicines, Institute of Materia Medica, Peking Union Medical College and Chinese Academy of Medical Sciences, Beijing 100050, China

Yan Zhang, Liying Wu, Zhongsu Ma, Jia Cheng and Jingbo Liu
College of Food Science and Engineering, Jilin University, Changchun 130062, China

El Sayed H. El Ashry
Chemistry Department, Faculty of Science, Alexandria University, P. O. Box 426, Alexandria 21321, Egypt

Yeldez El Kilany
Chemistry Department, Faculty of Applied Science, Umm Al-Qura University, Makkah 21955, Saudi Arabia
Chemistry Department, Faculty of Science, Alexandria University, P. O. Box 426, Alexandria 21321, Egypt

Nariman M. Nahas and Nadia Al-Qurashi
Chemistry Department, Faculty of Applied Science, Umm Al-Qura University, Makkah 21955, Saudi Arabia

Assem Barakat
Department of Chemistry, College of Science, King Saud University, P. O. Box 2455, Riyadh 11451, Saudi Arabia
Chemistry Department, Faculty of Science, Alexandria University, P. O. Box 426, Alexandria 21321, Egypt

Hazem A. Ghabbour and Hoong-Kun Fun
Department of Pharmaceutical Chemistry, College of Pharmacy, King Saud University, P. O. Box 2457, Riyadh 11451, Saudi Arabia

Alexander Boreham, Jens Pikkemaat and Pierre Volz
Institut für Experimentalphysik, Freie Universität Berlin, Arnimallee 14, 14195 Berlin, Germany

Robert Brodwolf and Ulrike Alexiev
Institut für Experimentalphysik, Freie Universität Berlin, Arnimallee 14, 14195 Berlin, Germany
Helmholtz Virtual Institute—Multifunctional Biomaterials for Medicine, Helmholtz-Zentrum Geesthacht, Kantstr. 55, 14513 Teltow, Germany

Christian Kuehne
Institut für Laboratoriumsmedizin, Klinische Chemie und Pathobiochemie, Charité—Universitätsmedizin Berlin, Augustenburger Platz 1, 13353 Berlin, Germany

Kai Licha
Mivenion GmbH, Robert-Koch-Platz 4, 10115 Berlin,
Germany

Rainer Haag
Helmholtz Virtual Institute — Multifunctional Biomaterials
for Medicine, Helmholtz-Zentrum Geesthacht, Kantstr.
55, 14513 Teltow, Germany
Institut für Chemie und Biochemie, Freie Universität
Berlin, Takustrasse 3, 14195 Berlin, Germany

Jens Dernedde
Helmholtz Virtual Institute — Multifunctional Biomaterials
for Medicine, Helmholtz-Zentrum Geesthacht, Kantstr.
55, 14513 Teltow, Germany
Institut für Laboratoriumsmedizin, Klinische Chemie und
Pathobiochemie, Charité — Universitätsmedizin Berlin,
Augustenburger Platz 1, 13353 Berlin, Germany

www.ingramcontent.com/pod-product-compliance
Lightning Source LLC
Chambersburg PA
CBHW080505200326
41458CB00012B/4092